LEARNING STATISTICS WITH JAMOVI

Learning Statistics with jamovi
A Tutorial for Beginners in Statistical Analysis

Danielle J. Navarro and David R. Foxcroft

©2025 David R. Foxcroft and Danielle J. Navarro

This work is licensed under an Attribution-ShareAlike 4.0 International (CC BY-SA 4.0).

This license allows you to copy and redistribute, transform, and build upon the material for any purpose, even commercially. providing attribution is made to the authors (but not in any way that suggests that they endorse you or your use of the work). Attribution should include the following information:

Danielle J. Navarro and David R. Foxcroft, *Learning Statistics with jamovi: A Tutorial for Beginners in Statistical Analysis*. Cambridge, UK: Open Book Publishers, 2025, https://doi.org/10.11647/OBP.0333

Further details about CC BY-SA licenses are available at https://creativecommons.org/licenses/by-sa/4.0/

All external links were active at the time of publication unless otherwise stated and have been archived via the Internet Archive Wayback Machine at https://archive.org/web

Digital material and resources associated with this volume are available at https://doi.org/10.11647/OBP.0333#resources

ISBN Paperback: 978-1-80064-937-8

ISBN Hardback: 978-1-80064-938-5

ISBN Digital (PDF): 978-1-80064-939-2

DOI: 10.11647/OBP.0333

Table of contents

Preface   1

I   Beginnings   3

1   Why do we learn statistics   5
- 1.1   On the psychology of statistics   5
  - 1.1.1   The curse of belief bias   6
- 1.2   The cautionary tale of Simpson's paradox   8
- 1.3   Statistics in psychology   11
- 1.4   Statistics in everyday life   13
- 1.5   There's more to research methods than statistics   13

2   A brief introduction to research design   15
- 2.1   Introduction to psychological measurement   15
  - 2.1.1   Some thoughts about psychological measurement   15
  - 2.1.2   Operationalisation: defining your measurement   17
- 2.2   Scales of measurement   18
  - 2.2.1   Nominal scale   18
  - 2.2.2   Ordinal scale   19
  - 2.2.3   Interval scale   20
  - 2.2.4   Ratio scale   21
  - 2.2.5   Continuous versus discrete variables   21
  - 2.2.6   Some complexities   22
- 2.3   Assessing the reliability of a measurement   23
- 2.4   The "role" of variables: predictors and outcomes   25
- 2.5   Experimental and non-experimental research   25
  - 2.5.1   Experimental research   26
  - 2.5.2   Non-experimental research   26
- 2.6   Assessing the validity of a study   27
  - 2.6.1   Internal validity   28
  - 2.6.2   External validity   28
  - 2.6.3   Construct validity   29
  - 2.6.4   Face validity   30
  - 2.6.5   Ecological validity   30
- 2.7   Confounders, artefacts and other threats to validity   31
  - 2.7.1   History effects   32
  - 2.7.2   Maturation effects   32
  - 2.7.3   Repeated testing effects   33

		2.7.4	Selection bias	33

      2.7.4   Selection bias 33
      2.7.5   Differential attrition 33
      2.7.6   Non-response bias 34
      2.7.7   Regression to the mean 35
      2.7.8   Experimenter bias 36
      2.7.9   Demand effects and reactivity 37
      2.7.10 Placebo effects 37
      2.7.11 Situation, measurement and sub-population effects 38
      2.7.12 Fraud, deception and self-deception 38
  2.8   Summary 40

II An introduction to jamovi   43

3 Getting started with jamovi   **45**
  3.1   Installing jamovi 46
      3.1.1   Starting up jamovi 46
  3.2   Analyses 47
  3.3   The spreadsheet 48
      3.3.1   Variables 48
      3.3.2   Computed variables 49
      3.3.3   Copy and paste 50
      3.3.4   Syntax mode 50
  3.4   Loading data in jamovi 51
      3.4.1   Importing data from csv files 51
  3.5   Importing unusual data files 52
      3.5.1   Loading data from text files 53
      3.5.2   Loading data from SPSS (and other statistics packages) 53
      3.5.3   Loading Excel files 54
  3.6   Changing data from one level to another 54
  3.7   Installing add-on modules into jamovi 54
  3.8   Quitting jamovi 55
  3.9   Summary 55

III Working with data   57

4 Descriptive statistics   **59**
  4.1   Measures of central tendency 60
      4.1.1   The mean 61
      4.1.2   Calculating the mean in jamovi 61
      4.1.3   The median 62
      4.1.4   Mean or median? What's the difference? 63
      4.1.5   A real-life example 64
      4.1.6   Mode 65
  4.2   Measures of variability 67
      4.2.1   Range 68
      4.2.2   Interquartile range 68
      4.2.3   Mean absolute deviation 69
      4.2.4   Variance 70

		4.2.5	Standard deviation	73
		4.2.6	Which measure to use?	73
	4.3	Skew and kurtosis	74
	4.4	Descriptive statistics separately for each group	77
	4.5	Standard scores	78
	4.6	Summary	80

5 Drawing graphs — 81
5.1 Histograms 82
5.2 Boxplots 84
5.2.1 Violin plots 85
5.2.2 Drawing multiple boxplots 87
5.2.3 Using box plots to detect outliers 87
5.3 Bar graphs 89
5.4 Saving image files using jamovi 90
5.5 Summary 91

6 Pragmatic matters — 93
6.1 Tabulating and cross-tabulating data 94
6.1.1 Creating tables for single variables 94
6.1.2 Adding percentages to a contingency table 95
6.2 Logical expressions in jamovi 95
6.2.1 Assessing mathematical truths 97
6.2.2 Logical operations 97
6.2.3 Applying logical operation to text 99
6.3 Transforming and recoding a variable 100
6.3.1 Creating a transformed variable 101
6.3.2 Collapsing a variable into a smaller number of discrete levels or categories 103
6.3.3 Creating a transformation that can be applied to multiple variables 104
6.4 A few more mathematical functions and operations 107
6.4.1 Logarithms and exponentials 107
6.5 Extracting a subset of the data 108
6.6 Summary 110

IV Statistical theory — 111

Prelude — 113
On the limits of logical reasoning 113
Learning without making assumptions is a myth 116

7 Introduction to probability — 117
7.1 How are probability and statistics different? 118
7.2 What does probability mean? 119
7.2.1 The frequentist view 120
7.2.2 The Bayesian view 123
7.2.3 What's the difference? And who is right? 123
7.3 Basic probability theory 124

		7.3.1 Introducing probability distributions	124
	7.4	The binomial distribution	127
		7.4.1 Introducing the binomial	127
	7.5	The normal distribution	129
		7.5.1 Probability density	133
	7.6	Other useful distributions	134
	7.7	Summary	136

8 Estimating unknown quantities from a sample — 139

	8.1	Samples, populations and sampling	139
		8.1.1 Defining a population	140
		8.1.2 Simple random samples	141
		8.1.3 Most samples are not simple random samples	143
		8.1.4 How much does it matter if you don't have a simple random sample?	144
		8.1.5 Population parameters and sample statistics	145
	8.2	The law of large numbers	147
	8.3	Sampling distributions and the central limit theorem	148
		8.3.1 Sampling distribution of the mean	149
		8.3.2 Sampling distributions exist for any sample statistic!	151
		8.3.3 The central limit theorem	151
	8.4	Estimating population parameters	154
		8.4.1 Estimating the population mean	155
		8.4.2 Estimating the population standard deviation	156
	8.5	Estimating a confidence interval	160
		8.5.1 Interpreting a confidence interval	161
		8.5.2 Calculating confidence intervals in jamovi	163
	8.6	Summary	163

9 Hypothesis testing — 165

	9.1	A menagerie of hypotheses	165
		9.1.1 Research hypotheses versus statistical hypotheses	166
		9.1.2 Null hypotheses and alternative hypotheses	168
	9.2	Two types of errors	169
	9.3	Test statistics and sampling distributions	170
	9.4	Making decisions	171
		9.4.1 Critical regions and critical values	172
		9.4.2 A note on statistical "significance"	174
		9.4.3 The difference between one-sided and two-sided tests	174
	9.5	The p-value of a test	175
		9.5.1 A softer view of decision making	175
		9.5.2 The probability of extreme data	176
		9.5.3 A common mistake	177
	9.6	Reporting the results of a hypothesis test	177
		9.6.1 The issue	177
		9.6.2 Two proposed solutions	178
	9.7	Running the hypothesis test in practice	179
	9.8	Effect size, sample size and power	180
		9.8.1 The power function	181
		9.8.2 The power function	183

		9.8.3 Increasing the power of your study	184
9.9	Some issues to consider		186
	9.9.1	Neyman versus Fisher	186
	9.9.2	Bayesians versus frequentists	187
	9.9.3	Traps	187
9.10	Summary		188

V Statistical tools 191

10 Categorical data analysis 193

10.1	The χ^2 (chi-square) goodness-of-fit test	193
	10.1.1 The cards data	194
	10.1.2 The null hypothesis and the alternative hypothesis	195
	10.1.3 The "goodness-of-fit" test statistic	196
	10.1.4 The sampling distribution of the GOF statistic	198
	10.1.5 Degrees of freedom	199
	10.1.6 Testing the null hypothesis	200
	10.1.7 Doing the test in jamovi	201
	10.1.8 Specifying a different null hypothesis	202
	10.1.9 How to report the results of the test	203
10.2	The χ^2 test of independence (or association)	205
	10.2.1 Constructing our hypothesis test	206
	10.2.2 Doing the test in jamovi	208
10.3	The continuity correction	210
10.4	Effect size	210
10.5	Assumptions of the test(s)	211
10.6	The Fisher exact test	212
10.7	The McNemar test	213
	10.7.1 Doing the McNemar test in jamovi	215
10.8	What's the difference between McNemar and independence?	216
10.9	Summary	218

11 Comparing two means 219

11.1	The one-sample z-test	219
	11.1.1 The inference problem that the test addresses	220
	11.1.2 Constructing the hypothesis test	221
	11.1.3 A worked example, by hand	224
	11.1.4 Assumptions of the z-test	225
11.2	The one-sample t-test	225
	11.2.1 Introducing the t-test	226
	11.2.2 Doing the test in jamovi	227
	11.2.3 Assumptions of the one sample t-test	229
11.3	The independent samples t-test (Student test)	229
	11.3.1 The data	229
	11.3.2 Introducing the test	230
	11.3.3 A "pooled estimate" of the standard deviation	231
11.4	Completing the test	232
	11.4.1 Doing the test in jamovi	233
	11.4.2 Positive and negative t-values	234

	11.4.3 Assumptions of the Student t-test 235
11.5	The independent samples t-test (Welch test) 236
	11.5.1 Doing the Welch test in jamovi 237
	11.5.2 Assumptions of the Welch test 238
11.6	The paired-samples t-test 238
	11.6.1 The data 238
	11.6.2 What is the paired samples t-test? 240
	11.6.3 Doing the test in jamovi 242
11.7	One-sided tests 242
11.8	Effect size 244
	11.8.1 Cohen's d from one sample 245
	11.8.2 Cohen's d from a Student's t-test 245
	11.8.3 Cohen's d from a paired-samples test 246
11.9	Checking the normality of a sample 247
	11.9.1 QQ plots 247
	11.9.2 QQ plots for independent and paired t-tests 248
	11.9.3 Shapiro-Wilk tests 248
	11.9.4 Example 251
11.10	Testing non-normal data 251
	11.10.1 Two sample Mann-Whitney U test 252
	11.10.2 One sample Wilcoxon test 253
11.11	Summary 254

12 Correlation and linear regression   255

12.1	Correlations 255
	12.1.1 The data 255
	12.1.2 The strength and direction of a relationship 256
	12.1.3 The correlation coefficient 258
	12.1.4 Calculating correlations in jamovi 258
	12.1.5 Interpreting a correlation 258
	12.1.6 Spearman's rank correlations 260
12.2	Scatterplots 263
	12.2.1 More elaborate options 265
12.3	What is a linear regression model? 265
12.4	Estimating a linear regression model 268
	12.4.1 Linear regression in jamovi 270
	12.4.2 Interpreting the estimated model 270
12.5	Multiple linear regression 271
	12.5.1 Doing it in jamovi 271
12.6	Quantifying the fit of the regression model 273
	12.6.1 The R^2 value 273
	12.6.2 The relationship between regression and correlation 274
	12.6.3 The adjusted R^2 value 274
12.7	Hypothesis tests for regression models 275
	12.7.1 Testing the model as a whole 275
	12.7.2 Tests for individual coefficients 275
	12.7.3 Running the hypothesis tests in jamovi 276
12.8	Regarding regression coefficients 277
	12.8.1 Confidence intervals for the coefficients 278
	12.8.2 Calculating standardised regression coefficients 278

12.9 Assumptions of regression	279
12.10 Model checking	280
12.10.1 Three kinds of residuals	281
12.10.2 Checking the linearity of the relationship	281
12.10.3 Checking the normality of the residuals	283
12.10.4 Checking equality of variance	285
12.10.5 Checking for collinearity	286
12.10.6 Outliers and anomalous data	286
12.11 Model selection	290
12.11.1 Backward elimination	291
12.11.2 Forward selection	292
12.11.3 A caveat	292
12.11.4 Comparing two regression models	293
12.12 Summary	294

13 Comparing several means (one-way ANOVA)   295

13.1 An illustrative data set	295
13.2 How ANOVA works	297
13.2.1 Two formulas for the variance of Y	297
13.2.2 From variances to sums of squares	299
13.2.3 From sums of squares to the F-test	300
13.2.4 A worked example	301
13.3 Running an ANOVA in jamovi	305
13.3.1 Using jamovi to specify your ANOVA	305
13.4 Effect size	306
13.5 Multiple comparisons and post hoc tests	307
13.5.1 Running "pairwise" t-tests	308
13.5.2 Corrections for multiple testing	308
13.5.3 Bonferroni corrections	309
13.5.4 Holm corrections	309
13.5.5 Writing up the post hoc test	310
13.6 The assumptions of one-way ANOVA	311
13.6.1 Checking the homogeneity of variance assumption	311
13.6.2 Running the Levene test in jamovi	312
13.6.3 Removing the homogeneity of variance assumption	312
13.6.4 Checking the normality assumption	313
13.6.5 Removing the normality assumption	314
13.6.6 The logic behind the Kruskal-Wallis test	315
13.6.7 Additional details	315
13.6.8 How to run the Kruskal-Wallis test in jamovi	315
13.7 Repeated measures one-way ANOVA	316
13.7.1 Repeated measures ANOVA in jamovi	317
13.8 The Friedman non-parametric repeated measures ANOVA test	321
13.9 On the relationship between ANOVA and the Student t-test	322
13.10 Summary	322

14 Factorial ANOVA   325

14.1 Factorial ANOVA 1: balanced designs, focus on main effects	325
14.1.1 What hypotheses are we testing?	326
14.1.2 Running the analysis in jamovi	329

- 14.1.3 How are the sum of squares calculated? 331
- 14.1.4 What are our degrees of freedom? 332
- 14.1.5 Factorial ANOVA versus one-way ANOVAs 332
- 14.1.6 What kinds of outcomes does this analysis capture? 333
- 14.2 Factorial ANOVA 2: balanced designs, interpreting interactions 333
  - 14.2.1 What exactly is an interaction effect? 336
  - 14.2.2 Degrees of freedom for the interaction 336
  - 14.2.3 Running the ANOVA in jamovi 337
  - 14.2.4 Interpreting the results 337
- 14.3 Effect size 339
  - 14.3.1 Estimated group means 340
- 14.4 Assumption checking 342
  - 14.4.1 Homogeneity of variance 343
  - 14.4.2 Normality of residuals 343
- 14.5 Analysis of covariance (ANCOVA) 343
  - 14.5.1 Running ANCOVA in jamovi 345
- 14.6 ANOVA as a linear model 348
  - 14.6.1 Some data 348
  - 14.6.2 ANOVA with binary factors as a regression model 349
  - 14.6.3 How to encode non binary factors as contrasts 353
  - 14.6.4 The equivalence between ANOVA and regression for non-binary factors 355
  - 14.6.5 Degrees of freedom as parameter counting! 357
- 14.7 Different ways to specify contrasts 358
  - 14.7.1 Treatment contrasts 359
  - 14.7.2 Helmert contrasts 359
  - 14.7.3 Sum to zero contrasts 360
  - 14.7.4 Optional contrasts in jamovi 361
- 14.8 Post hoc tests 361
- 14.9 The method of planned comparisons 364
- 14.10 Factorial ANOVA 3: unbalanced designs 365
  - 14.10.1 The *coffee* data 366
  - 14.10.2 "Standard ANOVA" does not exist for unbalanced designs . . . 367
  - 14.10.3 Type I sum of squares 367
  - 14.10.4 Type III sum of squares 370
  - 14.10.5 Type II sum of squares 372
  - 14.10.6 Effect sizes (and non-additive sums of squares) 375
- 14.11 Summary 376

15 Factor Analysis 377
- 15.1 Exploratory Factor Analysis 377
  - 15.1.1 Checking assumptions 379
  - 15.1.2 What is EFA good for? 379
  - 15.1.3 EFA in jamovi 380
  - 15.1.4 Writing up an EFA 390
- 15.2 Principal Component Analysis 391
- 15.3 Confirmatory Factor Analysis 392
  - 15.3.1 CFA in jamovi 394
  - 15.3.2 Reporting a CFA 400
- 15.4 Multi-Trait Multi-Method CFA 401

		15.4.1 MTMM CFA in jamovi	405
	15.5	Internal consistency reliability analysis	407
		15.5.1 Reliability analysis in jamovi	409
	15.6	Summary	411

VI Endings, alternatives and prospects 413

16 Bayesian statistics 415
    16.1 Probabilistic reasoning by rational agents 415
        16.1.1 Priors: what you believed before 416
        16.1.2 Likelihoods: theories about the data 416
        16.1.3 The joint probability of data and hypothesis 417
        16.1.4 Updating beliefs using Bayes' rule 419
    16.2 Bayesian hypothesis tests 420
        16.2.1 The Bayes factor 421
        16.2.2 Interpreting Bayes factors 422
    16.3 Why be a Bayesian? 423
        16.3.1 Statistics that mean what you think they mean 423
        16.3.2 Evidentiary standards you can believe 424
        16.3.3 The p-value is a lie. 425
        16.3.4 Is it really this bad? 429
    16.4 Bayesian t-tests 430
        16.4.1 Independent samples t-test 430
        16.4.2 Paired samples t-test 431
    16.5 Summary 432

Epilogue 433
    The undiscovered statistics 433
        Omissions within the topics covered 433
        Statistical models missing from the book 434
        Other ways of doing inference 437
        Miscellaneous topics 439
    Learning the basics, and learning them in jamovi 441

References 443

Chapter notes 447

About the team 475

Preface

In this textbook we cover the contents of an introductory statistics class, as typically taught to undergraduate psychology, health or social science students. The book covers how to get started in jamovi as well as giving an introduction to data manipulation. From a statistical perspective, the book discusses descriptive statistics and graphing first, followed by chapters on probability theory, sampling and estimation, and null hypothesis testing. After introducing the theory, the book covers the analysis of contingency tables, correlation, t-tests, regression, ANOVA and factor analysis. Bayesian statistics are touched on at the end of the book.

This book is an adaptation of DJ Navarro (2018). Learning statistics with R: A tutorial for psychology students and other beginners. (Version 0.6). https://learningstatisticswithr.com/.

The jamovi version of this book was first released in 2018, as version 0.65. Versions 0.70 and 0.75 were released in subsequent years with corrections and additions; details of the changes in earlier versions of the book can be found in the preface to version 0.75: https://github.com/user-attachments/files/18124061/learning-statistics-with-jamovi-0.75.pdf. In that time, many people have contacted us asking for a hard copy version of the book. To achieve this, and to preserve the open source attributes of the book and materials, we have worked with Open Book Publishers in Cambridge, UK, to release this updated version. Open Book Publishers are the leading independent open access publisher of academic research in the Humanities and Social Sciences in the UK. They are award-winning, not-for-profit, run by scholars, and committed to making high-quality research freely available to readers around the world.

> **i** Note
>
> A pdf version of the book is available for free download from Open Book Publishers (https://www.openbookpublishers.com/books/10.11647/obp.0333). All the data files you need can be accessed for free within jamovi via an add-on module in the jamovi library. Or you can download the files from https://www.learnstatswithjamovi.com.

If you spot any mistakes, or have any suggestions, please do let us know by raising an issue at https://github.com/davidfoxcroft/lsj-book/issues.

David Foxcroft
January 1st, 2025

Part I

Beginnings

Chapter 1

Why do we learn statistics

Thou shalt not answer questionnaires
Or quizzes upon World Affairs,
Nor with compliance
Take any test. Thou shalt not sit
With statisticians nor commit
A social science
– W.H. Auden[1]

1.1 On the psychology of statistics

To the surprise of many students, statistics is a fairly significant part of a psychological education. To the surprise of no-one, statistics is very rarely the *favourite* part of one's psychological education. After all, if you really loved the idea of doing statistics, you'd probably be enrolled in a statistics class right now, not a psychology class. So, not surprisingly, there's a pretty large proportion of the student base that isn't happy about the fact that psychology has so much statistics in it. In view of this, I thought that the right place to start might be to answer some of the more common questions that people have about stats.

A big part of this issue at hand relates to the very idea of statistics. What is it? What's it there for? And why are scientists so bloody obsessed with it? These are all good questions, when you think about it. So let's start with the last one. As a group, scientists seem to be bizarrely fixated on running statistical tests on everything. In fact, we use statistics so often that we sometimes forget to explain to people why we do. It's a kind of article of faith among scientists – and especially social scientists – that your findings can't be trusted until you've done some stats. Undergraduate students might be forgiven for thinking that we're all completely mad, because no-one takes the time to answer one very simple question:

Why do you do statistics? Why don't scientists just use **common sense**?

It's a naive question in some ways, but most good questions are. There's a lot of good answers to it,[2] but for my money, the best answer is a really simple one: we don't trust ourselves enough. We worry that we're human, and susceptible to all of the biases, temptations and frailties that humans suffer from. Much of statistics is basically a safeguard. Using "common sense" to evaluate evidence means trusting gut instincts, relying on verbal arguments and on using the raw power of human reason to come up with the right answer. Most scientists don't think this approach is likely to work.

In fact, come to think of it, this sounds a lot like a psychological question to me, and since I do work in a psychology department, it seems like a good idea to dig a little deeper here. Is it really plausible to think that this "common sense" approach is very trustworthy? Verbal arguments have to be constructed in language, and all languages have biases – some things are harder to say than others, and not necessarily because they're false (e.g., quantum electrodynamics is a good theory, but hard to explain in words). The instincts of our "gut" aren't designed to solve scientific problems, they're designed to handle day-to-day inferences – and given that biological evolution is slower than cultural change, we should say that they're designed to solve the day-to-day problems for a *different world* than the one we live in. Most fundamentally, reasoning sensibly requires people to engage in "induction", making wise guesses and going beyond the immediate evidence of the senses to make generalisations about the world. If you think that you can do that without being influenced by various distractors, well, I have a bridge in London I'd like to sell you. Heck, as the next section shows, we can't even solve "deductive" problems (ones where no guessing is required) without being influenced by our pre-existing biases.

1.1.1 The curse of belief bias

People are mostly pretty smart. We're smarter than the other species that we share the planet with (though many people might disagree). Our minds are quite amazing things, and we seem to be capable of the most incredible feats of thought and reason. That doesn't make us perfect though. And among the many things that psychologists have shown over the years is that we really do find it hard to be neutral, to evaluate evidence impartially and without being swayed by pre-existing biases. A good example of this is the **belief bias effect** in logical reasoning: if you ask people to decide whether a particular argument is logically valid (i.e., the conclusion would be true if the premises were true), we tend to be influenced by the believability of the conclusion, even when we shouldn't. For instance, here's a valid argument where the conclusion is believable:

> All cigarettes are expensive (Premise 1)
> Some addictive things are inexpensive (Premise 2)
> Therefore, some addictive things are not cigarettes (Conclusion)

And here's a valid argument where the conclusion is not believable:

> All addictive things are expensive (Premise 1)
> Some cigarettes are inexpensive (Premise 2)
> Therefore, some cigarettes are not addictive (Conclusion)

The logical *structure* of argument #2 is identical to the structure of argument #1, and they're both valid. However, in the second argument, there are good reasons to think that premise 1 is incorrect, and as a result it's probably the case that the conclusion is also incorrect. But that's entirely irrelevant to the topic at hand; an argument is deductively valid if the conclusion is a logical consequence of the premises. That is, a valid argument doesn't have to involve true statements.

On the other hand, here's an invalid argument that has a believable conclusion:

> All addictive things are expensive (Premise 1)
> Some cigarettes are inexpensive (Premise 2)
> Therefore, some addictive things are not cigarettes (Conclusion)

And finally, an invalid argument with an unbelievable conclusion:

> All cigarettes are expensive (Premise 1)
> Some addictive things are inexpensive (Premise 2)
> Therefore, some cigarettes are not addictive (Conclusion)

Now, suppose that people really are perfectly able to set aside their pre-existing biases about what is true and what isn't, and purely evaluate an argument on its logical merits. We'd expect 100% of people to say that the valid arguments are valid, and 0% of people to say that the invalid arguments are valid. So if you ran an experiment looking at this, you'd expect to see data as in Table 1.1.

Table 1.1: Validity of arguments

	conclusion feels true	conclusion feels false
argument is valid	100% say "valid"	100% say "valid"
argument is invalid	0% say "valid"	0% say "valid"

If the psychological data looked like this (or even a good approximation to this), we might feel safe in just trusting our gut instincts. That is, it'd be perfectly okay just to let scientists evaluate data based on their common sense, and not bother with all this murky statistics stuff. However, you guys have taken psych classes, and by now you probably know where this is going.

In a classic study, J. St. B. T. Evans et al. (1983) ran an experiment looking at exactly this. What they found is that when pre-existing biases (i.e., beliefs) were in agreement with the structure of the data, everything went the way you'd hope (Table 1.2). Not perfect, but that's pretty good. But look what happens when our intuitive feelings about the truth of the conclusion run against the logical structure of the argument, see (Table 1.3).

Oh dear, that's not as good. Apparently, when people are presented with a strong argument that contradicts our pre-existing beliefs, we find it pretty hard to even perceive it to be a strong argument (people only did so 46% of the time). Even worse, when people are presented with a weak argument that agrees with our pre-existing biases, almost no-one can see that the argument is weak (people got that one wrong 92% of the time!).[3]

Table 1.2: Pre-existing biases and argument validity

	conclusion feels true	conclusion feels false
argument is valid	92% say "valid"	
argument is invalid		8% say "valid"

Table 1.3: Intuition and argument validity

	conclusion feels true	conclusion feels false
argument is valid	92% say "valid"	46% say "valid"
argument is invalid	92% say "valid"	8% say "valid"

If you think about it, it's not as if these data are horribly damning. Overall, people did do better than chance at compensating for their prior biases, since about 60% of people's judgements were correct (you'd expect 50% by chance). Even so, if you were a professional "evaluator of evidence", and someone came along and offered you a magic tool that improves your chances of making the right decision from 60% to (say) 95%, you'd probably jump at it, right? Of course you would. Thankfully, we actually do have a tool that can do this. But it's not magic, it's statistics. So that's reason #1 why scientists love statistics. It's just too easy for us to "believe what we want to believe". So instead, if we want to "believe in the data", we're going to need a bit of help to keep our personal biases under control. That's what statistics does, it helps keep us honest.

1.2 The cautionary tale of Simpson's paradox

The following is a true story (I think!). In 1973, the University of California, Berkeley had some worries about the admissions of students into their postgraduate courses. Specifically, the thing that caused the problem was the gender breakdown of their admissions (Table 1.4).

Table 1.4: Berkeley students by gender

	Number of applicants	Percent admitted
Males	8442	44%
Females	4321	35%

Given this, they were worried about being sued![4] Given that there were nearly 13,000 applicants, a difference of 9% in admission rates between males and females is just way too big to be a coincidence. Pretty compelling data, right? And if I were to say to you that these data *actually* reflect a weak bias in favour of women (sort of!), you'd probably think that I was either crazy or sexist.

Oddly, it's actually sort of true. When people started looking more carefully at the admissions data they told a rather different story (Bickel et al., 1975). Specifically, when they looked at it on a department by department basis, it turned out that most of the departments actually had a slightly *higher* success rate for female applicants than for male applicants. Table 1.5 shows the admission figures for the six largest departments (with the names of the departments removed for privacy reasons):

Table 1.5: Berkeley students by gender for six largest departments

	Males		Females	
Department	Applicants	Percent admitted	Applicants	Percent admitted
A	825	62%	108	82%
B	560	63%	25	68%
C	325	37%	593	34%
D	417	33%	375	35%
E	191	28%	393	24%
F	272	6%	341	7%

Remarkably, most departments had a *higher* rate of admissions for females than for males! Yet the overall rate of admission across the university for females was lower than for males. How can this be? How can both of these statements be true at the same time?

Here's what's going on. Firstly, notice that the departments are not equal to one another in terms of their admission percentages: some departments (e.g., A, B) tended to admit a high percentage of the qualified applicants, whereas others (e.g., F) tended to reject most of the candidates, even if they were high quality. So, among the six departments shown above, notice that department A is the most generous, followed by B, C, D, E and F in that order. Next, notice that males and females tended to apply to different departments. If we rank the departments in terms of the total number of male applicants, we get **A**>**B**>D>C>F>E (the "easy" departments are in bold). On the whole, males tended to apply to the departments that had high admission rates.

Now compare this to how the female applicants distributed themselves. Ranking the departments in terms of the total number of female applicants produces a quite different ordering C>E>D>F>**A**>**B**. In other words, what these data seem to be suggesting is that the female applicants tended to apply to "harder" departments. And in fact, if we look at Figure 1.1 we see that this trend is systematic, and quite striking. This effect is known as **Simpson's paradox**. It's not common, but it does happen in real life, and most people are very surprised by it when they first encounter it, and many people refuse to even believe that it's real. It is very real. And while there are lots of very subtle statistical lessons buried in there, I want to use it to make a much more important point: doing research is hard, and there are lots of subtle, counter-intuitive traps lying in wait for the unwary. That's reason #2 why scientists love statistics, and why we teach research methods. Because science is hard, and the truth is sometimes cunningly hidden in the nooks and crannies of complicated data.

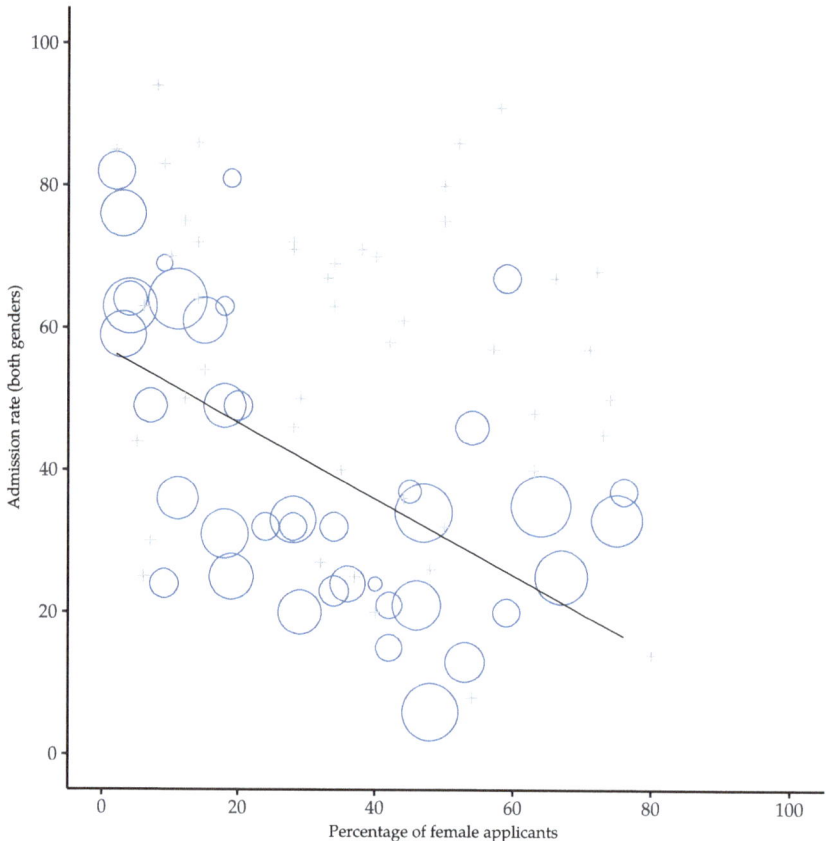

Figure 1.1: The Berkeley 1973 college admissions data. This figure plots the admission rate for the 85 departments that had at least one female applicant, as a function of the percentage of applicants that were female. The plot is a redrawing of Figure 1 from Bickel et al. (1975). Circles plot departments with more than 40 applicants; the area of the circle is proportional to the total number of applicants. The crosses plot departments with fewer than 40 applicants

Before leaving this topic entirely, I want to point out something else really critical that is often overlooked in a research methods class. Statistics only solves *part* of the problem. Remember that we started all this with the concern that Berkeley's admissions processes might be unfairly biased against female applicants. When we looked at the "aggregated" data, it did seem like the university was discriminating against women, but when we "disaggregate" and looked at the individual behaviour of all the departments, it turned out that the actual departments were, if anything, slightly biased in favour of women. The gender bias in total admissions was caused by the fact that women tended to self-select for harder departments. From a legal perspective, that would probably put the university in the clear. Postgraduate admissions are determined at the level of the individual department, and there are good reasons to do that. At the level of individual departments the decisions are more or less unbiased (the weak bias in favour of females at that level is small, and not consistent across departments). Since the university can't dictate which departments people choose to apply to, and the decision making takes place at the level of the department it can hardly be held accountable for any biases that those choices produce.

That was the basis for my somewhat glib remarks earlier, but that's not exactly the whole story, is it? After all, if we're interested in this from a more sociological and psychological perspective, we might want to ask *why* there are such strong gender differences in applications. Why do males tend to apply to engineering more often than females, and why is this reversed for the English department? And why is it the case that the departments that tend to have a female-application bias tend to have lower overall admission rates than those departments that have a male-application bias? Might this not still reflect a gender bias, even though every single department is itself unbiased? It might. Suppose, hypothetically, that males preferred to apply to "hard sciences" and females prefer "humanities". And suppose further that the reason for why the humanities departments have low admission rates is because the government doesn't want to fund the humanities (Ph.D. places, for instance, are often tied to government funded research projects). Does that constitute a gender bias? Or just an unenlightened view of the value of the humanities? What if someone at a high level in the government cut the humanities funds because they felt that the humanities are "useless chick stuff". That seems pretty blatantly gender biased. None of this falls within the purview of statistics, but it matters to the research project. If you're interested in the overall structural effects of subtle gender biases, then you probably want to look at both the aggregated and disaggregated data. If you're interested in the decision making process at Berkeley itself then you're probably only interested in the disaggregated data.

In short there are a lot of critical questions that you can't answer with statistics, but the answers to those questions will have a huge impact on how you analyse and interpret data. And this is the reason why you should always think of statistics as a tool to help you learn about your data. No more and no less. It's a powerful tool to that end, but there's no substitute for careful thought.

1.3 Statistics in psychology

I hope that the discussion above helped explain why science in general is so focused on statistics. But I'm guessing that you have a lot more questions about what role statistics plays in psychology, and specifically why psychology classes always devote so many lectures to stats. So here's my attempt to answer a few of them...

Why does psychology have so much statistics?

To be perfectly honest, there's a few different reasons, some of which are better than others. The most important reason is that psychology is a statistical science. What I mean by that is that the "things" that we study are *people*. Real, complicated, gloriously messy, infuriatingly perverse people. The "things" of physics include objects like electrons, and while there are all sorts of complexities that arise in physics, electrons don't have minds of their own. They don't have opinions, they don't differ from each other in weird and arbitrary ways, they don't get bored in the middle of an experiment, and they don't get angry at the experimenter and then deliberately try to sabotage the data set (not that I've ever done that!). At a fundamental level psychology is harder than physics.[5] Basically, we teach statistics to you as psychologists because you need to be better at stats than physicists. There's actually a saying used sometimes in physics, to the effect that "if your experiment needs statistics, you should have done a better experiment". They have the luxury of being able to say that because their objects of study are pathetically simple in comparison to the vast mess that confronts social scientists. And it's not just psychology. Most social sciences are desperately reliant on statistics. Not because we're bad experimenters, but because we've picked a harder problem to solve. We teach you stats because you really, really need it.

Can't someone else do the statistics?

To some extent, but not completely. It's true that you don't need to become a fully trained statistician just to do psychology, but you do need to reach a certain level of statistical competence. In my view, there's three reasons that every psychological researcher ought to be able to do basic statistics:

- Firstly, there's the fundamental reason: statistics is deeply intertwined with research design. If you want to be good at designing psychological studies, you need to at the very least understand the basics of stats.
- Secondly, if you want to be good at the psychological side of the research, then you need to be able to understand the psychological literature, right? But almost every paper in the psychological literature reports the results of statistical analyses. So if you really want to understand the psychology, you need to be able to understand what other people did with their data. And that means understanding a certain amount of statistics.
- Thirdly, there's a big practical problem with being dependent on other people to do all your statistics: statistical analysis is *expensive*. If you ever get bored and want to look up how much the Australian government charges for university fees, you'll notice something interesting: statistics is designated as a "national priority" category, and so the fees are much, much lower than for any other area of study. This is because there's a massive shortage of statisticians out there. So, from your perspective as a psychological researcher, the laws of supply and demand aren't exactly on your side here! As a result, in almost any real-life situation where you want to do psychological research, the cruel facts will be that you don't have enough money to afford a statistician. So the economics of the situation mean that you have to be pretty self-sufficient.

Note that a lot of these reasons generalise beyond researchers. If you want to be a practicing psychologist and stay on top of the field, it helps to be able to read the scientific literature, which relies pretty heavily on statistics.

I don't care about jobs, research or clinical work. Do I need statistics?

Okay, now you're just messing with me. Still, I think it should matter to you too. Statistics should matter to you in the same way that statistics should matter to *everyone*. We live in the 21st century, and data are *everywhere*. Frankly, given the world in which we live these days, a basic knowledge of statistics is pretty damn close to a survival tool! Which is the topic of the next section.

1.4 Statistics in everyday life

> *We are drowning in information,
> but we are starved for knowledge*
> – Various authors, original probably John Naisbitt

When I started writing up my lecture notes I took the 20 most recent news articles posted to the ABC News website. Of those articles, eight of them included a discussion of a statistical topic and six of those made a mistake. The most common error was failing to report baseline data (e.g., the article mentions that 5% of people in situation X have some characteristic Y, but doesn't say how common the characteristic is for everyone else!). The point I'm trying to make here isn't that journalists are bad at statistics (though they almost always are), it's that a basic knowledge of statistics is very helpful for trying to figure out when someone else is either making a mistake or even lying to you. In fact, one of the biggest things that a knowledge of statistics does to you is cause you to get angry at the newspaper or the internet on a far more frequent basis. You can find a good example of this in Section 4.1.5 in Chapter 4. In later versions of this book I'll try to include more anecdotes along those lines.

1.5 There's more to research methods than statistics

So far, most of what I've talked about is statistics, and so you'd be forgiven for thinking that statistics is all I care about. To be fair, you wouldn't be far wrong, but research methodology is a broader concept than statistics. So most research methods courses will cover topics that relate much more to the pragmatics of research design, and in particular the issues that you encounter when trying to do research with humans. However, about 99% of student fears relate to the statistics part of the course, so I've focused on the stats in this discussion, and hopefully I've convinced you that statistics matter, and more importantly, should not to be feared. That said, it's typical for introductory research methods classes to be very stats heavy. This is not (usually) because the lecturers are evil people. Quite the contrary, in fact. Introductory classes focus a lot on the statistics because you almost always find yourself needing statistics before you need other research methods training. Why? Because almost all your assignments in other classes will rely on statistical training, to a greater extent than they rely on other methodological tools. It's not common for undergraduate assignments to require you to design your own study from the ground up (in which case you would need to know a lot about research design), but it *is* common for assignments to ask you to analyse and interpret data that were collected in a study that someone else designed (in which case you need statistics). In that sense, from the perspective of enabling you to do well in all your other classes, statistics is more urgent.

But note that "urgent" is different from "important" – they both matter. I really do want to stress that research design is just as important as data analysis, and this book does spend some time on it. However, while statistics has a kind of universality, and provides a set of core tools that are useful for most types of psychological and social research, the research methods side isn't quite so universal. There are some general principles that everyone should think about, but a lot of research design is idiosyncratic and is specific to the area of research that you want to engage in.

Chapter 2

A brief introduction to research design

To consult the statistician after an experiment is finished is often merely to ask him to conduct a post mortem examination. He can perhaps say what the experiment died of.
– Sir Ronald Fisher[6]

In this chapter, we're going to start thinking about the basic ideas that go into designing a study, collecting data, checking whether your data collection works, and so on. It won't give you enough information to allow you to design studies of your own, but it will give you a lot of the basic tools that you need to assess the studies done by other people. However, since the focus of this book is much more on data analysis than on data collection, I'm only giving a very brief overview. Note that this chapter is "special" in two ways. Firstly, it's much more psychology specific than the later chapters. Secondly, it focuses much more heavily on the scientific problem of research methodology, and much less on the statistical problem of data analysis. Nevertheless, the two problems are related to one another, so it's traditional for stats textbooks to discuss the problem in a little detail. This chapter relies heavily on Campbell & Stanley (1963) and Stevens (1946) for the discussion of scales of measurement.

2.1 Introduction to psychological measurement

The first thing to understand is data collection can be thought of as a kind of **measurement**. That is, what we're trying to do here is measure something about human behaviour or the human mind. What do I mean by "measurement"?

2.1.1 Some thoughts about psychological measurement

Measurement itself is a subtle concept, but basically it comes down to finding some way of assigning numbers, or labels, or some other kind of well-defined descriptions, to "stuff". So, any of the following would count as a psychological measurement:

- My **age** is *33* years.
- I *do not* **like anchovies**.
- My **chromosomal gender** is *male*.
- My **self-identified gender** is *female*.

In the short list above, the **bolded part** is "the thing to be measured", and the *italicised part* is "the measurement itself". In fact, we can expand on this a little bit, by thinking about the set of possible measurements that could have arisen in each case:

- My **age** (in years) could have been *0, 1, 2, 3 …*, etc. The upper bound on what my age could possibly be is a bit fuzzy, but in practice you'd be safe in saying that the largest possible age is *150*, since no human has ever lived that long.
- When asked if I **like anchovies**, I might have said that *I do*, or *I do not*, or *I have no opinion*, or *I sometimes do*.
- My **chromosomal gender** is almost certainly going to be *male* (XY) or *female* (XX), but there are a few other possibilities. I could also have *Klinfelter's syndrome* (XXY), which is more similar to male than to female. And I imagine there are other possibilities too.
- My **self-identified** gender is also very likely to be male or female, but it doesn't have to agree with my chromosomal gender. I may also choose to identify with *neither*, or to explicitly call myself *transgender*.

As you can see, for some things (like age) it seems fairly obvious what the set of possible measurements should be, whereas for other things it gets a bit tricky. But I want to point out that even in the case of someone's age it's much more subtle than this. For instance, in the example above I assumed that it was okay to measure age in years. But if you're a developmental psychologist, that's way too crude, and so you often measure age in *years and months* (if a child is 2 years and 11 months this is usually written as "2;11"). If you're interested in newborns you might want to measure age in *days since birth*, maybe even *hours since birth*. In other words, the way in which you specify the allowable measurement values is important.

Looking at this a bit more closely, you might also realise that the concept of "age" isn't actually all that precise. In general, when we say "age" we implicitly mean "the length of time since birth". But that's not always the right way to do it. Suppose you're interested in how newborn babies control their eye movements. If you're interested in kids that young, you might also start to worry that "birth" is not the only meaningful point in time to care about. If Baby Alice is born 3 weeks premature and Baby Bianca is born 1 week late, would it really make sense to say that they are the "same age" if we encountered them "2 hours after birth"? In one sense, yes. By social convention we use birth as our reference point for talking about age in everyday life, since it defines the amount of time the person has been operating as an independent entity in the world. But from a scientific perspective that's not the only thing we care about. When we think about the biology of human beings, it's often useful to think of ourselves as organisms that have been growing and maturing since conception, and from that perspective Alice and Bianca aren't the same age at all. So you might want to define the concept of "age" in two different ways: the length of time since conception and the length of time since birth. When dealing with adults it won't make much difference, but when dealing with newborns it might.

Moving beyond these issues, there's the question of methodology. What specific "measurement method" are you going to use to find out someone's age? As before, there are lots of different possibilities:

- You could just ask people "how old are you?" The method of self-report is fast, cheap and easy. But it only works with people old enough to understand the question, and some people lie about their age.
- You could ask an authority (e.g., a parent) "how old is your child?" This method is fast, and when dealing with kids it's not all that hard since the parent is almost always around. It doesn't work as well if you want to know "age since conception", since a lot of parents can't say for sure when conception took place. For that, you might need a different authority (e.g., an obstetrician).
- You could look up official records, for example birth or death certificates. This is a time consuming and frustrating endeavour, but it has its uses (e.g., if the person is now dead).

2.1.2 Operationalisation: defining your measurement

All of the ideas discussed in the previous section relate to the concept of **operationalisation**. To be a bit more precise about the idea, operationalisation is the process by which we take a meaningful but somewhat vague concept and turn it into a precise measurement. The process of operationalisation can involve several different things:

- Being precise about what you are trying to measure. For instance, does "age" mean "time since birth" or "time since conception" in the context of your research?

- Determining what method you will use to measure it. Will you use self-report to measure age, ask a parent, or look up an official record? If you're using self-report, how will you phrase the question?

- Defining the set of allowable values that the measurement can take. Note that these values don't always have to be numerical, though they often are. When measuring age the values are numerical, but we still need to think carefully about what numbers are allowed. Do we want age in years, years and months, days, or hours? For other types of measurements (e.g., gender) the values aren't numerical. But, just as before, we need to think about what values are allowed. If we're asking people to self-report their gender, what options do we allow them to choose between? Is it enough to allow only "male" or "female"? Do you need an "other" option? Or should we not give people specific options and instead let them answer in their own words? And if you open up the set of possible values to include all verbal response, how will you interpret their answers?

Operationalisation is a tricky business, and there's no "one, true way" to do it. The way in which you choose to operationalise the informal concept of "age" or "gender" into a formal measurement depends on what you need to use the measurement for. Often you'll find that the community of scientists who work in your area have some fairly well-established ideas for how to go about it. In other words, operationalisation needs to be thought through on a case-by-case basis. Nevertheless, while there are a lot of issues that are specific to each individual research project, there are some aspects to it that are pretty general.

Before moving on I want to take a moment to clear up our terminology, and in the process introduce one more term. Here are four different things that are closely related to each other:

- **A theoretical construct.** This is the thing that you're trying to take a measurement of, like "age", "gender" or an "opinion". A theoretical construct can't be directly observed, and often they're actually a bit vague.
- **A measure.** The measure refers to the method or the tool that you use to make your observations. A question in a survey, a behavioural observation or a brain scan could all count as a measure.
- **An operationalisation.** The term "operationalisation" refers to the logical connection between the measure and the theoretical construct, or to the process by which we try to derive a measure from a theoretical construct.
- **A variable.** Finally, a new term. A variable is what we end up with when we apply our measure to something in the world. That is, variables are the actual "data" that we end up with in our data sets.

In practice, even scientists tend to blur the distinction between these things, but it's very helpful to try to understand the differences.

2.2 Scales of measurement

As the previous section indicates, the outcome of a psychological measurement is called a variable. But not all variables are of the same qualitative type and so it's useful to understand what types there are. A very useful concept for distinguishing between different types of variables is what's known as **scales of measurement**.

2.2.1 Nominal scale

A **nominal scale** variable (also referred to as a **categorical** variable) is one in which there is no particular relationship between the different possibilities. For these kinds of variables it doesn't make any sense to say that one of them is "bigger" or "better" than any other one, and it absolutely doesn't make any sense to average them. The classic example for this is "eye colour". Eyes can be blue, green or brown, amongst other possibilities, but none of them is any "bigger" than any other one. As a result, it would feel really weird to talk about an "average eye colour". Similarly, gender is nominal too: male isn't better or worse than female. Neither does it make sense to try to talk about an "average gender". In short, nominal scale variables are those for which the only thing you can say about the different possibilities is that they are different. That's it.

Let's take a slightly closer look at this. Suppose I was doing research on how people commute to and from work. One variable I would have to measure would be what kind of transportation people use to get to work. This "transport type" variable could have quite a few possible values, including: "train", "bus", "car", "bicycle". For now, let's suppose that these four are the only possibilities. Then imagine that I ask 100 people how they got to work today, with this result (Table 2.1).

Table 2.1: How did 100 people get to work today

Transportation	Number of people
(1) Train	12
(2) Bus	30
(3) Car	48
(4) Bicycle	10

So, what's the average transportation type? Obviously, the answer here is that there isn't one. It's a silly question to ask. You can say that travel by car is the most popular method, and travel by train is the least popular method, but that's about all. Similarly, notice that the order in which I list the options isn't very interesting. I could have chosen to display the data like in Table 2.2.

Table 2.2: How did 100 people get to work today, a different view

Transportation	Number of people
(3) Car	48
(1) Train	12
(4) Bicycle	10
(2) Bus	30

…and nothing really changes.

2.2.2 Ordinal scale

Ordinal scale variables have a bit more structure than nominal scale variables, but not by a lot. An ordinal scale variable is one in which there is a natural, meaningful way to order the different possibilities, but you can't do anything else. The usual example given of an ordinal variable is "finishing position in a race". You *can* say that the person who finished first was faster than the person who finished second, but you *do not* know how much faster. As a consequence we know that 1st > 2nd, and we know that 2nd > 3rd, but the difference between 1st and 2nd might be much larger than the difference between 2nd and 3rd.

Here's a more psychologically interesting example. Suppose I'm interested in people's attitudes to climate change. I then go and ask some people to pick the statement (from four listed statements) that most closely matches their beliefs:

1. Temperatures are rising because of human activity
2. Temperatures are rising but we don't know why
3. Temperatures are rising but not because of humans

4. Temperatures are not rising

Notice that these four statements actually do have a natural ordering, in terms of "the extent to which they agree with the current science". Statement 1 is a close match, statement 2 is a reasonable match, statement 3 isn't a very good match, and statement 4 is in strong opposition to current science. So, in terms of the thing I'm interested in (the extent to which people endorse the science), I can order the items as $1 > 2 > 3 > 4$. Since this ordering exists, it would be very weird to list the options like this…

1. Temperatures are rising but not because of humans
2. Temperatures are rising because of human activity
3. Temperatures are not rising
4. Temperatures are rising but we don't know why

…because it seems to violate the natural "structure" to the question.

So, let's suppose I asked 100 people these questions, and got the answers shown in Table 2.3.

Table 2.3: Attitudes to climate change

Response	Number
(1) Temperatures are rising because of human activity	51
(2) Temperatures are rising but we do not know why	20
(3) Temperatures are rising but not because of humans	10
(4) Temperatures are not rising	19

When analysing these data it seems quite reasonable to try to group (1), (2) and (3) together, and say that 81 out of 100 people were willing to at *least partially* endorse the science. And it's also quite reasonable to group (2), (3) and (4) together and say that 49 out of 100 people registered *at least some disagreement* with the dominant scientific view. However, it would be entirely bizarre to try to group (1), (2) and (4) together and say that 90 out of 100 people said… what? There's nothing sensible that allows you to group those responses together at all.

That said, notice that while we *can* use the natural ordering of these items to construct sensible groupings, what we can't do is average them. For instance, in my simple example here, the "average" response to the question is 1.97. If you can tell me what that means I'd love to know, because it seems like gibberish to me!

2.2.3 Interval scale

In contrast to nominal and ordinal scale variables, **interval scale** and ratio scale variables are variables for which the numerical value is genuinely meaningful. In the case of interval scale variables the *differences* between the numbers are interpretable, but the variable doesn't have a "natural" zero value. A good example of an interval scale variable is measuring temperature in degrees celsius. For instance, if it was 15° yesterday and 18° today, then the 3° difference between the two is genuinely meaningful. Moreover, that 3° difference is *exactly the same* as the 3° difference between 7° and 10°. In short, addition and subtraction are meaningful for interval scale variables.[7]

However, notice that the 0° does not mean "no temperature at all". It actually means "the temperature at which water freezes", which is pretty arbitrary. As a consequence it becomes pointless to try to multiply and divide temperatures. It is wrong to say that 20° is twice as hot as 10°, just as it is weird and meaningless to try to claim that 20° is negative two times as hot as -10°.

Again, lets look at a more psychological example. Suppose I'm interested in looking at how the attitudes of first-year university students have changed over time. Obviously, I'm going to want to record the year in which each student started. This is an interval scale variable. A student who started in 2003 did arrive 5 years before a student who started in 2008. However, it would be completely daft for me to divide 2008 by 2003 and say that the second student started "1.0024 times later" than the first one. That doesn't make any sense at all.

2.2.4 Ratio scale

The fourth and final type of variable to consider is a **ratio scale** variable, in which zero really means zero, and it's okay to multiply and divide. A good psychological example of a ratio scale variable is response time (RT). In a lot of tasks it's very common to record the amount of time somebody takes to solve a problem or answer a question, because it's an indicator of how difficult the task is. Suppose that Alan takes 2.3 seconds to respond to a question, whereas Ben takes 3.1 seconds. As with an interval scale variable, addition and subtraction are both meaningful here. Ben really did take 3.1 - 2.3 = 0.8 seconds longer than Alan did. However, notice that multiplication and division also make sense here too: Ben took 3.1/2.3 = 1.35 times as long as Alan did to answer the question. And the reason why you can do this is that, for a ratio scale variable such as RT, "zero seconds" really does mean "no time at all".

2.2.5 Continuous versus discrete variables

There's a second kind of distinction that you need to be aware of, regarding what types of variables you can run into. This is the distinction between continuous variables and discrete variables (Table 2.4). The difference between these is as follows:

- A **continuous variable** is one in which, for any two values that you can think of, it's always logically possible to have another value in between.
- A **discrete variable** is, in effect, a variable that isn't continuous. For a discrete variable it's sometimes the case that there's nothing in the middle.

These definitions probably seem a bit abstract, but they're pretty simple once you see some examples. For instance, response time is continuous. If Alan takes 3.1 seconds and Ben takes 2.3 seconds to respond to a question, then Cameron's response time will lie in between if he took 3.0 seconds. And of course it would also be possible for David to take 3.031 seconds to respond, meaning that his RT would lie in between Cameron's and Alan's. And while in practice it might be impossible to measure RT that precisely, it's certainly possible in principle. Because we can always find a new value for RT in between any two other ones we regard RT as a continuous measure.

Table 2.4: The relationship between the scales of measurement and the discrete/continuity distinction. Cells with a tick mark correspond to things that are possible

	continuous	discrete
nominal		✓
ordinal		✓
interval	✓	✓
ratio	✓	✓

Discrete variables occur when this rule is violated. For example, nominal scale variables are always discrete. There isn't a type of transportation that falls "in between" trains and bicycles, not in the strict mathematical way that 2.3 falls in between 2 and 3. So transportation type is discrete. Similarly, ordinal scale variables are always discrete. Although "2nd place" does fall between "1st place" and "3rd place", there's nothing that can logically fall in between "1st place" and "2nd place". Interval scale and ratio scale variables can go either way. As we saw above, response time (a ratio scale variable) is continuous. Temperature in degrees celsius (an interval scale variable) is also continuous. However, the year you went to school (an interval scale variable) is discrete. There's no year in between 2002 and 2003. The number of questions you get right on a true-or-false test (a ratio scale variable) is also discrete. Since a true-or-false question doesn't allow you to be "partially correct", there's nothing in between 5/10 and 6/10. Table 2.4 summarises the relationship between the scales of measurement and the discrete/continuity distinction. Cells with a tick mark correspond to things that are possible. I'm trying to hammer this point home, because (a) some textbooks get this wrong, and (b) people very often say things like "discrete variable" when they mean "nominal scale variable". It's very unfortunate.

2.2.6 Some complexities

Okay, I know you're going to be shocked to hear this, but the real world is much messier than this little classification scheme suggests. Very few variables in real life actually fall into these nice neat categories, so you need to be kind of careful not to treat the scales of measurement as if they were hard and fast rules. It doesn't work like that. They're guidelines, intended to help you think about the situations in which you should treat different variables differently. Nothing more.

So let's take a classic example, maybe *the* classic example, of a psychological measurement tool: the **Likert scale**. The humble Likert scale is the bread and butter tool of all survey design. You yourself have filled out hundreds, maybe thousands, of them and odds are you've even used one yourself. Suppose we have a survey question that looks like this:

> Which of the following best describes your opinion of the statement that "all pirates are freaking awesome"?

and then the options presented to the participant are these:

1. Strongly disagree
2. Disagree
3. Neither agree nor disagree
4. Agree
5. Strongly agree

This set of items is an example of a 5-point Likert scale, in which people are asked to choose among one of several (in this case 5) clearly ordered possibilities, generally with a verbal descriptor given in each case. However, it's not necessary that all items are explicitly described. This is a perfectly good example of a 5-point Likert scale too:

1. Strongly disagree
2.
3.
4.
5. Strongly agree

Likert scales are very handy, if somewhat limited, tools. The question is what kind of variable are they? They're obviously discrete, since you can't give a response of 2.5. They're obviously not nominal scale, since the items are ordered; and they're not ratio scale either, since there's no natural zero.

But are they ordinal scale or interval scale? One argument says that we can't really prove that the difference between "strongly agree" and "agree" is of the same size as the difference between "agree" and "neither agree nor disagree". In fact, in everyday life it's pretty obvious that they're not the same at all. So this suggests that we ought to treat Likert scales as ordinal variables. On the other hand, in practice most participants do seem to take the whole "on a scale from 1 to 5" part fairly seriously, and they tend to act as if the differences between the five response options were fairly similar to one another. As a consequence, a lot of researchers treat Likert scale data as interval scale.[8] It's not interval scale, but in practice it's close enough that we usually think of it as being **quasi-interval scale**.

2.3 Assessing the reliability of a measurement

At this point we've thought a little bit about how to operationalise a theoretical construct and thereby create a psychological measure. And we've seen that by applying psychological measures we end up with variables, which can come in many different types. At this point, we should start discussing the obvious question: is the measurement any good? We'll do this in terms of two related ideas: *reliability* and *validity*. Put simply, the **reliability** of a measure tells you how precisely you are measuring something, whereas the **validity** of a measure tells you how accurate the measure is. In this section we'll talk about reliability; we'll talk about validity in the section on Assessing the validity of a study.

Reliability is actually a very simple concept. It refers to the repeatability or consistency of your measurement. The measurement of my weight by means of a "bathroom scale" is very reliable. If I step on and off the scales over and over again, it'll keep giving me the same answer. Measuring my intelligence by means of "asking my mum" is very unreliable. Some days she tells me I'm a bit thick, and other days she tells me I'm a complete idiot. Notice that this concept of reliability is different to the question of whether the measurements are correct (the correctness of a measurement relates to it's validity). If I'm holding a sack of potatoes when I step on and off the bathroom scales the measurement will still be reliable: it will always give me the same answer. However, this highly reliable answer doesn't match up to my true weight at all, therefore it's wrong. In technical terms, this is a reliable but invalid measurement. Similarly, whilst my mum's estimate of my intelligence is a bit unreliable, she might be right. Maybe I'm just not too bright, and so while her estimate of my intelligence fluctuates pretty wildly from day to day, it's basically right. That would be an unreliable but valid measure. Of course, if my mum's estimates are too unreliable it's going to be very hard to figure out which one of her many claims about my intelligence is actually the right one. To some extent, then, a very unreliable measure tends to end up being invalid for practical purposes; so much so that many people would say that reliability is necessary (but not sufficient) to ensure validity.

Okay, now that we're clear on the distinction between reliability and validity, let's have a think about the different ways in which we might measure reliability:

- **Test-retest reliability**. This relates to consistency over time. If we repeat the measurement at a later date do we get the same answer?
- **Inter-rater reliability**. This relates to consistency across people. If someone else repeats the measurement (e.g., someone else rates my intelligence) will they produce the same answer?
- **Parallel forms reliability**. This relates to consistency across theoretically-equivalent measurements. If I use a different set of bathroom scales to measure my weight does it give the same answer?
- **Internal consistency reliability**. If a measurement is constructed from lots of different parts that perform similar functions (e.g., a personality questionnaire result is added up across several questions) do the individual parts tend to give similar answers. We'll look at this particular form of reliability later in the book, in Section 15.5.

Not all measurements need to possess all forms of reliability. For instance, educational assessment can be thought of as a form of measurement. One of the subjects that I teach, *Computational Cognitive Science*, has an assessment structure that has a research component and an exam component (plus other things). The exam component is *intended* to measure something different from the research component, so the assessment as a whole has low internal consistency. However, within the exam there are several questions that are intended to (approximately) measure the same things, and those tend to produce similar outcomes. So the exam on its own has a fairly high internal consistency. Which is as it should be. You should only demand reliability in those situations where you want to be measuring the same thing!

2.4 The "role" of variables: predictors and outcomes

I've got one last piece of terminology that I need to explain to you before moving away from variables. Normally, when we do some research we end up with lots of different variables. Then, when we analyse our data, we usually try to explain some of the variables in terms of some of the other variables. It's important to keep the two roles "thing doing the explaining" and "thing being explained" distinct. So let's be clear about this now. First, we might as well get used to the idea of using mathematical symbols to describe variables, since it's going to happen over and over again. Let's denote the "to be explained" variable Y, and denote the variables "doing the explaining" as X_1, X_2, etc.

When we are doing an analysis we have different names for X and Y, since they play different roles in the analysis. The classical names for these roles are **independent variable** (IV) and **dependent variable** (DV). The IV is the variable that you use to do the explaining (i.e., X) and the DV is the variable being explained (i.e., Y). The logic behind these names goes like this: if there really is a relationship between X and Y then we can say that Y depends on X, and if we have designed our study "properly" then X isn't dependent on anything else. However, I personally find those names horrible. They're hard to remember and they're highly misleading because (a) the IV is never actually "independent of everything else", and (b) if there's no relationship then the DV doesn't actually depend on the IV. And in fact, because I'm not the only person who thinks that IV and DV are just awful names, there are a number of alternatives that I find more appealing. The terms that I'll use in this book are **predictors** and **outcomes**. The idea here is that what you're trying to do is use X (the predictors) to make guesses about Y (the outcomes).[9] This is summarised in Table 2.5.

Table 2.5: Variable distinctions

role of the variable	classical name	modern name
"to be explained"	dependent variable (DV)	outcome
"to do the explaining"	independent variable (IV)	predictor

2.5 Experimental and non-experimental research

One of the big distinctions that you should be aware of is the distinction between "experimental research" and "non-experimental research". When we make this distinction, what we're really talking about is the degree of control that the researcher exercises over the people and events in the study.

2.5.1 Experimental research

The key feature of **experimental research** is that the researcher controls all aspects of the study, especially what participants experience during the study. In particular, the researcher manipulates or varies the predictor variables (IVs) but allows the outcome variable (DV) to vary naturally. The idea here is to deliberately vary the predictors (IVs) to see if they have any causal effects on the outcomes. Moreover, in order to ensure that there's no possibility that something other than the predictor variables is causing the outcomes, everything else is kept constant or is in some other way "balanced", to ensure that they have no effect on the results. In practice, it's almost impossible to *think* of everything else that might have an influence on the outcome of an experiment, much less keep it constant. The standard solution to this is **randomisation**. That is, we randomly assign people to different groups, and then give each group a different treatment (i.e., assign them different values of the predictor variables). We'll talk more about randomisation later, but for now it's enough to say that what randomisation does is minimise (but not eliminate) the possibility that there are any systematic difference between groups.

Let's consider a very simple, completely unrealistic and grossly unethical example. Suppose you wanted to find out if smoking causes lung cancer. One way to do this would be to find people who smoke and people who don't smoke and look to see if smokers have a higher rate of lung cancer. This is *not* a proper experiment, since the researcher doesn't have a lot of control over who is and isn't a smoker. And this really matters. For instance, it might be that people who choose to smoke cigarettes also tend to have poor diets, or maybe they tend to work in asbestos mines, or whatever. The point here is that the groups (smokers and non-smokers) actually differ on lots of things, not just smoking. So it might be that the higher incidence of lung cancer among smokers is caused by something else, and not by smoking per se. In technical terms these other things (e.g., diet) are called "confounders", and we'll talk about those in just a moment.

In the meantime, let's consider what a proper experiment might look like. Recall that our concern was that smokers and non-smokers might differ in lots of ways. The solution, as long as you have no ethics, is to control who smokes and who doesn't. Specifically, if we randomly divide young non-smokers into two groups and force half of them to become smokers, then it's very unlikely that the groups will differ in any respect other than the fact that half of them smoke. That way, if our smoking group gets cancer at a higher rate than the non-smoking group, we can feel pretty confident that (a) smoking does cause cancer and (b) we're murderers.

2.5.2 Non-experimental research

Non-experimental research is a broad term that covers "any study in which the researcher doesn't have as much control as they do in an experiment". Obviously, control is something that scientists like to have, but as the previous example illustrates there are lots of situations in which you can't or shouldn't try to obtain that control. Since it's grossly unethical (and almost certainly criminal) to force people to smoke in order to find out if they get cancer, this is a good example of a situation in which you really shouldn't try to obtain experimental control. But there are other reasons too. Even leaving aside the ethical issues, our "smoking experiment" does have a few other issues. For instance, when I suggested that we "force" half of the people to become smokers, I was talking about *starting* with a sample of non-smokers, and then forcing them to become smokers. While this sounds like the kind of solid, evil experimental design that a mad scientist would love, it might not be a very sound way of investigating the effect in the real world. For instance, suppose that smoking only causes lung cancer when people have poor diets, and suppose also that people who normally smoke do tend to have poor diets. However, since the "smokers" in our experiment aren't "natural" smokers (i.e., we forced non-smokers to become smokers, but they didn't take on all of the other normal, real-life characteristics that smokers might tend to possess) they probably have better diets. As such, in this silly example they wouldn't get lung cancer and our experiment will fail, because it violates the structure of the "natural" world (the technical name for this is an "artefactual" result).

One distinction worth making between two types of non-experimental research is the difference between **quasi-experimental research** and **case studies**. The example I discussed earlier, in which we wanted to examine incidence of lung cancer among smokers and non-smokers without trying to control who smokes and who doesn't, is a quasi-experimental design. That is, it's the same as an experiment but we don't control the predictors (IVs). We can still use statistics to analyse the results, but we have to be a lot more careful and circumspect.

The alternative approach, case studies, aims to provide a very detailed description of one or a few instances. In general, you can't use statistics to analyse the results of case studies and it's usually very hard to draw any general conclusions about "people in general" from a few isolated examples. However, case studies are very useful in some situations. Firstly, there are situations where you don't have any alternative. Neuropsychology has this issue a lot. Sometimes, you just can't find a lot of people with brain damage in a specific brain area, so the only thing you can do is describe those cases that you do have in as much detail and with as much care as you can. However, there's also some genuine advantages to case studies. Because you don't have as many people to study you have the ability to invest lots of time and effort trying to understand the specific factors at play in each case. This is a very valuable thing to do. As a consequence, case studies can complement the more statistically-oriented approaches that you see in experimental and quasi-experimental designs. We won't talk much about case studies in this book, but they are nevertheless very valuable tools!

2.6 Assessing the validity of a study

More than any other thing, a scientist wants their research to be "valid". The conceptual idea behind **validity** is very simple. Can you trust the results of your study? If not, the study is invalid. However, whilst it's easy to state, in practice it's much harder to check validity than it is to check reliability. And in all honesty, there's no precise, clearly agreed upon notion of what validity actually is. In fact, there are lots of different kinds of validity, each of which raises its own issues. And not all forms of validity are relevant to all studies. I'm going to talk about five different types of validity:

- Internal validity.
- External validity.
- Construct validity.

- Face validity.
- Ecological validity.

First, a quick guide as to what matters here. (1) Internal and external validity are the most important, since they tie directly to the fundamental question of whether your study really works. (2) Construct validity asks whether you're measuring what you think you are. (3) Face validity isn't terribly important except insofar as you care about "appearances". (4) Ecological validity is a special case of face validity that corresponds to a kind of appearance that you might care about a lot.

2.6.1 Internal validity

Internal validity refers to the extent to which you are able to draw the correct conclusions about the causal relationships between variables. It's called "internal" because it refers to the relationships between things "inside" the study. Let's illustrate the concept with a simple example. Suppose you're interested in finding out whether a university education makes you write better. To do so, you get a group of first year students, ask them to write a 1000 word essay, and count the number of spelling and grammatical errors they make. Then you find some third-year students, who obviously have had more of a university education than the first-years, and repeat the exercise. And let's suppose it turns out that the third-year students produce fewer errors. And so you conclude that a university education improves writing skills. Right? Except that the big problem with this experiment is that the third-year students are older and they've had more experience with writing things. So it's hard to know for sure what the causal relationship is. Do older people write better? Or people who have had more writing experience? Or people who have had more education? Which of the above is the true cause of the superior performance of the third-years? Age? Experience? Education? You can't tell. This is an example of a failure of internal validity, because your study doesn't properly tease apart the causal relationships between the different variables.

2.6.2 External validity

External validity relates to the **generalisability** or **applicability** of your findings. That is, to what extent do you expect to see the same pattern of results in "real life" as you saw in your study. To put it a bit more precisely, any study that you do in psychology will involve a fairly specific set of questions or tasks, will occur in a specific environment, and will involve participants that are drawn from a particular subgroup (disappointingly often it is college students!). So, if it turns out that the results don't actually generalise or apply to people and situations beyond the ones that you studied, then what you've got is a lack of external validity.

The classic example of this issue is the fact that a very large proportion of studies in psychology will use undergraduate psychology students as the participants. Obviously, however, the researchers don't care *only* about psychology students. They care about people in general. Given that, a study that uses only psychology students as participants always carries a risk of lacking external validity. That is, if there's something "special" about psychology students that makes them different to the general population in some relevant respect, then we may start worrying about a lack of external validity.

That said, it is absolutely critical to realise that a study that uses only psychology students does not necessarily have a problem with external validity. I'll talk about this again later, but it's such a common mistake that I'm going to mention it here. The external validity of a study is threatened by the choice of population if (a) the population from which you sample your participants is very narrow (e.g., psychology students), and (b) the narrow population that you sampled from is systematically different from the general population in some respect that is relevant to the *psychological phenomenon that you intend to study*. The italicised part is the bit that lots of people forget. It is true that psychology undergraduates differ from the general population in lots of ways, and so a study that uses only psychology students may have problems with external validity. However, if those differences aren't very relevant to the phenomenon that you're studying, then there's nothing to worry about. To make this a bit more concrete here are two extreme examples:

- You want to measure "attitudes of the general public towards psychotherapy", but all of your participants are psychology students. This study would almost certainly have a problem with external validity.
- You want to measure the effectiveness of a visual illusion, and your participants are all psychology students. This study is unlikely to have a problem with external validity.

Having just spent the last couple of paragraphs focusing on the choice of participants, since that's a big issue that everyone tends to worry most about, it's worth remembering that external validity is a broader concept. The following are also examples of things that might pose a threat to external validity, depending on what kind of study you're doing:

- People might answer a "psychology questionnaire" in a manner that doesn't reflect what they would do in real life.
- Your lab experiment on (say) "human learning" has a different structure to the learning problems people face in real life.

2.6.3 Construct validity

Construct validity is basically a question of whether you're measuring what you want to be measuring. A measurement has good construct validity if it is actually measuring the correct theoretical construct, and bad construct validity if it doesn't. To give a very simple (if ridiculous) example, suppose I'm trying to investigate the rates with which university students cheat on their exams. And the way I attempt to measure it is by asking the cheating students to stand up in the lecture theatre so that I can count them. When I do this with a class of 300 students 0 people claim to be cheaters. So I therefore conclude that the proportion of cheaters in my class is 0%. Clearly this is a bit ridiculous. But the point here is not that this is a very deep methodological example, but rather to explain what construct validity is. The problem with my measure is that while I'm trying to measure "the proportion of people who cheat" what I'm actually measuring is "the proportion of people stupid enough to own up to cheating, or bloody minded enough to pretend that they do". Obviously, these aren't the same thing! So my study has gone wrong, because my measurement has very poor construct validity.

2.6.4 Face validity

Face validity simply refers to whether or not a measure "looks like" it's doing what it's supposed to, nothing more. If I design a test of intelligence, and people look at it and they say "no, that test doesn't measure intelligence", then the measure lacks face validity. It's as simple as that. Obviously, face validity isn't very important from a pure scientific perspective. After all, what we care about is whether or not the measure *actually* does what it's supposed to do, not whether it *looks like* it does what it's supposed to do. As a consequence, we generally don't care very much about face validity. That said, the concept of face validity serves three useful pragmatic purposes:

- Sometimes, an experienced scientist will have a "hunch" that a particular measure won't work. While these sorts of hunches have no strict evidentiary value, it's often worth paying attention to them. Because often times people have knowledge that they can't quite verbalise, there might be something to worry about even if you can't quite say why. In other words, when someone you trust criticises the face validity of your study, it's worth taking the time to think more carefully about your design to see if you can think of reasons why it might go awry. Mind you, if you don't find any reason for concern, then you should probably not worry. After all, face validity really doesn't matter very much.
- Often (very often), completely uninformed people will also have a "hunch" that your research is crap. And they'll criticise it on the internet or something. On close inspection you may notice that these criticisms are actually focused entirely on how the study "looks", but not on anything deeper. The concept of face validity is useful for gently explaining to people that they need to substantiate their arguments further.
- Expanding on the last point, if the beliefs of untrained people are critical (e.g., this is often the case for applied research where you actually want to convince policy makers of something or other) then you have to care about face validity. Simply because, whether you like it or not, a lot of people will use face validity as a proxy for real validity. If you want the government to change a law on scientific psychological grounds, then it won't matter how good your studies "really" are. If they lack face validity you'll find that politicians ignore you. Of course, it's somewhat unfair that policy often depends more on appearance than fact, but that's how things go.

2.6.5 Ecological validity

Ecological validity is a different notion of validity, which is similar to external validity, but less important. The idea is that, in order to be ecologically valid, the entire set up of the study should closely approximate the real-world scenario that is being investigated. In a sense, ecological validity is a kind of face validity. It relates mostly to whether the study "looks" right, but with a bit more rigour to it. To be ecologically valid the study has to look right in a fairly specific way. The idea behind it is the intuition that a study that is ecologically valid is more likely to be externally valid. It's no guarantee, of course. But the nice thing about ecological validity is that it's much easier to check whether a study is ecologically valid than it is to check whether a study is externally valid. A simple example would be eyewitness identification studies. Most of these studies tend to be done in a university setting, often with a fairly simple array of faces to look at, rather than a line up. The length of time between seeing the "criminal" and being asked to identify the suspect in the "line up" is usually shorter. The "crime" isn't real so there's no chance of the witness being scared, and there are no police officers present so there's not as much chance of feeling pressured. These things all mean that the study definitely lacks ecological validity. They might (but might not) mean that it also lacks external validity.

2.7 Confounders, artefacts and other threats to validity

If we look at the issue of validity in the most general fashion the two biggest worries that we have are *confounders* and *artefacts*. These two terms are defined in the following way:

- **Confounder:** A confounder is an additional, often unmeasured variable[10] that turns out to be related to both the predictors and the outcome. The existence of confounders threatens the internal validity of the study because you can't tell whether the predictor causes the outcome, or if the confounding variable causes it.
- **Artefact:** A result is said to be "artefactual" if it only holds in the special situation that you happened to test in your study. The possibility that your result is an artefact poses a threat to your external validity, because it raises the possibility that you can't generalise or apply your results to the actual population that you care about.

As a general rule confounders are a bigger concern for non-experimental studies, precisely because they're not proper experiments. By definition, you're leaving lots of things uncontrolled, so there's a lot of scope for confounders being present in your study. Experimental research tends to be much less vulnerable to confounders. The more control you have over what happens during the study, the more you can prevent confounders from affecting the results. With random allocation, for example, confounders are distributed randomly, and evenly, between different groups.

However, there are always swings and roundabouts and when we start thinking about artefacts rather than confounders the shoe is very firmly on the other foot. For the most part, artefactual results tend to be more of a concern for experimental studies than for non-experimental studies. To see this, it helps to realise that the reason that a lot of studies are non-experimental is precisely because what the researcher is trying to do is examine human behaviour in a more naturalistic context. By working in a more real-world context you lose experimental control (making yourself vulnerable to confounders), but because you tend to be studying human psychology "in the wild" you reduce the chances of getting an artefactual result. Or, to put it another way, when you take psychology out of the wild and bring it into the lab (which we usually have to do to gain our experimental control), you always run the risk of accidentally studying something different to what you wanted to study.

Be warned though. The above is a rough guide only. It's absolutely possible to have confounders in an experiment, and to get artefactual results with non-experimental studies. This can happen for all sorts of reasons, not least of which is experimenter or researcher error. In practice, it's really hard to think everything through ahead of time and even very good researchers make mistakes.

Although there's a sense in which almost any threat to validity can be characterised as a confounder or an artefact, they're pretty vague concepts. So let's have a look at some of the most common examples.

2.7.1 History effects

History effects refer to the possibility that specific events may occur during the study that might influence the outcome measure. For instance, something might happen in between a pretest and a post-test. Or in-between testing participant 23 and participant 24. Alternatively, it might be that you're looking at a paper from an older study that was perfectly valid for its time, but the world has changed enough since then that the conclusions are no longer trustworthy. Examples of things that would count as history effects are:

- You're interested in how people think about risk and uncertainty. You started your data collection in December 2010. But finding participants and collecting data takes time, so you're still finding new people in February 2011. Unfortunately for you (and even more unfortunately for others), the Queensland floods occurred in January 2011 causing billions of dollars of damage and killing many people. Not surprisingly, the people tested in February 2011 express quite different beliefs about handling risk than the people tested in December 2010. Which (if any) of these reflects the "true" beliefs of participants? I think the answer is probably both. The Queensland floods genuinely changed the beliefs of the Australian public, though possibly only temporarily. The key thing here is that the "history" of the people tested in February is quite different to people tested in December.

- You're testing the psychological effects of a new anti-anxiety drug. So what you do is measure anxiety before administering the drug (e.g., by self-report, and taking physiological measures). Then you administer the drug, and afterwards you take the same measures. In the interim however, because your lab is in Los Angeles, there's an earthquake which increases the anxiety of the participants.

2.7.2 Maturation effects

As with history effects, **maturational effects** are fundamentally about change over time. However, maturation effects aren't in response to specific events. Rather, they relate to how people change on their own over time. We get older, we get tired, we get bored, etc. Some examples of maturation effects are:

- When doing developmental psychology research you need to be aware that children grow up quite rapidly. So, suppose that you want to find out whether some educational trick helps with vocabulary size among 3 year olds. One thing that you need to be aware of is that the vocabulary size of children that age is growing

at an incredible rate (multiple words per day) all on its own. If you design your study without taking this maturational effect into account, then you won't be able to tell if your educational trick works.
- When running a very long experiment in the lab (say, something that lasts for three hours) it's very likely that people will begin to get bored and tired, and that this maturational effect will cause performance to decline regardless of anything else going on in the experiment.

2.7.3 Repeated testing effects

An important type of history effect is the effect of **repeated testing**. Suppose I want to take two measurements of some psychological construct (e.g., anxiety). One thing I might be worried about is if the first measurement has an effect on the second measurement. In other words, this is a history effect in which the "event" that influences the second measurement is the first measurement itself! This is not at all uncommon. Examples of this include:

- Learning and practice: e.g., "intelligence" at time 2 might appear to go up relative to time 1 because participants learned the general rules of how to solve "intelligence-test-style" questions during the first testing session.
- Familiarity with the testing situation: e.g., if people are nervous at time 1, this might make performance go down. But after sitting through the first testing situation they might calm down a lot precisely because they've seen what the testing looks like.
- Auxiliary changes caused by testing: e.g., if a questionnaire assessing mood is boring then mood rating at measurement time 2 is more likely to be "bored" precisely because of the boring measurement made at time 1.

2.7.4 Selection bias

Selection bias is a pretty broad term. Suppose that you're running an experiment with two groups of participants where each group gets a different "treatment", and you want to see if the different treatments lead to different outcomes. However, suppose that, despite your best efforts, you've ended up with a gender imbalance across groups (say, group A has 80% females and group B has 50% females). It might sound like this could never happen but, trust me, it can. This is an example of a selection bias, in which the people "selected into" the two groups have different characteristics. If any of those characteristics turns out to be relevant (say, your treatment works better on females than males) then you're in a lot of trouble.

2.7.5 Differential attrition

When thinking about the effects of attrition, it is sometimes helpful to distinguish between two different types. The first is **homogeneous attrition**, in which the attrition effect is the same for all groups, treatments or conditions. In the example I gave above, the attrition would be homogeneous if (and only if) the easily bored participants are dropping out of all of the conditions in my experiment at about the same rate. In general, the main effect of homogeneous attrition is likely to be that it makes your sample unrepresentative. As such, the biggest worry that you'll have is that the generalisability of the results decreases. In other words, you lose external validity.

The second type of attrition is **heterogeneous attrition**, in which the attrition effect is different for different groups. More often called **differential attrition**, this is a kind of selection bias that is caused by the study itself. Suppose that, for the first time ever in the history of psychology, I manage to find the perfectly balanced and representative sample of people. I start running "Dani's incredibly long and tedious experiment" on my perfect sample but then, because my study is incredibly long and tedious, lots of people start dropping out. I can't stop this. Participants absolutely have the right to stop doing any experiment, any time, for whatever reason they feel like, and as researchers we are morally (and professionally) obliged to remind people that they do have this right. So, suppose that "Dani's incredibly long and tedious experiment" has a very high drop out rate. What do you suppose the odds are that this drop out is random? Answer: zero. Almost certainly the people who remain are more conscientious, more tolerant of boredom, etc., than those who leave. To the extent that (say) conscientiousness is relevant to the psychological phenomenon that I care about, this attrition can decrease the validity of my results.

Here's another example. Suppose I design my experiment with two conditions. In the "treatment" condition, the experimenter insults the participant and then gives them a questionnaire designed to measure obedience. In the "control" condition, the experimenter engages in a bit of pointless chitchat and then gives them the questionnaire. Leaving aside the questionable scientific merits and dubious ethics of such a study, let's have a think about what might go wrong here. As a general rule, when someone insults me to my face I tend to get much less co-operative. So, there's a pretty good chance that a lot more people are going to drop out of the treatment condition than the control condition. And this drop out isn't going to be random. The people most likely to drop out would probably be the people who don't care all that much about the importance of obediently sitting through the experiment. Since the most bloody minded and disobedient people all left the treatment group but not the control group, we've introduced a confounder: the people who actually took the questionnaire in the treatment group were already more likely to be dutiful and obedient than the people in the control group. In short, in this study insulting people doesn't make them more obedient. It makes the more disobedient people leave the experiment! The internal validity of this experiment is completely shot.

2.7.6 Non-response bias

Non-response bias is closely related to selection bias and to differential attrition. The simplest version of the problem goes like this. You mail out a survey to 1000 people but only 300 of them reply. The 300 people who replied are almost certainly not a random subsample. People who respond to surveys are systematically different to people who don't. This introduces a problem when trying to generalise from those 300 people who replied to the population at large, since you now have a very non-random sample. The issue of non-response bias is more general than this, though. Among the (say) 300 people that did respond to the survey, you might find that not everyone answers every question. If (say) 80 people chose not to answer one of your questions, does this introduce problems? As always, the answer is maybe. If the question that wasn't answered was on the last page of the questionnaire, and those 80 surveys were returned with the last page missing, there's a good chance that the missing data isn't a big deal; probably the pages just fell off. However, if the question that 80 people didn't answer was the most confrontational or invasive personal question in the questionnaire, then almost certainly you've got a problem. In essence, what you're dealing with here is what's called the problem of **missing data**. If the data that is missing was "lost" randomly, then it's not a big problem. If it's missing systematically, then it can be a big problem.

2.7.7 Regression to the mean

Regression to the mean refers to any situation where you select data based on an extreme value on some measure. Because the variable has natural variation it almost certainly means that when you take a subsequent measurement the later measurement will be less extreme than the first one, purely by chance.

Here's an example. Suppose I'm interested in whether a psychology education has an adverse effect on very smart kids. To do this, I find the 20 Psychology I students with the best high school grades and look at how well they're doing at university. It turns out that they're doing a lot better than average, but they're not topping the class at university even though they did top their classes at high school. What's going on? The natural first thought is that this must mean that the psychology classes must be having an adverse effect on those students. However, while that might very well be the explanation, it's more likely that what you're seeing is an example of "regression to the mean". To see how it works, let's take a moment to think about what is required to get the best mark in a class, regardless of whether that class be at high school or at university. When you've got a big class there are going to be lots of very smart people enrolled. To get the best mark you have to be very smart, work very hard, and be a bit lucky. The exam has to ask just the right questions for your idiosyncratic skills, and you have to avoid making any dumb mistakes (we all do that sometimes) when answering them. And that's the thing, whilst intelligence and hard work are transferable from one class to the next, luck isn't. The people who got lucky in high school won't be the same as the people who get lucky at university. That's the very definition of "luck". The consequence of this is that when you select people at the very extreme values of one measurement (the top 20 students), you're selecting for hard work, skill and luck. But because the luck doesn't transfer to the second measurement (only the skill and work), these people will all be expected to drop a little bit when you measure them a second time (at university). So their scores fall back a little bit, back towards everyone else. This is regression to the mean.

Regression to the mean is surprisingly common. For instance, if two very tall people have kids their children will tend to be taller than average but not as tall as the parents. The reverse happens with very short parents. Two very short parents will tend to have short children, but nevertheless those kids will tend to be taller than the parents. It can also be extremely subtle. For instance, there have been studies done that suggested that people learn better from negative feedback than from positive feedback. However, the way that people tried to show this was to give people positive reinforcement whenever they did good, and negative reinforcement when they did bad. And what you see is that after the positive reinforcement people tended to do worse, but after the negative reinforcement they tended to do better. But notice that there's a selection bias here! When people do very well, you're selecting for "high" values, and so you should expect, because of regression to the mean, that performance on the next trial should be worse regardless of whether reinforcement is given. Similarly, after a bad trial, people will tend to improve all on their own. The apparent superiority of negative feedback is an artefact caused by regression to the mean (see Kahneman & Tversky (1973), for discussion).

2.7.8 Experimenter bias

Experimenter bias can come in multiple forms. The basic idea is that the experimenter, despite the best of intentions, can accidentally end up influencing the results of the experiment by subtly communicating the "right answer" or the "desired behaviour" to the participants. Typically, this occurs because the experimenter has special knowledge that the participant does not, for example the right answer to the questions being asked or knowledge of the expected pattern of performance for the condition that the participant is in. The classic example of this happening is the case study of "Clever Hans", which dates back to 1907 (Pfungst, 1911). Clever Hans was a horse that apparently was able to read and count and perform other human like feats of intelligence. After Clever Hans became famous, psychologists started examining his behaviour more closely. It turned out that, not surprisingly, Hans didn't know how to do maths. Rather, Hans was responding to the human observers around him, because the humans did know how to count and the horse had learned to change its behaviour when people changed theirs.

The general solution to the problem of experimenter bias is to engage in double blind studies, where neither the experimenter nor the participant knows which condition the participant is in or knows what the desired behaviour is. This provides a very good solution to the problem, but it's important to recognise that it's not quite ideal, and hard to pull off perfectly. For instance, the obvious way that I could try to construct a double blind study is to have one of my Ph.D. students (one who doesn't know anything about the experiment) run the study. That feels like it should be enough. The only person (me) who knows all the details (e.g., correct answers to the questions, assignments of participants to conditions) has no interaction with the participants, and the person who does all the talking to people (the Ph.D. student) doesn't know anything. Except for the reality that the last part is very unlikely to be true. In order for the Ph.D. student to run the study effectively they need to have been briefed by me, the researcher. And, as it happens, the Ph.D. student also knows me and knows a bit about my general beliefs about people and psychology (e.g., I tend to think humans are much smarter than psychologists give them credit for). As a result of all this, it's almost impossible for the experimenter to avoid knowing a little bit about what expectations I have. And even a little bit of knowledge can have an effect. Suppose the experimenter accidentally conveys the fact that the participants are expected to do well in this task. Well, there's a thing called the "Pygmalion effect", where if you expect great things of people they'll tend to rise to the occasion. But if you expect them to fail then they'll do that too. In other words, the expectations become a self-fulfilling prophecy.

2.7.9 Demand effects and reactivity

When talking about experimenter bias, the worry is that the experimenter's knowledge or desires for the experiment are communicated to the participants, and that these can change people's behaviour (Rosenthal, 1966). However, even if you manage to stop this from happening, it's almost impossible to stop people from knowing that they're part of a psychological study. And the mere fact of knowing that someone is watching or studying you can have a pretty big effect on behaviour. This is generally referred to as **reactivity** or **demand effects**. The basic idea is captured by the Hawthorne effect: people alter their performance because of the attention that the study focuses on them. The effect takes its name from a study that took place in the "Hawthorne Works" factory outside of Chicago (see Adair (1984)). This study, from the 1920s, looked at the effects of factory lighting on worker productivity. But, importantly, change in worker behaviour occurred because the workers knew they were being studied, rather than any effect of factory lighting.

To get a bit more specific about some of the ways in which the mere fact of being in a study can change how people behave, it helps to think like a social psychologist and look at some of the roles that people might adopt during an experiment but might not adopt if the corresponding events were occurring in the real world:

- The *good participant* tries to be too helpful to the researcher. He or she seeks to figure out the experimenter's hypotheses and confirm them.
- The *negative participant* does the exact opposite of the good participant. He or she seeks to break or destroy the study or the hypothesis in some way.
- The *faithful participant* is unnaturally obedient. He or she seeks to follow instructions perfectly, regardless of what might have happened in a more realistic setting.
- The *apprehensive participant* gets nervous about being tested or studied, so much so that his or her behaviour becomes highly unnatural, or overly socially desirable.

2.7.10 Placebo effects

The **placebo effect** is a specific type of demand effect that we worry a lot about. It refers to the situation where the mere fact of being treated causes an improvement in outcomes. The classic example comes from clinical trials. If you give people a completely chemically inert drug and tell them that it's a cure for a disease, they will tend to get better faster than people who aren't treated at all. In other words, it is people's belief that they are being treated that causes the improved outcomes, not the drug.

However, the current consensus in medicine is that true placebo effects are quite rare and most of what was previously considered placebo effect is in fact some combination of natural healing (some people just get better on their own), regression to the mean and other quirks of study design. Of interest to psychology is that the strongest evidence for at least some placebo effect is in self-reported outcomes, most notably in treatment of pain (Hróbjartsson & Gøtzsche, 2010).

2.7.11 Situation, measurement and sub-population effects

In some respects, these terms are a catch-all term for "all other threats to external validity". They refer to the fact that the choice of sub-population from which you draw your participants, the location, timing and manner in which you run your study (including who collects the data) and the tools that you use to make your measurements might all be influencing the results. Specifically, the worry is that these things might be influencing the results in such a way that the results won't generalise to a wider array of people, places and measures.

2.7.12 Fraud, deception and self-deception

> *It is difficult to get a man to understand something, when his salary depends on his not understanding it.*
> – Upton Sinclair

There's one final thing I feel I should mention. While reading what the textbooks often have to say about assessing the validity of a study I couldn't help but notice that they seem to make the assumption that the researcher is honest. I find this hilarious. While the vast majority of scientists are honest, in my experience at least, some are not.[11] Not only that, as I mentioned earlier, scientists are not immune to belief bias. It's easy for a researcher to end up deceiving themselves into believing the wrong thing, and this can lead them to conduct subtly flawed research and then hide those flaws when they write it up. So you need to consider not only the (probably unlikely) possibility of outright fraud, but also the (probably quite common) possibility that the research is unintentionally "slanted". I opened a few standard textbooks and didn't find much of a discussion of this problem, so here's my own attempt to list a few ways in which these issues can arise:

- **Data fabrication**. Sometimes, people just make up the data. This is occasionally done with "good" intentions. For instance, the researcher believes that the fabricated data do reflect the truth, and may actually reflect "slightly cleaned up" versions of actual data. On other occasions, the fraud is deliberate and malicious. Some high-profile examples where data fabrication has been alleged or shown include Cyril Burt (a psychologist who is thought to have fabricated some of his data), Andrew Wakefield (who has been accused of fabricating his data connecting the MMR vaccine to autism) and Hwang Woo-suk (who falsified a lot of his data on stem cell research).
- **Hoaxes**. Hoaxes share a lot of similarities with data fabrication, but they differ in the intended purpose. A hoax is often a joke, and many of them are intended to be (eventually) discovered. Often, the point of a hoax is to discredit someone or some field. There's quite a few well known scientific hoaxes that have occurred over the years (e.g., Piltdown man) and some were deliberate attempts to discredit particular fields of research (e.g., the Sokal affair).
- **Data misrepresentation**. While fraud gets most of the headlines, it's much more common in my experience to see data being misrepresented. When I say this I'm not referring to newspapers getting it wrong (which they do, almost always). I'm referring to the fact that often the data don't actually say what the researchers

think they say. My guess is that, almost always, this isn't the result of deliberate dishonesty but instead is due to a lack of sophistication in the data analyses. For instance, think back to the example of Simpson's paradox that I discussed in the beginning of this book. It's very common to see people present "aggregated" data of some kind and sometimes, when you dig deeper and find the raw data yourself you find that the aggregated data tell a different story to the disaggregated data. Alternatively, you might find that some aspect of the data is being hidden, because it tells an inconvenient story (e.g., the researcher might choose not to refer to a particular variable). There's a lot of variants on this, many of which are very hard to detect.

- **Study "misdesign"**. Okay, this one is subtle. Basically, the issue here is that a researcher designs a study that has built-in flaws and those flaws are never reported in the paper. The data that are reported are completely real and are correctly analysed, but they are produced by a study that is actually quite wrongly put together. The researcher really wants to find a particular effect and so the study is set up in such a way as to make it "easy" to (artefactually) observe that effect. One sneaky way to do this, in case you're feeling like dabbling in a bit of fraud yourself, is to design an experiment in which it's obvious to the participants what they're "supposed" to be doing, and then let reactivity work its magic for you. If you want you can add all the trappings of double blind experimentation but it won't make a difference since the study materials themselves are subtly telling people what you want them to do. When you write up the results the fraud won't be obvious to the reader. What's obvious to the participant when they're in the experimental context isn't always obvious to the person reading the paper. Of course, the way I've described this makes it sound like it's always fraud. Probably there are cases where this is done deliberately, but in my experience the bigger concern is with unintentional misdesign. The researcher believes and so the study just happens to end up with a built-in flaw, and that flaw then magically erases itself when the study is written up for publication.

- **Data mining and post hoc hypothesising**. Another way in which the authors of a study can more or less misrepresent the data is by engaging in what's referred to as "data mining" (see Gelman and Loken 2014, for a broader discussion of this as part of the "garden of forking paths" in statistical analysis). As we'll discuss later, if you keep trying to analyse your data in lots of different ways, you'll eventually find something that "looks" like a real effect but isn't. This is referred to as "data mining". It used to be quite rare because data analysis used to take weeks, but now that everyone has very powerful statistical software on their computers it's becoming very common. Data mining per se isn't "wrong", but the more that you do it the bigger the risk you're taking. The thing that is wrong, and I suspect is very common, is unacknowledged data mining. That is, the researcher runs every possible analysis known to humanity, finds the one that works, and then pretends that this was the only analysis that they ever conducted. Worse yet, they often "invent" a hypothesis after looking at the data to cover up the data mining. To be clear. It's not wrong to change your beliefs after looking at the data, and to reanalyse your data using your new "post hoc" hypotheses. What is wrong (and I suspect common) is failing to acknowledge what you've done. If you acknowledge that you did it then other researchers are able to take your behaviour into account. If you don't, then they can't. And that makes your behaviour deceptive. Bad!

- **Publication bias and self-censoring**. Finally, a pervasive bias is "non-reporting" of negative results. This is almost impossible to prevent. Journals don't publish every article that is submitted to them. They prefer to publish articles that find "something". So, if 20 people run an experiment looking at whether reading Finnegans Wake causes insanity in humans, and 19 of them find that it doesn't, which one do you think is going to get published? Obviously, it's the one study that did find that Finnegans Wake causes insanity.[12] This is an example of a publication bias. Since no-one ever published the 19 studies that didn't find an effect, a naive reader would never know that they existed. Worse yet, most researchers "internalise" this bias and end up self-censoring their research. Knowing that negative results aren't going to be accepted for publication, they never even try to report them. As a friend of mine says "for every experiment that you get published, you also have 10 failures". And she's right. The catch is, while some (maybe most) of those studies are failures for boring reasons (e.g., you stuffed something up) others might be genuine "null" results that you ought to acknowledge when you write up the "good" experiment. And telling which is which is often hard to do. A good place to start is a paper by Ioannidis (2005) with the depressing title "Why most published research findings are false". I'd also suggest taking a look at work by Kühberger et al. (2014) presenting statistical evidence that this actually happens in psychology.

There's probably a lot more issues like this to think about, but that'll do to start with. What I really want to point out is the blindingly obvious truth that real-world science is conducted by actual humans, and only the most gullible of people automatically assume that everyone else is honest and impartial. Actual scientists aren't usually that naive, but for some reason the world likes to pretend that we are, and the textbooks we usually write seem to reinforce that stereotype.

2.8 Summary

This chapter isn't really meant to provide a comprehensive discussion of psychological research methods. It would require another volume just as long as this one to do justice to the topic. However, in real life statistics and study design are so tightly intertwined that it's very handy to discuss some of the key topics. In this chapter, I've briefly discussed the following topics:

- Introduction to psychological measurement. What does it mean to operationalise a theoretical construct? What does it mean to have variables and take measurements?
- Scales of measurement and types of variables. Remember that there are two different distinctions here. There's the difference between discrete and continuous data, and there's the difference between the four different scale types (nominal, ordinal, interval and ratio).
- Assessing the reliability of a measurement. If I measure the "same" thing twice, should I expect to see the same result? Only if my measure is reliable. But what does it mean to talk about doing the "same" thing? Well, that's why we have different types of reliability. Make sure you remember what they are.

- The "role" of variables: predictors and outcomes. What roles do variables play in an analysis? Can you remember the difference between predictors and outcomes? Dependent and independent variables? Etc.
- Experimental and non-experimental research designs. What makes an experiment an experiment? Is it a nice white lab coat, or does it have something to do with researcher control over variables?
- Assessing the validity of a study. Does your study measure what you want it to? How might things go wrong? And is it my imagination, or was that a very long list of possible ways in which things can go wrong?

All this should make clear to you that study design is a critical part of research methodology. I built this chapter from the classic little book by Campbell & Stanley (1963), but there are of course a large number of textbooks out there on research design. Spend a few minutes with your favourite search engine and you'll find dozens.

Part II

An introduction to jamovi

Chapter 3

Getting started with jamovi

Robots are nice to work with.
– Roger Zelazny[13]

In this chapter I'll discuss how to get started in jamovi. I'll briefly talk about how to download and install jamovi, but most of the chapter will be focused on getting you started with finding your way around the jamovi graphical user interface (GUI). Our goal in this chapter is not to learn any statistical concepts: we're just trying to learn the basics of how jamovi works and get comfortable interacting with the system. To do this we'll spend a bit of time looking at data sets and variables. In doing so, you'll get something of a feel for what it's like to work in jamovi.

However, before going into any of the specifics, it's worth talking a little about why you might want to use jamovi at all. Given that you're reading this you've probably got your own reasons. However, if those reasons are "because that's what my stats class uses", it might be worth explaining a little why your lecturer has chosen to use jamovi for the class. Of course, I don't really know why *other* people choose jamovi so I'm really talking about why I use it.

- It's sort of obvious but worth saying anyway: doing your statistics on a computer is faster, easier and more powerful than doing statistics by hand. Computers excel at mindless repetitive tasks, and a lot of statistical calculations are both mindless and repetitive. For most people the only reason to ever do statistical calculations with pencil and paper is for learning purposes. In my class I do occasionally suggest doing some calculations that way, but the only real value to it is pedagogical. It does help you to get a "feel" for statistics to do some calculations yourself, so it's worth doing it once. But only once!
- Doing statistics in a conventional spreadsheet (e.g., Microsoft Excel) is generally a bad idea in the long run. Although many people likely feel more familiar with them, spreadsheets are very limited in terms of what analyses they allow you do. If you get into the habit of trying to do your real-life data analysis using spreadsheets then you've dug yourself into a very deep hole.
- Avoiding proprietary software is a very good idea. There are a lot of commercial packages out there that you can buy, some of which I like and some of which I

don't. They're usually very glossy in their appearance, and generally very powerful (much more powerful than spreadsheets). However, they're also very expensive: usually, the company sells "student versions" (crippled versions of the real thing) very cheaply; they sell full powered "educational versions" at a price that makes me wince; and they sell commercial licences with a staggeringly high price tag. The business model here is to suck you in during your student days and then leave you dependent on their tools when you go out into the real world. It's hard to blame them for trying, but personally I'm not in favour of shelling out thousands of dollars if I can avoid it. And you can avoid it. If you make use of packages like jamovi that are open source and free you never get trapped having to pay exorbitant licensing fees.

- Something that you might not appreciate now, but will love later on if you do anything involving data analysis, is the fact that jamovi is basically a sophisticated front end for the free R statistical programming language. When you download and install R you get all the basic "packages" and those are very powerful on their own. However, because R is so open and so widely used, it's become something of a standard tool in statistics and so lots of people write their own packages that extend the system. And these are freely available too. One of the consequences of this, I've noticed, is that if you look at recent advanced data analysis textbooks then a lot of them use R.

Those are the main reasons I use jamovi. It's not without its flaws, though. It's relatively new[14] so there is not a huge set of textbooks and other resources to support it, and it has a few annoying quirks that we're all pretty much stuck with, but on the whole I think the strengths outweigh the weakness; more so than any other option I've encountered so far.

3.1 Installing jamovi

Okay, enough with the sales pitch. Let's get started. Just as with any piece of software, jamovi needs to be installed on a "computer", which is a magical box that does cool things and delivers free ponies. Or something along those lines; I may be confusing computers with the iPad marketing campaigns. Anyway, jamovi is freely distributed online and you can download it from the jamovi homepage, which is https://www.jamovi.org/.

At the top of the page under the heading "Download", you'll see separate links for Windows users, Mac users and Linux users. If you follow the relevant link you'll see that the online instructions are pretty self-explanatory. At the time of writing, the current version of jamovi is 2.3, but they usually issue updates every few months, so you'll probably have a newer version.[15]

3.1.1 Starting up jamovi

One way or another, regardless of what operating system you're using, it's time to open jamovi and get started. When first starting jamovi you will be presented with a user interface which looks something like Figure 3.1.

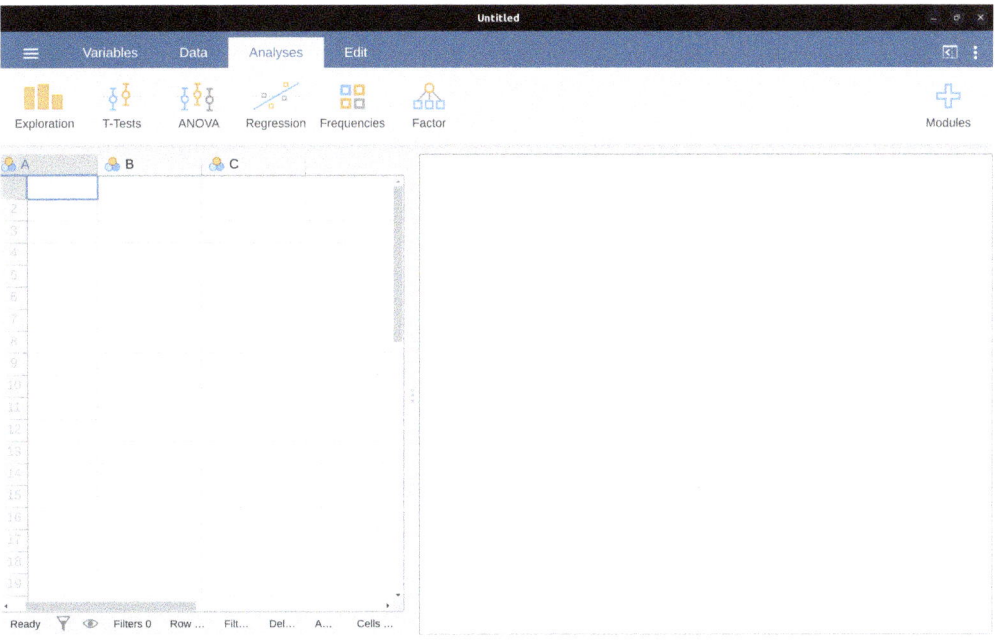

Figure 3.1: jamovi starts up!

To the left is the spreadsheet view, and to the right is where the results of statistical tests appear. Down the middle is a bar separating these two regions and this can be dragged to the left or the right to change their sizes.

It is possible to simply begin typing values into the jamovi spreadsheet as you would in any other spreadsheet software. Alternatively, existing data sets in the csv (.csv) file format can be opened in jamovi. Additionally, you can easily import SPSS, SAS, STATA and JASP files directly into jamovi. To open a file select the 'File'[16] tab (three horizontal lines signify this tab) at the top left hand corner, select 'Open' and then choose from the files listed on 'Browse' depending on whether you want to open an example or a file stored on your computer.

3.2 Analyses

Analyses can be selected from the analysis ribbon or menu along the top. Selecting an analysis will present an 'Options panel' for that particular analysis, allowing you to assign different variables to different parts of the analysis, and select different options. At the same time, the results for the analysis will appear in the right 'Results panel' and will update in real time as you make changes to the options.

When you have the analysis set up correctly you can dismiss the analysis options by clicking the arrow to the top right of the optional panel. If you wish to return to these options, you can click on the results that were produced. In this way, you can return to any analysis that you (or say, a colleague) created earlier.

If you decide you no longer need a particular analysis, you can remove it with the results context menu. Right-clicking on the analysis results will bring up a menu and by selecting 'Analysis' and then 'Remove' the analysis can be removed. But more on this later. First, let's take a more detailed look at the spreadsheet view.

3.3 The spreadsheet

In jamovi data is represented in a spreadsheet with each column representing a 'variable' and each row representing a 'case' or 'participant'.

3.3.1 Variables

The most commonly used variables in jamovi are 'Data variables', these variables simply contain data either loaded from a data file, or 'typed in' by the user. Data variables can be one of several measurement levels (Figure 3.2).

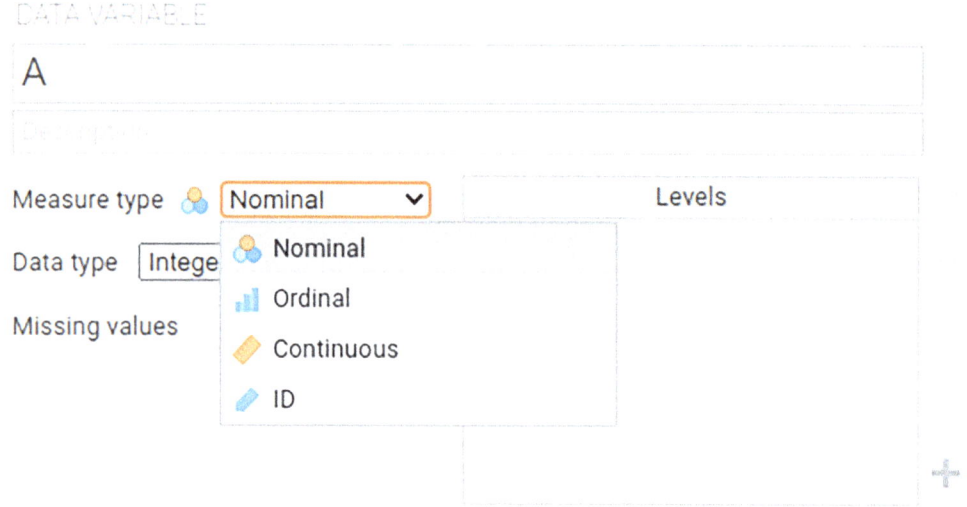

Figure 3.2: measurement levels

These levels are designated by the symbol in the header of the variable's column. The ID variable type is unique to jamovi. It's intended for variables that contain identifiers that you would almost never want to analyse. For example, a person's name, or a participant ID. Specifying an ID variable type can improve performance when interacting with very large data sets.

Nominal variables are for categorical variables which are text labels, for example a column called 'gender' with the values 'male' and 'female' would be nominal. So would a person's name. Nominal variable values can also have a numeric value. These variables are used most often when importing data which codes values with numbers rather than text. For example, a column in a data set may contain the values 1 for males, and 2 for females. It is possible to add nice 'human-readable' labels to these values with the variable editor (more on this later).

Ordinal variables are like nominal variables, except the values have a specific order. An example is a Likert scale with 3 being 'strongly agree' and -3 being 'strongly disagree'.

Continuous variables are variables which exist on a continuous scale. Examples might be height or weight. This is also referred to as 'Interval' or 'Ratio scale'.

In addition, you can also specify different data types: variables have a data type of either 'Text', 'Integer' or 'Decimal'.

When starting with a blank spreadsheet and typing values in the variable type will change automatically depending on the data you enter. This is a good way to get a feel for which variable types go with which sorts of data. Similarly, when opening a data file jamovi will try and guess the variable type from the data in each column. In both cases this automatic approach may not be correct, and it may be necessary to manually specify the variable type with the variable editor.

The variable editor can be opened by selecting 'Setup' from the data tab or by double-clicking on the variable column header. The variable editor allows you to change the name of the variable and, for data variables, the variable type, the order of the levels, and the label displayed for each level. Changes can be applied by clicking the 'tick' to the top right. The variable editor can be dismissed by clicking the 'Hide' arrow.

New variables can be inserted or appended to the data set using the 'Add' button from the data ribbon. The 'Add' button also allows the addition of computed variables.

3.3.2 Computed variables

Computed Variables are those which take their value by performing a computation on other variables. Computed Variables can be used for a range of purposes, including log transforms, *z*-scores, sum-scores, negative scoring and means.

Computed variables can be added to the data set with the 'Add' button available on the data tab. This will produce a formula box where you can specify the formula. The usual arithmetic operators are available. Some examples of formulas are:

A + B LOG10(len) MEAN(A, B) (len - VMEAN(len)) / VSTDEV(len)

In order, these are the sum of A and B, a log (base 10) transform of len, the mean of A and B, and the *z*-score of the variable len.[17] Figure 3.3 shows the jamovi screen for the new variable computed as the *z*-score of len (from the 'Tooth Growth' example data set).

3.3.2.1 V-functions

Several functions are already available in jamovi and available from the drop down box labelled fx. A number of functions appear in pairs, one prefixed with a V and the other not. V functions perform their calculation on a variable as a whole, where as non-V functions perform their calculation row by row. For example, MEAN(A, B) will produce the mean of A and B for each row. Where as VMEAN(A) gives the mean of all the values in A.

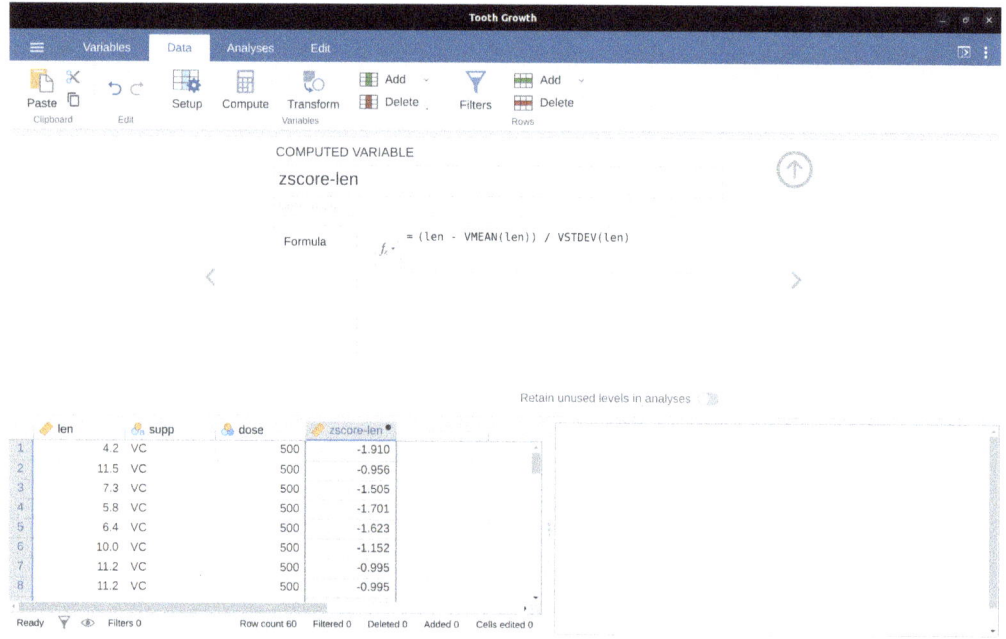

Figure 3.3: A newly computed variable, the z-score of 'dose'

3.3.3 Copy and paste

jamovi produces nice American Psychological Association (APA) formatted tables and attractive plots. It is often useful to be able to copy and paste these, perhaps into a Word document, or into an email to a colleague. To copy results right click on the object of interest and from the menu select exactly what you want to copy. The menu allows you to choose to copy only the image or the entire analysis. Selecting 'Copy' copies the content to the clipboard and this can be pasted into other programs in the usual way. You can practice this later on when we do some analyses.

3.3.4 Syntax mode

jamovi also provides an 'R Syntax mode'.[18] In this mode jamovi produces equivalent R code for each analysis. To change to syntax mode, select the 'Application' menu to the top right of jamovi (a button with three vertical dots) and click the 'Syntax mode' checkbox there. You can turn off syntax mode by clicking this a second time.

In syntax mode analyses continue to operate as before but now they produce R syntax, and "ascii output" like an R session. Like all results objects in jamovi, you can right click on these items (including the R syntax) and copy and paste them, for example into an R session. At present, the provided R syntax does not include the data import step and so this must be performed manually in R. There are many resources explaining how to import data into R and if you are interested we recommend you take a look at these; just search on the interweb.

3.4 Loading data in jamovi

There are several different types of files that are likely to be relevant to us when doing data analysis. There are two in particular that are especially important from the perspective of this book:

- *jamovi files* are those with a .omv file extension. This is the standard kind of file that jamovi uses to store data, and variables and analyses.

- *Comma separated value (csv) files* are those with a .csv file extension. These are just regular old text files and they can be opened with many different software programs. It's quite typical for people to store data in csv files, precisely because they're so simple.

There are also several other kinds of data file that you might want to import into jamovi. For instance, you might want to open Microsoft Excel spreadsheets (.xls files), or data files that have been saved in the native file formats for other statistics software, such as SPSS or SAS. Whichever file formats you are using, it's a good idea to create a folder or folders especially for your jamovi data sets and analyses and to make sure you keep these backed up regularly.

3.4.1 Importing data from csv files

One quite commonly used data format is the humble "comma separated value" file, also called a csv file, and usually bearing the file extension .csv. csv files are just plain old-fashioned text files and what they store is basically just a table of data. This is illustrated in Figure 3.4, which shows a file called *booksales.csv* that I've created. As you can see, each row represents the book sales data for one month. The first row doesn't contain actual data though, it has the names of the variables.

It's easy to open csv files in jamovi. From the top left menu (the button with three parallel lines) choose 'Open' and browse to where you have stored the csv file on your computer. If you're on a Mac, it'll look like the usual Finder window that you use to choose a file; on Windows it looks like an Explorer window. An example of what it looks like on a Mac is shown in Figure 3.5. I'm assuming that you're familiar with your own computer, so you should have no problem finding the csv file that you want to import! Find the one you want, then click on the 'Open' button.

There are a few things that you can check to make sure that the data gets imported correctly:

- Heading. Does the first row of the file contain the names for each variable – a 'header' row? The *booksales.csv* file has a header, so that's a yes.
- Decimal. What character is used to specify the decimal point? In English-speaking countries this is almost always a period (i.e., .). That's not universally true though, many European countries use a comma.
- Quote. What character is used to denote a block of text? That's usually going to be a double quote mark ("). It is for the *booksales.csv* file.

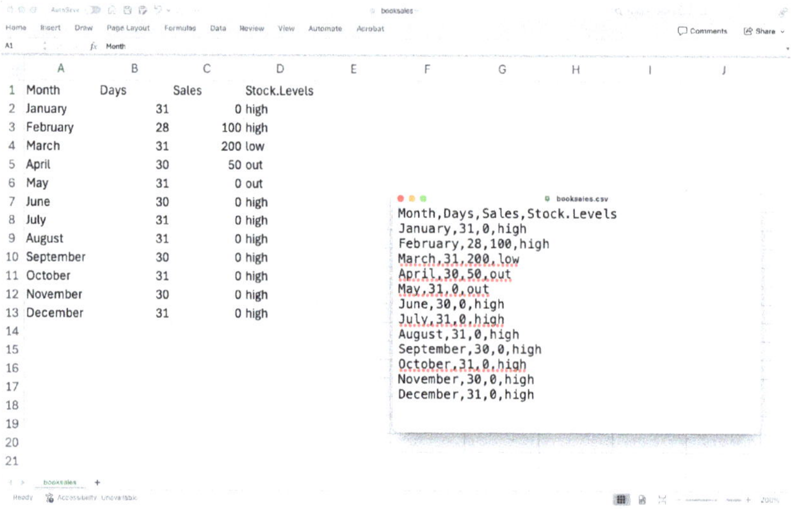

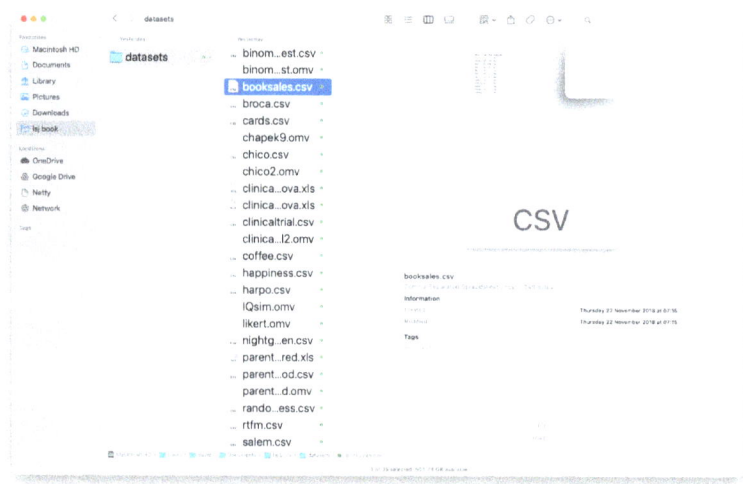

Figure 3.4: The *booksales.csv* data file. On the left I have opened the file using a spreadsheet program, which shows that the file is basically a table. On the right the same file is open in a standard text editor (the TextEdit program on a Mac), which shows how the file is formatted. The entries in the table are separated by commas

Figure 3.5: A dialog box on a Mac asking you to select the csv file jamovi should try to import. Mac users will recognise this immediately, as it is the usual way in which a Mac asks you to find a file. Windows users will not see this, instead they will see the usual explorer window that Windows always gives you when it wants you to select a file

3.5 Importing unusual data files

Throughout this book I've assumed that your data are stored as a jamovi .omv file or as a "properly" formatted csv file. However, in real life that's not a terribly plausible assumption to make so I'd better talk about some of the other possibilities that you might run into.

3.5.1 Loading data from text files

The first thing I should point out is that if your data are saved as a text file but aren't quite in the proper csv format then there's still a pretty good chance that jamovi will be able to open it. You just need to try it and see if it works. Sometimes though you will need to change some of the formatting. The ones that I've often found myself needing to change are:

- header. A lot of the time when you're storing data as a csv file the first row actually contains the column names and not data. If that's not true then it's a good idea to open up the csv file in a spreadsheet programme such as Open Office and add the header row manually.
- sep. As the name "comma separated value" indicates, the values in a row of a csv file are usually separated by commas. This isn't universal, however. In Europe the decimal point is typically written as , instead of . and as a consequence it would be somewhat awkward to use , as the separator. Therefore it is not unusual to use ; instead of , as the separator. At other times, I've seen a TAB character used.
- quote. It's conventional in csv files to include a quoting character for textual data. As you can see by looking at the *booksales.csv* file, this is usually a double quote character, ". But sometimes there is no quoting character at all, or you might see a single quote mark ' used instead.
- skip. It's actually very common to receive csv files in which the first few rows have nothing to do with the actual data. Instead, they provide a human readable summary of where the data came from, or maybe they include some technical info that doesn't relate to the data.
- missing values. Often you'll get given data with missing values. For one reason or another, some entries in the table are missing. The data file needs to include a "special" value to indicate that the entry is missing. By default jamovi assumes that this value is 99,[19] for both numeric and text data, so you should make sure that, where necessary, all missing values in the csv file are replaced with 99 (or -9999; whichever you choose) before opening / importing the file into jamovi. Once you have opened / imported the file into jamovi all the missing values are converted to blank or greyed out cells in the jamovi spreadsheet view. You can also change the missing value for each variable as an option in the Data - Setup view.

3.5.2 Loading data from SPSS (and other statistics packages)

The commands listed above are the main ones we'll need for data files in this book. But in real life we have many more possibilities. For example, you might want to read data files in from other statistics programs. Since SPSS is probably the most widely used statistics package in psychology, it's worth mentioning that jamovi can also import SPSS data files (file extension .sav). Just follow the instructions above for how to open a csv file, but this time navigate to the .sav file you want to import. For SPSS files, jamovi will regard all values as missing if they are regarded as "system missing" files in SPSS. The 'Default missings' value does not seem to work as expected when importing SPSS files, so be aware of this – you might need another step: import the SPSS file into jamovi, then export as a csv file before re-opening in jamovi.[20]

And that's pretty much it, at least as far as SPSS goes. As far as other statistical software goes, jamovi can also directly open / import SAS and STATA files.

3.5.3 Loading Excel files

A different problem is posed by Excel files. Despite years of yelling at people for sending data to me encoded in a proprietary data format, I get sent a lot of Excel files. The way to handle Excel files is to open them up first in Excel or another spreadsheet programme that can handle Excel files, and then export the data as a csv file before opening / importing the csv file into jamovi.

3.6 Changing data from one level to another

Sometimes you want to change the variable level. This can happen for all sorts of reasons. Sometimes when you import data from files, it can come to you in the wrong format. Numbers sometimes get imported as nominal, text values. Dates may get imported as text. Participant ID values can sometimes be read as continuous: nominal values can sometimes be read as ordinal or even continuous. There's a good chance that sometimes you'll want to convert a variable from one measurement level into another one. Or, to use the correct term, you want to **coerce** the variable from one class into another.

Earlier we saw how to specify different variable levels, and if you want to change a variable's measurement level then you can do this in the jamovi data view for that variable. Just click the check box for the measurement level you want – continuous, ordinal, or nominal.

3.7 Installing add-on modules into jamovi

A really great feature of jamovi is the ability to install add-on modules from the jamovi library. These add-on modules have been developed by the jamovi community, i.e., jamovi users and developers who have created special software add-ons that do other, usually more advanced, analyses that go beyond the capabilities of the base jamovi program.

To install add-on modules, just click on the large + in the top right of the jamovi window, select "jamovi-library" and then browse through the various add-on modules that are available. Choose the one(s) you want, and then install them, as in Figure 3.6. It's that easy. The newly installed modules can then be accessed from the "Analyses" button bar. Try it…useful add-on modules to install include "scatr" (added under "Descriptives"), R_j and **of course** the data files for this book: "lsj-data".

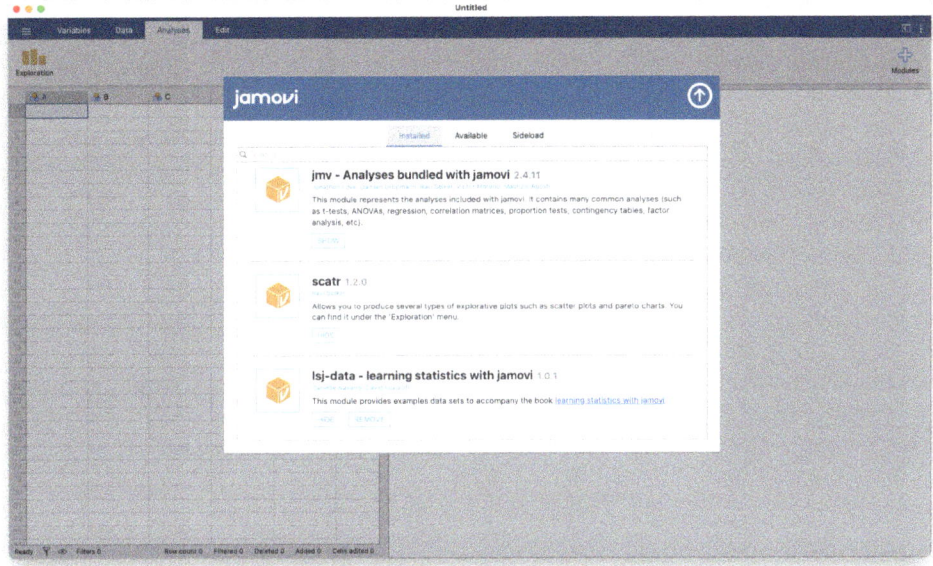

Figure 3.6: Installing add-on modules in jamovi

3.8 Quitting jamovi

There's one last thing I should cover in this chapter: how to quit jamovi. It's not hard, just close the program the same way you would any other program. However, what you might want to do before you quit is save your work! There are two parts to this: saving any changes to the data set, and saving the analyses that you ran.

It is good practice to save any changes to the data set as a *new* data set. That way you can always go back to the original data. To save any changes in jamovi, select 'Export'...'Data' from the main jamovi menu (button with three horizontal bars in the top left) and create a new file name for the changed data set.

Alternatively, you can save *both* the changed data and any analyses you have undertaken by saving as a jamovi file. To do this, from the main jamovi menu select 'Save as' and type in a file name for this 'jamovi file (.omv)'. Remember to save the file in a location where you can find it again later. I usually create a new folder for specific data sets and analyses.

3.9 Summary

Every book that tries to teach a new statistical software program to novices has to cover roughly the same topics, and in roughly the same order. Ours is no exception, and so in the grand tradition of doing it just the same way everyone else did it, this chapter covered the following topics:

- Installing jamovi. We downloaded and installed jamovi, and started it up.
- Analyses. We very briefly oriented to the part of jamovi where analyses are done and results appear, but then deferred this until later in the book.
- The spreadsheet. We spent more time looking at the spreadsheet part of jamovi, and considered different variable types, and how to compute new variables.
- Loading data in jamovi. We also saw how to load data files in jamovi.
- Importing unusual data files. Then we figured out how to open other data files, from different file types.
- Changing data from one level to another. And saw that sometimes we need to coerce data from one type to another.
- Installing add-on modules into jamovi. Installing add-on modules from the jamovi community really extends jamovi capabilities.
- Quitting jamovi. Finally, we looked at good practice in terms of saving your data set and analyses when you have finished and are about to quit jamovi.

We still haven't arrived at anything that resembles data analysis. Maybe the next chapter will get us a bit closer!

Part III

Working with data

Chapter 4

Descriptive statistics

Any time that you get a new data set to look at one of the first tasks that you have to do is find ways of summarising the data in a compact, easily understood fashion. This is what **descriptive statistics** (as opposed to inferential statistics) is all about. In fact, to many people the term "statistics" is synonymous with descriptive statistics. It is this topic that we'll consider in this chapter, but before going into any details, let's take a moment to get a sense of why we need descriptive statistics. To do this, let's open the *aflsmall_margins.csv* file and see what variables are stored in the file, see Figure 4.1.

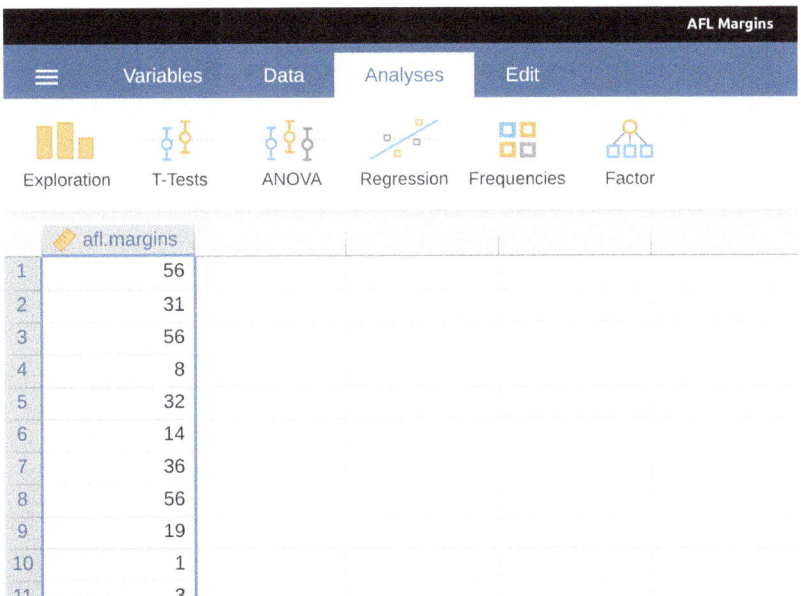

Figure 4.1: A screenshot of jamovi showing the variables stored in the *aflsmall_margins.csv* file

In fact, there is just one variable here, *afl.margins*. We'll focus a bit on this variable in this chapter, so I'd better tell you what it is. Unlike most of the data sets in this book, this is actually real data, relating to the Australian Football League (AFL).[21] The *afl.margins* variable contains the winning margin (number of points) for all 176 home and away games played during the 2010 season.

This output doesn't make it easy to get a sense of what the data are actually saying. Just "looking at the data" isn't a terribly effective way of understanding data. In order to get some idea about what the data are actually saying we need to calculate some descriptive statistics (this chapter) and draw some nice pictures (Chapter 5). Since the descriptive statistics are the easier of the two topics I'll start with those, but nevertheless I'll show you a histogram of the *afl.margins* data since it should help you get a sense of what the data we're trying to describe actually look like, see Figure 4.2. We'll talk a lot more about how to draw histograms in Section 5.1 in the next chapter. For now, it's enough to look at the histogram and note that it provides a fairly interpretable representation of the *afl.margins* data.

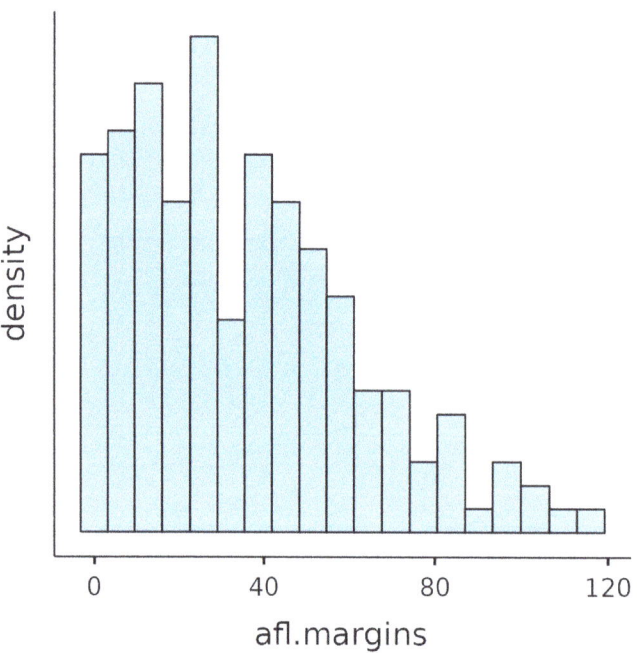

Figure 4.2: A histogram of the AFL 2010 winning margin data (the *afl.margins* variable). As you might expect, the larger the winning margin the less frequently you tend to see it

4.1 Measures of central tendency

Drawing pictures of the data, as I did in Figure 4.2, is an excellent way to convey the "gist" of what the data is trying to tell you. It's often extremely useful to try to condense the data into a few simple "summary" statistics. In most situations, the first thing that you'll want to calculate is a measure of **central tendency**. That is, you'd like to know something about where the "average" or "middle" of your data lies. The three most commonly used measures are the mean, median and mode. I'll explain each of these in turn, and then discuss when each of them is useful.

4.1.1 The mean

The **mean** of a set of observations is just a normal, old-fashioned average. Add all of the values up, and then divide by the total number of values. The first five AFL winning margins were 56, 31, 56, 8 and 32, so the mean of these observations is just:

$$\frac{56 + 31 + 56 + 8 + 32}{5} = \frac{183}{5} = 36.60$$

Of course, this definition of the mean isn't news to anyone. Averages (i.e., means) are used so often in everyday life that this is pretty familiar stuff. However, since the concept of a mean is something that everyone already understands, I'll use this as an excuse to start introducing some of the mathematical notation that statisticians use to describe this calculation, and talk about how the calculations would be done in jamovi.

The first piece of notation to introduce is N, which we'll use to refer to the number of observations that we're averaging (in this case $N = 5$). Next, we need to attach a label to the observations themselves. It's traditional to use X for this, and to use subscripts to indicate which observation we're actually talking about. That is, we'll use X_1 to refer to the first observation, X_2 to refer to the second observation, and so on all the way up to X_N for the last one. Or, to say the same thing in a slightly more abstract way, we use X_i to refer to the i-th observation. Just to make sure we're clear on the notation, Table 4.1 lists the 5 observations in the *afl.margins* variable, along with the mathematical symbol used to refer to it and the actual value that the observation corresponds to.

Table 4.1: Observations in the *afl.margins* variable

the observation	its symbol	the observed value
winning margin, game 1	X_1	56 points
winning margin, game 2	X_2	31 points
winning margin, game 3	X_3	56 points
winning margin, game 4	X_4	8 points
winning margin, game 5	X_5	32 points

[Additional technical detail[22]]

4.1.2 Calculating the mean in jamovi

Okay, that's the maths. So how do we get the magic computing box to do the work for us? When the number of observations starts to become large it's much easier to do these sorts of calculations using a computer. To calculate the mean using all the data we can use jamovi. The first step is to click on the 'Exploration' button and then click 'Descriptives'. Then you can highlight the *afl.margins* variable and click the 'Right arrow' to move it across into the 'Variables box'. As soon as you do that a Table appears on the right-hand side of the screen containing default 'Descriptives' information; see Figure 4.3.

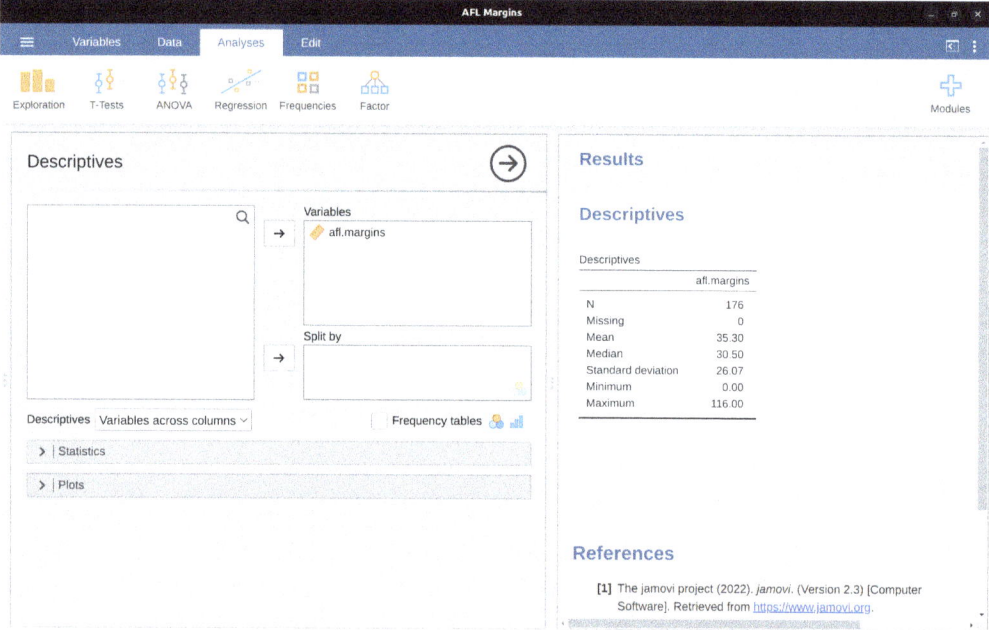

Figure 4.3: Default descriptives for the AFL 2010 winning margin data (the *afl.margins* variable)

As you can see in Figure 4.3, the mean value for the *afl.margins* variable is 35.30. Other information presented includes the total number of observations (N=176), the number of missing values (none), and the median, minimum and maximum values for the variable.

4.1.3 The median

The second measure of central tendency that people use a lot is the **median**, and it's even easier to describe than the mean. The median of a set of observations is just the middle value. As before let's imagine we were interested only in the first 5 AFL winning margins: 56, 31, 56, 8 and 32. To figure out the median we sort these numbers into ascending order: 8, 31, **32**, 56, 56.

From inspection, it's obvious that the median value of these 5 observations is 32 since that's the middle one in the sorted list (I've put it in bold to make it even more obvious). Easy stuff. But what should we do if we are interested in the first 6 games rather than the first 5? Since the sixth game in the season had a winning margin of 14 points, our sorted list is now 8, 14, **31**, **32**, 56, 56.

And there are two middle numbers, 31 and 32. The median is defined as the average of those two numbers, which is of course 31.5. As before, it's very tedious to do this by hand when you've got lots of numbers. In real life, of course, no-one actually calculates the median by sorting the data and then looking for the middle value. In real life we use a computer to do the heavy lifting for us, and jamovi has provided us with a median value of 30.50 for the *afl.margins* variable (Figure 4.3).

4.1.4 Mean or median? What's the difference?

Knowing how to calculate means and medians is only a part of the story. You also need to understand what each one is saying about the data, and what that implies for when you should use each one. This is illustrated in Figure 4.4. The mean is kind of like the "centre of gravity" of the data set, whereas the median is the "middle value" in the data. What this implies, as far as which one you should use, depends a little on what type of data you've got and what you're trying to achieve. As a rough guide:

- If your data are nominal scale you probably shouldn't be using either the mean or the median. Both the mean and the median rely on the idea that the numbers assigned to values are meaningful. If the numbering scheme is arbitrary then it's probably best to use the Mode instead.
- If your data are ordinal scale you're more likely to want to use the median than the mean. The median only makes use of the order information in your data (i.e., which numbers are bigger) but doesn't depend on the precise numbers involved. That's exactly the situation that applies when your data are ordinal scale. The mean, on the other hand, makes use of the precise numeric values assigned to the observations, so it's not really appropriate for ordinal data.
- For interval and ratio scale data either one is generally acceptable. Which one you pick depends a bit on what you're trying to achieve. The mean has the advantage that it uses all the information in the data (which is useful when you don't have a lot of data). But it's very sensitive to extreme, outlying values.

Let's expand on that last part a little. One consequence is that there are systematic differences between the mean and the median when the histogram is asymmetric (Skew and kurtosis). This is illustrated in Figure 4.4. Notice that the median (right-hand side) is located closer to the "body" of the histogram, whereas the mean (left-hand side) gets dragged towards the "tail" (where the extreme values are). To give a concrete example, suppose Bob (income $50,000), Kate (income $60,000) and Jane (income $65,000) are sitting at a table. The average income at the table is $58,333 and the median income is $60,000. Then Bill sits down with them (income $100,000,000). The average income has now jumped to $25,043,750 but the median rises only to $62,500. If you're interested in looking at the overall income at the table the mean might be the right answer. But if you're interested in what counts as a typical income at the table the median would be a better choice here.

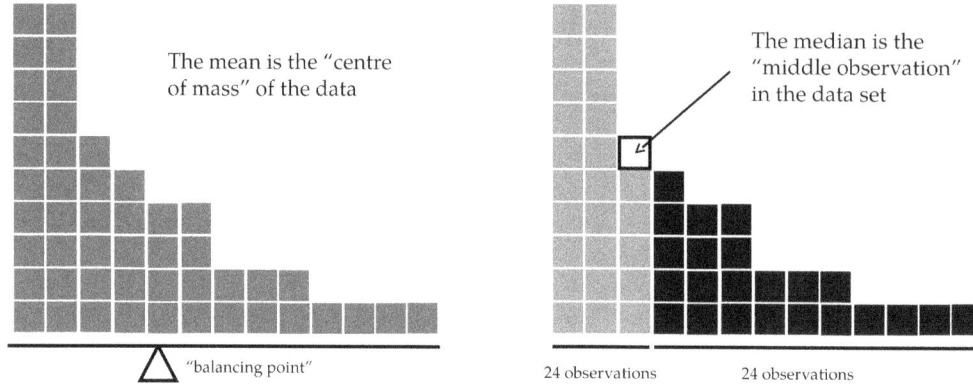

Figure 4.4: An illustration of the difference between how the mean and the median should be interpreted. The mean is basically the 'centre of gravity' of the data set. If you imagine that the histogram of the data is a solid object, then the point on which you could balance it (as if on a see-saw) is the mean. In contrast, the median is the middle observation, with half of the observations smaller and half of the observations larger

4.1.5 A real-life example

To try to get a sense of why you need to pay attention to the differences between the mean and the median let's consider a real life example. Since I tend to mock journalists for their poor scientific and statistical knowledge, I should give credit where credit is due. This is an excellent article by Michael Janda from the ABC news website[23] from 24 September, 2010:

> Senior Commonwealth Bank executives have travelled the world in the past couple of weeks with a presentation showing how Australian house prices, and the key price to income ratios, compare favourably with similar countries. "Housing affordability has actually been going sideways for the last five to six years," said Craig James, the chief economist of the bank's trading arm, CommSec.

This probably comes as a huge surprise to anyone with a mortgage, or who wants a mortgage, or pays rent, or isn't completely oblivious to what's been going on in the Australian housing market over the last several years. Back to the article:

> CBA has waged its war against what it believes are housing doomsayers with graphs, numbers and international comparisons. In its presentation, the bank rejects arguments that Australia's housing is relatively expensive compared to incomes. It says Australia's house price to household income ratio of 5.6 in the major cities, and 4.3 nationwide, is comparable to many other developed nations. It says San Francisco and New York have ratios of 7, Auckland's is 6.7, and Vancouver comes in at 9.3.

More excellent news! Except, the article goes on to make the observation that:

> Many analysts say that has led the bank to use misleading figures and comparisons. If you go to page four of CBA's presentation and read the source information at the bottom of the graph and table, you would notice there is an additional source on the international comparison – Demographia. However, if the Commonwealth Bank had also used Demographia's analysis of Australia's house price to income ratio, it would have come up with a figure closer to 9 rather than 5.6 or 4.3.

That's, um, a rather serious discrepancy. One group of people say 9, another says 4-5. Should we just split the difference and say the truth lies somewhere in between? Absolutely not! This is a situation where there is a right answer and a wrong answer. Demographia is correct, and the Commonwealth Bank is wrong. As the article points out:

> [An] obvious problem with the Commonwealth Bank's domestic price to income figures is they compare average incomes with median house prices (unlike the Demographia figures that compare median incomes to median prices). The median is the mid-point, effectively cutting out the highs and lows, and that means the average is generally higher when it comes to incomes and asset prices, because it includes the earnings of Australia's wealthiest people. To put it another way: the Commonwealth Bank's figures count Ralph Norris' multi-million dollar pay packet on the income side, but not his (no doubt) very expensive house in the property price figures, thus understating the house price to income ratio for middle-income Australians.

Couldn't have put it better myself. The way that Demographia calculated the ratio is correct. The way that the Bank did it is incorrect. As for why an extremely quantitatively sophisticated organisation such as a major bank made such an elementary mistake, well… I can't say for sure since I have no special insight into their thinking. But the article itself does happen to mention the following facts, which may or may not be relevant:

> [As] Australia's largest home lender, the Commonwealth Bank has one of the biggest vested interests in house prices rising. It effectively owns a massive swathe of Australian housing as security for its home loans as well as many small business loans.

My, my.

4.1.6 Mode

The mode of a sample is very simple. It is the value that occurs most frequently. We can illustrate the mode using a different AFL variable: who has played in the most finals? Open the aflsmall finalists file and take a look at the *afl.finalists* variable, see Figure 4.5.

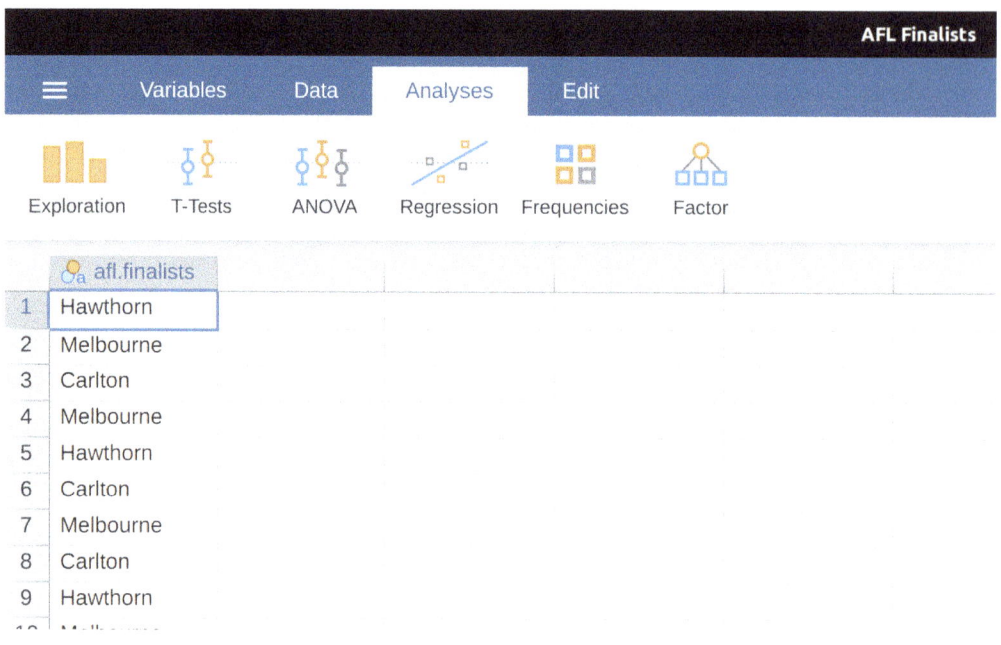

Figure 4.5: A screenshot of jamovi showing the variables stored in the aflsmall_finalists.csv file

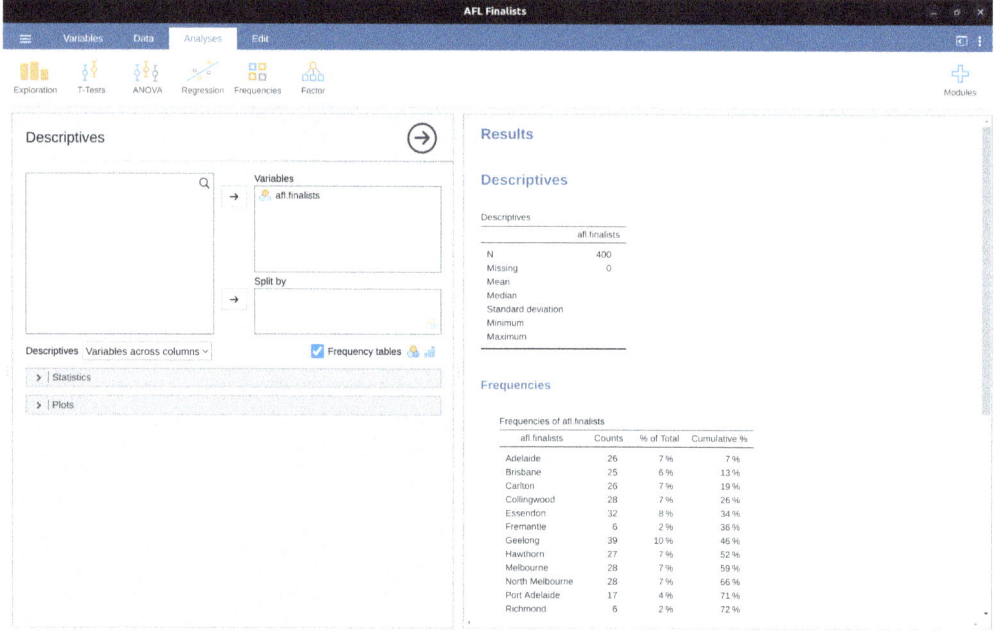

Figure 4.6: A screenshot of jamovi showing the frequency table for the *afl.finalists* variable

This variable contains the names of all 400 teams that played in all 200 finals matches played during the period 1987 to 2010. What we could do is read through all 400 entries and count the number of occasions on which each team name appears in our list of finalists, thereby producing a **frequency table**. However, that would be mindless and boring: exactly the sort of task that computers are great at. So let's use jamovi to do this for us. Under 'Exploration' – 'Descriptives' click the small check box labelled 'Frequency tables' and you should get something like Figure 4.6.

Now that we have our frequency table we can just look at it and see that, over the 24 years for which we have data, Geelong has played in more finals than any other team. Thus, the mode of the *afl.finalists* data is "Geelong". We can see that Geelong (39 finals) played in more finals than any other team during the 1987-2010 period. It's also worth noting that in the 'Descriptives' Table no results are calculated for mean, median, minimum or maximum. This is because the *afl.finalists* variable is a nominal text variable so it makes no sense to calculate these values.

One last point to make regarding the mode. Whilst the mode is most often calculated when you have nominal data, because means and medians are useless for those sorts of variables, there are some situations in which you really do want to know the mode of an ordinal, interval or ratio scale variable. For instance, let's go back to our *afl.margins* variable. This variable is clearly ratio scale (if it's not clear to you, it may help to re-read Section 2.2), and so in most situations the mean or the median is the measure of central tendency that you want. But consider this scenario: a friend of yours is offering a bet and they pick a football game at random. Without knowing who is playing you have to guess the exact winning margin. If you guess correctly you win $50. If you don't you lose $1. There are no consolation prizes for "almost" getting the right answer. You have to guess exactly the right margin. For this bet, the mean and the median are completely useless to you. It is the mode that you should bet on. To calculate the mode for the *afl.margins* variable in jamovi, go back to that data set and on the 'Exploration' – 'Descriptives' screen you will see you can expand the section marked 'Statistics'. Click on the checkbox marked 'Mode' and you will see the modal value presented in the 'Descriptives' Table, as in Figure 4.7. So the 2010 data suggest you should bet on a 3 point margin.

4.2 Measures of variability

The statistics that we've discussed so far all relate to central tendency. That is, they all talk about which values are "in the middle" or "popular" in the data. However, central tendency is not the only type of summary statistic that we want to calculate. The second thing that we really want is a measure of the **variability** of the data. That is, how "spread out" are the data? How "far" away from the mean or median do the observed values tend to be? For now, let's assume that the data are interval or ratio scale, and we'll continue to use the *afl.margins* data. We'll use this data to discuss several different measures of spread, each with different strengths and weaknesses.

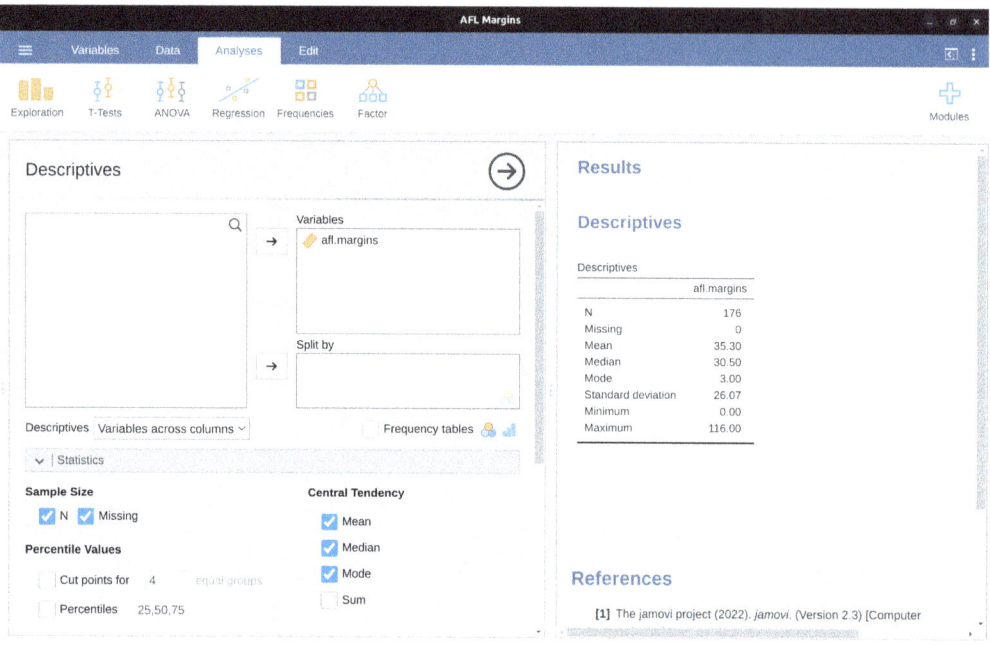

Figure 4.7: A screenshot of jamovi showing the modal value for the *afl.margins* variable

4.2.1 Range

The **range** of a variable is very simple. It's the biggest value minus the smallest value. For the AFL winning margins data the maximum value is 116 and the minimum value is 0. Although the range is the simplest way to quantify the notion of "variability", it's one of the worst. Recall from our discussion of the mean that we want our summary measure to be robust. If the data set has one or two extremely bad values in it we'd like our statistics to not be unduly influenced by these cases. For example, in a variable containing very extreme outliers

-100, 2, 3, 4, 5, 6, 7, 8, 9, 10

it is clear that the range is not robust. This variable has a range of 110 but if the outlier were removed we would have a range of only 8.

4.2.2 Interquartile range

The **interquartile range** (IQR) is like the range, but instead of the difference between the biggest and smallest value the difference between the 25th percentile and the 75th percentile is taken. If you don't already know what a **percentile** is, the 10th percentile of a data set is the smallest number x such that 10% of the data is less than x. In fact, we've already come across the idea. The median of a data set is its 50th percentile! In jamovi you can easily specify the 25th, 50th and 75th percentiles by clicking the checkbox 'Quartiles' in the 'Exploration' – 'Descriptives' – 'Statistics' screen.

And not surprisingly, in Figure 4.8 the 50th percentile is the same as the median value.

And, by noting that $50.50 - 12.75 = 37.75$, we can see that the interquartile range for the 2010 AFL winning margins data is 37.75. While it's obvious how to interpret the range it's a little less obvious how to interpret the IQR. The simplest way to think about it is like this: the interquartile range is the range spanned by the "middle half" of the data. That is, one quarter of the data falls below the 25th percentile and one quarter of the data is above the 75th percentile, leaving the "middle half" of the data lying in between the two. And the IQR is the range covered by that middle half.

Descriptives

	afl.margins
N	176
Missing	0
Mean	35.30
Median	30.50
Mode	3.00
Standard deviation	26.07
Minimum	0.00
Maximum	116.00
25th percentile	12.75
50th percentile	30.50
75th percentile	50.50

Figure 4.8: A screenshot of jamovi showing the Quartiles for the *afl.margins* variable

4.2.3 Mean absolute deviation

The two measures we've looked at so far, the range and the interquartile range, both rely on the idea that we can measure the spread of the data by looking at the percentiles of the data. However, this isn't the only way to think about the problem. A different approach is to select a meaningful reference point (usually the mean or the median) and then report the "typical" deviations from that reference point. What do we mean by "typical" deviation? Usually, this is the mean or median value of these deviations. In practice, this leads to two different measures: the "mean absolute deviation" (from the mean) and the "median absolute deviation" (from the median). From what I've read, the measure based on the median seems to be used in statistics and does seem to be the better of the two. But to be honest I don't think I've seen it used much in psychology. The measure based on the mean does occasionally show up in psychology though. In this section I'll talk about the first one, and I'll come back to talk about the second one later.

Since the previous paragraph might sound a little abstract, let's go through the **mean absolute deviation** from the mean a little more slowly. One useful thing about this measure is that the name actually tells you exactly how to calculate it. Let's think about our AFL winning margins data, and once again we'll start by pretending that there are only 5 games in total, with winning margins of 56, 31, 56, 8 and 32. Since our calculations rely on an examination of the deviation from some reference point (in this case the mean), the first thing we need to calculate is the mean, $\bar{X}$. For these five observations, our mean is $\bar{X} = 36.6$. The next step is to convert each of our observations X_i into a deviation score. We do this by calculating the difference between the observation X_i and the mean $\bar{X}$. That is, the deviation score is defined to be $X_i - \bar{X}$. For the first observation in our sample, this is equal to $56 - 36.6 = 19.4$. Okay, that's simple enough. The next step in the process is to convert these deviations to absolute deviations, and we do this by converting any negative values to positive ones. Mathematically, we would denote the absolute value of -3 as $|-3|$, and so we say that $|-3| = 3$. We use the absolute value here because we don't really care whether the value is higher than the mean or lower than the mean, we're just interested in how close it is to the mean. To help make this process as obvious as possible, Table 4.2 shows these calculations for all five observations.

Table 4.2: Measures of variability

English notation:	notation	value	deviation from mean	absolute deviation
	i	X_i	$X_i - \bar{X}$	$\|X_i - \bar{X}\|$
	1	56	19.4	19.4
	2	31	-5.6	5.6
	3	56	19.4	19.4
	4	8	-28.6	28.6
	5	32	-4.6	4.6

Now that we have calculated the absolute deviation score for every observation in the data set, all that we have to do to calculate the mean of these scores. Let's do that:

$$\frac{19.4 + 5.6 + 19.4 + 28.6 + 4.6}{5} = 15.52$$

And we're done. The mean absolute deviation for these five scores is 15.52.

[Additional technical detail[24]]

4.2.4 Variance

Although the average absolute deviation measure has its uses, it's not the best measure of variability to use. From a purely mathematical perspective there are some solid reasons to prefer squared deviations rather than absolute deviations. If we do that we obtain a measure called the **variance**, which has a lot of really nice statistical properties that I'm going to ignore,[25] and one massive psychological flaw that I'm going to make a big deal out of in a moment. The variance of a data set X is sometimes written as Var(X), but it's more commonly denoted s^2 (the reason for this will become clearer shortly).

[Additional technical detail[26]]

Now that we've got the basic idea, let's have a look at a concrete example. Once again, let's use the first five AFL games as our data. If we follow the same approach that we took last time, we end up with the information shown in Table 4.3.

Table 4.3: Measures of variability for the first five AFL games

English notation:	maths: i	value X_i	deviation from mean $X_i - \bar{X}$	absolute deviation $(X_i - \bar{X})^2$
	1	56	19.4	376.36
	2	31	-5.6	31.36
	3	56	19.4	376.36
	4	8	-28.6	817.96
	5	32	-4.6	21.16

That last column contains all of our squared deviations, so all we have to do is average them. If we do that by hand, using a calculator, we end up with a variance of 324.64. For the moment, don't worry about what a variance of 324.64 actually means. Instead le's talk a bit more about how to do the calculations in jamovi, because this will reveal something very weird. Start a new jamovi session by clicking on the main menu button (three horizontal lines in the top left corner and selecting 'New'. Now type in the first five values from the *afl.margins* data set in column A (56, 31, 56, 8, 32. Change the variable type to 'Continuous' and under 'Descriptives' click the 'Variance' check box, and you get the same values for variance as the one we calculated by hand (324.64). No, wait, you get a completely different answer (405.80) – see Figure 4.9.

That's just weird – is jamovi broken? As it happens, the answer is no.[27] jamovi is not making a mistake. In fact, it's very simple to explain what jamovi is doing here, but slightly trickier to explain why jamovi is doing it. So let's start with the "what". What jamovi is doing is evaluating a slightly different formula to the one I showed you above. Instead of averaging the squared deviations, which requires you to divide by the number of data points N, jamovi has chosen to divide by $N - 1$.

[Additional technical detail[28]]

So that's the *what*. The real question is why jamovi is dividing by $N - 1$ and not by N. After all, the variance is supposed to be the mean squared deviation, right? So shouldn't we be dividing by N, the actual number of observations in the sample? Well, yes, we should. However, as we'll discuss in Chapter 8, there's a subtle distinction between "describing a sample" and "making guesses about the population from which the sample came". Up to this point, it's been a distinction without a difference. Regardless of whether you're describing a sample or drawing inferences about the population, the mean is calculated exactly the same way. Not so for the variance, or the standard deviation, or for many other measures.

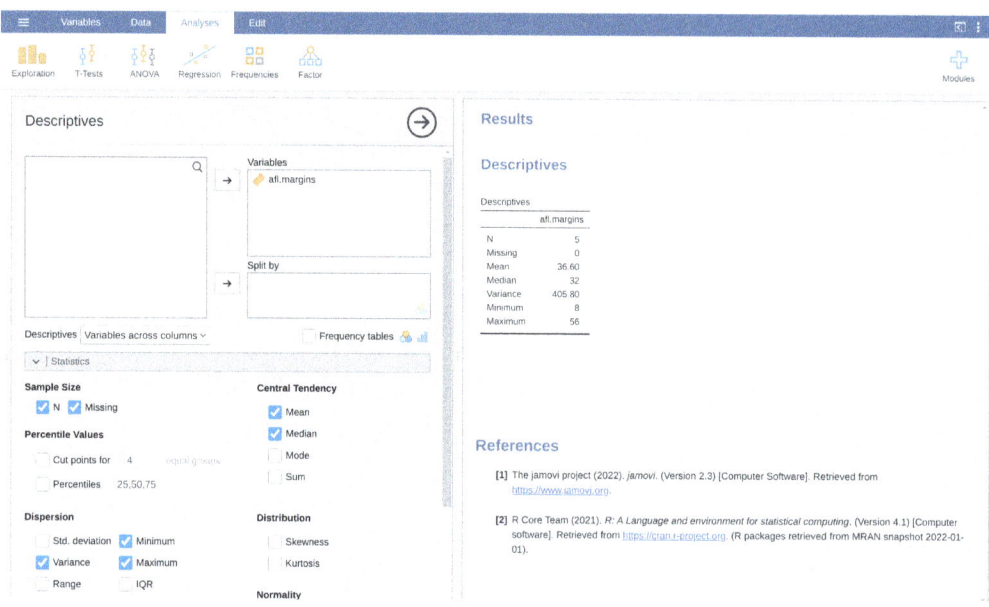

Figure 4.9: A screenshot of jamovi showing the Variance for the first 5 values of the afl.margins variable

What I outlined to you initially (i.e., take the actual average, and thus divide by N) assumes that you literally intend to calculate the variance of the sample. Most of the time, however, you're not terribly interested in the sample in and of itself. Rather, the sample exists to tell you something about the world. If so, you're actually starting to move away from calculating a "sample statistic" and towards the idea of estimating a "population parameter". However, I'm getting ahead of myself. For now, let's just take it on faith that jamovi knows what it's doing, and we'll revisit the question later on when we talk about estimation in Chapter 8.

Okay, one last thing. This section so far has read a bit like a mystery novel. I've shown you how to calculate the variance, described the weird "$N-1$" thing that jamovi does and hinted at the reason why it's there, but I haven't mentioned the single most important thing. How do you interpret the variance? Descriptive statistics are supposed to describe things, after all, and right now the variance is really just a gibberish number. Unfortunately, the reason why I haven't given you the human-friendly interpretation of the variance is that there really isn't one. This is the most serious problem with the variance. Although it has some elegant mathematical properties that suggest that it really is a fundamental quantity for expressing variation, it's completely useless if you want to communicate with an actual human. Variances are completely uninterpretable in terms of the original variable! All the numbers have been squared and they don't mean anything anymore. This is a huge issue. For instance, according to Table 4.3, the margin in game 1 was "376.36 points-squared higher than the average margin". This is *exactly* as stupid as it sounds, and so when we calculate a variance of 324.64 we're in the same situation. I've watched a lot of footy games, and at no time has anyone ever referred to "points squared". It's not a real unit of measurement, and since the variance is expressed in terms of this gibberish unit, it is totally meaningless to a human.

4.2.5 Standard deviation

Okay, suppose that you like the idea of using the variance because of those nice mathematical properties that I haven't talked about, but since you're a human and not a robot you'd like to have a measure that is expressed in the same units as the data itself (i.e., points, not points squared). What should you do? The solution to the problem is obvious! Take the square root of the variance, known as the **standard deviation**, also called the "root mean squared deviation", or RMSD. This solves our problem fairly neatly. Whilst nobody has a clue what "a variance of 324.68 points-squared" really means, it's much easier to understand "a standard deviation of 18.01 points" since it's expressed in the original units. It is traditional to refer to the standard deviation of a sample of data as s, though "sd" and "std dev." are also used at times.

[Additional technical detail[29]]

However, as you might have guessed from our discussion of the variance, what jamovi actually calculates is slightly different to the formula given above. Just like we saw with the variance, what jamovi calculates is a version that divides by $N - 1$ rather than N.

[Additional technical detail[30]]

Interpreting standard deviations is slightly more complex. Because the standard deviation is derived from the variance, and the variance is a quantity that has little to no meaning that makes sense to us humans, the standard deviation doesn't have a simple interpretation. As a consequence, most of us just rely on a simple rule of thumb. In general, you should expect 68% of the data to fall within 1 standard deviation of the mean, 95% of the data to fall within 2 standard deviation of the mean, and 99.7% of the data to fall within 3 standard deviations of the mean. This rule tends to work pretty well most of the time, but it's not exact. It's actually calculated based on an assumption that the histogram is symmetric and "bell shaped". As you can tell from looking at the AFL winning margins histogram in Figure 4.2, this isn't exactly true of our data! Even so, the rule is approximately correct. As it turns out, 65.3% of the *afl.margins* data fall within one standard deviation of the mean. This is shown visually in Figure 4.10.

4.2.6 Which measure to use?

We've discussed quite a few measures of spread: range, IQR, mean absolute deviation, variance and standard deviation; and hinted at their strengths and weaknesses. Here's a quick summary:

- *Range*. Gives you the full spread of the data. It's very vulnerable to outliers and as a consequence it isn't often used unless you have good reasons to care about the extremes in the data.
- *Interquartile range*. Tells you where the "middle half" of the data sits. It's pretty robust and complements the median nicely. This is used a lot.
- *Mean absolute deviation*. Tells you how far "on average" the observations are from the mean. It's very interpretable but has a few minor issues (not discussed here) that make it less attractive to statisticians than the standard deviation. Used sometimes, but not often.

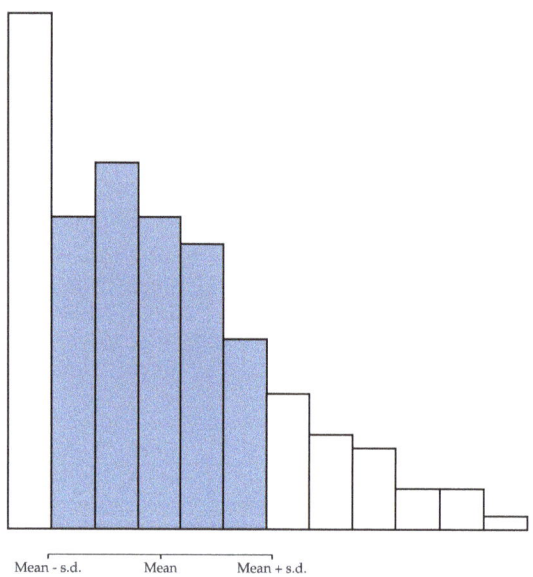

Figure 4.10: An illustration of the standard deviation from the AFL winning margins data. The shaded bars in the histogram show how much of the data fall within one standard deviation of the mean. In this case, 65.3% of the data set lies within this range, which is pretty consistent with the "approximately 68% rule" discussed in the main text

- *Variance.* Tells you the average squared deviation from the mean. It's mathematically elegant and is probably the "right" way to describe variation around the mean, but it's completely uninterpretable because it doesn't use the same units as the data. Almost never used except as a mathematical tool, but it's buried "under the hood" of a very large number of statistical tools.
- *Standard deviation.* This is the square root of the variance. It's fairly elegant mathematically and it's expressed in the same units as the data so it can be interpreted pretty well. In situations where the mean is the measure of central tendency, this is the default. This is by far the most popular measure of variation.

In short, the IQR and the standard deviation are easily the two most common measures used to report the variability of the data. But there are situations in which the others are used. I've described all of them in this book because there's a fair chance you'll run into most of these somewhere.

4.3 Skew and kurtosis

There are two more descriptive statistics that you will sometimes see reported in the psychological literature: skew and kurtosis. In practice, neither one is used anywhere near as frequently as the measures of central tendency and variability that we've been talking about. Skew is pretty important, so you do see it mentioned a fair bit, but I've actually never seen kurtosis reported in a scientific article to date.

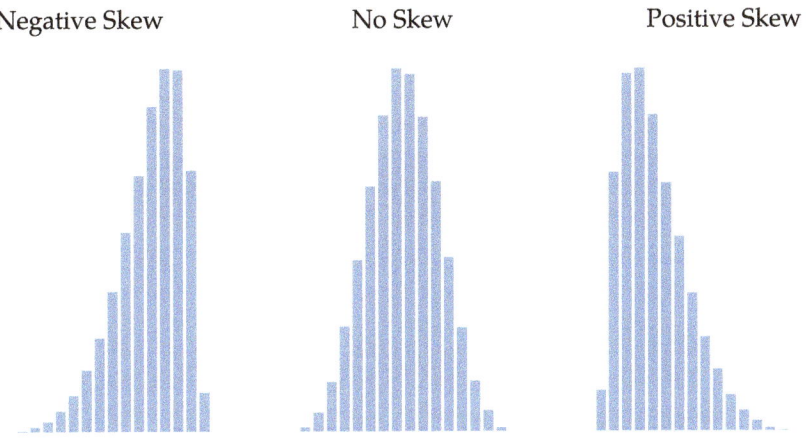

Figure 4.11: An illustration of skewness. On the left we have a negatively skewed data set, in the middle we have a data set with no skew, and on the right we have a positively skewed data set

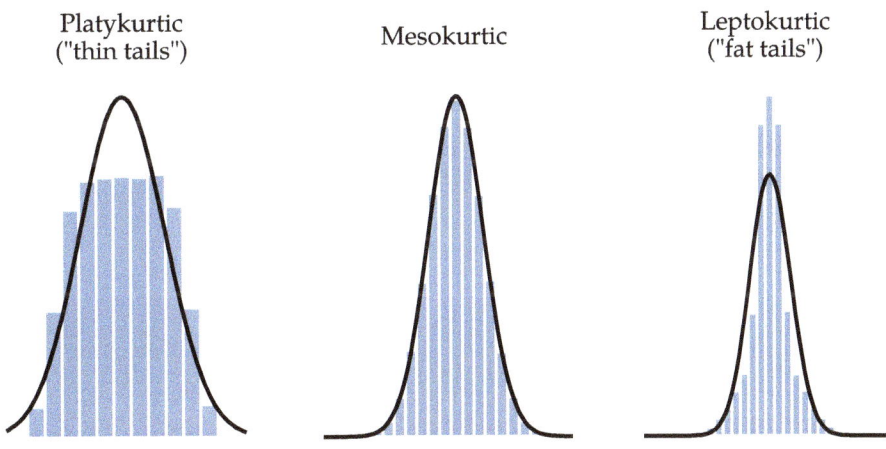

Figure 4.12: An illustration of kurtosis. On the left, we have a "platykurtic" distribution (kurtosis = -.95) meaning that the distribution has "thin" or flat tails. In the middle we have a "mesokurtic" distribution (kurtosis is almost exactly 0) which means that the tails are neither thin or fat. Finally, on the right, we have a "leptokurtic" distribution (kurtosis = 2.12) indicating that the distribution has "fat" tails. Note that kurtosis is measured with respect to a normal curve (black line)

Since it's the more interesting of the two, let's start by talking about the **skewness**. Skewness is basically a measure of asymmetry and the easiest way to explain it is by drawing some pictures. As Figure 4.11 illustrates, if the data tend to have a lot of extreme small values (i.e., the lower tail is "longer" than the upper tail) and not so many extremely large values (left panel) then we say that the data are *negatively skewed*. On the other hand, if there are more extremely large values than extremely small ones (right panel) we say that the data are positively skewed. That's the qualitative idea behind skewness. If there are relatively more values that are far greater than the mean, the distribution is positively skewed or right skewed, with a tail stretching to the right. Negative or left skew is the opposite. A symmetric distribution has a skewness of 0. The skewness value for a positively skewed distribution is positive, and a negative value for a negatively skewed distribution.

[Additional technical detail[31]]

Perhaps more helpfully, you can use jamovi to calculate skewness: it's a check box in the 'Statistics' options under 'Exploration' – 'Descriptives'. For the *afl.margins* variable, the skewness figure is 0.780. If you divide the skewness estimate by the Std. error for skewness you have an indication of how skewed the data is. Especially in small samples (N < 50), one rule of thumb suggests that a value of 2 or less can mean that the data is not very skewed, and a value of over 2 that there is sufficient skew in the data to possibly limit its use in some statistical analyses. Though there is no clear agreement on this interpretation. That said, this does indicate that the AFL winning margins data is somewhat skewed ($\frac{0.780}{0.183} = 4.262$).

The final measure that is sometimes referred to, though very rarely in practice, is the kurtosis of a data set. Put simply, kurtosis is a measure of how thin or fat the tails of a distribution are, as illustrated in Figure 4.12. By convention, we say that the "normal curve" (black lines) has zero kurtosis, so the degree of kurtosis is assessed relative to this curve. In this Figure, the data on the left have a pretty flat distribution, with thin tails, so the kurtosis is negative and we call the data platykurtic. The data on the right have a distribution with fat tails, so the kurtosis is positive and we say that the data is leptokurtic. But the data in the middle have neither thin or fat tails, so we say that it is mesokurtic and has kurtosis zero. This is summarised in Table 4.4.

Table 4.4: Thin to fat tails to illustrate kurtosis

English	informal term	kurtosis value
"tails too thin"	platykurtic	negative
"tails neither thin or fat"	mesokurtic	zero
"tails too fat"	leptokurtic	positive

[Additional technical detail[32]]

More to the point, jamovi has a check box for kurtosis just below the check box for skewness, and this gives a value for kurtosis of 0.101 with a standard error of 0.364. This means that the AFL winning margins data has only a small kurtosis, which is ok.

4.4 Descriptive statistics separately for each group

It is very commonly the case that you find yourself needing to look at descriptive statistics broken down by some grouping variable. This is pretty easy to do in jamovi. For instance, let's say I want to look at the descriptive statistics for some clinical trial data, broken down separately by therapy type. This is a new data set, one that you've never seen before. The data is stored in the *clinicaltrial.csv* file and we'll use it a lot later on in Chapter 13 (you can find a complete description of the data at the start of that chapter). Let's load it and see what we've got (Figure 4.13).

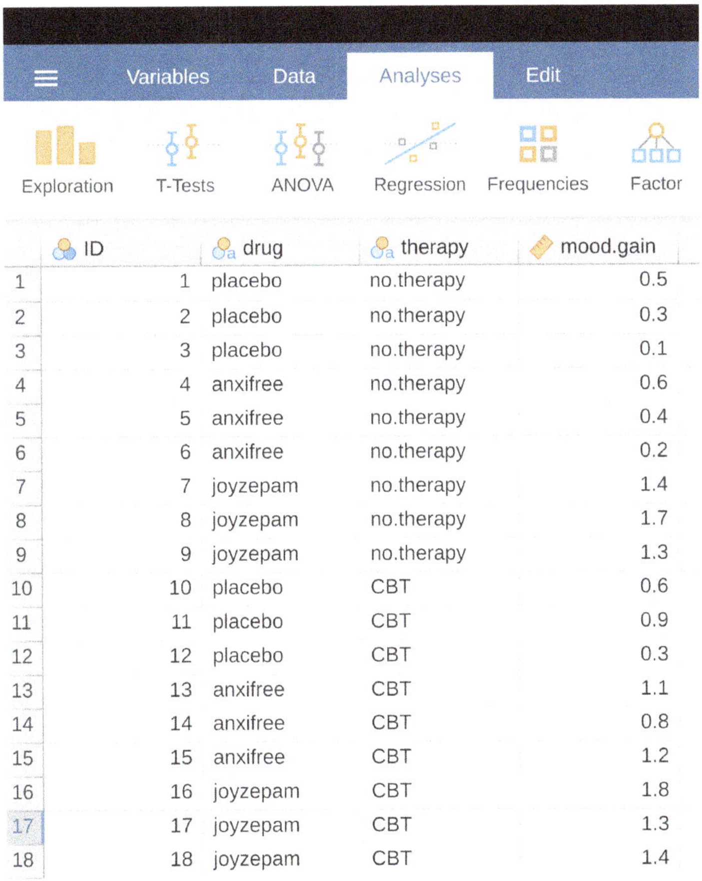

	ID	drug	therapy	mood.gain
1	1	placebo	no.therapy	0.5
2	2	placebo	no.therapy	0.3
3	3	placebo	no.therapy	0.1
4	4	anxifree	no.therapy	0.6
5	5	anxifree	no.therapy	0.4
6	6	anxifree	no.therapy	0.2
7	7	joyzepam	no.therapy	1.4
8	8	joyzepam	no.therapy	1.7
9	9	joyzepam	no.therapy	1.3
10	10	placebo	CBT	0.6
11	11	placebo	CBT	0.9
12	12	placebo	CBT	0.3
13	13	anxifree	CBT	1.1
14	14	anxifree	CBT	0.8
15	15	anxifree	CBT	1.2
16	16	joyzepam	CBT	1.8
17	17	joyzepam	CBT	1.3
18	18	joyzepam	CBT	1.4

Figure 4.13: A screenshot of jamovi showing the variables stored in the *clinicaltrial.csv* file

Evidently there were three drugs: a placebo, something called "anxifree" and something called "joyzepam", and there were 6 people administered each drug. There were 9 people treated using cognitive behavioural therapy (CBT) and 9 people who received no psychological treatment. And we can see from looking at the 'Descriptives' of the mood.gain variable that most people did show a mood gain ($mean = 0.88$), though without knowing what the scale is here it's hard to say much more than that. Still, that's not too bad. Overall I feel that I learned something from that.

We can also go ahead and look at some other descriptive statistics, and this time separately for each type of therapy. In jamovi, check Std. deviation, Skewness and Kurtosis in the 'Statistics' options. At the same time, transfer the therapy variable into the 'Split by' box, and you should get something like Figure 4.14.

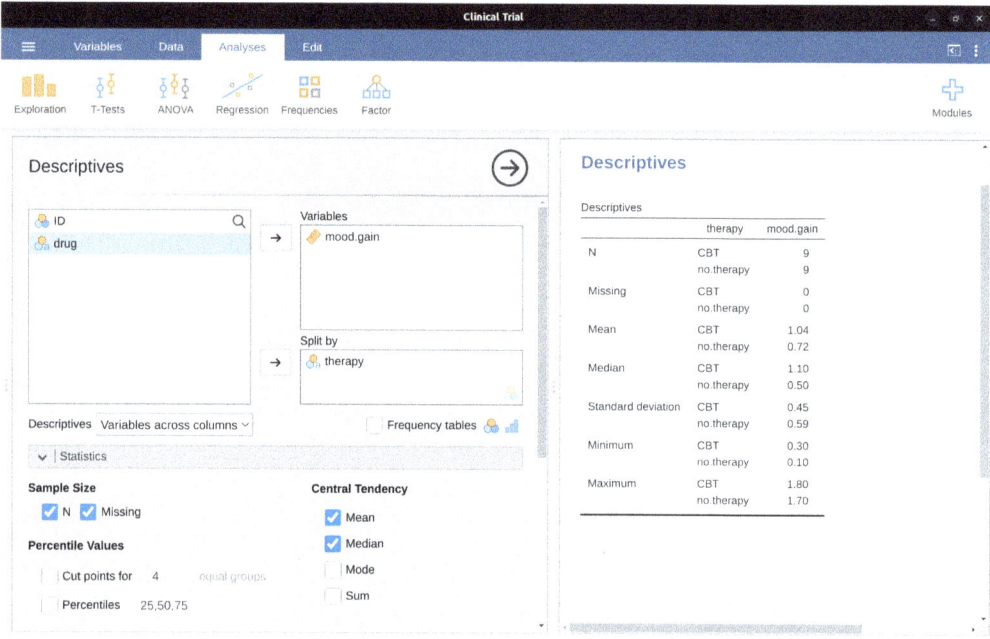

Figure 4.14: A screenshot of jamovi showing Descriptives split by therapy type

What if you have multiple grouping variables? Suppose you want to look at the average mood gain separately for all possible combinations of drug and therapy. It is possible to do this by adding another variable, drug, into the 'Split by' box. Easy peasy, though sometimes if you split too much there isn't enough data in each breakdown combination to make meaningful calculations. In this case jamovi tells you this by stating something like 'NaN' or 'Inf'.[33]

4.5 Standard scores

Suppose my friend is putting together a new questionnaire intended to measure "grumpiness". The survey has 50 questions which you can answer in a grumpy way or not. Across a big sample (hypothetically, let's imagine a million people or so!) the data are fairly normally distributed, with the mean grumpiness score being 17 out of 50 questions answered in a grumpy way, and the standard deviation is 5. In contrast, when I take the questionnaire I answer 35 out of 50 questions in a grumpy way. So, how grumpy am I? One way to think about it would be to say that I have grumpiness of $\frac{35}{50}$, so you might say that I'm 70% grumpy. But that's a bit weird, when you think about it. If my friend had phrased her questions a bit differently people might have answered them in a different way, so the overall distribution of answers could easily move up or down depending on the precise way in which the questions were asked. So, I'm only 70% grumpy *with respect to this set of survey questions*. Even if it's a very good questionnaire this isn't a very informative statement.

A simpler way around this is to describe my grumpiness by comparing me to other people. Shockingly, out of my friend's sample of $1,000,000$ people, only 159 people were as grumpy as me (that's not at all unrealistic, frankly) suggesting that I'm in the top 0.016% of people for grumpiness. This makes much more sense than trying to interpret the raw data. This idea, that we should describe my grumpiness in terms of the overall distribution of the grumpiness of humans, is the qualitative idea that standardisation attempts to get at. One way to do this is to do exactly what I just did and describe everything in terms of percentiles. However, the problem with doing this is that "it's lonely at the top". Suppose that my friend had only collected a sample of 1000 people (still a pretty big sample for the purposes of testing a new questionnaire, I'd like to add), and this time had gotten, let's say, a mean of 16 out of 50 with a standard deviation of 5. The problem is that almost certainly not a single person in that sample would be as grumpy as me.

However, all is not lost. A different approach is to convert my grumpiness score into a **standard score**, also referred to as a *z*-score. The standard score is defined as the number of standard deviations above the mean that my grumpiness score lies. To phrase it in "pseudomaths" the standard score is calculated like this:

$$\text{standard score} = \frac{\text{raw score} - mean}{\text{standard deviation}}$$

[Additional technical detail[34]]

So, going back to the grumpiness data, we can now transform Dani's raw grumpiness into a standardised grumpiness score:

$$z = \frac{35 - 17}{5} = 3.6$$

To interpret this value, recall the rough heuristic that I provided in Section 4.2.5 in which I noted that 99.7% of values are expected to lie within 3 standard deviations of the mean. So the fact that my grumpiness corresponds to a *z*-score of 3.6 indicates that I'm very grumpy indeed. In fact this suggests that I'm grumpier than 99.98% of people. Sounds about right.

In addition to allowing you to interpret a raw score in relation to a larger population (and thereby allowing you to make sense of variables that lie on arbitrary scales), standard scores serve a second useful function. Standard scores can be compared to one another in situations where the raw scores can't. Suppose, for instance, my friend also had another questionnaire that measured extraversion using a 24 item questionnaire. The overall mean for this measure turns out to be 13 with standard deviation 4, and I scored a 2. As you can imagine, it doesn't make a lot of sense to try to compare my raw score of 2 on the extraversion questionnaire to my raw score of 35 on the grumpiness questionnaire. The raw scores for the two variables are "about" fundamentally different things, so this would be like comparing apples to oranges.

What about the standard scores? Well, this is a little different. If we calculate the standard scores we get $(z = \frac{(35-17)}{5} = 3.6)$ for grumpiness and $(z = \frac{(2-13)}{4} = -2.75)$

for extraversion. These two numbers can be compared to each other.[35] I'm much less extraverted than most people ($z = -2.75$) and much grumpier than most people ($z = 3.6$). But the extent of my unusualness is much more extreme for grumpiness, since 3.6 is a bigger number than 2.75. Because each standardised score is a statement about where an observation falls relative to its own population, it is possible to compare standardised scores across completely different variables.

4.6 Summary

Calculating some basic descriptive statistics is one of the very first things you do when analysing real data, and descriptive statistics are much simpler to understand than inferential statistics, so like every other statistics textbook I've started with descriptives. In this chapter, we talked about the following topics:

- Measures of central tendency. Broadly speaking, central tendency measures tell you where the data are. There's three measures that are typically reported in the literature: the mean, median and mode.
- Measures of variability. In contrast, measures of variability tell you about how "spread out" the data are. The key measures are: range, standard deviation, and interquartile range.
- Skew and kurtosis. We also looked at assymetry in a variable's distribution (skew) and thin or fat tailed distributions (kurtosis).
- Descriptive statistics separately for each group. Since this book focuses on doing data analysis in jamovi, we spent a bit of time talking about how descriptive statistics are computed for different subgroups.
- Standard scores. The z-score is a slightly unusual beast. It's not quite a descriptive statistic, and not quite an inference. Make sure you understand this section. It'll come up again later.

In the next chapter we'll move on to a discussion of how to draw pictures! Everyone loves a pretty picture, right? But before we do, I want to end on an important point. A traditional first course in statistics spends only a small proportion of the class on descriptive statistics, maybe one or two lectures at most. The vast majority of the lecturer's time is spent on inferential statistics because that's where all the hard stuff is. That makes sense, but it hides the practical everyday importance of choosing good descriptives.

Chapter 5

Drawing graphs

Above all else show the data.
– Edward Tufte[36]

Visualising data is one of the most important tasks facing the data analyst. It's important for two reasons. First, for drawing "presentation graphics" – displaying data in a clean, visually appealing way makes it easier for readers to understand what you're trying to tell them. Second, and perhaps more important, is that drawing graphs helps you to understand the data. To that end it's important that "exploratory graphics" help you learn about the data as part of your analysis.

To give a sense of the importance of this chapter, I'll start with a classic illustration of just how powerful a good graph can be. Figure 5.1 is a redrawing of one of the most famous data visualisations of all time – John Snow's 1854 map of cholera deaths. The map is elegant in its simplicity. A street map helps orient the viewer, overlayed with a large number of small squares, each one representing the location of a cholera case. The larger dots show the location of water pumps, labelled by name. Even the most casual inspection of the graph makes it clear that the source of the outbreak is almost certainly the Broad Street pump. Upon viewing this graph Dr Snow arranged to have the handle removed from the pump, thus ending the outbreak that had killed over 500 people. Such is the power of a good data visualisation.

There are two goals in this chapter. First, to discuss several fairly standard graphs that we use a lot when analysing and presenting data, and second to show you how to create these graphs in jamovi. The graphs themselves tend to be pretty straightforward, so in one respect this chapter is pretty simple. Where people usually struggle is learning how to produce graphs, and especially learning how to produce good graphs. Fortunately, learning how to draw graphs in jamovi is reasonably simple as long as you're not too picky about what your graph looks like. What I mean when I say this is that jamovi has a lot of very good default graphs, or plots, that most of the time produce a clean, high-quality graphic. However, on those occasions when you do want to do something non-standard, or if you need to make highly specific changes to the figure, then the graphics functionality in jamovi is not yet capable of supporting advanced work or detail editing.

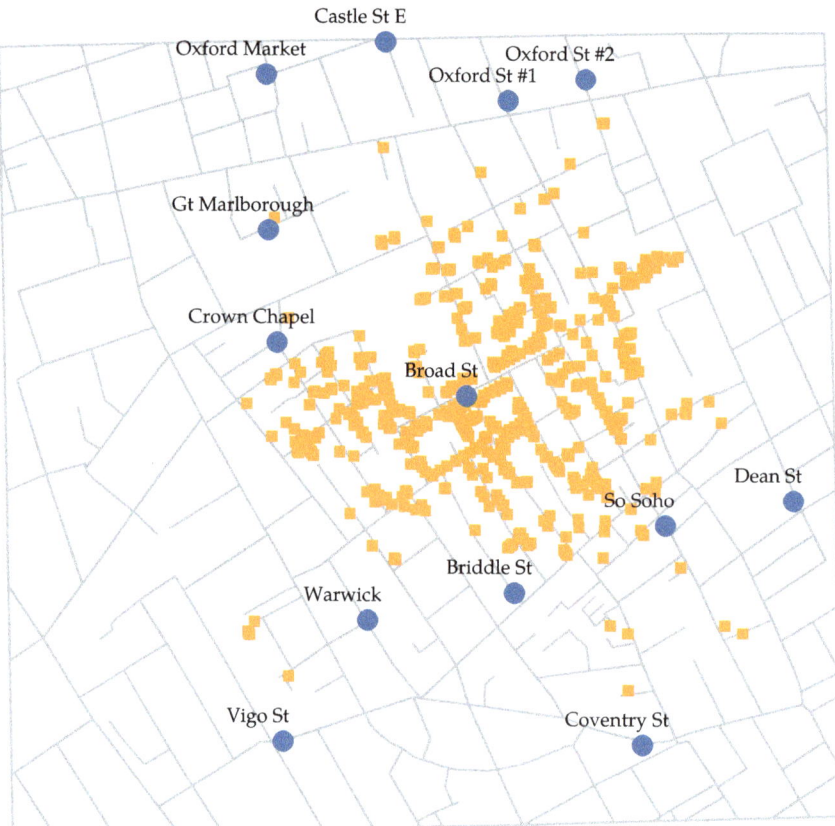

Figure 5.1: A stylised redrawing of John Snow's original cholera map of London. Each small orange square represents the location of a cholera death and each blue circle shows the location of a water pump. As the plot makes clear, the cholera outbreak is centred very closely on the Broad St pump

5.1 Histograms

Let's begin with the humble **histogram**. Histograms are one of the simplest and most useful ways of visualising data. They make most sense when you have an interval or ratio scale variable (e.g., the *afl.margins* data from Chapter 4) and you want to get an overall impression of the variable. Most of you probably know how histograms work, since they're so widely used, but for the sake of completeness I'll describe them. All you do is divide up the possible values into **bins** and then count the number of observations that fall within each bin. This count is referred to as the frequency or density of the bin and is displayed as a vertical bar. In the AFL winning margins data there are 33 games in which the winning margin was less than 10 points and it is this fact that is represented by the height of the leftmost bar that we showed earlier in Chapter 4, Figure 4.2. With these earlier graphs we used an advanced plotting package in R which, for now, is beyond the capability of jamovi. But jamovi gets us close, and drawing this histogram in jamovi is pretty straightforward. Open up the 'plots' options under 'Exploration' – 'Descriptives' and click the 'histogram' check box, as in Figure 5.2. jamovi defaults to labelling the y-axis as 'density' and the x-axis with the variable name. The **bins** are selected automatically, and there is no scale, or count, information on the y-axis unlike the previous Figure 4.2. But this does not matter too much because after all what we are really interested in is our impression of the shape of the distribution: is it normally distributed or is there a skew or kurtosis? Our first impressions of these characteristics come from drawing a **histogram**.

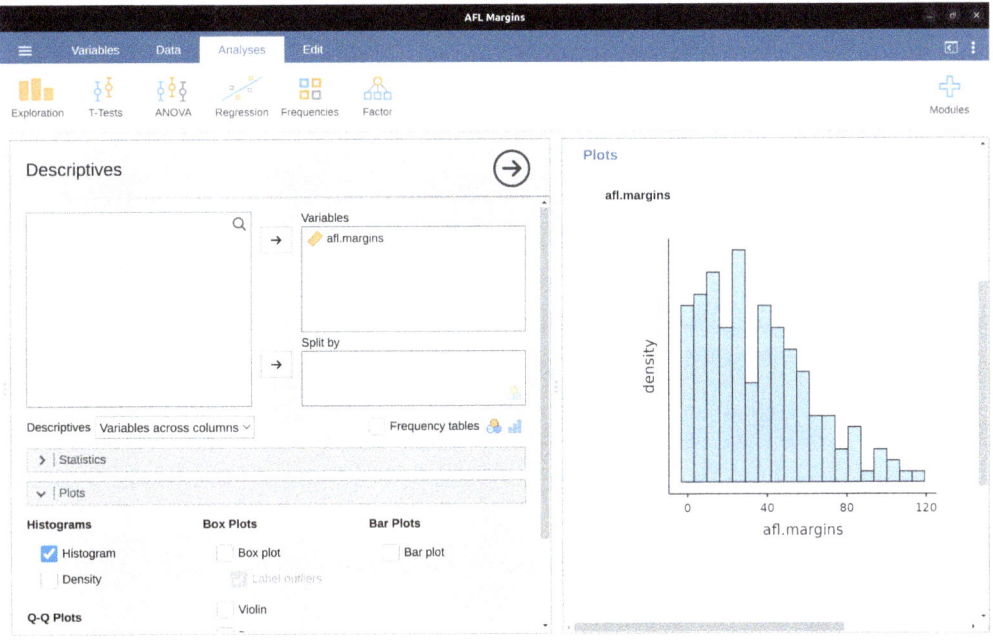

Figure 5.2: jamovi screen showing the histogram check box

One additional feature that jamovi provides is the ability to plot a 'Density' curve. You can do this by clicking the 'Density' check box under the 'Plots' options (and unchecking 'Histogram'), and this gives us the plot shown in Figure 5.3. A density plot visualises the distribution of data over a continuous interval or time period. This chart is a variation of a histogram that uses **kernel smoothing** to plot values, allowing for smoother distributions by smoothing out the noise. The peaks of a density plot help display where values are concentrated over the interval. An advantage density plots have over histograms is that they are better at determining the distribution shape because they're not affected by the number of bins used (each bar used in a typical histogram). A histogram comprising of only 4 bins wouldn't produce a distinguishable enough shape of distribution as a 20-bin histogram would. However, with density plots, this isn't an issue.

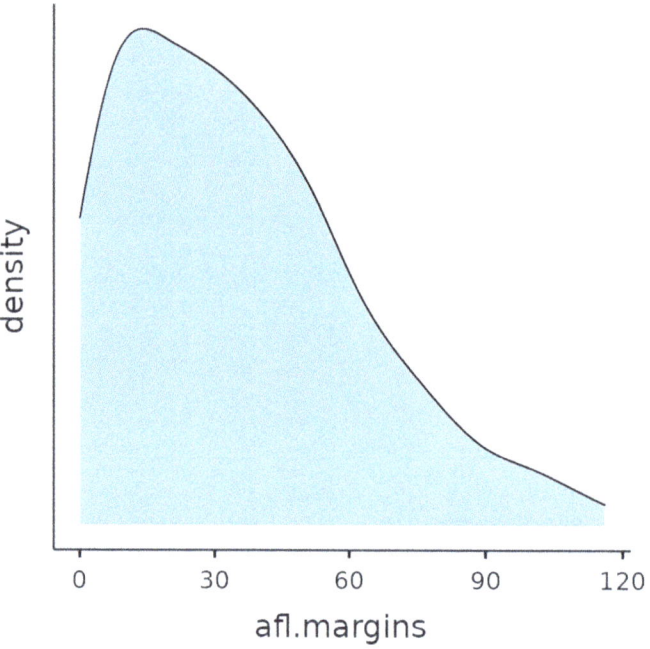

Figure 5.3: A density plot of the *afl.margins* variable plotted in jamovi

Although this image would need a lot of cleaning up in order to make a good presentation graphic (i.e., one you'd include in a report), it nevertheless does a pretty good job of describing the data. In fact, the big strength of a histogram or density plot is that (properly used) it does show the entire spread of the data, so you can get a pretty good sense about what it looks like. The downside to histograms is that they aren't very compact. Unlike some of the other plots I'll talk about it's hard to cram 20-30 histograms into a single image without overwhelming the viewer. And of course, if your data are nominal scale then histograms are useless.

5.2 Boxplots

Another alternative to histograms is a **boxplot**, sometimes called a "box and whiskers" plot. Like histograms they're most suited to interval or ratio scale data. The idea behind a boxplot is to provide a simple visual depiction of the median, the interquartile range, and the range of the data. And because they do so in a fairly compact way boxplots have become a very popular statistical graphic, especially during the exploratory stage of data analysis when you're trying to understand the data yourself. Let's have a look at how they work, again using the *afl.margins* data as our example.

The easiest way to describe what a boxplot looks like is just to draw one. Click on the 'Box plot' check box and you will get the plot shown in Figure 5.4. jamovi has drawn the most basic boxplot possible.

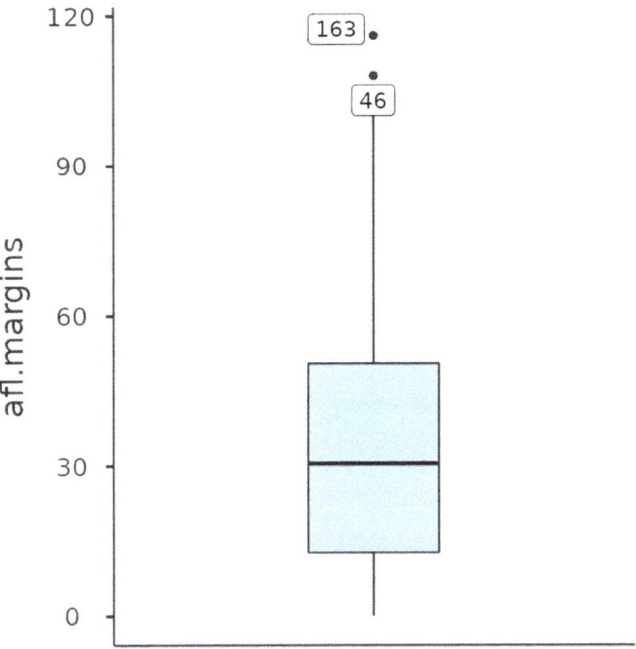

Figure 5.4: A box plot of the *afl.margins* variable plotted in jamovi

When you look at this plot this is how you should interpret it: the thick line in the middle of the box is the median; the box itself spans the range from the 25th percentile to the 75th percentile; and the "whiskers" go out to the most extreme data point that doesn't exceed a certain bound. By default, this value is 1.5 times the interquartile range (IQR), calculated as 25th percentile $-(1.5 \times IQR)$ for the lower boundary, and 75th percentile $+(1.5 \times IQR)$ for the upper boundary. Any observation whose value falls outside this range is plotted as a circle or dot instead of being covered by the whiskers, and is commonly referred to as an **outlier**. For our *afl.margins* data there are two observations that fall outside this range, and these observations are plotted as dots (the upper boundary is 107, and looking over the data column in the spreadsheet there are two observations with values higher than this, 46 and 163, so these are the dots).

5.2.1 Violin plots

A variation to the traditional box plot is the violin plot. Violin plots are similar to box plots except that they also show the kernel probability density of the data at different values. Typically, violin plots will include a marker for the median of the data and a box indicating the interquartile range, as in standard box plots. In jamovi you can achieve this sort of functionality by checking both the 'Violin' and the 'Box plot' check boxes. See Figure 5.5, which also has the 'Data' check box turned on to show the actual data points on the plot. This does tend to make the graph a bit too busy though, in my opinion. Clarity is simplicity, so in practice it might be better to just use a simple box plot.

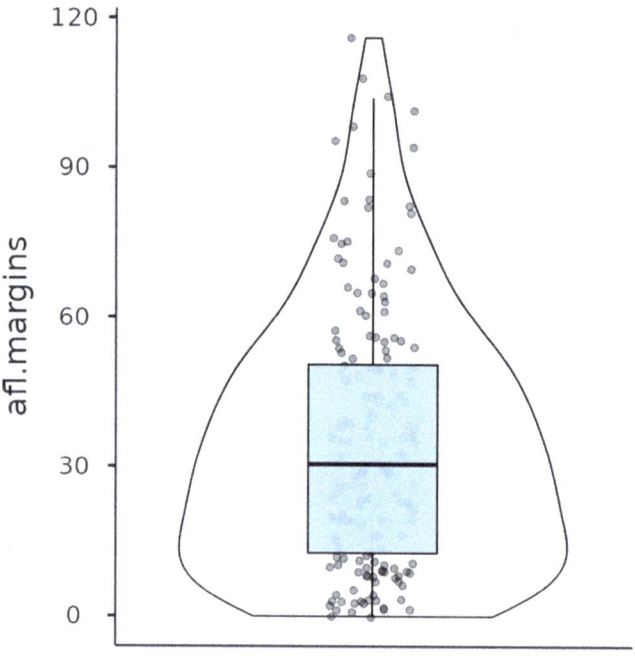

Figure 5.5: A violin plot of the *afl.margins* variable plotted in jamovi, also showing a box plot and data points

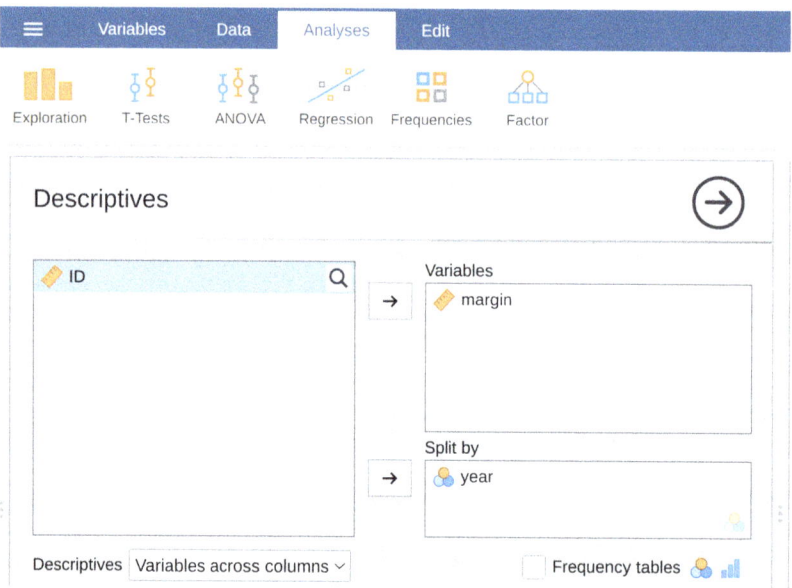

Figure 5.6: jamovi screen shot showing the 'Split by' window

5.2.2 Drawing multiple boxplots

One last thing. What if you want to draw multiple boxplots at once? Suppose, for instance, I wanted separate boxplots showing the *afl.margins* not just for 2010 but for every year between 1987 and 2010. To do that the first thing we'll have to do is find the data. These are stored in the **aflmarginbyyear.csv** file. So let's load it into jamovi and see what is in it. You will see that it is a pretty big data set. It contains 4296 games and the variables that we're interested in. What we want to do is have jamovi draw boxplots for the margin variable, but plotted separately for each year. The way to do this is to move the year variable across into the 'Split by' box, as in Figure 5.6.

The result is shown in Figure 5.7. This version of the box plot, split by year, gives a sense of why it's sometimes useful to choose box plots instead of histograms. It's possible to get a good sense of what the data look like from year to year without getting overwhelmed with too much detail. Now imagine what would have happened if I'd tried to cram 24 histograms into this space: no chance at all that the reader is going to learn anything useful.

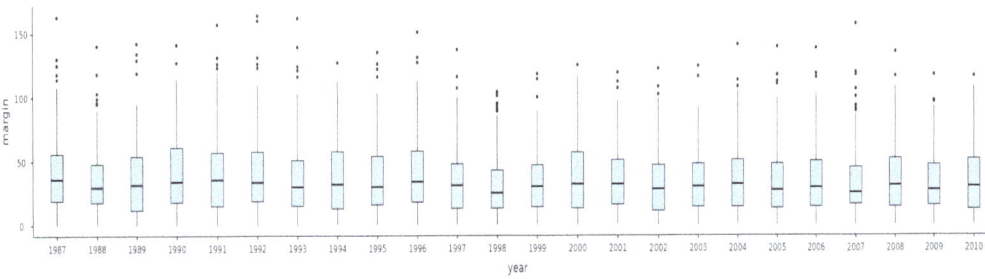

Figure 5.7: Multiple boxplots plotted in jamovi, for the margin by year variables

5.2.3 Using box plots to detect outliers

As the boxplot automatically separates out observations that lie outside a certain range, depicting them with a dot in jamovi, people often use them as an informal method for detecting **outliers**: observations that are "suspiciously" distant from the rest of the data. Here's an example. Suppose that I'd drawn the boxplot for the *afl.margins* data and it came up looking like Figure 5.8. It's pretty clear that something funny is going on with two of the observations. Apparently, there were two games in which the margin was over 300 points![37] That doesn't sound right to me. Now that I've become suspicious it's time to look a bit more closely at the data. In jamovi you can quickly find out which of these observations are suspicious and then you can go back to the raw data to see if there has been a mistake in data entry. One way to do this is to tell jamovi to label the outliers, by checking the box next to the 'Box plot' check box. This adds a row number label next to the outlier in the boxplot, so you can go look at that row and find the extreme value. Another, more flexible way, is to set up a filter so that only those observations with values over a certain threshold are included. In our example, the threshold is over 300, so that is the filter we will create. First, click on the 'Filters' button at the top of the jamovi window, and then type 'margin > 300' into the filter field, as in Figure 5.9.

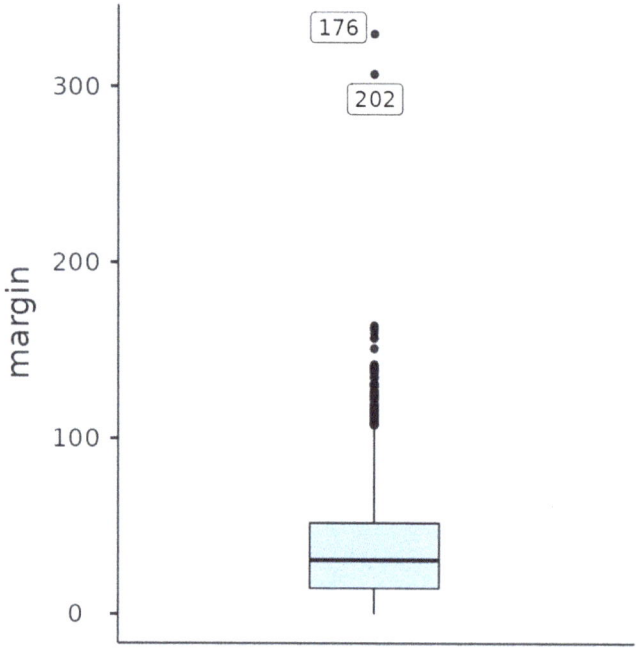

Figure 5.8: A boxplot showing two very suspicious outliers!

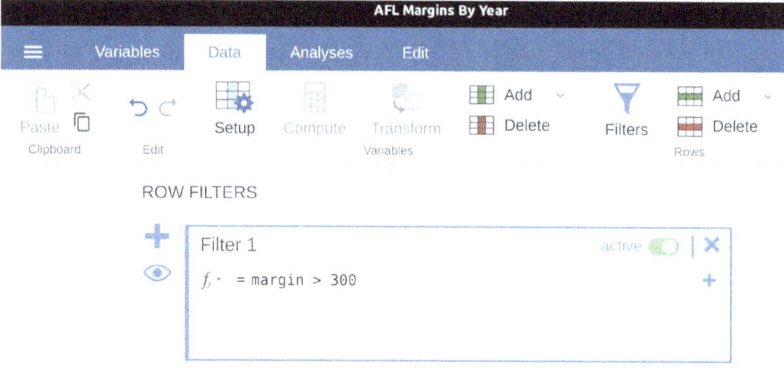

Figure 5.9: The jamovi filter screen

This filter creates a new column in the spreadsheet view where only those observations that pass the filter are included. One neat way to quickly identify which observations these are is to tell jamovi to produce a 'Frequency table' (in the 'Exploration' – 'Descriptives' window) for the ID variable (which must be a nominal variable otherwise the Frequency table is not produced). In Figure 5.10 you can see that the ID values for the observations where the margin was over 300 are 176 and 202. These are suspicious cases, or observations, where you should go back to the original data source to find out what is going on.

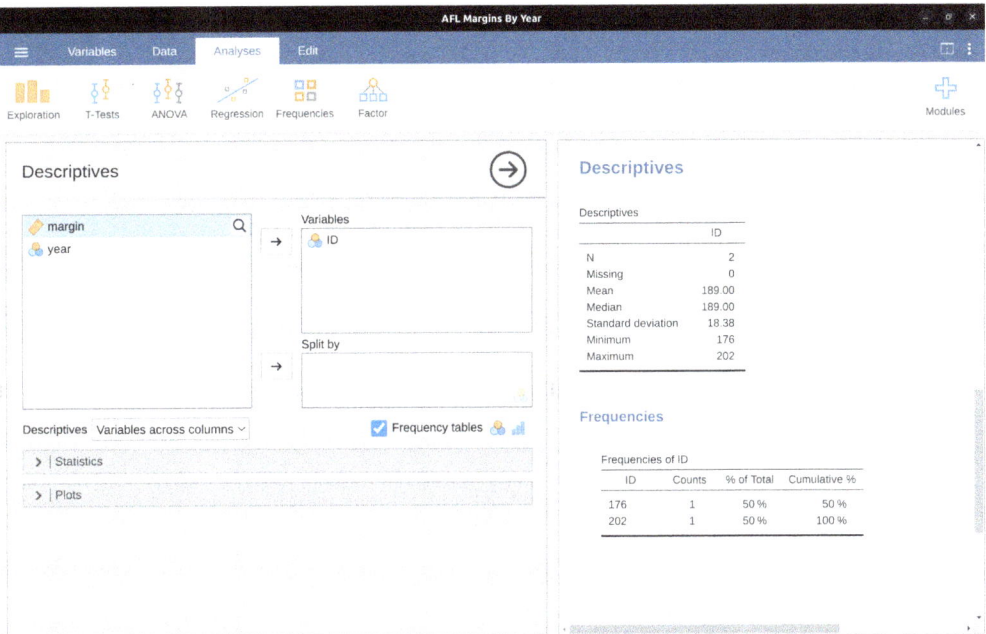

Figure 5.10: Frequency table for ID showing the ID numbers for the two suspicious outliers, 176 and 202

Usually you find that someone has just typed in the wrong number. Whilst this might seem like a silly example, I should stress that this kind of thing actually happens a lot. Real world data sets are often riddled with stupid errors, especially when someone had to type something into a computer at some point. In fact, there's actually a name for this phase of data analysis and in practice it can take up a huge chunk of our time: data cleaning. It involves searching for typing mistakes ("typos"), missing data and all sorts of other obnoxious errors in raw data files.

For less extreme values, even if they are flagged in a a boxplot as outliers, the decision about whether to include outliers or exclude them in any analysis depends heavily on why you think the data look they way they do and what you want to use the data for. You really need to exercise good judgement here. If the outlier looks legitimate to you, then keep it. In any case, I'll return to the topic again in Section 12.10 in Chapter 12.

5.3 Bar graphs

Another form of graph that you often want to plot is the **bar graph**. Let's use the *afl.finalists* data set with the *afl.finalists* variable that I introduced in Section 4.1.6. What I want to do is draw a bar graph that displays the number of finals that each team has played in over the time spanned by the *afl.finalists* data set. There are lots of teams, but I am particularly interested in just four: Brisbane, Carlton, Fremantle and Richmond. So the first step is to set up a filter so just those four teams are included in the bar graph. This is straightforward in jamovi and you can do it by using the 'Filters' function that we used previously. Open up the 'Filters' screen and type in the following exactly as written – including the single quote marks:

afl.finalists == 'Brisbane' or afl.finalists == 'Carlton' or afl.finalists == 'Fremantle' or afl.finalists == 'Richmond' [38]

When you have done this you will see, in the 'Data' view, that jamovi has filtered out all values apart from those we have specified. Next, open up the 'Exploration' – 'Descriptives' window and click on the 'Bar plot' check box (remember to move the 'afl.finalists' variable across into the 'Variables' box so that jamovi knows which variable to use). You should then get a bar graph, something like that shown in Figure 5.11.

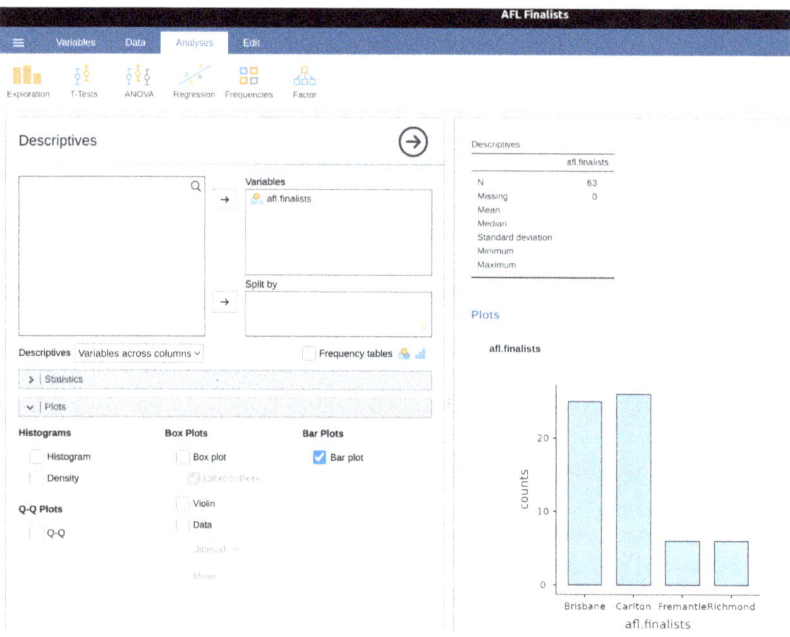

Figure 5.11: Filtering to include just four AFL teams, and drawing a bar plot in jamovi

5.4 Saving image files using jamovi

Hold on, you might be thinking. What's the good of being able to draw pretty pictures in jamovi if I can't save them and send them to friends to brag about how awesome my data is? How do I save the picture? Simples. Just right click on the plot image and export it to a file, either as 'png', 'eps', 'svg' or 'pdf'. These formats all produce nice images that you can then send to your friends, or include in your assignments or papers.

5.5 Summary

Perhaps I'm a simple-minded person, but I love pictures. Every time I write a new scientific paper one of the first things I do is sit down and think about what the pictures will be. In my head an article is really just a sequence of pictures linked together by a story. All the rest of it is just window dressing. What I'm really trying to say here is that the human visual system is a very powerful data analysis tool. Give it the right kind of information and it will supply a human reader with a massive amount of knowledge very quickly. Not for nothing do we have the saying "a picture is worth a thousand words". With that in mind, I think that this is one of the most important chapters in the book. The topics covered were:

- *Common plots*. Much of the chapter was focused on standard graphs that statisticians like to produce: Histograms, Boxplots and Bar graphs.
- Saving image files using jamovi. Importantly, we also covered how to export your pictures.

One final thing to point out. Whilst jamovi produces some really neat default graphics, editing the plots is currently not possible. For more advanced graphics and plotting capability the packages available in R are much more powerful. One of the most popular graphics systems is provided by the ggplot2 package (see https://ggplot2.tidyverse.org/), which is loosely based on *The grammar of graphics* (Wilkinson et al., 2006). It's not for novices. You need to have a pretty good grasp of R before you can start using it, and even then it takes a while to really get the hang of it. But when you're ready it's worth taking the time to teach yourself, because it's a much more powerful and cleaner system.

Chapter 6

Pragmatic matters

> *The garden of life never seems to confine itself to the plots philosophers have laid out for its convenience. Maybe a few more tractors would do the trick.*
> – Roger Zelazny[39]

This is a somewhat strange chapter, even by my standards. My goal in this chapter is to talk a bit more honestly about the realities of working with data than you'll see anywhere else in the book. The problem with real world data sets is that they are *messy*. Very often the data file that you start out with doesn't have the variables stored in the right format for the analysis you want to do. Sometimes there might be a lot of missing values in your data set. Sometimes you only want to analyse a subset of the data. Et cetera. In other words, there's a lot of **data manipulation** that you need to do just to get the variables in your data set into the format that you need it. The purpose of this chapter is to provide a basic introduction to these pragmatic topics. Although the chapter is motivated by the kinds of practical issues that arise when manipulating real data, I'll stick with the practice that I've adopted throughout most of the book and rely on very small, toy data sets that illustrate the underlying issue. Because this chapter is essentially a collection of techniques and doesn't tell a single coherent story, it may be useful to start with a list of topics:

- Tabulating and cross-tabulating data.
- Logical expressions in jamovi.
- Transforming and recoding a variable.
- A few more mathematical functions and operations.
- Extracting a subset of the data.

As you can see, the list of topics that the chapter covers is pretty broad, and there's a lot of content there. Even though this is one of the longest and hardest chapters in the book, I'm really only scratching the surface of several fairly different and important topics. My advice, as usual, is to read through the chapter once and try to follow as much of it as you can. Don't worry too much if you can't grasp it all at once, especially the later sections. The rest of the book is only lightly reliant on this chapter so you can get away with just understanding the basics. However, what you'll probably find is that later on you'll need to flick back to this chapter in order to understand some of the concepts that I refer to here.

6.1 Tabulating and cross-tabulating data

A very common task when analysing data is the construction of frequency tables, or crosstabulation of one variable against another. These tasks can be achieved in jamovi and I'll show you how in this section.

6.1.1 Creating tables for single variables

Let's start with a simple example. As the father of a small child I naturally spend a lot of time watching TV shows like *In the Night Garden*. In the *nightgarden.csv* file, I've transcribed a short section of the dialogue. The file contains two variables of interest, speaker and utterance. Open up this data set in jamovi and take a look at the data in the 'spreadsheet' view. You will see that the data looks something like this:

'speaker' variable: upsy-daisy upsy-daisy upsy-daisy upsy-daisy tombliboo tombliboo makka-pakka makka-pakka makka-pakka makka-pakka 'utterance' variable: pip pip onk onk ee oo pip pip onk onk

Looking at this it becomes very clear what happened to my sanity! With these as my data, one task I might find myself needing to do is construct a frequency count of the number of words each character speaks during the show. The jamovi 'Descriptives' screen has a check box called 'Frequency tables' which does just this, see Table 6.1.

Table 6.1: Frequency table for the speaker variable

levels	Counts	% of Total	Cumulative %
makka-pakka	4	40%	40%
tombliboo	2	20%	60%
upsy-daisy	4	40%	100%

The output here tells us on the first line that what we're looking at is a tabulation of the speaker variable. In the 'Levels' column it lists all the different speakers that exist in the data, and in the 'Counts' column it tells you how many times that speaker appears in the data. In other words, it's a frequency table.

In jamovi, the 'Frequency tables' check box will only produce a table for single variables. For a table of two variables, for example combining speaker and utterance so that we can see how many times each speaker said a particular utterance, we need a cross-tabulation or contingency table. In jamovi you can do this by selecting the 'Frequencies' – 'Contingency Tables' – 'Independent Samples' analysis, and moving the speaker variable into the 'Rows' box, and the utterances variable into the 'Columns' box. You then should have a contingency table like the one shown in Figure 6.1.

Don't worry about the "χ^2 Tests" table that is produced. We are going to cover this later on in Chapter 10. When interpreting the contingency table remember that these are counts, so the fact that the first row and second column of numbers corresponds to a value of 2 indicates that Makka-Pakka (row 1) says "onk" (column 2) twice in this data set.

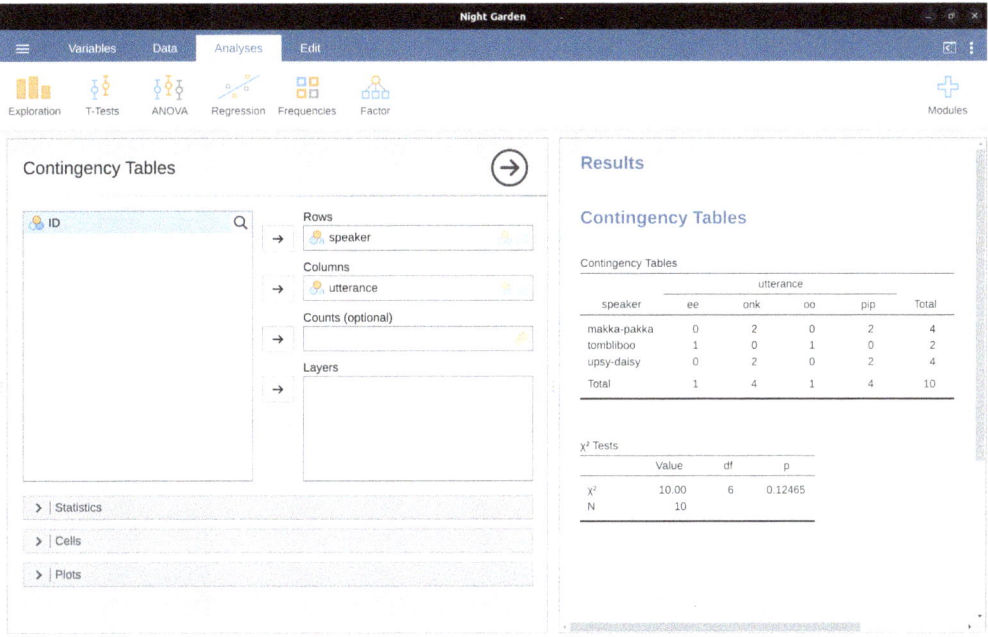

Figure 6.1: Contingency table for the speaker and utterances variables

6.1.2 Adding percentages to a contingency table

The contingency table shown in Figure 6.1 shows a table of raw frequencies. That is, a count of the total number of cases for different combinations of levels of the specified variables. However, often you want your data to be organised in terms of percentages as well as counts. You can find the check boxes for different percentages under the 'Cells' option in the 'Contingency Tables' window. First, click on the 'Row' check box and the Contingency Table in the output window will change to the one in Figure 6.2.

What we're looking at here is the percentage of utterances made by each character. In other words, 50% of Makka-Pakka's utterances are "pip", and the other 50% are "onk". Let's contrast this with the table we get when we calculate column percentages (uncheck 'Row' and check 'Column' in the Cells options window), see Figure 6.3. In this version, what we're seeing is the percentage of characters associated with each utterance. For instance, whenever the utterance "ee" is made (in this data set), 100% of the time it's a Tombliboo saying it.

6.2 Logical expressions in jamovi

A key concept that a lot of data transformations in jamovi rely on is the idea of a **logical value**. A logical value is an assertion about whether something is true or false. This is implemented in jamovi in a pretty straightforward way. There are two logical values, namely TRUE and FALSE. Despite the simplicity, logical values are very useful things. Let's see how they work.

Contingency Tables

Contingency Tables

speaker		utterance				Total
		ee	onk	oo	pip	
makka-pakka	Observed	0	2	0	2	4
	% within row	0 %	50 %	0 %	50 %	100 %
tombliboo	Observed	1	0	1	0	2
	% within row	50 %	0 %	50 %	0 %	100 %
upsy-daisy	Observed	0	2	0	2	4
	% within row	0 %	50 %	0 %	50 %	100 %
Total	Observed	1	4	1	4	10
	% within row	10 %	40 %	10 %	40 %	100 %

Figure 6.2: Contingency table for the speaker and utterances variables, with row percentages

Contingency Tables

Contingency Tables

speaker		utterance				Total
		ee	onk	oo	pip	
makka-pakka	Observed	0	2	0	2	4
	% within column	0 %	50 %	0 %	50 %	40 %
tombliboo	Observed	1	0	1	0	2
	% within column	100 %	0 %	100 %	0 %	20 %
upsy-daisy	Observed	0	2	0	2	4
	% within column	0 %	50 %	0 %	50 %	40 %
Total	Observed	1	4	1	4	10
	% within column	100 %	100 %	100 %	100 %	100 %

Figure 6.3: Contingency table for the speaker and utterances variables, with column percentages

6.2.1 Assessing mathematical truths

In George Orwell's classic book *1984* one of the slogans used by the totalitarian Party was "two plus two equals five". The idea being that the political domination of human freedom becomes complete when it is possible to subvert even the most basic of truths. It's a terrifying thought, especially when the protagonist Winston Smith finally breaks down under torture and agrees to the proposition. "Man is infinitely malleable", the book says. I'm pretty sure that this isn't true of humans[40] and it's definitely not true of jamovi. jamovi is not infinitely malleable, it has rather firm opinions on the topic of what is and isn't true, at least as regards basic mathematics. If I ask it to calculate $2 + 2$,[41] it always gives the same answer, and it's not 5!

Of course, so far jamovi is just doing the calculations. I haven't asked it to explicitly assert that $2 + 2 = 4$ is a true statement. If I want jamovi to make an explicit judgement, I can use a command like this: $2 + 2 == 4$.

What I've done here is use the **equality operator**, $==$, to force jamovi to make a "true or false" judgement.[42] Okay, let's see what jamovi thinks of the Party slogan, so type this into the compute new variable 'formula' box:

$$2 + 2 == 5$$

And what do you get? It should be a whole set of 'false' values in the spreadsheet column for your newly computed variable. It was worth having a look at what happens if I try to force jamovi to believe that two plus two is five by making a statement like $2 + 2 = 5$. I know that if I do this in another program, say R, then it throws up an error message. But wait, if you do this in jamovi you get a whole set of 'false' values. So what is going on? Well, it seems that jamovi is being pretty smart and realises that you are testing whether it is TRUE or FALSE that $2 + 2 = 5$, regardless of whether you use the correct **equality operator** $==$, or the equals sign $=$.

6.2.2 Logical operations

So now we've seen logical operations at work. But so far we've only seen the simplest possible example. You probably won't be surprised to discover that we can combine logical operations with other operations and functions in a more complicated way, like this: $3 \times 3 + 4 \times 4 == 5 \times 5$ or this $SQRT(25) == 5$

Not only that, but as Table 6.2 illustrates, there are several other logical operators that you can use corresponding to some basic mathematical concepts. Hopefully these are all pretty self-explanatory. For example, the **less than** operator $<$ checks to see if the number on the left is less than the number on the right. If it's less, then jamovi returns an answer of TRUE, but if the two numbers are equal, or if the one on the right is smaller, then jamovi returns an answer of FALSE.

In contrast, the **less than or equal to** operator $<=$ will do exactly what it says. It returns a value of TRUE if the number on the left-hand side is less than or equal to the number on the right-hand side. At this point I hope it's pretty obvious what the **greater than** operator $>$ and the **greater than or equal to** operator $>=$ do!

Table 6.2: Some logical operators

operation	operator	example input	answer
less than	<	2 < 3	TRUE
less than or equal to	<=	2 <= 2	TRUE
greater than	>	2 > 3	FALSE
greater than or equal to	>=	2 >= 2	TRUE
equal to	==	2 == 3	FALSE
not equal to	!=	2 != 3	TRUE

Next on the list of logical operators is the **not equal to** operator $!=$ which, as with all the others, does what it says it does. It returns a value of TRUE when things on either side are not identical to each other. Therefore, since $2 + 2$ isn't equal to 5, we would get "true" as the value for our newly computed variable. Try it and see:

$$2 + 2 != 5$$

We're not quite done yet. There are three more logical operations that are worth knowing about, listed in Table 6.3. These are the **not** operator $!$, the **and** operator *and*, and the **or** operator *or*. Like the other logical operators, their behaviour is more or less exactly what you'd expect given their names. For instance, if I ask you to assess the claim that "either $2 + 2 = 4$ or $2 + 2 = 5$" you'd say that it's true. Since it's an "either-or" statement, all we need is for one of the two parts to be true. That's what the *or* operator does.[43]

Table 6.3: Some more logical operators

operation	operator	example input	answer
not	NOT	NOT(1==1)	FALSE
or	or	(1==1) or (2==3)	TRUE
and	and	(1==1) and (2==3)	FALSE

On the other hand, if I ask you to assess the claim that "both $2 + 2 = 4$ and $2 + 2 = 5$" you'd say that it's false. Since this is an **and** statement we need both parts to be true. And that's what the *and* operator does:

$$(2 + 2 == 4) \text{ and } (2 + 2 == 5)$$

Finally, there's the **not** operator, which is simple but annoying to describe in English. If I ask you to assess my claim that "it is not true that $2 + 2 = 5$" then you would say that my claim is true, because actually my claim is that "$2 + 2 = 5$ is false". And I'm right.

If we write this in jamovi we use this:

$$NOT(2+2==5)$$

In other words, since $2+2==5$ is a FALSE statement, it must be the case that $NOT(2+2==5)$ is a TRUE one. Essentially, what we've really done is claim that "not false" is the same thing as "true". Obviously, this isn't really quite right in real life. But jamovi lives in a much more black or white world. For jamovi everything is either true or false. No shades of grey are allowed. Of course, in our $2+2=5$ example, we didn't really need to use the "not" operator NOT and the "equals to" operator $==$ as two separate operators. We could have just used the "not equals to" operator $!=$ like this:

$$2+2\;!=5$$

6.2.3 Applying logical operation to text

I also want to briefly point out that you can apply these logical operators to text as well as to logical data. It's just that we need to be a bit more careful in understanding how jamovi interprets the different operations. In this section I'll talk about how the equal to operator $==$ applies to text, since this is the most important one. Obviously, the not equal to operator $!=$ gives the exact opposite answers to $==$ so I'm implicitly talking about that one too, but I won't give specific commands showing the use of $!=$.

Okay, let's see how it works. In one sense, it's very simple. For instance, I can ask jamovi if the word "cat" is the same as the word "dog", like this:

"cat" $==$ "dog"

That's pretty obvious, and it's good to know that even jamovi can figure that out. Similarly, jamovi does recognise that a "cat" is a "cat":

"cat" $==$ "cat"

Again, that's exactly what we'd expect. However, what you need to keep in mind is that jamovi is not at all tolerant when it comes to grammar and spacing. If two strings differ in any way whatsoever, jamovi will say that they're not equal to each other, as with the following:

" cat" $==$ "cat"   "cat" $==$ "CAT"   "cat" $==$ "c a t"

You can also use other logical operators too. For instance jamovi also allows you to use the > and < operators to determine which of two text 'strings' comes first, alphabetically speaking. Sort of. Actually, it's a bit more complicated than that, but let's start with a simple example:

"cat" $<$ "dog"

In jamovi, this example evaluates to 'true'. This is because "cat" does does come before "dog" alphabetically, so jamovi judges the statement to be true. However, if we ask jamovi to tell us if "cat" comes before "anteater" then it will evaluate the expression as false. So far, so good. But text data is a bit more complicated than the dictionary suggests. What about "cat" and "CAT"? Which of these comes first? Try it and find out:

"CAT" < "cat"

This in fact evaluates to 'true'. In other words, jamovi assumes that uppercase letters come before lowercase ones. Fair enough. No-one is likely to be surprised by that. What you might find surprising is that jamovi assumes that all uppercase letters come before all lowercase ones. That is, while "anteater" < "zebra" is a true statement, and the uppercase equivalent "ANTEATER" < "ZEBRA" is also true, it is not true to say that "anteater" < "ZEBRA", as the following extract illustrates. Try this:

"anteater" < "ZEBRA"

This evaluates to 'false', and this may seem slightly counter-intuitive. With that in mind, it may help to have a quick look at Table 6.4 which lists various text characters in the order that jamovi processes them.

Table 6.4: Text characters in the order that jamovi processes them

!	"	#	$	%	&	'	(
)	*	+	,	-	.	/	0
1	2	3	4	5	6	7	8
9	:	;	<	=	>	?	@
A	B	C	D	E	F	G	H
I	J	K	L	M	N	O	P
Q	R	S	T	U	V	W	X
Y	Z	[	\	]	^	_	'
a	b	c	d	e	g	h	i
j	k	l	m	n	o	p	q
r	s	t	u	v	w	x	y
z	{			}				

6.3 Transforming and recoding a variable

It's not uncommon in real-world data analysis to find that one of your variables isn't quite equivalent to the variable that you really want. For instance, it's often convenient to take a continuous-valued variable (e.g., age) and break it up into a smallish number of categories (e.g., younger, middle, older). At other times, you may need to convert a numeric variable into a different numeric variable (e.g., you may want to analyse at the absolute value of the original variable). In this section I'll describe a few key ways you can do these things in jamovi.

6.3.1 Creating a transformed variable

The first trick to discuss is the idea of **transforming** a variable. Taken literally, anything you do to a variable is a transformation, but in practice what it usually means is that you apply a relatively simple mathematical function to the original variable in order to create a new variable that either (a) provides a better way of describing the thing you're actually interested in, or (b) is more closely in agreement with the assumptions of the statistical tests you want to do. Since, at this stage, I haven't talked about statistical tests or their assumptions, I'll show you an example based on the first case.

Suppose I've run a short study in which I ask 10 people a single question: On a scale of 1 (strongly disagree) to 7 (strongly agree), to what extent do you agree with the proposition that "Dinosaurs are awesome"?

Now let's load and look at the data. The data file *likert.omv* contains a single variable that contains raw Likert-scale responses for these 10 people. However, if you think about it, this isn't the best way to represent these responses. Because of the fairly symmetric way that we set up the response scale, there's a sense in which the midpoint of the scale should have been coded as 0 (no opinion), and the two endpoints should be "3 (strongly agree)" and "3 (strongly disagree)". By recoding the data in this way it's a bit more reflective of how we really think about the responses. The recoding here is pretty straightforward, we just subtract 4 from the raw scores. In jamovi you can do this by computing a new variable: click on the 'Data' – 'Compute' button and you will see that a new variable has been added to the spreadsheet. Let's call this new variable 'likert.centred' (go ahead and type that in) and then add the following in the formula box, like in Figure 6.4: 'likert.raw - 4'.

One reason why it might be useful to have the data in this format is that there are a lot of situations where you might prefer to analyse the strength of the opinion separately from the direction of the opinion. We can do two different transformations on this likert.centred variable in order to distinguish between these two different concepts. First, to compute an opinion.strength variable, we want to take the absolute value of the centred data (using the 'ABS' function).[44] In jamovi, create another new variable using the 'Compute' button. Name the variable opinion.strength and this time click on the fx button next to the 'Formula' box. This shows the different 'Functions' and 'Variables' that you can add to the 'Formula' box, so double click on 'ABS' and then double click on "likert.centred" and you will see that the 'Formula' box is populated with 'ABS(likert.centred)' and a new variable has been created in the spreadsheet view, as in Figure 6.5.

Second, to compute a variable that contains only the direction of the opinion and ignores the strength, we want to calculate the "sign" of the variable. In jamovi we can use the 'IF' function to do this. Create another new variable using the 'Compute' button, name this one 'opinion.sign', and then type the following into the function box:

IF(likert.centred == 0, 0, likert.centred / opinion.strength)

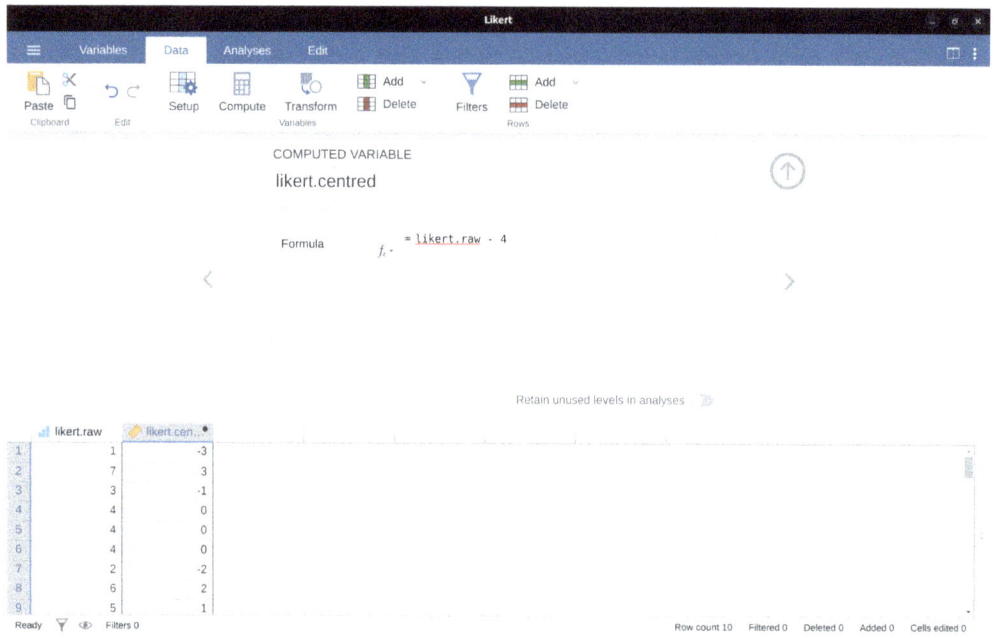

Figure 6.4: Creating a new computed variable in jamovi

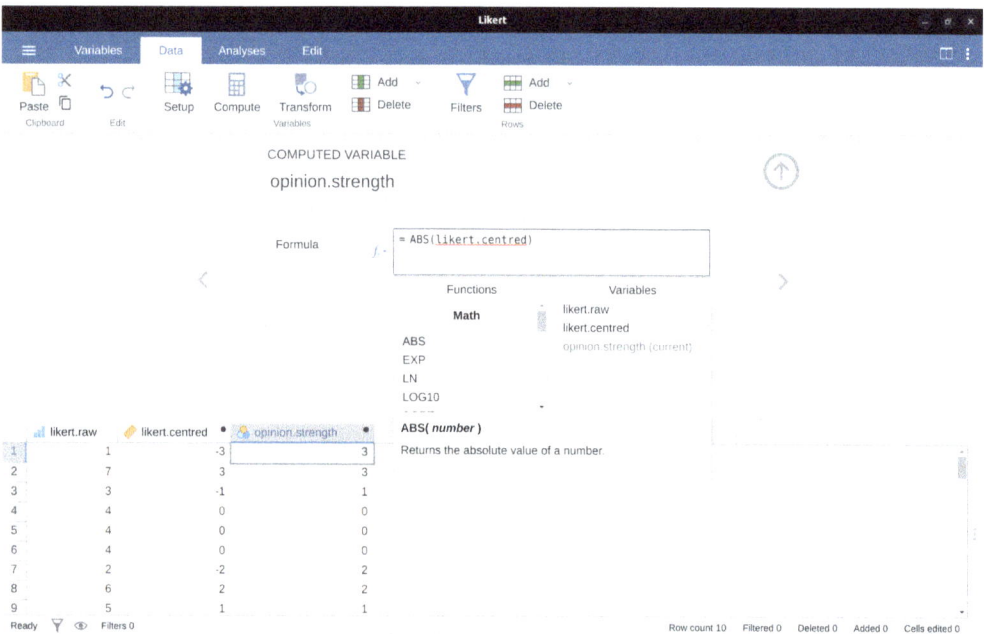

Figure 6.5: Using the f_x button to select functions and variables

When done, you'll see that all negative numbers from the 'likert.centred' variable are converted to -1, all positive numbers are converted to 1 and zero stays as 0, like so:

-1 1 -1 0 0 0 -1 1 1 1

Let's break down what this 'IF' command is doing. In jamovi there are three parts to an 'IF' statement, written as 'IF(expression, value, else)'. The first part, 'expression' can be a logical or mathematical statement. In our example, we have specified 'likert.centred == 0', which is TRUE for values where 'likert.centred' is zero. The next part, 'value', is the new value where the expression in part one is TRUE. In our example, we have said that for all those values where 'likert.centred' is zero, keep them zero. In the next part, 'else', we can enter another logical or mathematical statement to be used if part one evaluates to FALSE, i.e. where 'likert.centred' is not zero. In our example we have divided 'likert.centred' by 'opinion.strength' to give '-1' or '+1' depending of the sign of the original value in 'likert.centred'.[45]

And we're done. We now have three shiny new variables, all of which are useful transformations of the original 'likert.raw' data.

6.3.2 Collapsing a variable into a smaller number of discrete levels or categories

One pragmatic task that comes up quite often is the problem of collapsing a variable into a smaller number of discrete levels or categories. For instance, suppose I'm interested in looking at the age distribution of people at a social gathering:

60,58,24,26,34,42,31,30,33,2,9

In some situations it can be quite helpful to group these into a smallish number of categories. For example, we could group the data into three broad categories: young (0-20), adult (21-40) and older (41-60). This is a quite coarse-grained classification, and the labels that I've attached only make sense in the context of this data set (e.g., viewed more generally, a 42 year old wouldn't consider themselves as "older"). We can slice this variable up quite easily using the jamovi 'IF' function that we have already used. This time we have to specify nested 'IF' statements, meaning simply that IF the first logical expression is TRUE, insert a first value, but IF a second logical expression is TRUE, insert a second value, but IF a third logical expression is TRUE, then insert a third value. This can be written as:

IF(Age >= 0 and Age <= 20, 1, IF(Age >= 21 and Age <= 40, 2, IF(Age >= 41 and Age <= 60, 3)))

Note that there are three left parentheses used during the nesting, so the whole statement has to end with three right parentheses otherwise you will get an error message. The jamovi screenshot for this data manipulation, along with an accompanying frequency table, is shown in Figure 6.6.

It's important to take the time to figure out whether or not the resulting categories make any sense at all in terms of your research project. If they don't make any sense to you as meaningful categories, then any data analysis that uses those categories is likely to be just as meaningless. More generally, in practice I've noticed that people have a very strong desire to carve their (continuous and messy) data into a few (discrete and simple) categories, and then run analyses using the categorised data instead of the original data.[46] I wouldn't go so far as to say that this is an inherently bad idea, but it does have some fairly serious drawbacks at times, so I would advise some caution if you are thinking about doing it.

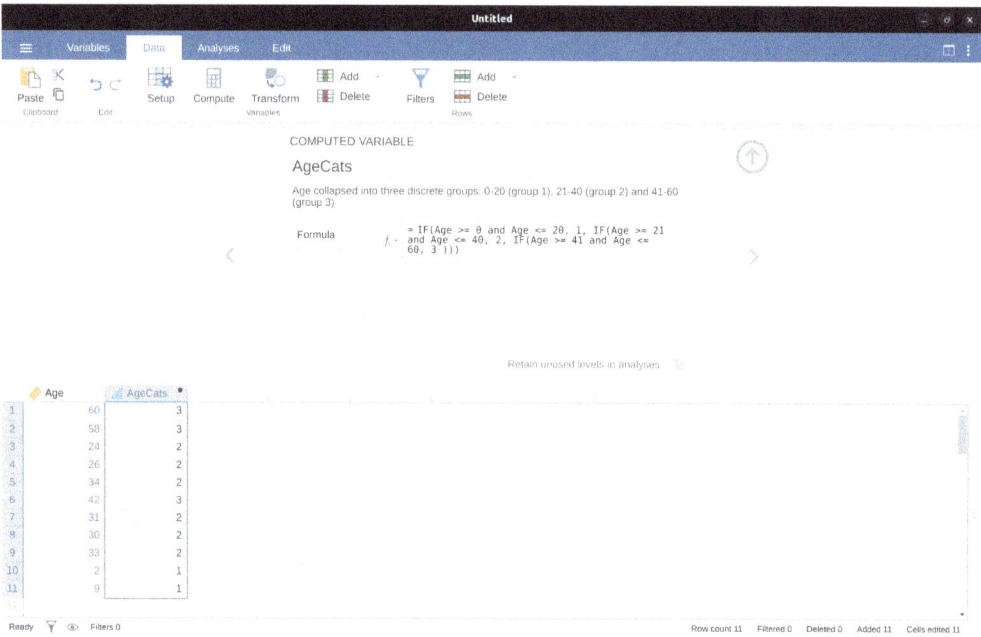

Figure 6.6: Collapsing a variable into a smaller number of discrete levels using the jamovi 'IF' function

6.3.3 Creating a transformation that can be applied to multiple variables

Sometimes you want to apply the same transformation to more than one variable, for example when you have multiple questionnaire items that all need to be recalculated or recoded in the same way. And one of the neat features in jamovi is that you can create a transformation, using the 'Data' – 'Transform' button, that can then be saved and applied to multiple variables. Let's go back to the first example above, using the data file *likert.omv* that contains a single variable with raw Likert-scale responses for 10 people. To create a transformation that you can save and then apply across multiple variables (assuming you had more variables like this in your data file), first in the spreadsheet editor select (i.e., click) the variable you want to use to initially create the transformation. In our example this is likert.raw. Next click the 'Transform' button in the jamovi 'Data' ribbon, and you'll see something like Figure 6.7.

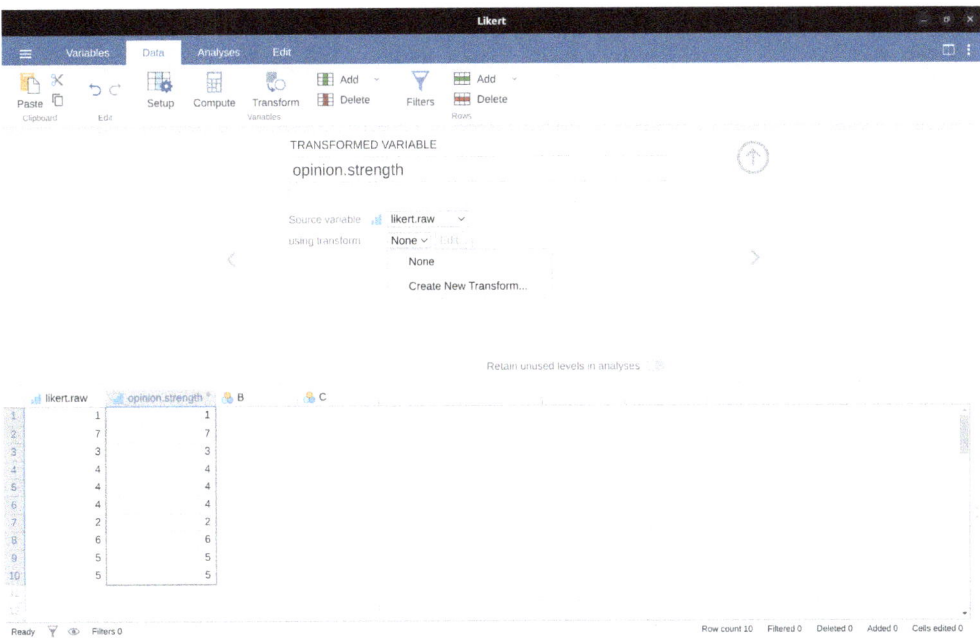

Figure 6.7: Creating a new variable transformation using the jamovi 'Transform' command

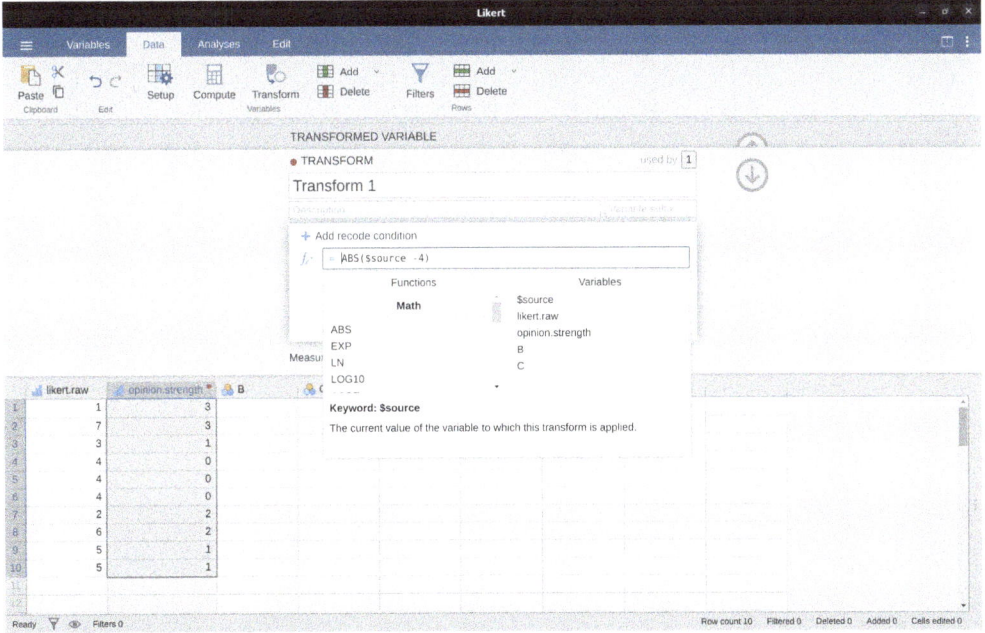

Figure 6.8: Specifying a transformation in jamovi, to be saved as the imaginatively named 'Transform 1'

Give your new variable a name, let's call it 'opinion.strength', and then click on the 'using transform' selection box and select 'Create New Transform…'. This is where you will create, and name, the transformation that can be re-applied to as many variables as you like. The transformation is automatically named for us as 'Transform 1' (imaginative, huh. You can change this if you like). Then type the expression "ABS($source - 4)" into the function text box, as in Figure 6.8, press Enter or Return on your keyboard and, hey presto, you have created a new transformation and applied it to the likert.raw variable! Good, eh. Note that instead of using the variable label in the expression, we have instead used '$source'. This is so that we can then use the same transformation with as many different variables as we like – jamovi requires you to use '$source' to refer to the source variable you are transforming. Your transformation has also been saved and can be re-used any time you like (providing you save the data set as an '.omv' file, otherwise you'll lose it!).

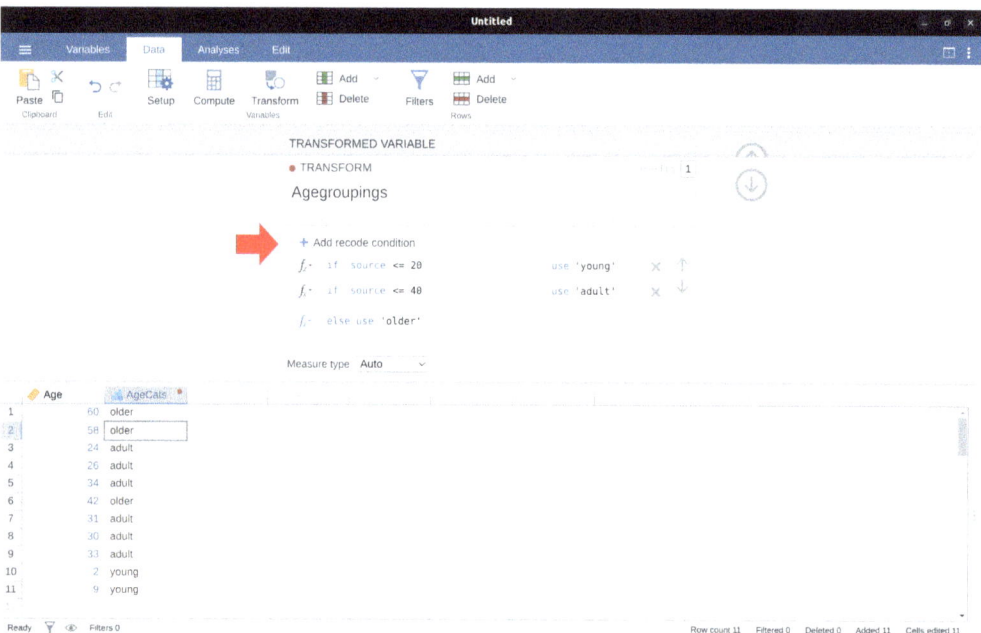

Figure 6.9: jamovi transformation into three age categories, using the 'Add condition' button

You can also create a transformation with the second example we looked at, the age distribution of people at a social gathering. Go on, you know you want to! Remember that we collapsed this variable into three groups: younger, adult and older. This time we will achieve the same thing, but using the jamovi 'Transform' – 'Add condition' button. With this data set (go back to it or create it again if you didn't save it) set up a new variable transformation. Call the transformed variable 'AgeCats' and the transformation you will create 'Agegroupings'. Then click on the big "+" sign next to the function box. This is the 'Add condition' button and I've stuck a big red arrow onto Figure 6.9 so you can see exactly where this is. Re-create the transformation shown in Figure 6.9 and when you are done, you will see the new values appear in the spreadsheet window. What's more, the age groupings transformation has been saved and can be re-applied any time you like. Ok, so I know that it's unlikely you will have more than one 'Age' variable, but you get the idea now of how to set up transformations in jamovi, so you can follow this idea with other sorts of variables. A typical scenario for this is when you have a questionnaire scale with, say, 20 items (variables) and each item was originally scored from 1 to 6 but, for some reason or quirk of the data you decide to recode all the items as 1 to 3. You can easily do this in jamovi by creating and then re-applying your transformation for each variable that you want to recode.

6.4 A few more mathematical functions and operations

In the section on Transforming and recoding a variable I discussed the ideas behind variable transformations and showed that a lot of the transformations that you might want to apply to your data are based on fairly simple mathematical functions and operations. In this section I want to return to that discussion and mention several other mathematical functions and arithmetic operations that are actually quite useful for a lot of real world data analysis. Table 6.5 gives a brief overview of the various mathematical functions I want to talk about here, or later.[47] Obviously this doesn't even come close to cataloguing the range of possibilities available; but it does cover a range of functions that are used regularly in data analysis and that are available in jamovi.

Table 6.5: Some mathematical operators

	function	example input	(answer)
square root	SQRT(x)	SQRT(25)	5
absolute value	ABS(x)	ABS(-23)	23
logarithm (base 10)	LOG10(x)	LOG10(1000)	3
logarithm (base e)	LN(x)	LN(1000)	6.91
exponentiation	EXP(x)	EXP(6.908)	1e+03
box-cox	BOXCOX(x, lamda)	BOXCOX(6.908, 3)	110

6.4.1 Logarithms and exponentials

As I've mentioned earlier, jamovi has a useful range of mathematical functions built into it and there really wouldn't be much point in trying to describe or even list all of them. For the most part, I've focused only on those functions that are strictly necessary for this book. However I do want to make an exception for logarithms and exponentials. Although they aren't needed anywhere else in this book, they are *everywhere* in statistics more broadly. And not only that, there are a *lot of* situations in which it is convenient to analyse the logarithm of a variable (i.e., to take a "log-transform" of the variable). I suspect that many (maybe most) readers of this book will have encountered logarithms and exponentials before, but from past experience I know that there's a substantial proportion of students who take a social science statistics class who haven't touched logarithms since high school, and would appreciate a bit of a refresher.

In order to understand logarithms and exponentials, the easiest thing to do is to actually calculate them and see how they relate to other simple calculations. There are three jamovi functions in particular that I want to talk about, namely LN(), LOG10() and EXP(). To start with, let's consider LOG10(), which is known as the "logarithm in base 10". The trick to understanding a **logarithm** is to understand that it's basically the "opposite" of taking a power. Specifically, the logarithm in base 10 is closely related to the powers of 10. So let's start by noting that 10-cubed is 1000. Mathematically, we would write this:

$$10^3 = 1000$$

The trick to understanding a logarithm is to recognise that the statement that "10 to the power of 3 is equal to 1000" is equivalent to the statement that "the logarithm (in base 10) of 1000 is equal to 3". Mathematically, we write this as:

$$\log_{10}(1000) = 3$$

Okay, since the LOG10() function is related to the powers of 10, you might expect that there are other logarithms (in bases other than 10) that are related to other powers too. And of course that's true: there's not really anything mathematically special about the number 10. You and I happen to find it useful because decimal numbers are built around the number 10, but the big bad world of mathematics scoffs at our decimal numbers. Sadly, the universe doesn't actually care how we write down numbers. Anyway, the consequence of this cosmic indifference is that there's nothing particularly special about calculating logarithms in base 10. You could, for instance, calculate your logarithms in base 2. Alternatively, a third type of logarithm, and one we see a lot more of in statistics than either base 10 or base 2, is called the **natural logarithm**, and corresponds to the logarithm in base e. Since you might one day run into it, I'd better explain what e is. The number e, known as **Euler's number**, is one of those annoying "irrational" numbers whose decimal expansion is infinitely long, and is considered one of the most important numbers in mathematics. The first few digits of e are:

$$e \approx 2.718282$$

There are quite a few situations in statistics that require us to calculate powers of e, though none of them appear in this book. Raising e to the power x is called the **exponential** of x, and so it's very common to see e^x written as exp(x). And so it's no surprise that jamovi has a function that calculates exponentials, called EXP(). Because the number e crops up so often in statistics, the natural logarithm (i.e., logarithm in base e) also tends to turn up. Mathematicians often write it as $\log_e(x)$ or ln(x). In fact, jamovi works the same way: the LN() function corresponds to the natural logarithm.

And with that, I think we've had quite enough exponentials and logarithms for this book!

6.5 Extracting a subset of the data

One very important kind of data handling is being able to extract a particular subset of the data. For instance, you might be interested only in analysing the data from one experimental condition, or you may want to look closely at the data from people over 50 years in age. To do this, the first step is getting jamovi to filter the subset of the data corresponding to the observations that you're interested in.

This section returns to the *nightgarden.csv* data set. If you're reading this whole chapter in one sitting, then you should already have this data set loaded into a jamovi window. For this section, let's focus on the two variables speaker and utterance (see Tabulating and cross-tabulating data if you've forgotten what those variables look like). Suppose that what I want to do is pull out only those utterances that were made by Makka-Pakka. To that end, we need to specify a filter in jamovi. First open up a filter window by clicking on 'Filters' on the main jamovi 'Data' toolbar. Then, in the 'Filter 1' text box, next to the '=' sign, type the following, including the single quote marks:

speaker == 'makka-pakka'

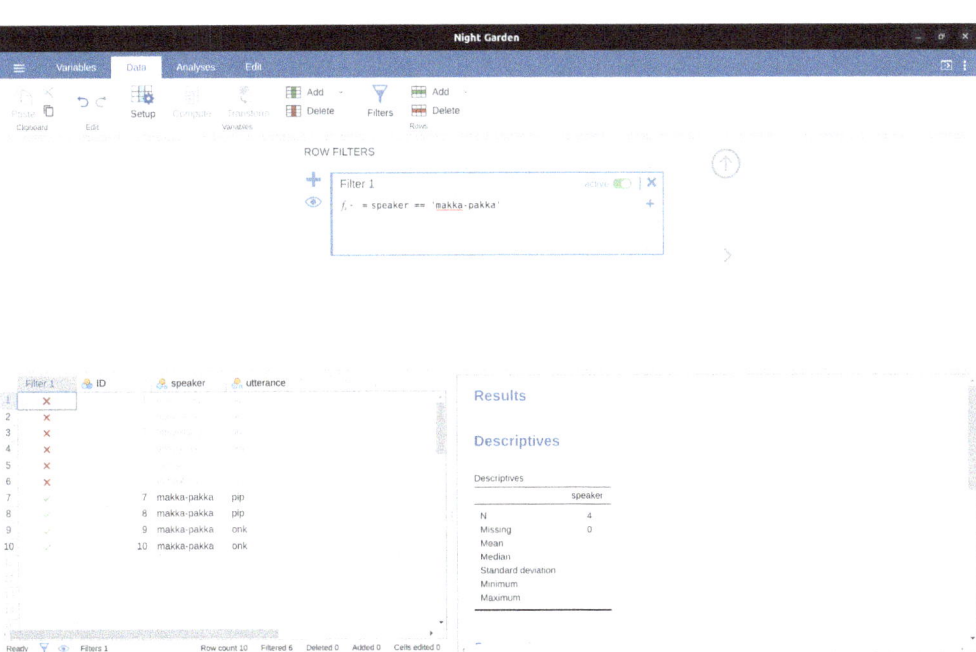

Figure 6.10: Creating a subset of the nightgarden data using the jamovi 'Filters' option

When you have done this, you will see that a new column has been added to the spreadsheet window (see Figure 6.10), labelled 'Filter 1', with the cases where speaker is not 'makka-pakka' greyed-out (i.e., filtered out) and, conversely, the cases where speaker is 'makka-pakka' have a green check mark indicating they are filtered in. You can test this by running 'Exploration' – 'Descriptives' – 'Frequency tables' for the speaker variable and seeing what that shows. Go on, try it!

Following on from this simple example, you can also build up more complex filters using logical expressions in jamovi. For instance, suppose I wanted to keep only those cases when the utterance is either "pip" or "oo". In this case in the 'Filter 1' text box, next to the '=' sign, you would type the following:

> utterance == 'pip' or utterance == 'oo'

6.6 Summary

Obviously, there's no real coherence to this chapter. It's just a grab bag of topics and tricks that can be handy to know about, so the best wrap up I can give here is just to repeat this list:

- Tabulating and cross-tabulating data.
- Logical expressions in jamovi.
- Transforming and recoding a variable.
- A few more mathematical functions and operations.
- Extracting a subset of the data.

Part IV

Statistical theory

Prelude

Part IV of the book is by far the most theoretical, focusing as it does on the theory of statistical inference. Over the next three chapters my goal is to give you an introduction to probability theory, sampling and estimation in Chapter 8 and statistical hypothesis testing in Chapter 9. Before we get started though, I want to say something about the big picture. Statistical inference is primarily about learning from data. The goal is no longer merely to describe our data but to use the data to draw conclusions about the world. To motivate the discussion I want to spend a bit of time talking about a philosophical puzzle known as the riddle of induction, because it speaks to an issue that will pop up over and over again throughout the book: statistical inference relies on assumptions. This sounds like a bad thing. In everyday life people say things like "you should never make assumptions", and psychology classes often talk about assumptions and biases as bad things that we should try to avoid. From personal experience I have learned never to say such things around philosophers!

On the limits of logical reasoning

> *The whole art of war consists in getting at what is on the other side of the hill, or, in other words, in learning what we do not know from what we do.*
> – Arthur Wellesley, 1st Duke of Wellington

This quote (https://www.bartleby.com/lit-hub/samuel-arthur-bent/duke-of-wellington/quote) came about as a consequence of a carriage ride across the countryside. Wellesley and his companion, J. W. Croker, were playing a guessing game, each trying to predict what would be on the other side of each hill. In every case it turned out that Wellesley was right and Croker was wrong. Many years later when Wellesley was asked about the game he explained that "the whole art of war consists in getting at what is on the other side of the hill". Indeed, war is not special in this respect. All of life is a guessing game of one form or another, and getting by on a day-to-day basis requires us to make good guesses. So let's play a guessing game of our own.

Suppose you and I are observing the Wellesley-Croker competition and after every three hills you and I have to predict who will win the next one, Wellesley or Croker. Let's say that W refers to a Wellesley victory and C refers to a Croker victory. After three hills, our data set looks like this: WWW

Our conversation goes like this:

> you: Three in a row doesn't mean much. I suppose Wellesley might be better

> at this than Croker, but it might just be luck. Still, I'm a bit of a gambler. I'll bet on Wellesley.
>
> me: I agree that three in a row isn't informative and I see no reason to prefer Wellesley's guesses over Croker's. I can't justify betting at this stage. Sorry. No bet for me.

Your gamble paid off: three more hills go by and Wellesley wins all three. Going into the next round of our game the score is 1-0 in favour of you and our data set looks like this: $WWW\ WWW$. I've organised the data into blocks of three so that you can see which batch corresponds to the observations that we had available at each step in our little side game. After seeing this new batch, our conversation continues:

> you: Six wins in a row for Duke Wellesley. This is starting to feel a bit suspicious. I'm still not certain, but I reckon that he's going to win the next one too.
>
> me: I guess I don't see that. Sure, I agree that Wellesley has won six in a row, but I don't see any logical reason why that means he'll win the seventh one. No bet. you: Do you really think so? Fair enough, but my bet worked out last time and I'm okay with my choice.

For a second time you were right, and for a second time I was wrong. Wellesley wins the next three hills, extending his winning record against Croker to 9-0. The data set available to us is now this: $WWW\ WWW\ WWW$. And our conversation goes like this:

> you: Okay, this is pretty obvious. Wellesley is way better at this game. We both agree he's going to win the next hill, right?
>
> me: Is there really any logical evidence for that? Before we started this game, there were lots of possibilities for the first 10 outcomes, and I had no idea which one to expect. $WWW\ WWW\ WWW\ W$ was one possibility, but so was $WCC\ CWC\ WWC\ C$ and $WWW\ WWW\ WWW\ C$ or even $CCC\ CCC\ CCC\ C$. Because I had no idea what would happen so I'd have said they were all equally likely. I assume you would have too, right? I mean, that's what it means to say you have "no idea", isn't it?
>
> you: I suppose so.
>
> me: Well then, the observations we've made logically rule out all possibilities except two: $WWW\ WWW\ WWW\ C$ or $WWW\ WWW\ WWW\ W$. Both of these are perfectly consistent with the evidence we've encountered so far, aren't they?
>
> you: Yes, of course they are. Where are you going with this?

me: So what's changed then? At the start of our game, you'd have agreed with me that these are equally plausible and none of the evidence that we've encountered has discriminated between these two possibilities. Therefore, both of these possibilities remain equally plausible and I see no logical reason to prefer one over the other. So yes, while I agree with you that Wellesley's run of 9 wins in a row is remarkable, I can't think of a good reason to think he'll win the 10th hill. No bet.

you: I see your point, but I'm still willing to chance it. I'm betting on Wellesley.

Wellesley's winning streak continues for the next three hills. The score in the Wellesley-Croker game is now 12-0, and the score in our game is now 3-0. As we approach the fourth round of our game, our data set is this: $WWW\ WWW\ WWW\ WWW$. And the conversation continues:

you: Oh yeah! Three more wins for Wellesley and another victory for me. Admit it, I was right about him! I guess we're both betting on Wellesley this time around, right?

me: I don't know what to think. I feel like we're in the same situation we were in last round, and nothing much has changed. There are only two legitimate possibilities for a sequence of 13 hills that haven't already been ruled out, $WWW\ WWW\ WWW\ WWW\ C$ and $WWW\ WWW\ WWW\ WWW\ W$. It's just like I said last time. If all possible outcomes were equally sensible before the game started, shouldn't these two be equally sensible now given that our observations don't rule out either one? I agree that it feels like Wellesley is on an amazing winning streak, but where's the logical evidence that the streak will continue?

you: I think you're being unreasonable. Why not take a look at our scorecard, if you need evidence? You're the expert on statistics and you've been using this fancy logical analysis, but the fact is you're losing. I'm just relying on common sense and I'm winning. Maybe you should switch strategies.

me: Hmm, that is a good point and I don't want to lose the game, but I'm afraid I don't see any logical evidence that your strategy is better than mine. It seems to me that if there were someone else watching our game, what they'd have observed is a run of three wins to you. Their data would look like this: YYY. Logically, I don't see that this is any different to our first round of watching Wellesley and Croker. Three wins to you doesn't seem like a lot of evidence, and I see no reason to think that your strategy is working out any better than mine. If I didn't think that WWW was good evidence then for Wellesley being better than Croker at their game, surely I have no reason now to think that YYY is good evidence that you're better at ours?

you: Okay, now I think you're being a jerk.

me: I don't see the logical evidence for that.

Learning without making assumptions is a myth

There are lots of different ways in which we could dissect this dialogue, but since this is a statistics book pitched at psychologists and not an introduction to the philosophy and psychology of reasoning, I'll keep it brief. What I've described above is sometimes referred to as the riddle of induction. It seems entirely reasonable to think that a 12-0 winning record by Wellesley is pretty strong evidence that he will win the 13th game, but it is not easy to provide a proper logical justification for this belief. On the contrary, despite the obviousness of the answer, it's not actually possible to justify betting on Wellesley without relying on some assumption that you don't have any logical justification for.

The riddle of induction is most associated with the philosophical work of David Hume and more recently Nelson Goodman, but you can find examples of the problem popping up in fields as diverse as literature (Lewis Carroll) and machine learning (the "no free lunch" theorem). There really is something weird about trying to "learn what we do not know from what we do know". The critical point is that assumptions and biases are unavoidable if you want to learn anything about the world. There is no escape from this, and it is just as true for statistical inference as it is for human reasoning. In the dialogue I was taking aim at your perfectly sensible inferences as a human being, but the common sense reasoning that you relied on is no different to what a statistician would have done. Your "common sense" half of the dialog relied on an implicit assumption that there exists some difference in skill between Wellesley and Croker, and what you were doing was trying to work out what that difference in skill level would be. My "logical analysis" rejects that assumption entirely. All I was willing to accept is that there are sequences of wins and losses and that I did not know which sequences would be observed. Throughout the dialogue I kept insisting that all logically possible data sets were equally plausible at the start of the Wellesely-Croker game, and the only way in which I ever revised my beliefs was to eliminate those possibilities that were factually inconsistent with the observations.

That sounds perfectly sensible on its own terms. In fact, it even sounds like the hallmark of good deductive reasoning. Like Sherlock Holmes, my approach was to rule out that which is impossible in the hope that what would be left is the truth. Yet as we saw, ruling out the impossible never led me to make a prediction. On its own terms everything I said in my half of the dialogue was entirely correct. An inability to make any predictions is the logical consequence of making "no assumptions". In the end I lost our game because you did make some assumptions and those assumptions turned out to be right. Skill is a real thing, and because you believed in the existence of skill you were able to learn that Wellesley had more of it than Croker. Had you relied on a less sensible assumption to drive your learning you might not have won the game.

Ultimately there are two things you should take away from this. First, as I've said, you cannot avoid making assumptions if you want to learn anything from your data. But second, once you realise that assumptions are necessary it becomes important to make sure you make the right ones! A data analysis that relies on few assumptions is not necessarily better than one that makes many assumptions, it all depends on whether those assumptions are good ones for your data. As we go through the rest of this book I'll often point out the assumptions that underpin a particular statistical technique, and how you can check whether those assumptions are sensible.

Chapter 7

Introduction to probability

> *[God] has afforded us only the twilight ... of Probability.*
> – John Locke

Up to this point in the book we've discussed some of the key ideas in experimental design, and we've talked a little about how you can summarise a data set. To a lot of people this is all there is to statistics: collecting all the numbers, calculating averages, drawing pictures, and putting them all in a report somewhere. Kind of like stamp collecting but with numbers. However, statistics covers much more than that. In fact, descriptive statistics is one of the smallest parts of statistics and one of the least powerful. The bigger and more useful part of statistics is that it provides information that lets you make inferences about data.

Once you start thinking about statistics in these terms, that statistics is there to help us draw inferences from data, you start seeing examples of it everywhere. For instance, here's a tiny extract from a newspaper article in the *Sydney Morning Herald* (30 Oct 2010):

> *"I have a tough job," the Premier said in response to a poll which found her government is now the most unpopular Labor administration in polling history, with a primary vote of just 23 per cent.*

This kind of remark is entirely unremarkable in the papers or in everyday life, but let's have a think about what it entails. A polling company has conducted a survey, usually a pretty big one because they can afford it. I'm too lazy to track down the original survey so let's just imagine that they called 1000 New South Wales (NSW) voters at random, and 230 (23%) of those claimed that they intended to vote for the Australian Labor Party (ALP). For the 2010 Federal election the Australian Electoral Commission reported 4,610,795 enrolled voters in NSW, so the opinions of the remaining 4,609,795 voters (about 99.98% of voters) remain unknown to us. Even assuming that no-one lied to the polling company the only thing we can say with 100% confidence is that the true ALP primary vote is somewhere between 230/4,610,795 (about 0.005%) and 4,610,025/4,610,795 (about 99.83%). So, on what basis is it legitimate for the polling company, the newspaper, and the readership to conclude that the ALP primary vote is only about 23%?

The answer to the question is pretty obvious. If I call 1000 people at random, and 230 of them say they intend to vote for the ALP, then it seems very unlikely that these are the only 230 people out of the entire voting public who actually intend to vote ALP. In other words, we assume that the data collected by the polling company is pretty representative of the population at large. But how representative? Would we be surprised to discover that the true ALP primary vote is actually 24%? 29%? 37%? At this point everyday intuition starts to break down a bit. No-one would be surprised by 24%, and everybody would be surprised by 37%, but it's a bit hard to say whether 29% is plausible. We need some more powerful tools than just looking at the numbers and guessing.

Inferential statistics provides the tools that we need to answer these sorts of questions, and since these kinds of questions lie at the heart of the scientific enterprise, they take up the lions share of every introductory course on statistics and research methods. However, the theory of statistical inference is built on top of **probability theory**. And it is to probability theory that we must now turn. This discussion of probability theory is basically background detail. There's not a lot of statistics per se in this chapter, and you don't need to understand this material in as much depth as the other chapters in this part of the book. Nevertheless, because probability theory does underpin so much of statistics, it's worth covering some of the basics.

7.1 How are probability and statistics different?

Before we start talking about probability theory, it's helpful to spend a moment thinking about the relationship between probability and statistics. The two disciplines are closely related but they're not identical. Probability theory is "the doctrine of chances". It's a branch of mathematics that tells you how often different kinds of events will happen. For example, all of these questions are things you can answer using probability theory:

- What are the chances of a fair coin coming up heads 10 times in a row?
- If I roll a six-sided dice twice, how likely is it that I'll roll two sixes?
- How likely is it that five cards drawn from a perfectly shuffled deck will all be hearts?
- What are the chances that I'll win the lottery?

Notice that all of these questions have something in common. In each case the "truth of the world" is known and my question relates to "what kind of events" will happen. In the first question I know that the coin is fair so there's a 50% chance that any individual coin flip will come up heads. In the second question I know that the chance of rolling a 6 on a single die is 1 in 6. In the third question I know that the deck is shuffled properly. And in the fourth question I know that the lottery follows specific rules. You get the idea. The critical point is that probabilistic questions start with a known **model** of the world, and we use that model to do some calculations. The underlying model can be quite simple. For instance, in the coin flipping example we can write down the model like this:

$$P(head) = 0.5$$

which you can read as "the probability of heads is 0.5". As we'll see later, in the same way that percentages are numbers that range from 0% to 100%, probabilities are just numbers that range from 0 to 1. When using this probability model to answer the first question I don't actually know exactly what's going to happen. Maybe I'll get 10 heads, like the question says. But maybe I'll get three heads. That's the key thing. In probability theory the model is known but the data are not.

So that's probability. What about statistics? Statistical questions work the other way around. In statistics we do not know the truth about the world. All we have is the data and it is from the data that we want to learn the truth about the world. Statistical questions tend to look more like these:

- If my friend flips a coin 10 times and gets 10 heads are they playing a trick on me?
- If five cards off the top of the deck are all hearts how likely is it that the deck was shuffled?
- If the lottery commissioner's spouse wins the lottery how likely is it that the lottery was rigged?

This time around the only thing we have are data. What I know is that I saw my friend flip the coin 10 times and it came up heads every time. And what I want to infer is whether or not I should conclude that what I just saw was actually a fair coin being flipped 10 times in a row, or whether I should suspect that my friend is playing a trick on me. The data I have look like this:

H H H H H H H H H H

and what I'm trying to do is work out which "model of the world" I should put my trust in. If the coin is fair then the model I should adopt is one that says that the probability of heads is 0.5, that is P(heads) = 0.5. If the coin is not fair then I should conclude that the probability of heads is not 0.5, which we would write as $P(heads) \neq 0.5$. In other words, the statistical inference problem is to figure out which of these probability models is right. Clearly, the statistical question isn't the same as the probability question, but they're deeply connected to one another. Because of this, a good introduction to statistical theory will start with a discussion of what probability is and how it works.

7.2 What does probability mean?

Let's start with the first of these questions. What is "probability"? It might seem surprising to you but while statisticians and mathematicians (mostly) agree on what the rules of probability are, there's much less of a consensus on what the word really means. It seems weird because we're all very comfortable using words like "chance", "likely", "possible" and "probable", and it doesn't seem like it should be a very difficult question to answer. But if you've ever had that experience in real life you might walk away from the conversation feeling like you didn't quite get it right, and that (like many everyday concepts) it turns out that you don't really know what it's all about.

So I'll have a go at it. Let's suppose I want to bet on a soccer game between two teams of robots, Arduino Arsenal and C Milan. After thinking about it, I decide that there is an 80% probability of Arduino Arsenal winning. What do I mean by that? Here are three possibilities:

- They're robot teams so I can make them play over and over again, and if I did that Arduino Arsenal would win 8 out of every 10 games on average.
- For any given game, I would agree that betting on this game is only "fair" if a $1 bet on C Milan gives a $5 payoff (i.e. I get my $1 back plus a $4 reward for being correct), as would a $4 bet on Arduino Arsenal (i.e., my $4 bet plus a $1 reward).
- My subjective "belief" or "confidence" in an Arduino Arsenal victory is four times as strong as my belief in a C Milan victory.

Each of these seems sensible. However, they're not identical and not every statistician would endorse all of them. The reason is that there are different statistical ideologies (yes, really!) and depending on which one you subscribe to, you might say that some of those statements are meaningless or irrelevant. In this section I give a brief introduction the two main approaches that exist in the literature. These are by no means the only approaches, but they're the two big ones.

7.2.1 The frequentist view

The first of the two major approaches to probability, and the more dominant one in statistics, is referred to as the **frequentist view** and it defines probability as a **long-run frequency**. Suppose we were to try flipping a fair coin over and over again. By definition this is a coin that has $P(H) = 0.5$. What might we observe? One possibility is that the first 20 flips might look like this:

T,H,H,H,H,T,T,H,H,H,H,T,H,H,T,T,T,T,T,H

In this case 11 of these 20 coin flips (55%) came up heads. Now suppose that I'd been keeping a running tally of the number of heads (which I'll call N_H) that I've seen, across the first N flips, and calculate the proportion of heads $\frac{N_H}{N}$ every time. Table 7.1 shows what I'd get (I did literally flip coins to produce this!).

Notice that at the start of the sequence the *proportion* of heads fluctuates wildly, starting at .00 and rising as high as .80. Later on, one gets the impression that it dampens out a bit, with more and more of the values actually being pretty close to the "right" answer of .50. This is the frequentist definition of probability in a nutshell. Flip a fair coin over and over again, and as N grows large (approaches infinity, denoted $N \to \infty$) the proportion of heads will converge to 50%. There are some subtle technicalities that the mathematicians care about, but qualitatively speaking that's how the frequentists define probability. Unfortunately, I don't have an infinite number of coins or the infinite patience required to flip a coin an infinite number of times. However, I do have a computer and computers excel at mindless repetitive tasks. So I asked my computer to simulate flipping a coin 1000 times and then drew a picture of what happens to the proportion $\frac{N_H}{N}$ as N increases. Actually, I did it four times just to make sure it wasn't a fluke. The results are shown in Figure 7.1. As you can see, the proportion of observed heads eventually stops fluctuating and settles down. When it does, the number at which it finally settles is the true probability of heads.

Table 7.1: Coin flips and proportion of heads

number of flips	number of heads	proportion
1	0	0.00
2	1	0.50
3	2	0.67
4	3	0.75
5	4	0.80
6	4	0.67
7	4	0.57
8	5	0.63
9	6	0.67
10	7	0.70
11	8	0.73
12	8	0.67
13	9	0.69
14	10	0.71
15	10	0.67
16	10	0.63
17	10	0.59
18	10	0.56
19	10	0.53
20	11	0.55

The frequentist definition of probability has some desirable characteristics. First, it is objective. The probability of an event is *necessarily* grounded in the world. The only way that probability statements can make sense is if they refer to (a sequence of) events that occur in the physical universe.[48] Secondly, it is unambiguous. Any two people watching the same sequence of events unfold, trying to calculate the probability of an event, must inevitably come up with the same answer.

However, it also has undesirable characteristics. Infinite sequences don't really exist in the physical world. Suppose you picked up a coin from your pocket and started to flip it. Every time it lands it impacts on the ground and each impact wears the coin down a bit. Eventually the coin will be destroyed. So, one might ask whether it really makes sense to pretend that an "infinite" sequence of coin flips is even a meaningful concept, or an objective one. We can't say that an "infinite sequence" of events is a real thing in the physical universe, because the physical universe doesn't allow infinite anything. More seriously, the frequentist definition has a narrow scope. There are lots of things out there that human beings are happy to assign probability to in everyday language, but cannot (even in theory) be mapped onto a hypothetical sequence of events. For instance, if a meteorologist comes on TV and says "the probability of rain in Adelaide on 2 November 2048 is 60%" we humans are happy to accept this. But it's not clear how to define this in frequentist terms. There's only one city of Adelaide, and only one 2 November 2048. There's no infinite sequence of events here, just a one-off thing.

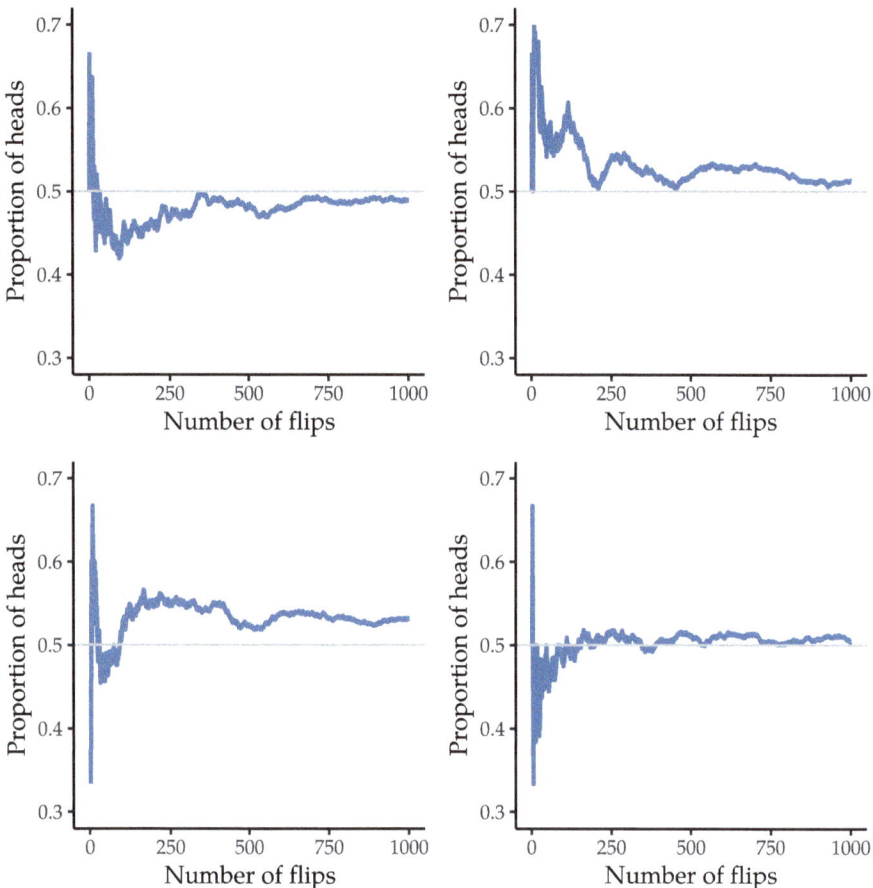

Figure 7.1: An illustration of how frequentist probability works. If you flip a fair coin over and over again the proportion of heads that you have seen eventually settles down and converges to the true probability of 0.5. Each panel shows four different simulated experiments. In each case we pretend we flipped a coin 1000 times and kept track of the proportion of flips that were heads as we went along. Although none of these sequences actually ended up with an exact value of .5, if we had extended the experiment for an infinite number of coin flips they would have

Frequentist probability genuinely *forbids* us from making probability statements about a single event. From the frequentist perspective it will either rain tomorrow or it will not. There is no "probability" that attaches to a single non-repeatable event. Now, it should be said that there are some very clever tricks that frequentists can use to get around this. One possibility is that what the meteorologist means is something like "There is a category of days for which I predict a 60% chance of rain, and if we look only across those days for which I make this prediction, then on 60% of those days it will actually rain". It's very weird and counter-intuitive to think of it this way, but you do see frequentists do this sometimes. And it will come up later in this book (e.g. in Section 8.5).

7.2.2 The Bayesian view

The Bayesian view of probability is often called the subjectivist view, and although it has been a minority view among statisticians it has been steadily gaining traction for the last several decades. There are many flavours of Bayesianism, making it hard to say exactly what "the" Bayesian view is. The most common way of thinking about subjective probability is to define the probability of an event as the **degree of belief** that an intelligent and rational agent assigns to that truth of that event. From that perspective, probabilities don't exist in the world but rather in the thoughts and assumptions of people and other intelligent beings.

However, in order for this approach to work we need some way of operationalising "degree of belief". One way that you can do this is to formalise it in terms of "rational gambling", though there are many other ways. Suppose that I believe that there's a 60% probability of rain tomorrow. If someone offers me a bet that if it rains tomorrow then I win $5, but if it doesn't rain I lose $5. Clearly, from my perspective, this is a pretty good bet. On the other hand, if I think that the probability of rain is only 40% then it's a bad bet to take. So we can operationalise the notion of a "subjective probability" in terms of what bets I'm willing to accept.

What are the advantages and disadvantages to the Bayesian approach? The main advantage is that it allows you to assign probabilities to any event you want to. You don't need to be limited to those events that are repeatable. The main disadvantage (to many people) is that we can't be purely objective. Specifying a probability requires us to specify an entity that has the relevant degree of belief. This entity might be a human, an alien, a robot, or even a statistician. But there has to be an intelligent agent out there that believes in things. To many people this is uncomfortable, it seems to make probability arbitrary. Whilst the Bayesian approach requires that the agent in question be rational (i.e., obey the rules of probability), it does allow everyone to have their own beliefs. I can believe the coin is fair and you don't have to, even though we're both rational. The frequentist view doesn't allow any two observers to attribute different probabilities to the same event. When that happens then at least one of them must be wrong. The Bayesian view does not prevent this from occurring. Two observers with different background knowledge can legitimately hold different beliefs about the same event. In short, where the frequentist view is sometimes considered to be too narrow (forbids lots of things that that we want to assign probabilities to), the Bayesian view is sometimes thought to be too broad (allows too many differences between observers).

7.2.3 What's the difference? And who is right?

Now that you've seen each of these two views independently it's useful to make sure you can compare the two. Go back to the hypothetical robot soccer game at the start of the section. What do you think a frequentist and a Bayesian would say about these three statements? Which statement would a frequentist say is the correct definition of probability? Which one would a Bayesian opt for? Would some of these statements be meaningless to a frequentist or a Bayesian? If you've understood the two perspectives you should have some sense of how to answer those questions.

Okay, assuming you understand the difference then you might be wondering which of them is *right*? Honestly, I don't know that there is a right answer. As far as I can tell there's nothing mathematically incorrect about the way frequentists think about sequences of events, and there's nothing mathematically incorrect about the way that Bayesians define the beliefs of a rational agent. In fact, when you dig down into the details Bayesians and frequentists actually agree about a lot of things. Many frequentist methods lead to decisions that Bayesians agree a rational agent would make. Many Bayesian methods have very good frequentist properties.

For the most part, I'm a pragmatist so I'll use any statistical method that I trust. As it turns out, that makes me prefer Bayesian methods for reasons I'll explain towards the end of the book. But I'm not fundamentally opposed to frequentist methods. Not everyone is quite so relaxed. For instance, consider Sir Ronald Fisher, one of the towering figures of 20th century statistics and a vehement opponent to all things Bayesian, whose paper on the mathematical foundations of statistics referred to Bayesian probability as "an impenetrable jungle [that] arrests progress towards precision of statistical concepts" (Fisher, 1922b, p. 311). Or the psychologist Paul Meehl, who suggests that relying on frequentist methods could turn you into "a potent but sterile intellectual rake who leaves in his merry path a long train of ravished maidens but no viable scientific offspring" (Meehl, 1967, p. 114). The history of statistics, as you might gather, is not devoid of entertainment.

In any case, whilst I personally prefer the Bayesian view, the majority of statistical analyses are based on the frequentist approach. My reasoning is pragmatic. The goal of this book is to cover roughly the same territory as a typical undergraduate stats class in psychology, and if you want to understand the statistical tools used by most psychologists you'll need a good grasp of frequentist methods. I promise you that this isn't wasted effort. Even if you end up wanting to switch to the Bayesian perspective, you really should read through at least one book on the "orthodox" frequentist view. Besides, I won't completely ignore the Bayesian perspective. Every now and then I'll add some commentary from a Bayesian point of view, and I'll revisit the topic in more depth in Chapter 16.

7.3 Basic probability theory

Ideological arguments between Bayesians and frequentists notwithstanding, it turns out that people mostly agree on the rules that probabilities should obey. There are lots of different ways of arriving at these rules. The most commonly used approach is based on the work of Andrey Kolmogorov, one of the great Soviet mathematicians of the 20th century. I won't go into a lot of detail, but I'll try to give you a bit of a sense of how it works. And in order to do so I'm going to have to talk about my trousers.

7.3.1 Introducing probability distributions

One of the disturbing truths about my life is that I only own five pairs of trousers. Even sadder, I've given them names: I call them X_1, X_2, X_3, X_4 and X_5. Now, on any given day, I pick out exactly one pair of trousers to wear. If I were to describe this situation using the language of probability theory, I would refer to each pair of trousers (i.e., each X) as an elementary event. The key characteristic of **elementary events** is that every time we make an observation (e.g., every time I put on a pair of trousers) then the outcome will be one and only one of these events. I always wear exactly one pair of trousers so my trousers satisfy this constraint. Similarly, the set of all possible events is called a **sample space**. Granted, some people would call it a "wardrobe", but that's because they're refusing to think about my trousers in probabilistic terms.

Okay, now that we have a sample space (a wardrobe), which is built from lots of possible elementary events (trousers), what we want to do is assign a **probability** of one of these elementary events. For an event X, the probability of that event $P(X)$ is a number that lies between 0 and 1. The bigger the value of $P(X)$, the more likely the event is to occur. So, for example, if $P(X) = 0$ it means the event X is impossible (i.e., I never wear those trousers). On the other hand, if $P(X) = 1$ it means that event X is certain to occur (i.e., I always wear those trousers). For probability values in the middle it means that I sometimes wear those trousers. For instance, if $P(X) = 0.5$ it means that I wear those trousers half of the time.

At this point, we're almost done. The last thing we need to recognise is that "something always happens". Every time I put on trousers, I really do end up wearing trousers. What this somewhat trite statement means, in probabilistic terms, is that the probabilities of the elementary events need to add up to 1. This is known as the **law of total probability**. More importantly, if these requirements are satisfied then what we have is a **probability distribution**. For example, Table 7.2 shows an example of a probability distribution. Each of the events has a probability that lies between 0 and 1, and if we add up the probability of all events they sum to 1. Awesome. We can even draw a nice bar graph (see Section 5.3) to visualise this distribution, as shown in Figure 7.2.

Table 7.2: A probability distribution for trouser wearing

Which trousers?	Label	Probability
Blue jeans	X_1	$P(X_1) = .5$
Grey jeans	X_2	$P(X_2) = .3$
Black jeans	X_3	$P(X_3) = .1$
Black suit	X_4	$P(X_4) = 0$
Blue tracksuit	X_5	$P(X_5) = .1$

And, at this point, we've all achieved something. You've learned what a probability distribution is, and I've finally managed to find a way to create a graph that focuses entirely on my trousers. Everyone wins! The only other thing that I need to point out is that probability theory allows you to talk about **non elementary events** as well as elementary ones. The easiest way to illustrate the concept is with an example. In the trousers example it's perfectly legitimate to refer to the probability that I wear jeans. In this scenario, the "Dani wears jeans" event is said to have happened as long as the elementary event that actually did occur is one of the appropriate ones. In this case "blue jeans", "black jeans" or "grey jeans". In mathematical terms we defined the "jeans" event E to correspond to the set of elementary events $(X1, X2, X3)$. If any of these elementary events occurs then E is also said to have occurred. Having decided to write down the definition of the E this way, it's pretty straightforward to state what the probability P(E) and, since the probabilities of blue, grey and black jeans respectively are .5, .3 and .1, the probability that I wear jeans is equal to .9. is: we just add everything up.

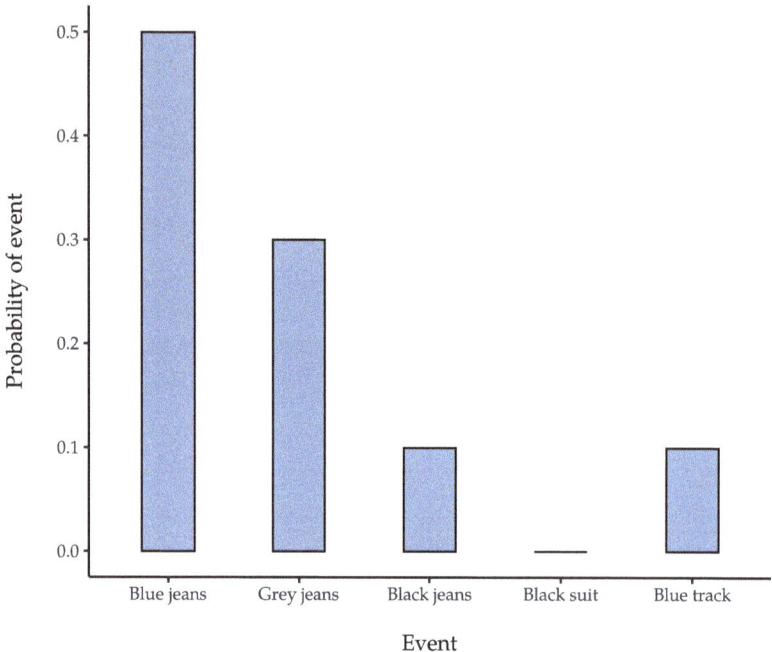

Figure 7.2: A visual depiction of the "trousers" probability distribution. There are five "elementary events", corresponding to the five pairs of trousers that I own. Each event has some probability of occurring – this probability is a number between 0 to 1. The sum of these probabilities is 1

In this particular case:

$$P(E) = P(X_1) + P(X_2) + P(X_3)$$

You might be thinking that this is all terribly obvious and simple and you'd be right. All we've really done is wrap some basic mathematics around a few common sense intuitions. However, from these simple beginnings it's possible to construct some extremely powerful mathematical tools. I'm definitely not going to go into the details in this book, but what I will do is list, in Table 7.3, some of the other rules that probabilities satisfy. These rules can be derived from the simple assumptions that I've outlined above, but since we don't actually use these rules for anything in this book I won't do so here.

Table 7.3: Some rules that probabilities satisfy

English	Notation	Formula
not A	$P(\neg A)$	$1 - P(A)$
A or B	$P(A \cup B)$	$P(A) + P(B) - P(A \cap B)$
A and B	$P(A \cap B)$	$P(A\|B)P(B)$

7.4 The binomial distribution

As you might imagine, probability distributions vary enormously. However, they aren't all equally important. In fact, the vast majority of the content in this book relies on just five distributions: the binomial distribution, the normal distribution, the t-distribution, the χ^2 ("chi-square") distribution and the F-distribution. Given this, what I'll do over the next few sections is provide a brief introduction to all five of these, paying special attention to the binomial and the normal. I'll start with the binomial distribution since it's the simplest of the five.

7.4.1 Introducing the binomial

The theory of probability originated in the attempt to describe how games of chance work, so it seems fitting that our discussion of the **binomial distribution** should involve a discussion of rolling dice and flipping coins. Let's imagine a simple "experiment". In my hot little hand I'm holding 20 identical six-sided dice. On one face of each die there's a picture of a skull, the other five faces are all blank. If I proceed to roll all 20 dice, what's the probability that I'll get exactly 4 skulls? Assuming that the dice are fair, we know that the chance of any one die coming up skulls is 1 in 6. To say this another way, the skull probability for a single die is approximately .167. This is enough information to answer our question, so let's have a look at how it's done.

As usual, we'll want to introduce some names and some notation. We'll let N denote the number of dice rolls in our experiment, which is often referred to as the **size parameter** of our binomial distribution. We'll also use θ to refer to the the probability that a single die comes up skulls, a quantity that is usually called the **success probability** of the binomial.[49] Finally, we'll use X to refer to the results of our experiment, namely the number of skulls I get when I roll the dice. Since the actual value of X is due to chance we refer to it as a **random variable**. In any case, now that we have all this terminology and notation we can use it to state the problem a little more precisely. The quantity that we want to calculate is the probability that $X = 4$ given that we know that $\theta = .167$ and $N = 20$. The general "form" of the thing I'm interested in calculating could be written as:

$$P(X|\theta, N)$$

and we're interested in the special case where $X = 4, \theta = .167$ and $N = 20$.

[Additional technical detail[50]]

Yeah, yeah. I know what you're thinking: notation, notation, notation. Really, who cares? Very few readers of this book are here for the notation, so I should probably move on and talk about how to use the binomial distribution. I've included the formula for the binomial distribution in a note,[51] since some readers may want to play with it themselves, but since most people probably don't care that much and because we don't need the formula in this book, I won't talk about it in any detail. Instead, I just want to show you what the binomial distribution looks like.

To that end, Figure 7.3 plots the binomial probabilities for all possible values of X for our dice rolling experiment, from $X = 0$ (no skulls) all the way up to $X = 20$ (all skulls). Note that this is basically a bar chart, and is no different to the "trousers probability" plot I drew in Figure 7.2. On the horizontal axis we have all the possible events, and on the vertical axis we can read off the probability of each of those events. So, the probability of rolling 4 skulls out of 20 is about 0.20 (the actual answer is 0.2022036, as we'll see in a moment). In other words, you'd expect that to happen about 20% of the times you repeated this experiment.

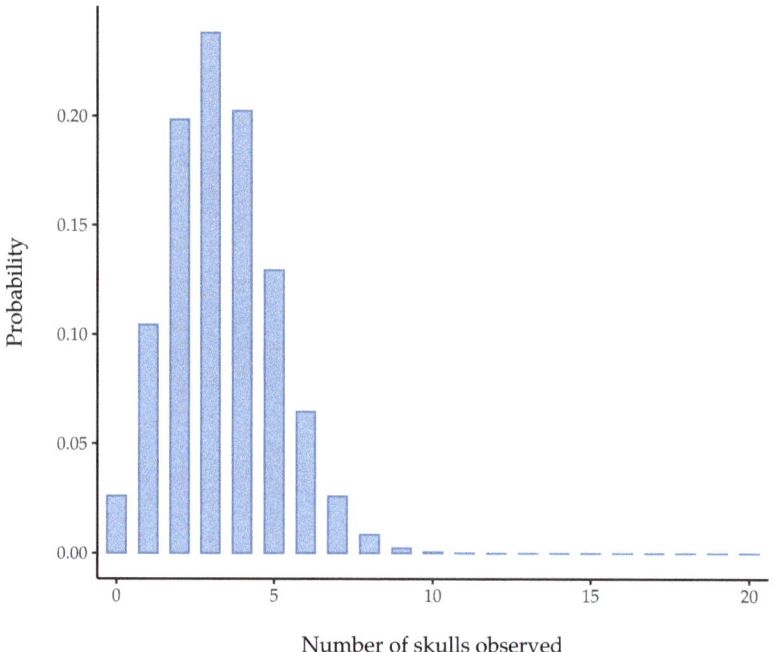

Figure 7.3: The binomial distribution with size parameter of $N = 20$ and an underlying success probability of $\theta = \frac{1}{6}$. Each vertical bar depicts the probability of one specific outcome (i.e., one possible value of X). Because this is a probability distribution, each of the probabilities must be a number between 0 and 1, and the heights of the bars must sum to 1 as well

To give you a feel for how the binomial distribution changes when we alter the values of θ and N, let's suppose that instead of rolling dice I'm actually flipping coins. This time around, my experiment involves flipping a fair coin repeatedly and the outcome that I'm interested in is the number of heads that I observe. In this scenario, the success probability is now $\theta = \frac{1}{2}$. Suppose I were to flip the coin $N = 20$ times. In this example, I've changed the success probability but kept the size of the experiment the same. What does this do to our binomial distribution? Well, as Figure 7.4 shows, the main effect of this is to shift the whole distribution, as you'd expect. Okay, what if we flipped a coin $N = 100$ times? Well, in that case we get Figure 7.4 (b). The distribution stays roughly in the middle but there's a bit more variability in the possible outcomes.

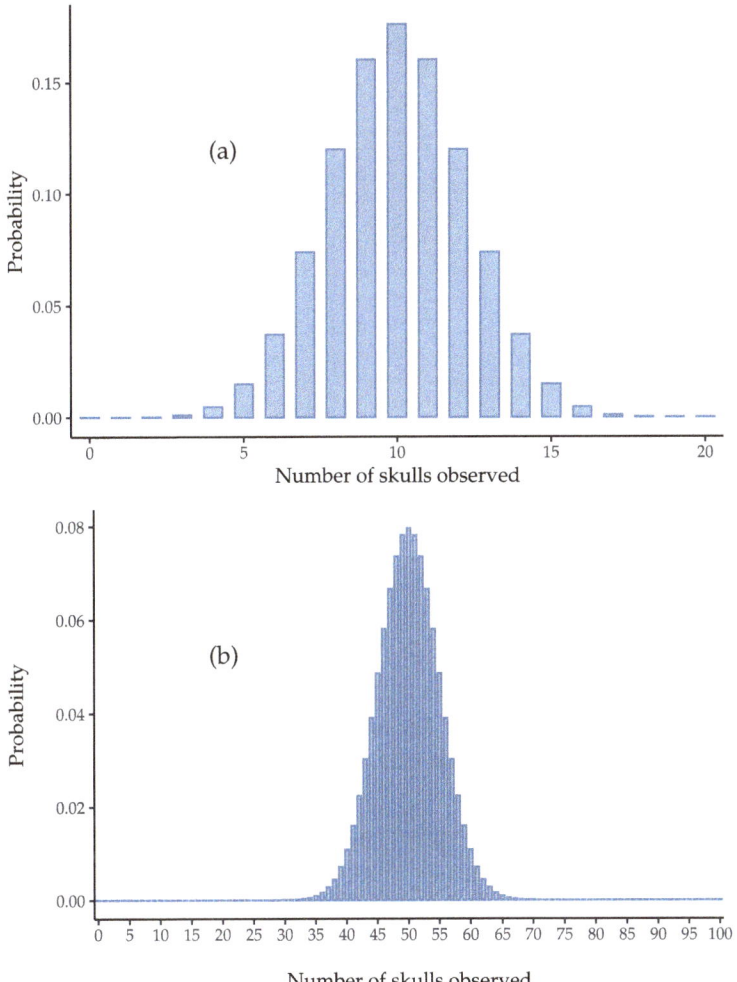

Figure 7.4: Two binomial distributions, involving a scenario in which I flip a fair coin so the underlying success probability is $\theta = \frac{1}{2}$. In panel (a), I flipped the coin $N = 20$ times. In panel (b) the coin was flipped $N = 100$ times

7.5 The normal distribution

While the binomial distribution is conceptually the simplest distribution to understand, it's not the most important one. That particular honour goes to the normal distribution, also referred to as "the bell curve" or a "Gaussian distribution". A **normal distribution** is described using two parameters: the mean of the distribution μ and the standard deviation of the distribution σ. The notation that we sometimes use to say that a variable X is normally distributed is as follows:

$$X \sim Normal(\mu, \sigma)$$

[Additional technical detail[52]]

Let's try to get a sense for what it means for a variable to be normally distributed. To that end, have a look at Figure 7.5 which plots a normal distribution with mean $\mu = 0$ and standard deviation $\sigma = 1$.

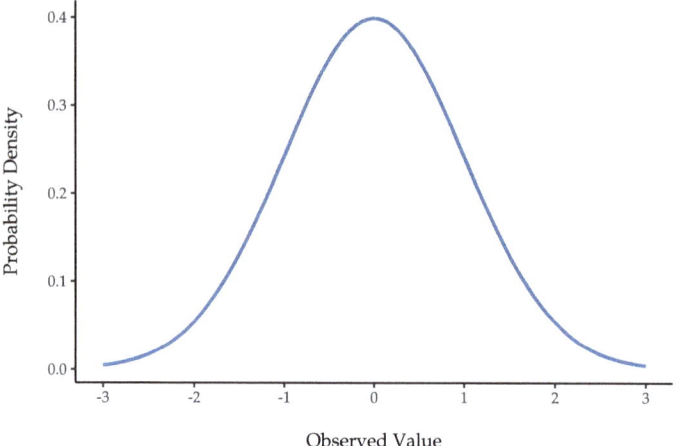

Figure 7.5: The normal distribution with mean $\mu = 0$ and standard deviation $\sigma = 1$. The x-axis corresponds to the value of some variable, and the y-axis tells us something about how likely we are to observe that value. However, notice that the y-axis is labelled *Probability Density* and not *Probability*

You can see where the name "bell curve" comes from; it looks a bit like a bell. Notice that, unlike the plots that I drew to illustrate the binomial distribution, the picture of the normal distribution in Figure 7.5 shows a smooth curve instead of "histogram-like" bars. This isn't an arbitrary choice, the normal distribution is continuous whereas the binomial is discrete.[53] For instance, in the die rolling example from the last section it was possible to get 3 skulls or 4 skulls, but impossible to get 3.9 skulls. The figures that I drew in the previous section reflected this fact. In Figure 7.3, for instance, there's a bar located at $X = 3$ and another one at $X = 4$ but there's nothing in between. Continuous quantities don't have this constraint. For instance, suppose we're talking about the weather. The temperature on a pleasant Spring day could be 23 degrees, 24 degrees, 23.9 degrees, or anything in between since temperature is a continuous variable. And so a normal distribution might be quite appropriate for describing Spring temperatures.[54]

With this in mind, let's see if we can't get an intuition for how the normal distribution works. First, let's have a look at what happens when we play around with the parameters of the distribution. To that end, Figure 7.6 plots normal distributions that have different means but have the same standard deviation. As you might expect, all of these distributions have the same "width". The only difference between them is that they've been shifted to the left or to the right. In every other respect they're identical. In contrast, if we increase the standard deviation while keeping the mean constant, the peak of the distribution stays in the same place but the distribution gets wider, as you can see in Figure 7.7. Notice, though, that when we widen the distribution the height of the peak shrinks. This has to happen, in the same way that the heights of the bars that we used to draw a discrete binomial distribution have to sum to 1, the total area under the curve for the normal distribution must equal 1.

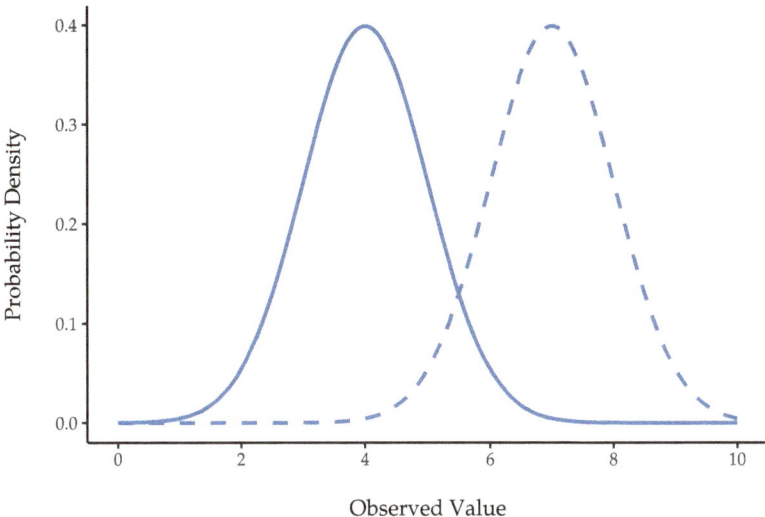

Figure 7.6: An illustration of what happens when you change the mean of a normal distribution. The solid line depicts a normal distribution with a mean of $\mu = 4$. The dashed line shows a normal distribution with a mean of $\mu = 7$. In both cases, the standard deviation is $\sigma = 1$. Not surprisingly, the two distributions have the same shape, but the dashed line is shifted to the right

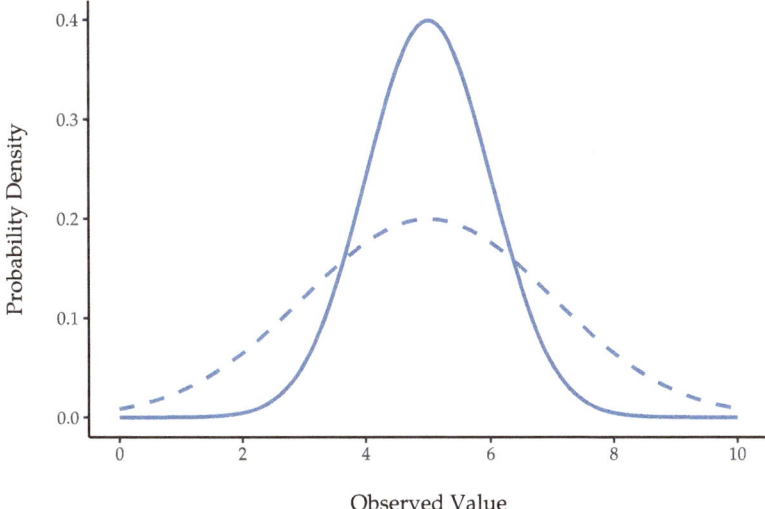

Figure 7.7: An illustration of what happens when you change the the standard deviation of a normal distribution. Both distributions plotted in this figure have a mean of $\mu = 5$, but they have different standard deviations. The solid line plots a distribution with standard deviation $\sigma = 1$, and the dashed line shows a distribution with standard deviation $\sigma = 2$. As a consequence, both distributions are "centred" on the same spot, but the dashed line is wider than the solid one

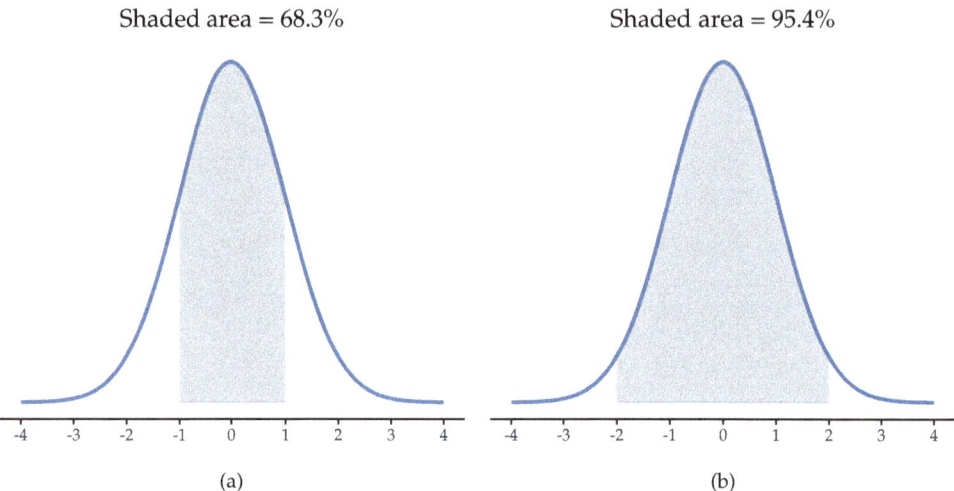

Figure 7.8: The area under the curve tells you the probability that an observation falls within a particular range. The solid lines plot normal distributions with mean $\mu = 0$ and standard deviation $\sigma = 1$. The shaded areas illustrate "areas under the curve" for two important cases. In panel (a), we can see that there is a 68.3% chance that an observation will fall within one standard deviation of the mean. In panel (b), we see that there is a 95.4% chance that an observation will fall within two standard deviations of the mean

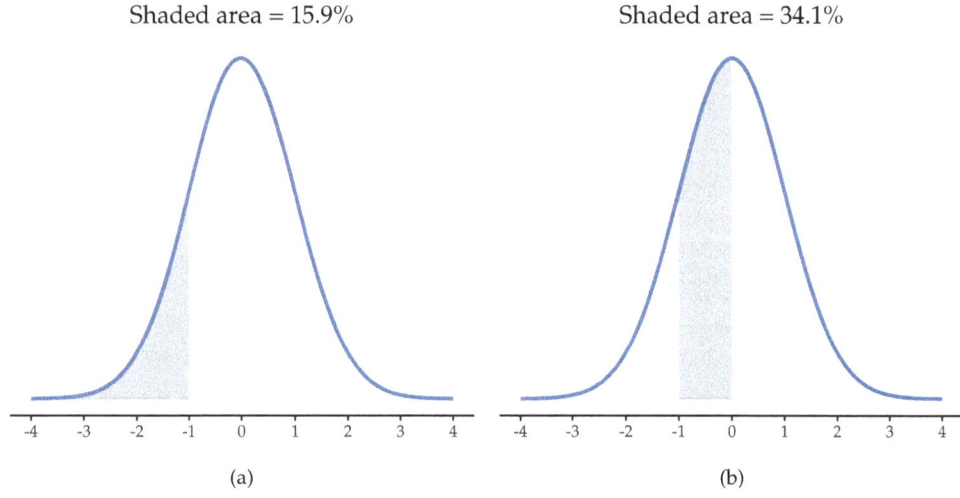

Figure 7.9: Two more examples of the "area under the curve idea". There is a 15.9% chance that an observation is one standard deviation below the mean or smaller (panel (a)), and a 34.1% chance that the observation is somewhere between one standard deviation below the mean and the mean (panel (b)). Notice that if you add these two numbers together you get 15.9% + 34.1% = 50%. For normally distributed data, there is a 50% chance that an observation falls below the mean. And of course that also implies that there is a 50% chance that it falls above the mean

Before moving on, I want to point out one important characteristic of the normal distribution. Irrespective of what the actual mean and standard deviation are, 68.3% of the area falls within 1 standard deviation of the mean. Similarly, 95.4% of the distribution falls within 2 standard deviations of the mean, and (99.7%) of the distribution is within 3 standard deviations. This idea is illustrated in Figure 7.8; see also Figure 7.9.

7.5.1 Probability density

There's something I've been trying to hide throughout my discussion of the normal distribution, something that some introductory textbooks omit completely. They might be right to do so. This "thing" that I'm hiding is weird and counter-intuitive even by the admittedly distorted standards that apply in statistics. Fortunately, it's not something that you need to understand at a deep level in order to do basic statistics. Rather, it's something that starts to become important later on when you move beyond the basics. So, if it doesn't make complete sense, don't worry too much, but try to make sure that you follow the gist of it.

Throughout my discussion of the normal distribution there's been one or two things that don't quite make sense. Perhaps you noticed that the y-axis in these figures is labelled "Probability Density" rather than density. Maybe you noticed that I used $p(X)$ instead of $P(X)$ when giving the formula for the normal distribution.

As it turns out, what is presented here isn't actually a probability, it's something else. To understand what that something is you have to spend a little time thinking about what it really means to say that X is a continuous variable. Let's say we're talking about the temperature outside. The thermometer tells me it's 23 degrees, but I know that's not really true. It's not exactly 23 degrees. Maybe it's 23.1 degrees, I think to myself. But I know that that's not really true either because it might actually be 23.09 degrees. But I know that... well, you get the idea. The tricky thing with genuinely continuous quantities is that you never really know exactly what they are.

Now think about what this implies when we talk about probabilities. Suppose that tomorrow's maximum temperature is sampled from a normal distribution with mean 23 and standard deviation 1. What's the probability that the temperature will be exactly 23 degrees? The answer is "zero", or possibly "a number so close to zero that it might as well be zero". Why is this? It's like trying to throw a dart at an infinitely small dart board. No matter how good your aim, you'll never hit it. In real life you'll never get a value of exactly 23. It'll always be something like 23.1 or 22.99998 or suchlike. In other words, it's completely meaningless to talk about the probability that the temperature is exactly 23 degrees. However, in everyday language if I told you that it was 23 degrees outside and it turned out to be 22.9998 degrees you probably wouldn't call me a liar. Because in everyday language "23 degrees" usually means something like "somewhere between 22.5 and 23.5 degrees". And while it doesn't feel very meaningful to ask about the probability that the temperature is exactly 23 degrees, it does seem sensible to ask about the probability that the temperature lies between 22.5 and 23.5, or between 20 and 30, or any other range of temperatures.

The point of this discussion is to make clear that when we're talking about continuous distributions it's not meaningful to talk about the probability of a specific value. However, what we can talk about is the probability that the value lies within a particular range of values. To find out the probability associated with a particular range what you need to do is calculate the "area under the curve". We've seen this concept already, in Figure 7.8 the shaded areas shown depict genuine probabilities (e.g., in Figure 7.8) it shows the probability of observing a value that falls within 1 standard deviation of the mean).

Okay, so that explains part of the story. I've explained a little bit about how continuous probability distributions should be interpreted (i.e., area under the curve is the key thing). But what does the formula for $p(x)$ that I described earlier actually mean? Obviously, $P(x)$ doesn't describe a probability, but what is it? The name for this quantity $P(x)$ is a **probability density**, and in terms of the plots we've been drawing it corresponds to the height of the curve. The densities themselves aren't meaningful in and of themselves, but they're "rigged" to ensure that the area under the curve is always interpretable as genuine probabilities. To be honest, that's about as much as you really need to know for now.[55]

7.6 Other useful distributions

The normal distribution is the distribution that statistics makes most use of (for reasons to be discussed shortly), and the binomial distribution is a very useful one for lots of purposes. But the world of statistics is filled with probability distributions, some of which we'll run into in passing. In particular, the three that will appear in this book are the t-distribution, the χ^2 distribution and the F-distribution. I won't give formulas for any of these, or talk about them in too much detail, but I will show you some pictures: Figure 7.10, Figure 7.11 and Figure 7.12.

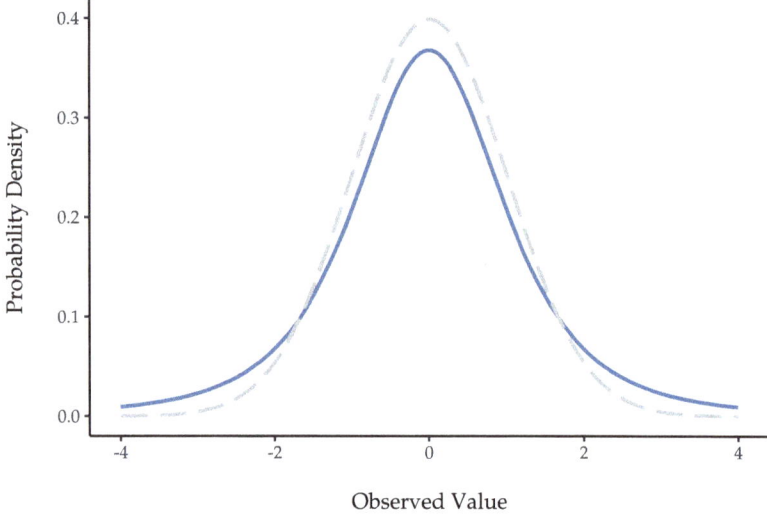

Figure 7.10: A t-distribution with 3 degrees of freedom (solid line). It looks similar to a normal distribution, but it is not quite the same. For comparison purposes I have plotted a standard normal distribution as the dashed line

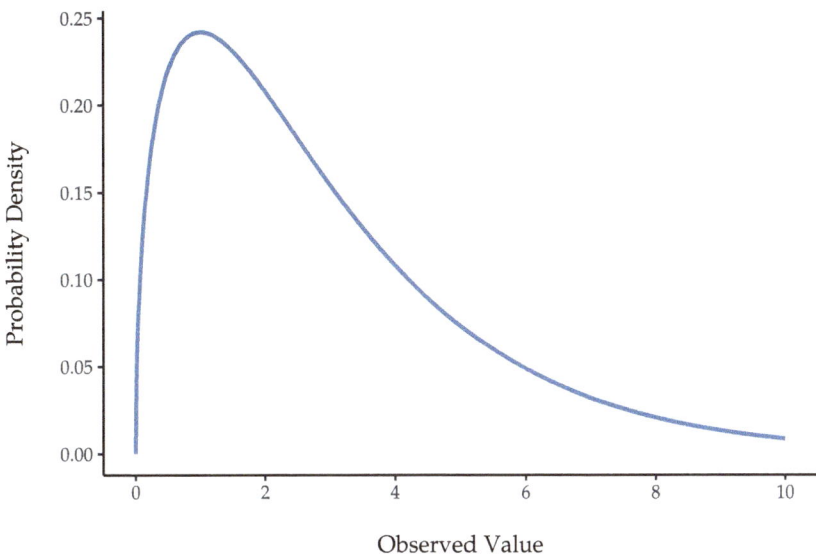

Figure 7.11: χ^2 distribution with 3 degrees of freedom. Notice that the observed values must always be greater than zero, and that the distribution is pretty skewed. These are the key features of a chi-square distribution

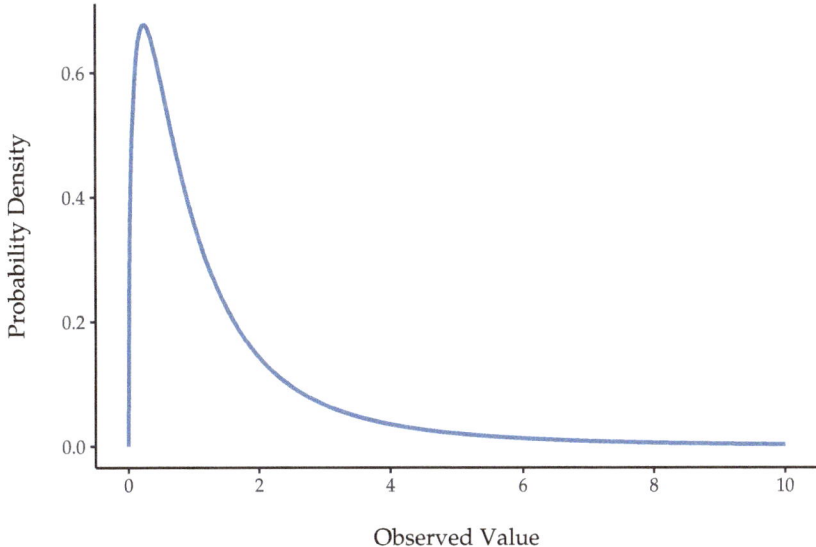

Figure 7.12: An F-distribution with 3 and 5 degrees of freedom. Qualitatively speaking, it looks pretty similar to a chi-square distribution, but they are not quite the same in general

The t-distribution is a continuous distribution that looks very similar to a normal distribution, see Figure 7.10. Note that the "tails" of the t-distribution are "heavier" (i.e., extend further outwards) than the tails of the normal distribution). That's the important difference between the two. This distribution tends to arise in situations where you think that the data actually follow a normal distribution, but you don't know the mean or standard deviation. We'll run into this distribution again in Chapter 11.

The χ^2 distribution is another distribution that turns up in lots of different places. The situation in which we'll see it is when doing categorical data analysis in Chapter 10, but it's one of those things that actually pops up all over the place. When you dig into the maths (and who doesn't love doing that?), it turns out that the main reason why the χ^2 distribution turns up all over the place is that if you have a bunch of variables that are normally distributed, square their values and then add them up (a procedure referred to as taking a "sum of squares"), this sum has a χ^2 distribution. You'd be amazed how often this fact turns out to be useful. Anyway, here's what a χ^2 distribution looks like: Figure 7.11.

The F-distribution looks a bit like a χ^2 distribution, and it arises whenever you need to compare two χ^2 distributions to one another. Admittedly, this doesn't exactly sound like something that any sane person would want to do, but it turns out to be very important in real-world data analysis. Remember when I said that χ^2 turns out to be the key distribution when we're taking a "sum of squares"? Well, what that means is if you want to compare two different "sums of squares", you're probably talking about something that has an F-distribution. Of course, as of yet I still haven't given you an example of anything that involves a sum of squares, but I will in Chapter 13. And that's where we'll run into the F-distribution. Oh, and there's a picture in Figure 7.12.

Okay, time to wrap this section up. We've seen three new distributions: χ^2, t and F. They're all continuous distributions, and they're all closely related to the normal distribution. The main thing for our purposes is that you grasp the basic idea that these distributions are all deeply related to one another, and to the normal distribution. Later on in this book we're going to run into data that are normally distributed, or at least assumed to be normally distributed. What I want you to understand right now is that, if you make the assumption that your data are normally distributed, you shouldn't be surprised to see χ^2, t and F-distributions popping up all over the place when you start trying to do your data analysis.

7.7 Summary

In this chapter we've talked about probability. We've talked about what probability means and why statisticians can't agree on what it means. We talked about the rules that probabilities have to obey. And we introduced the idea of a probability distribution and spent a good chunk of the chapter talking about some of the more important probability distributions that statisticians work with. The section-by-section breakdown looks like this:

- Probability theory versus statistics: How are probability and statistics different?
- The frequentist view versus The Bayesian view of probability.
- Basic probability theory.

- The binomial distribution, The normal distribution, and Other useful distributions.

As you'd expect, my coverage is by no means exhaustive. Probability theory is a large branch of mathematics in its own right, entirely separate from its application to statistics and data analysis. As such, there are thousands of books written on the subject and universities generally offer multiple classes devoted entirely to probability theory. Even the "simpler" task of documenting standard probability distributions is a big topic. I've described five standard probability distributions in this chapter, but sitting on my bookshelf I have a 45-chapter book called "Statistical Distributions" (M. Evans et al., 2011) that lists a lot more than that. Fortunately for you, very little of this is necessary. You're unlikely to need to know dozens of statistical distributions when you go out and do real-world data analysis, and you definitely won't need them for this book, but it never hurts to know that there's other possibilities out there.

Picking up on that last point, there's a sense in which this whole chapter is something of a digression. Many undergraduate psychology classes on statistics skim over this content very quickly (I know mine did), and even the more advanced classes will often "forget" to revisit the basic foundations of the field. Most academic psychologists would not know the difference between probability and density, and until recently very few would have been aware of the difference between Bayesian and frequentist probability. However, I think it's important to understand these things before moving onto the applications. For example, there are a lot of rules about what you're "allowed" to say when doing statistical inference and many of these can seem arbitrary and weird. However, they start to make sense if you understand that there is this Bayesian vs. frequentist distinction. Similarly, in Chapter 11 we're going to talk about something called the t-test, and if you really want to have a grasp of the mechanics of the t-test it really helps to have a sense of what a t-distribution actually looks like. You get the idea, I hope.

Chapter 8

Estimating unknown quantities from a sample

At the start of the last chapter I highlighted the critical distinction between descriptive statistics and *inferential statistics*. As discussed in Chapter 4, the role of descriptive statistics is to concisely summarise what we *do* know. In contrast, the purpose of inferential statistics is to "learn what we do not know from what we do". Now that we have a foundation in probability theory we are in a good position to think about the problem of statistical inference. What kinds of things would we like to learn about? And how do we learn them? These are the questions that lie at the heart of inferential statistics, and they are traditionally divided into two "big ideas": estimation and hypothesis testing. The goal in this chapter is to introduce the first of these big ideas, estimation theory, but I'm going to witter on about sampling theory first because estimation theory doesn't make sense until you understand sampling. As a consequence, this chapter divides naturally into two parts, the first three sections are focused on sampling theory, and the last two sections make use of sampling theory to discuss how statisticians think about estimation.

8.1 Samples, populations and sampling

In the Prelude to part IV I discussed the riddle of induction and highlighted the fact that all learning requires you to make assumptions. Accepting that this is true, our first task is to come up with some fairly general assumptions about data that make sense. This is where **sampling theory** comes in. If probability theory is the foundation upon which all statistical theory builds, sampling theory is the frame around which you can build the rest of the house. Sampling theory plays a huge role in specifying the assumptions upon which your statistical inferences rely. And in order to talk about "making inferences" the way statisticians think about it we need to be a bit more explicit about what it is that we're drawing inferences *from* (the sample) and what it is that we're drawing inferences *about* (the population).

In almost every situation of interest what we have available to us as researchers is a **sample** of data. We might have run an experiment with some number of participants, a polling company might have phoned some number of people to ask questions about voting intentions, and so on. In this way the data set available to us is finite and incomplete. We can't possibly get every person in the world to do our experiment, for example a polling company doesn't have the time or the money to ring up every voter in the country. In our earlier discussion of descriptive statistics in Chapter 4 this sample was the only thing we were interested in. Our only goal was to find ways of describing, summarising and graphing that sample. This is about to change.

8.1.1 Defining a population

A sample is a concrete thing. You can open up a data file and there's the data from your sample. A **population**, on the other hand, is a more abstract idea. It refers to the set of all possible people, or all possible observations, that you want to draw conclusions about and is generally *much bigger* than the sample. In an ideal world the researcher would begin the study with a clear idea of what the population of interest is, since the process of designing a study and testing hypotheses with the data does depend on the population about which you want to make statements.

Sometimes it's easy to state the population of interest. For instance, in the "polling company" example that opened the chapter the population consisted of all voters enrolled at the time of the study, millions of people. The sample was a set of 1000 people who all belong to that population. In most studies the situation is much less straightforward. In a typical psychological experiment determining the population of interest is a bit more complicated. Suppose I run an experiment using 100 undergraduate students as my participants. My goal, as a cognitive scientist, is to try to learn something about how the mind works. So, which of the following would count as "the population":

- All of the undergraduate psychology students at the University of Adelaide?
- Undergraduate psychology students in general, anywhere in the world?
- Australians currently living?
- Australians of similar ages to my sample?
- Anyone currently alive?
- Any human being, past, present or future?
- Any intelligent being?

Each of these defines a real group of mind-possessing entities, all of which might be of interest to me as a cognitive scientist, and it's not at all clear which one ought to be the true population of interest. As another example, consider the Wellesley-Croker game that we discussed in the Prelude to Part IV. The sample here is a specific sequence of 12 wins and 0 losses for Wellesley. What is the population? Again, it's not obvious what the population is.

- All outcomes until Wellesley and Croker arrived at their destination?
- All outcomes if Wellesley and Croker had played the game for the rest of their lives?
- All outcomes if Wellseley and Croker lived forever and played the game until the world ran out of hills?
- All outcomes if we created an infinite set of parallel universes and the Wellesely/Croker pair made guesses about the same 12 hills in each universe?

8.1.2 Simple random samples

Irrespective of how I define the population, the critical point is that the sample is a subset of the population and our goal is to use our knowledge of the sample to draw inferences about the properties of the population. The relationship between the two depends on the procedure by which the sample was selected. This procedure is referred to as a **sampling method** and it is important to understand why it matters.

To keep things simple, let's imagine that we have a bag containing 10 chips. Each chip has a unique letter printed on it so we can distinguish between the 10 chips. The chips come in two colours, black and white. This set of chips is the population of interest and it is depicted graphically on the left of Figure 8.1. As you can see from looking at the picture there are 4 black chips and 6 white chips, but of course in real life we wouldn't know that unless we looked in the bag. Now imagine you run the following "experiment": you shake up the bag, close your eyes, and pull out 4 chips without putting any of them back into the bag. First out comes the a chip (black), then the c chip (white), then j (white) and then finally b (black). If you wanted you could then put all the chips back in the bag and repeat the experiment, as depicted on the right-hand side of Figure 8.1. Each time you get different results but the procedure is identical in each case. The fact that the same procedure can lead to different results each time is what we refer to as a *random process*.[56] However, because we shook the bag before pulling any chips out, it seems reasonable to think that every chip has the same chance of being selected. A procedure in which every member of the population has the same chance of being selected is called a **simple random sample**. The fact that we did not put the chips back in the bag after pulling them out means that you can't observe the same thing twice, and in such cases the observations are said to have been sampled **without replacement**.

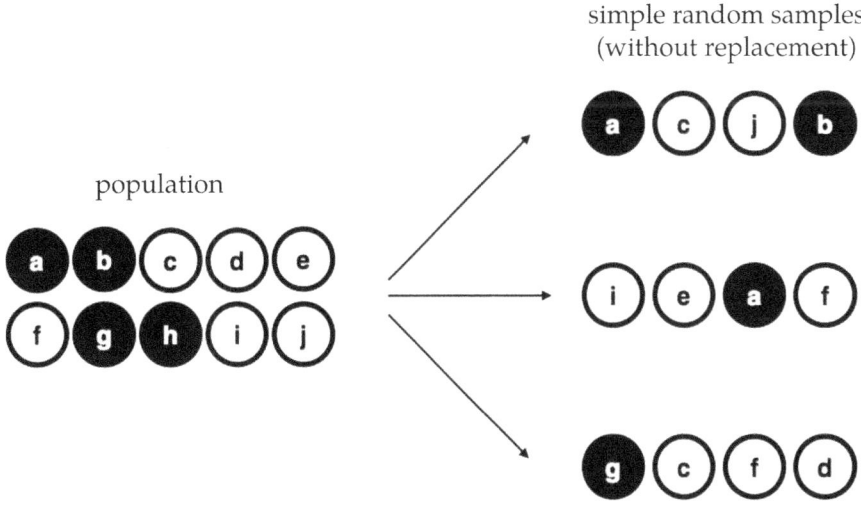

Figure 8.1: Simple random sampling without replacement from a finite population

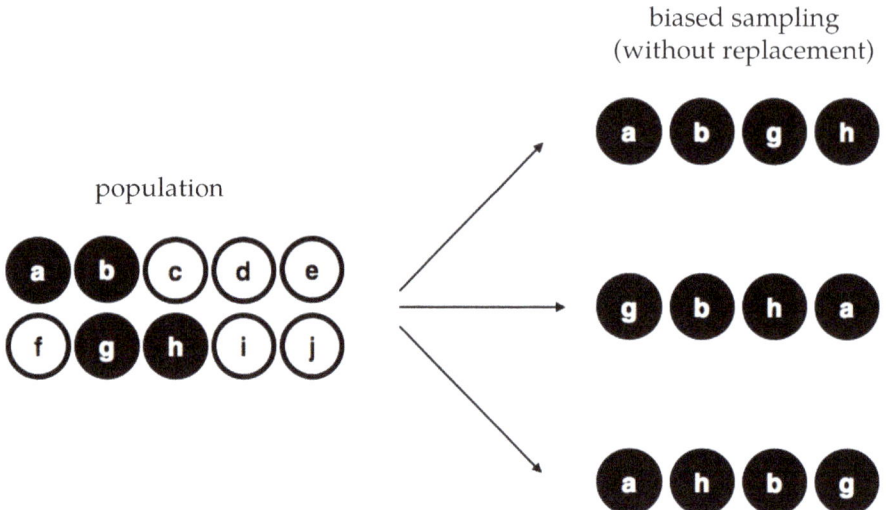

Figure 8.2: Biased sampling without replacement from a finite population

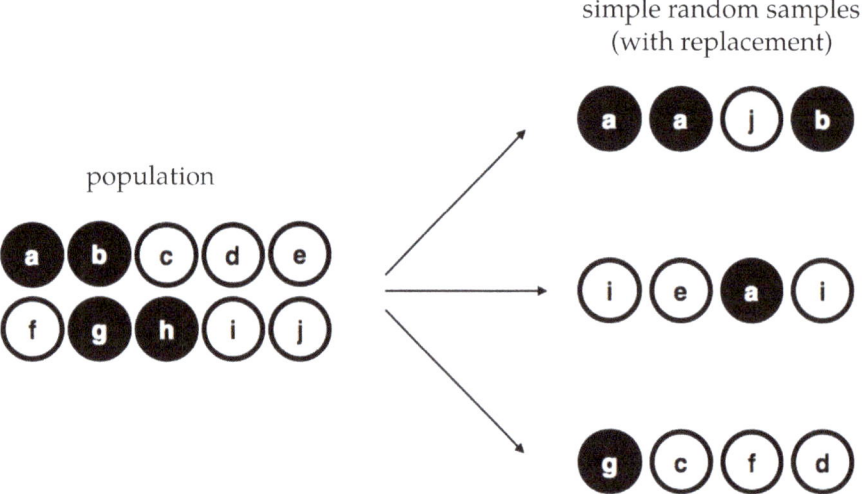

Figure 8.3: Simple random sampling *with* replacement from a finite population

To help make sure you understand the importance of the sampling procedure, consider an alternative way in which the experiment could have been run. Suppose that my five-year old son had opened the bag and decided to pull out four black chips without putting any of them back in the bag. This biased sampling scheme is depicted in Figure 8.2. Now consider the evidential value of seeing 4 black chips and 0 white chips. Clearly it depends on the sampling scheme, does it not? If you know that the sampling scheme is biased to select only black chips then a sample that consists of only black chips doesn't tell you very much about the population! For this reason statisticians really like it when a data set can be considered a simple random sample, because it makes the data analysis *much* easier. A third procedure is worth mentioning. This time around we close our eyes, shake the bag, and pull out a chip. This time, however, we record the observation and then put the chip back in the bag. Again we close our eyes, shake the bag, and pull out a chip. We then repeat this procedure until we have 4 chips. Data sets generated in this way are still simple random samples, but because we put the chips back in the bag immediately after drawing them it is referred to as a sample **with replacement**. The difference between this situation and the first one is that it is possible to observe the same population member multiple times, as illustrated in Figure 8.3.

In my experience, most psychology experiments tend to be sampling without replacement, because the same person is not allowed to participate in the experiment twice. However, most statistical theory is based on the assumption that the data arise from a simple random sample **with replacement**. In real life this very rarely matters. If the population of interest is large (e.g., has more than 10 entities!) the difference between sampling with and without replacement is too small to be concerned with. The difference between simple random samples and biased samples, on the other hand, is not such an easy thing to dismiss.

8.1.3 Most samples are not simple random samples

As you can see from looking at the list of possible populations that I showed above, it is almost impossible to obtain a simple random sample from most populations of interest. When I run experiments I'd consider it a minor miracle if my participants turned out to be a random sampling of the undergraduate psychology students at Adelaide university, even though this is by far the narrowest population that I might want to generalise to. A thorough discussion of other types of sampling schemes is beyond the scope of this book, but to give you a sense of what's out there I'll list a few of the more important ones.

- *Stratified sampling*. Suppose your population is (or can be) divided into several different sub-populations, or strata. Perhaps you're running a study at several different sites, for example. Instead of trying to sample randomly from the population as a whole, you instead try to collect a separate random sample from each of the strata. Stratified sampling is sometimes easier to do than simple random sampling, especially when the population is already divided into the distinct strata. It can also be more efficient than simple random sampling, especially when some of the sub-populations are rare. For instance, when studying schizophrenia it would be much better to divide the population into two[57] strata (schizophrenic and not-schizophrenic) and then sample an equal number of people from each group. If you selected people randomly you would get so few schizophrenic people in the sample that your study would be useless. This specific kind of of stratified sampling is referred to as oversampling because it makes a deliberate attempt to over-represent rare groups
- *Snowball sampling* is a technique that is especially useful when sampling from a "hidden" or hard to access population and is especially common in social sciences. For instance, suppose the researchers want to conduct an opinion poll among transgender people. The research team might only have contact details for a few trans folks, so the survey starts by asking them to participate (stage 1).

At the end of the survey the participants are asked to provide contact details for other people who might want to participate. In stage 2 those new contacts are surveyed. The process continues until the researchers have sufficient data. The big advantage to snowball sampling is that it gets you data in situations that might otherwise be impossible to get any. On the statistical side, the main disadvantage is that the sample is highly non-random, and non-random in ways that are difficult to address. On the real life side, the disadvantage is that the procedure can be unethical if not handled well, because hidden populations are often hidden for a reason. I chose transgender people as an example here to highlight this issue. If you weren't careful you might end up outing people who don't want to be outed (very, very bad form), and even if you don't make that mistake it can still be intrusive to use people's social networks to study them. It's certainly very hard to get people's informed consent before contacting them, yet in many cases the simple act of contacting them and saying "hey we want to study you" can be hurtful. Social networks are complex things, and just because you can use them to get data doesn't always mean you should.

- *Convenience sampling* is more or less what it sounds like. The samples are chosen in a way that is convenient to the researcher, and not selected at random from the population of interest. Snowball sampling is one type of convenience sampling, but there are many others. A common example in psychology are studies that rely on undergraduate psychology students. These samples are generally non-random in two respects. First, reliance on undergraduate psychology students automatically means that your data are restricted to a single sub-population. Second, the students usually get to pick which studies they participate in, so the sample is a self selected subset of psychology students and not a randomly selected subset. In real life most studies are convenience samples of one form or another. This is sometimes a severe limitation, but not always.

8.1.4 How much does it matter if you don't have a simple random sample?

Okay, so real world data collection tends not to involve nice simple random samples. Does that matter? A little thought should make it clear to you that it can matter if your data are not a simple random sample. Just think about the difference between Figure 8.1 and Figure 8.2. However, it's not quite as bad as it sounds. Some types of biased samples are entirely unproblematic. For instance, when using a stratified sampling technique you actually know what the bias is because you created it deliberately, often to *increase* the effectiveness of your study, and there are statistical techniques that you can use to adjust for the biases you've introduced (not covered in this book!). So in those situations it's not a problem.

More generally though, it's important to remember that random sampling is a means to an end, and not the end in itself. Let's assume you've relied on a convenience sample, and as such you can assume it's biased. A bias in your sampling method is only a problem if it causes you to draw the wrong conclusions. When viewed from that perspective, I'd argue that we don't need the sample to be randomly generated in *every* respect, we only need it to be random with respect to the psychologically-relevant phenomenon of interest. Suppose I'm doing a study looking at working memory capacity. In study 1, I actually have the ability to sample randomly from all human beings currently alive, with one exception: I can only sample people born on a Monday. In study 2, I am able to sample randomly from the Australian population. I want to generalise my results to the population of all living humans. Which study is better? The answer, obviously, is study 1. Why? Because we have no reason to think that being "born on a Monday" has any interesting relationship to working memory capacity. In contrast, I can think of several reasons why "being Australian" might matter. Australia is a wealthy, industrialised country with a very well-developed education system. People growing up in that system will have had life experiences much more similar to the experiences of the people who designed the tests for working memory capacity. This shared experience might easily translate into similar beliefs about how to "take a test", a shared assumption about how psychological experimentation works, and so on. These things might actually matter. For instance, "test taking" style might have taught the Australian participants how to direct their attention exclusively on fairly abstract test materials much more than people who haven't grown up in a similar environment. This could therefore lead to a misleading picture of what working memory capacity is.

There are two points hidden in this discussion. First, when designing your own studies, it's important to think about what population you care about and try hard to sample in a way that is appropriate to that population. In practice, you're usually forced to put up with a "sample of convenience" (e.g., psychology lecturers sample psychology students because that's the least expensive way to collect data, and our coffers aren't exactly overflowing with gold), but if so you should at least spend some time thinking about what the dangers of this practice might be. Second, if you're going to criticise someone else's study because they've used a sample of convenience rather than laboriously sampling randomly from the entire human population, at least have the courtesy to offer a specific theory as to how this might have distorted the results.

8.1.5 Population parameters and sample statistics

Okay. Setting aside the thorny methodological issues associated with obtaining a random sample, let's consider a slightly different issue. Up to this point we have been talking about populations the way a scientist might. To a psychologist a population might be a group of people. To an ecologist a population might be a group of bears. In most cases the populations that scientists care about are concrete things that actually exist in the real world. Statisticians, however, are a funny lot. On the one hand, they are interested in real-world data and real science in the same way that scientists are. On the other hand, they also operate in the realm of pure abstraction in the way that mathematicians do. As a consequence, statistical theory tends to be a bit abstract in how a population is defined. In much the same way that psychological researchers operationalise our abstract theoretical ideas in terms of concrete measurements (Section 2.1), statisticians operationalise the concept of a "population" in terms of mathematical objects that they know how to work with. You've already come across these objects in Chapter 7. They're called probability distributions.

The idea is quite simple. Let's say we're talking about IQ scores. To a psychologist the population of interest is a group of actual humans who have IQ scores. A statistician "simplifies" this by operationally defining the population as the probability distribution depicted in Figure 8.4 (a). IQ tests are designed so that the average IQ is 100, the standard deviation of IQ scores is 15, and the distribution of IQ scores is normal. These values are referred to as the **population parameters** because they are characteristics of the entire population. That is, we say that the population mean μ is 100 and the population standard deviation σ is 15.

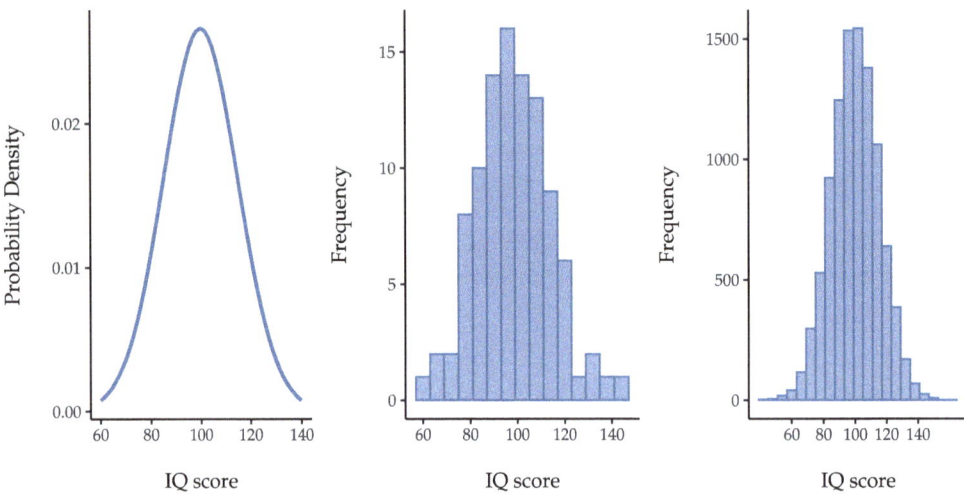

Figure 8.4: The population distribution of IQ scores (panel (a)) and two samples drawn randomly from it. In panel (b) we have a sample of 100 observations, and panel (c) we have a sample of 10,000 observations

Now suppose I run an experiment. I select 100 people at random and administer an IQ test, giving me a simple random sample from the population. My sample would consist of a collection of numbers like this:

106 101 98 80 74 ... 107 72 100

Each of these IQ scores is sampled from a normal distribution with mean 100 and standard deviation 15. So if I plot a histogram of the sample I get something like the one shown in Figure 8.4 (b). As you can see, the histogram is roughly the right shape but it's a very crude approximation to the true population distribution shown in Figure 8.4 (a). When I calculate the mean of my sample, I get a number that is fairly close to the population mean 100 but not identical. In this case, it turns out that the people in my sample have a mean IQ of 98.5, and the standard deviation of their IQ scores is 15.9. These **sample statistics** are properties of my data set, and although they are fairly similar to the true population values they are not the same. In general, sample statistics are the things you can calculate from your data set and the population parameters are the things you want to learn about. Later on in this chapter I'll talk about Estimating population parameters using your sample statistics and also Estimating a confidence interval but before we get to that there's a few more ideas in sampling theory that you need to know about

8.2 The law of large numbers

In the previous section I showed you the results of one fictitious IQ experiment with a sample size of $N = 100$. The results were somewhat encouraging as the true population mean is 100 and the sample mean of 98.5 is a pretty reasonable approximation to it. In many scientific studies that level of precision is perfectly acceptable, but in other situations you need to be a lot more precise. If we want our sample statistics to be much closer to the population parameters, what can we do about it? The obvious answer is to collect more data. Suppose that we ran a much larger experiment, this time measuring the IQs of 10,000 people. We can simulate the results of this experiment using jamovi. The *IQsim.omv* file is a jamovi data file. In this file I have generated 10,000 random numbers sampled from a normal distribution for a population with $mean = 100$ and $sd = 15$. This was done by computing a new variable using the '= NORM(100,15)' function. In Figure 8.5 a histogram and density plot shows that this larger sample is a better approximation to the true population distribution than the smaller one. This is reflected in the sample statistics. The mean IQ for the larger sample is 99.68 and the standard deviation is 14.90. These values are now very close to the true population.

I feel a bit silly saying this, but the thing I want you to take away from this is that large samples generally give you better information. I feel silly saying it because it's so obvious that it shouldn't need to be said. In fact, it's such an obvious point that when Jacob Bernoulli, one of the founders of probability theory, formalised this idea back in 1713 he was kind of a jerk about it. Here's how he described the fact that we all share this intuition:

> *For even the most stupid of men, by some instinct of nature, by himself and without any instruction (which is a remarkable thing), is convinced that the more observations have been made, the less danger there is of wandering from one's goal* (Stigler, 1986, p. 65).

Okay, so the passage comes across as a bit condescending (not to mention sexist), but his main point is correct. It really does feel obvious that more data will give you better answers. The question is, why is this so? Not surprisingly, this intuition that we all share turns out to be correct, and statisticians refer to it as the **law of large numbers**. The law of large numbers is a mathematical law that applies to many different sample statistics but the simplest way to think about it is as a law about averages. The sample mean is the most obvious example of a statistic that relies on averaging (because that's what the mean is... an average), so let's look at that. When applied to the sample mean what the law of large numbers states is that as the sample gets larger, the sample mean tends to get closer to the true population mean. Or, to say it a little bit more precisely, as the sample size "approaches" infinity (written as $N \longrightarrow \infty$), the sample mean approaches the population mean $\bar{X} \longrightarrow \mu$)[58]

I don't intend to subject you to a proof that the law of large numbers is true, but it's one of the most important tools for statistical theory. The law of large numbers is the thing we can use to justify our belief that collecting more and more data will eventually lead us to the truth. For any particular data set the sample statistics that we calculate from it will be wrong, but the law of large numbers tells us that if we keep collecting more data those sample statistics will tend to get closer and closer to the true population parameters.

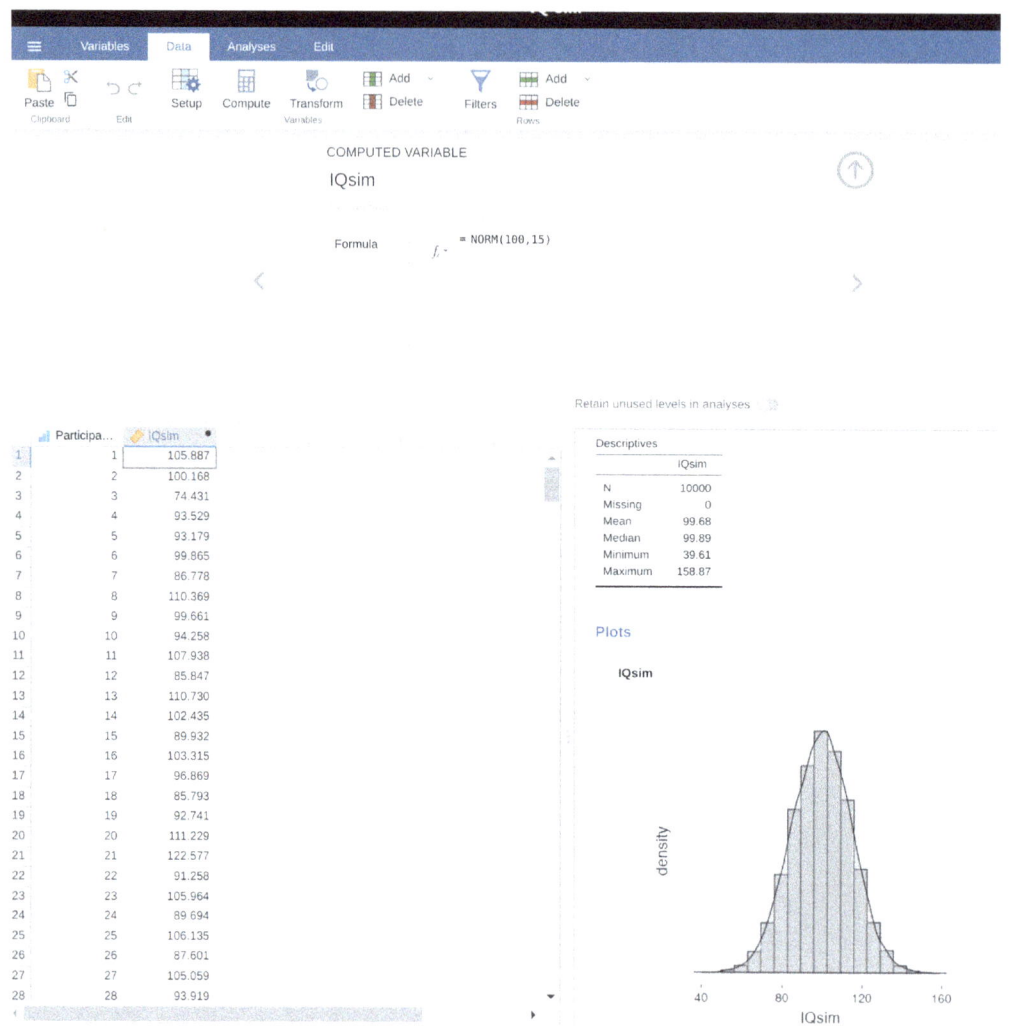

Figure 8.5: A random sample drawn from a normal distribution using jamovi

8.3 Sampling distributions and the central limit theorem

The law of large numbers is a very powerful tool but it's not going to be good enough to answer all our questions. Among other things, all it gives us is a "long run guarantee". In the long run, if we were somehow able to collect an infinite amount of data, then the law of large numbers guarantees that our sample statistics will be correct. But as John Maynard Keynes famously argued in economics, a long run guarantee is of little use in real life.

> [The] long run is a misleading guide to current affairs. In the long run we are all dead. Economists set themselves too easy, too useless a task, if in tempestuous seasons they can only tell us, that when the storm is long past, the ocean is flat again. (Keynes, 1923, p. 80).

As in economics, so too in psychology and statistics. It is not enough to know that we will eventually arrive at the right answer when calculating the sample mean. Knowing that an infinitely large data set will tell me the exact value of the population mean is cold comfort when my actual data set has a sample size of $N = 100$. In real life, then, we must know something about the behaviour of the sample mean when it is calculated from a more modest data set!

8.3.1 Sampling distribution of the mean

With this in mind, let's abandon the idea that our studies will have sample sizes of 10,000 and consider instead a very modest experiment indeed. This time around we'll sample $N = 5$ people and measure their IQ scores. As before, I can simulate this experiment in jamovi = NORM(100,15) function, but I only need 5 participant IDs this time, not 10,000. These are the five numbers that jamovi generated:

90 82 94 99 110

The mean IQ in this sample turns out to be exactly 95. Not surprisingly, this is much less accurate than the previous experiment. Now imagine that I decided to **replicate** the experiment. That is, I repeat the procedure as closely as possible and I randomly sample 5 new people and measure their IQ. Again, jamovi allows me to simulate the results of this procedure, and generates these five numbers:

78 88 111 111 117

This time around, the mean IQ in my sample is 101. If I repeat the experiment 10 times I obtain the results shown in Table 8.1, and as you can see the sample mean varies from one replication to the next.

Table 8.1: Ten replications of the IQ experiment, each with a sample size of ($N = 5$)

	Person 1	Person 2	Person 3	Person 4	Person 5	Sample Mean
Rep. 1	90	82	94	99	110	95.0
Rep. 2	78	88	111	111	117	101.0
Rep. 3	111	122	91	98	86	101.6
Rep. 4	98	96	119	99	107	103.8
Rep. 5	105	113	103	103	98	104.4
Rep. 6	81	89	93	85	114	92.4
Rep. 7	100	93	108	98	133	106.4
Rep. 8	107	100	105	117	85	102.8
Rep. 9	86	119	108	73	116	100.4
Rep. 10	95	126	112	120	76	105.8

Now suppose that I decided to keep going in this fashion, replicating this "five IQ scores" experiment over and over again. Every time I replicate the experiment I write down the sample mean. Over time, I'd be amassing a new data set, in which every experiment generates a single data point. The first 10 observations from my data set are the sample means listed in Table 8.1, so my data set starts out like this:

95.0 101.0 101.6 103.8 104.4 ...

What if I continued like this for 10,000 replications, and then drew a histogram. Well that's exactly what I did, and you can see the results in Figure 8.6. As this picture illustrates, the average of 5 IQ scores is usually between 90 and 110. But more importantly, what it highlights is that if we replicate an experiment over and over again, what we end up with is a distribution of sample means! (Table 8.1). This distribution has a special name in statistics, it's called the **sampling distribution of the mean**.

Sampling distributions are another important theoretical idea in statistics, and they're crucial for understanding the behaviour of small samples. For instance, when I ran the very first "five IQ scores" experiment, the sample mean turned out to be 95. What the sampling distribution in Figure 8.6 tells us, though, is that the "five IQ scores" experiment is not very accurate. If I repeat the experiment, the sampling distribution tells me that I can expect to see a sample mean anywhere between 80 and 120.

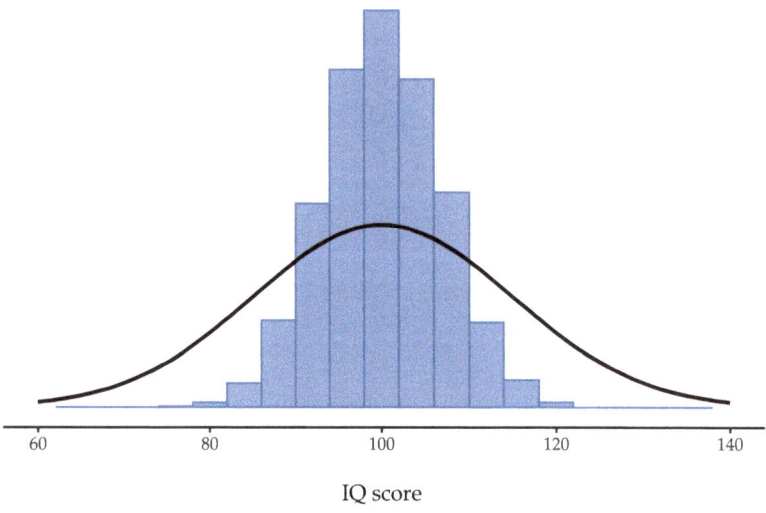

Figure 8.6: The sampling distribution of the mean for the "five IQ scores experiment". If you sample 5 people at random and calculate their average IQ you will almost certainly get a number between 80 and 120, even though there are quite a lot of individuals who have IQs above 120 or below 80. For comparison, the black line plots the population distribution of IQ scores

8.3.2 Sampling distributions exist for any sample statistic!

One thing to keep in mind when thinking about sampling distributions is that any sample statistic you might care to calculate has a sampling distribution. For example, suppose that each time I replicated the "five IQ scores" experiment I wrote down the largest IQ score in the experiment. This would give me a data set that started out like this:

110 117 122 119 113 ...

Doing this over and over again would give me a very different sampling distribution, namely the sampling distribution of the maximum. The sampling distribution of the maximum of 5 IQ scores is shown in Figure 8.7. Not surprisingly, if you pick 5 people at random and then find the person with the highest IQ score, they're going to have an above average IQ. Most of the time you'll end up with someone whose IQ is measured in the 100 to 140 range.

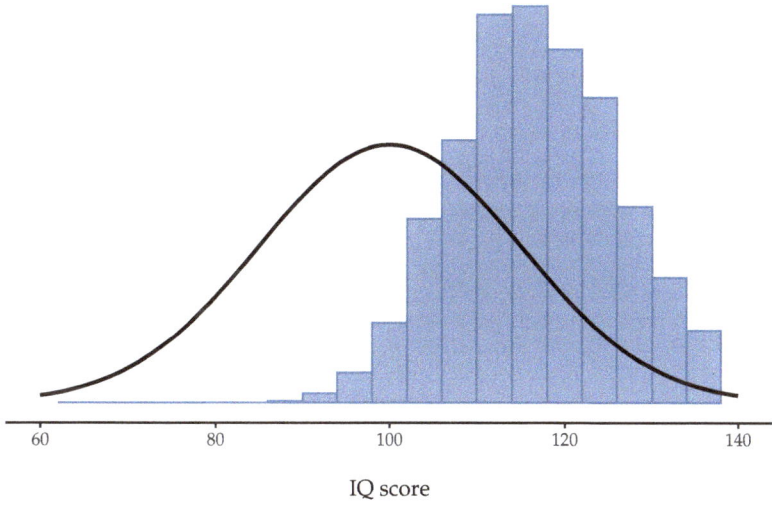

Figure 8.7: The sampling distribution of the maximum for the "five IQ scores experiment". If you sample 5 people at random and select the one with the highest IQ score you will probably see someone with an IQ between 100 and 140

8.3.3 The central limit theorem

At this point I hope you have a pretty good sense of what sampling distributions are, and in particular what the sampling distribution of the mean is. In this section I want to talk about how the sampling distribution of the mean changes as a function of sample size. Intuitively, you already know part of the answer. If you only have a few observations, the sample mean is likely to be quite inaccurate. If you replicate a small experiment and recalculate the mean you'll get a very different answer. In other words, the sampling distribution is quite wide. If you replicate a large experiment and recalculate the sample mean you'll probably get the same answer you got last time, so the sampling distribution will be very narrow. You can see this visually in Figure 8.8, showing that the bigger the sample size, the narrower the sampling distribution gets: in panel (a), each data set contained only a single observation, so the mean of each sample is the IQ score of just one person. As a consequence, the sampling distribution of the mean is of course identical to the population distribution of IQ scores. However, when we raise the sample size to 2 the mean of any one sample tends to be closer to the population mean than the IQ score of any one person, and so the histogram (i.e., the sampling distribution) is a bit narrower than the population distribution. By the time we raise the sample size to 10 (panel (c)), we can see that the distribution of sample means tend to be fairly tightly clustered around the true population mean. We can quantify this effect by calculating the standard deviation of the sampling distribution, which is referred to as the **standard error**. The standard error of a statistic is often denoted SE, and since we're usually interested in the standard error of the sample mean, we often use the acronym SEM. As you can see just by looking at the picture, as the sample size N increases, the SEM decreases.

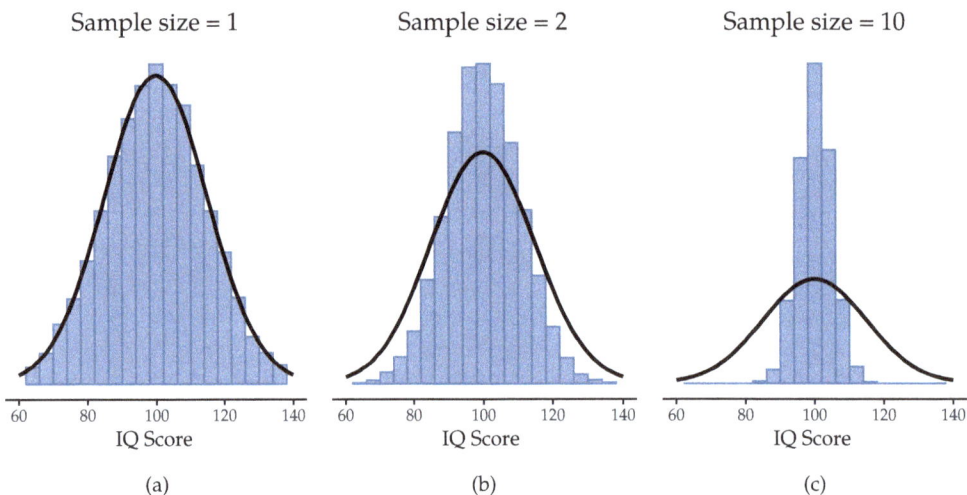

Figure 8.8: An illustration of the how sampling distribution of the mean depends on sample size. In each panel I generated 10,000 samples of IQ data and calculated the mean IQ observed within each of these data sets. The histograms in these plots show the distribution of these means (i.e., the sampling distribution of the mean). Each individual IQ score was drawn from a normal distribution with mean 100 and standard deviation 15, which is shown as the solid black line.

Okay, so that's one part of the story. However, there's something I've been glossing over so far. All my examples up to this point have been based on the "IQ scores" experiments, and because IQ scores are roughly normally distributed I've assumed that the population distribution is normal. What if it isn't normal? What happens to the sampling distribution of the mean? The remarkable thing is this, no matter what shape your population distribution is, as N increases the sampling distribution of the mean starts to look more like a normal distribution. To give you a sense of this I ran some simulations. To do this, I started with the "ramped" distribution shown in the histogram in Figure 8.9. As you can see by comparing the triangular shaped histogram to the bell curve plotted by the black line, the population distribution doesn't look very much like a normal distribution at all. Next, I simulated the results of a large number of experiments. In each experiment I took $N = 2$ samples from this distribution, and then calculated the sample mean. Figure 8.9 (b) plots the histogram of these sample means (i.e., the sampling distribution of the mean for $N = 2$). This time, the histogram produces a χ^2-shaped distribution. It's still not normal, but it's a lot closer to the black line than the population distribution in Figure 8.9 (a). When I increase the sample size to $N = 4$, the sampling distribution of the mean is very close to normal (Figure 8.9 (c)), and by the time we reach a sample size of $N = 8$ it's almost perfectly normal. In other words, as long as your sample size isn't tiny, the sampling distribution of the mean will be approximately normal no matter what your population distribution looks like!

On the basis of these figures, it seems like we have evidence for all of the following claims about the sampling distribution of the mean.

- The mean of the sampling distribution is the same as the mean of the population.
- The standard deviation of the sampling distribution (i.e., the standard error) gets smaller as the sample size increases.
- The shape of the sampling distribution becomes normal as the sample size increases.

As it happens, not only are all of these statements true, there is a very famous theorem in statistics that proves all three of them, known as the **central limit theorem**. Among other things, the central limit theorem tells us that if the population distribution has mean μ and standard deviation σ, then the sampling distribution of the mean also has mean μ and the standard error of the mean is:

$$SEM = \frac{\sigma}{\sqrt{N}}$$

Because we divide the population standard deviation σ by the square root of the sample size N, the SEM gets smaller as the sample size increases. It also tells us that the shape of the sampling distribution becomes normal.[59]

This result is useful for all sorts of things. It tells us why large experiments are more reliable than small ones, and because it gives us an explicit formula for the standard error it tells us how much more reliable a large experiment is. It tells us why the normal distribution is, well, normal. In real experiments, many of the things that we want to measure are actually averages of lots of different quantities (e.g., arguably, "general" intelligence as measured by IQ is an average of a large number of "specific" skills and abilities), and when that happens, the averaged quantity should follow a normal distribution. Because of this mathematical law, the normal distribution pops up over and over again in real data.

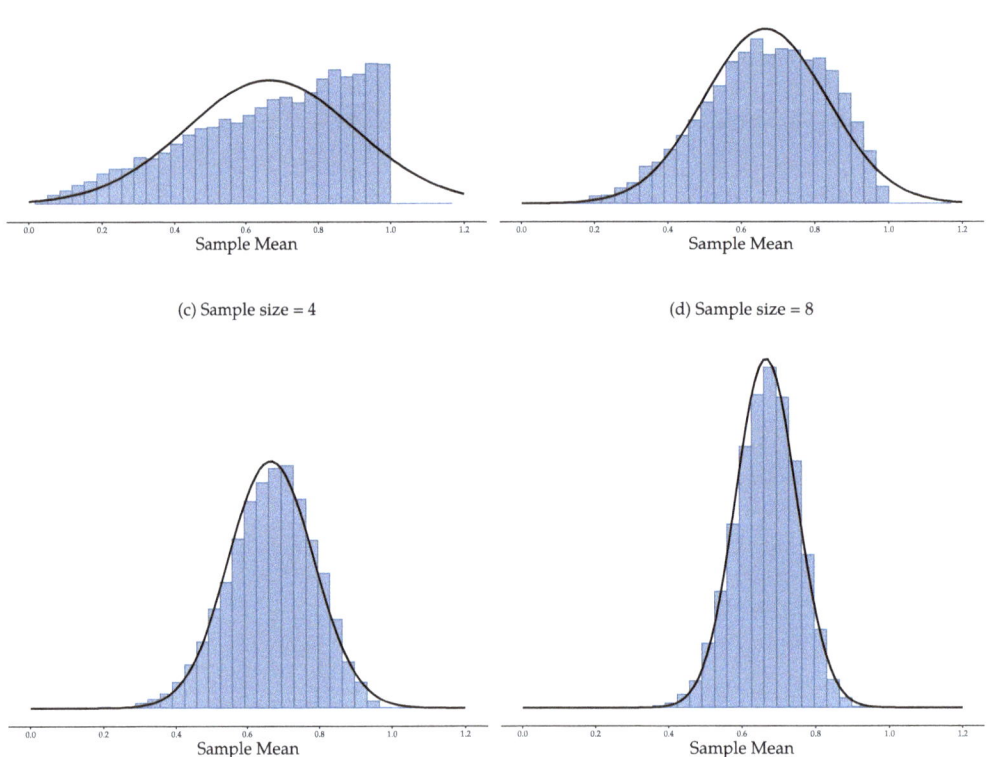

Figure 8.9: A demonstration of the central limit theorem. In panel (a), we have a non-normal population distribution, and panels (b)-(d) show the sampling distribution of the mean for samples of size 2,4 and 8 for data drawn from the distribution in panel (a). As you can see, even though the original population distribution is non-normal the sampling distribution of the mean becomes pretty close to normal by the time you have a sample of even four observations

8.4 Estimating population parameters

In all the IQ examples in the previous sections we actually knew the population parameters ahead of time. As every undergraduate gets taught in their very first lecture on the measurement of intelligence, IQ scores are defined to have mean 100 and standard deviation 15. However, this is a bit of a lie. How do we know that IQ scores have a true population mean of 100? Well, we know this because the people who designed the tests have administered them to very large samples, and have then "rigged" the scoring rules so that their sample has mean 100. That's not a bad thing of course; it's an important part of designing a psychological measurement. However, it's important to keep in mind that this theoretical mean of 100 only attaches to the population that the test designers used to design the tests. Good test designers will actually go to some lengths to provide "test norms" that can apply to lots of different populations (e.g., different age groups, nationalities etc).

This is very handy, but of course almost every research project of interest involves looking at a different population of people to those used in the test norms. For instance, suppose you wanted to measure the effect of low level lead poisoning on cognitive functioning in Port Pirie, a South Australian industrial town with a lead smelter. Perhaps you decide that you want to compare IQ scores among people in Port Pirie to a comparable sample in Whyalla, a South Australian industrial town with a steel refinery.[60] Regardless of which town you're thinking about, it doesn't make a lot of sense simply to assume that the true population mean IQ is 100. No-one has, to my knowledge, produced sensible norming data that can automatically be applied to South Australian industrial towns. We're going to have to **estimate** the population parameters from a sample of data. So how do we do this?

8.4.1 Estimating the population mean

Suppose we go to Port Pirie and 100 of the locals are kind enough to sit through an IQ test. The average IQ score among these people turns out to be $\bar{X} = 98.5$. So what is the true mean IQ for the entire population of Port Pirie? Obviously, we don't know the answer to that question. It could be 97.2, but it could also be 103.5. Our sampling isn't exhaustive so we cannot give a definitive answer. Nevertheless, if I was forced at gunpoint to give a "best guess" I'd have to say 98.5. That's the essence of statistical estimation: giving a best guess.

In this example estimating the unknown population parameter is straightforward. I calculate the sample mean and I use that as my **estimate of the population mean**. It's pretty simple, and in the next section I'll explain the statistical justification for this intuitive answer. However, for the moment what I want to do is make sure you recognise that the sample statistic and the estimate of the population parameter are conceptually different things. A sample statistic is a description of your data, whereas the estimate is a guess about the population. With that in mind, statisticians often use different notation to refer to them. For instance, if the true population mean is denoted μ, then we would use $\hat{\mu}$ to refer to our estimate of the population mean. In contrast, the sample mean is denoted $\bar{X}$ or sometimes m. However, in simple random samples the estimate of the population mean is identical to the sample mean. If I observe a sample mean of $\bar{X} = 98.5$ then my estimate of the population mean is also $\hat{\mu} = 98.5$. To help keep the notation clear, here's a handy table (Table 8.2).

Table 8.2: Notation for the mean

Symbol	What is it?	Do we know what it is?
$\hat{X}$	Sample mean	Yes, calculated from the raw data
μ	True population mean	Almost never known for sure
$\hat{\mu}$	Estimate of the population mean	Yes, identical to the sample mean in simple random samples

8.4.2 Estimating the population standard deviation

So far, estimation seems pretty simple, and you might be wondering why I forced you to read through all that stuff about sampling theory. In the case of the mean our estimate of the population parameter (i.e. $\hat{\mu}$) turned out to be identical to the corresponding sample statistic (i.e. $\bar{X}$). However, that's not always true. To see this, let's have a think about how to construct an **estimate of the population standard deviation**, which we'll denote $\hat{\sigma}$. What shall we use as our estimate in this case? Your first thought might be that we could do the same thing we did when estimating the mean, and just use the sample statistic as our estimate. That's almost the right thing to do, but not quite.

Here's why. Suppose I have a sample that contains a single observation. For this example, it helps to consider a sample where you have no intuitions at all about what the true population values might be, so let's use something completely fictitious. Suppose the observation in question measures the cromulence of my shoes. It turns out that my shoes have a cromulence of 20. So here's my sample:

$$20$$

This is a perfectly legitimate sample, even if it does have a sample size of $N = 1$. It has a sample mean of 20, and because every observation in this sample is equal to the sample mean (obviously!) it has a sample standard deviation of 0. As a description of the *sample* this seems quite right, the sample contains a single observation and therefore there is no variation observed within the sample. A sample standard deviation of $s = 0$ is the right answer here. But as an estimate of the *population* standard deviation it feels completely insane, right? Admittedly, you and I don't know anything at all about what "cromulence" is, but we know something about data. The only reason that we don't see any variability in the *sample* is that the sample is too small to display any variation! So, if you have a sample size of $N = 1$ it feels like the right answer is just to say "no idea at all".

Notice that you *don't* have the same intuition when it comes to the sample mean and the population mean. If forced to make a best guess about the population mean it doesn't feel completely insane to guess that the population mean is 20. Sure, you probably wouldn't feel very confident in that guess because you have only the one observation to work with, but it's still the best guess you can make.

Let's extend this example a little. My data set now has $N = 2$ observations of the cromulence of shoes, and the complete sample now looks like this:

$$20, 22$$

This time around, our sample is just large enough for us to be able to observe some variability: two observations is the bare minimum number needed for any variability to be observed! For our new data set, the sample mean is $\bar{X} = 21$, and the sample standard deviation is $s = 1$. What intuitions do we have about the population? Again, as far as the population mean goes, the best guess we can possibly make is the sample mean. If forced to guess we'd probably guess that the population mean cromulence is 21. What about the standard deviation? This is a little more complicated. The sample standard deviation is only based on two observations, and if you're at all like me you probably have the intuition that, with only two observations we haven't given the population "enough of a chance" to reveal its true variability to us. It's not just that we suspect that the estimate is wrong, after all with only two observations we expect it to be wrong to some degree. The worry is that the error is systematic. Specifically, we suspect that the sample standard deviation is likely to be smaller than the population standard deviation.

This intuition feels right, but it would be nice to demonstrate this somehow. There are in fact mathematical proofs that confirm this intuition, but unless you have the right mathematical background they don't help very much. Instead, what I'll do is simulate the results of some experiments. With that in mind, let's return to our IQ studies. Suppose the true population mean IQ is 100 and the standard deviation is 15. First I'll conduct an experiment in which I measure $N = 2$ IQ scores and I'll calculate the sample standard deviation. If I do this over and over again, and plot a histogram of these sample standard deviations, what I have is the sampling distribution of the standard deviation. I've plotted this distribution in Figure 8.10. Even though the true population standard deviation is 15 this experiment would, on average, produce an estimated standard deviation of only 8.4 – well below the true value! In other words, the sample standard deviation is a biased estimate of the population standard deviation Notice that this is a very different result to what we found in Figure 8.8 (b) when we plotted the sampling distribution of the mean, where the population mean is 100 and the average of the sample means is also 100.

Now let's extend the simulation. Instead of restricting ourselves to the situation where $N = 2$, let's repeat the exercise for sample sizes from 1 to 10. If we plot the average sample mean and average sample standard deviation as a function of sample size, you get the results shown in Figure 8.11. For the figure I generated 10,000 simulated data sets with 1 observation each, 10,000 more with 2 observations, and so on up to a sample size of 10. Each data set consisted of fake IQ data, that is the data were normally distributed with a true population mean of 100 and standard deviation 15. On average, the sample means turn out to be 100, regardless of sample size (panel a), and is equal to the population mean. It is an **unbiased estimator**, which is essentially the reason why your best estimate for the population mean is the sample mean.[61] The plot on the right (panel b) is quite different: on average, the sample standard deviation s is smaller than the population standard deviation σ, especially for small sample sizes. It is a **biased estimator**. In other words, if we want to make a "best guess" $\hat{\sigma}$ about the value of the population standard deviation $\hat{\sigma}$ we should make sure our guess is a little bit larger than the sample standard deviation s.

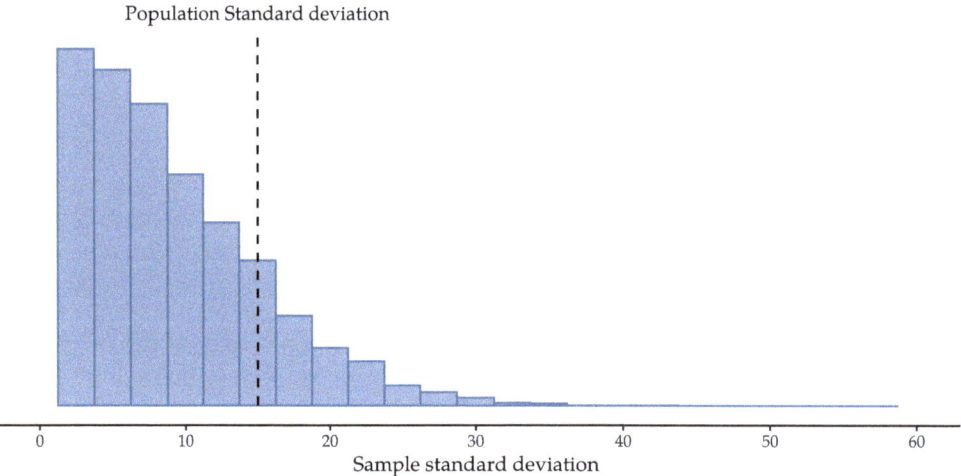

Figure 8.10: The sampling distribution of the sample standard deviation for a "two IQ scores" experiment. The true population standard deviation is 15 (dashed line), but as you can see from the histogram the vast majority of experiments will produce a much smaller sample standard deviation than this

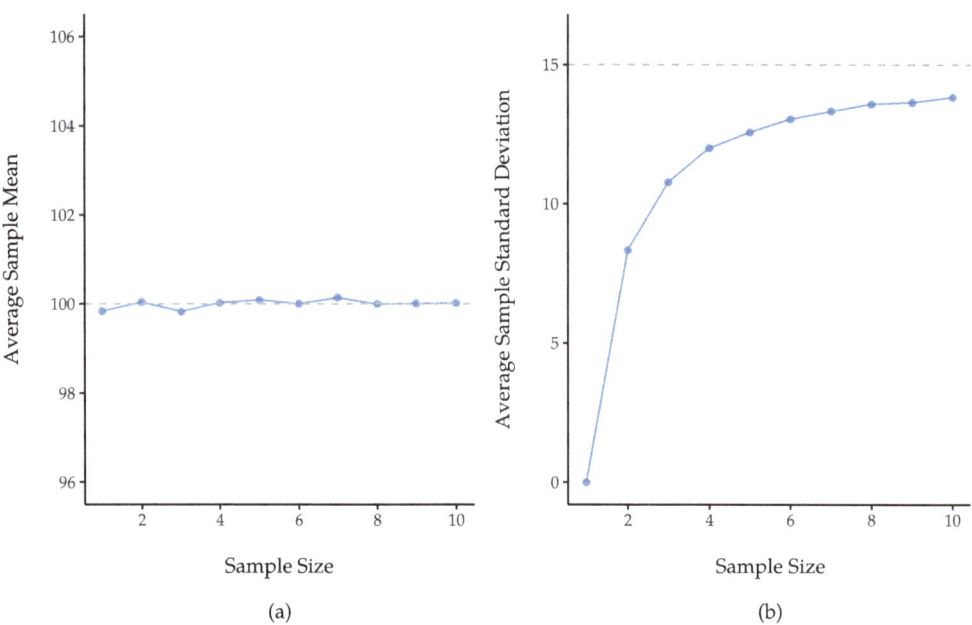

Figure 8.11: An illustration of the fact that the sample mean is an unbiased estimator of the population mean (panel a), but the sample standard deviation is a biased estimator of the population standard deviation (panel b)

The fix to this systematic bias turns out to be very simple. Here's how it works. Before tackling the standard deviation let's look at the variance. If you recall from the section on Estimating population parameters, the sample variance is defined to be the average of the squared deviations from the sample mean. That is:

$$s^2 = \frac{1}{N} \sum_{i=1}^{N} (X_i - \bar{X})^2$$

The sample variance s^2 is a biased estimator of the population variance σ^2. But as it turns out, we only need to make a tiny tweak to transform this into an unbiased estimator. All we have to do is divide by $N-1$ rather than by N.

This is an unbiased estimator of the population variance σ. Moreover, this finally answers the question we raised in Estimating population parameters. Why did jamovi give us slightly different answers for variance? It's because jamovi calculates $\hat{\sigma}^2$ not s^2, that's why. A similar story applies for the standard deviation. If we divide by $N-1$ rather than N our estimate of the population standard deviation is unbiased, and when we use jamovi's built in standard deviation function, what it's doing is calculating $\hat{\sigma}$ not s.[62]

One final point. In practice, a lot of people tend to refer to $\hat{\sigma}$ (i.e., the formula where we divide by $N-1$) as the sample standard deviation. Technically, this is incorrect. The sample standard deviation should be equal to s (i.e., the formula where we divide by N). These aren't the same thing, either conceptually or numerically. One is a property of the sample, the other is an estimated characteristic of the population. However, in almost every real life application what we actually care about is the estimate of the population parameter, and so people always report $\hat{\sigma}$ rather than s. This is the right number to report, of course. It's just that people tend to get a little bit imprecise about terminology when they write it up, because "sample standard deviation" is shorter than "estimated population standard deviation". It's no big deal, and in practice I do the same thing everyone else does. Nevertheless, I think it's important to keep the two concepts separate. It's never a good idea to confuse "known properties of your sample" with "guesses about the population from which it came". The moment you start thinking that s and $\hat{\sigma}$ are the same thing, you start doing exactly that.

To finish this section off, here's another couple of tables to help keep things clear (Table 8.3 and Table 8.4).

Table 8.3: Notation for standard deviation

Symbol	What is it?	Do we know what it is?
s	Sample standard deviation	Yes, calculated from the raw data
σ	Population standard deviation	Almost never known for sure
$\hat{\sigma}$	Estimate of the population standard deviation	Yes, but not the same as the sample standard deviation

Table 8.4: Notation for variance

Symbol	What is it?	Do we know what it is?
s^2	Sample variance	Yes, calculated from the raw data
σ^2	Population variance	Almost never known for sure
$\hat{\sigma}^2$	Estimate of the population variance	Yes, but not the same as the sample variance

8.5 Estimating a confidence interval

Statistics means never having to say you're certain
– Unknown origin[63]

Up to this point in this chapter, I've outlined the basics of sampling theory which statisticians rely on to make guesses about population parameters on the basis of a sample of data. As this discussion illustrates, one of the reasons we need all this sampling theory is that every data set leaves us with some uncertainty, so our estimates are never going to be perfectly accurate. The thing that has been missing from this discussion is an attempt to quantify the amount of uncertainty that attaches to our estimate. It's not enough to be able guess that, say, the mean IQ of undergraduate psychology students is 115 (yes, I just made that number up). We also want to be able to say something that expresses the degree of certainty that we have in our guess. For example, it would be nice to be able to say that there is a 95% chance that the true mean lies between 109 and 121. The name for this is a **confidence interval** for the mean.

Armed with an understanding of sampling distributions, constructing a confidence interval for the mean is actually pretty easy. Here's how it works. Suppose the true population mean is μ and the standard deviation is σ. I've just finished running my study that has N participants, and the mean IQ among those participants is $\bar{X}$. We know from our discussion of The central limit theorem that the sampling distribution of the mean is approximately normal. We also know from our discussion of the normal distribution in Section 7.5 that there is a 95% chance that a normally-distributed quantity will fall within about two standard deviations of the true mean.

To be more precise, the more correct answer is that there is a 95% chance that a normally distributed quantity will fall within 1.96 standard deviations of the true mean. Next, recall that the standard deviation of the sampling distribution is referred to as the standard error, and the standard error of the mean is written as SEM. When we put all these pieces together, we learn that there is a 95% probability that the sample mean $\bar{X}$ that we have actually observed lies within 1.96 standard errors of the population mean.

Of course, there's nothing special about the number 1.96. It just happens to be the multiplier you need to use if you want a 95% confidence interval. If I'd wanted a 70% confidence interval, I would have used 1.04 as the magic number rather than 1.96.

[Additional technical detail[64]]

8.5.1 Interpreting a confidence interval

The hardest thing about confidence intervals is understanding what they mean. Whenever people first encounter confidence intervals, the first instinct is almost always to say that "there is a 95% probability that the true mean lies inside the confidence interval". It's simple and it seems to capture the common sense idea of what it means to say that I am "95% confident". Unfortunately, it's not quite right. The intuitive definition relies very heavily on your own personal beliefs about the value of the population mean. I say that I am 95% confident because those are my beliefs. In everyday life that's perfectly okay, but if you remember back to the the section What does probability mean?, you'll notice that talking about personal belief and confidence is a Bayesian idea. However, confidence intervals are not Bayesian tools. Like everything else in this chapter, confidence intervals are frequentist tools, and if you are going to use frequentist methods then it's not appropriate to attach a Bayesian interpretation to them. If you use frequentist methods, you must adopt frequentist interpretations! Okay, so if that's not the right answer, what is? Remember what we said about frequentist probability. The only way we are allowed to make "probability statements" is to talk about a sequence of events, and to count up the frequencies of different kinds of events. From that perspective, the interpretation of a 95% confidence interval must have something to do with replication. Specifically, if we replicated the experiment over and over again and computed a 95% confidence interval for each replication, then 95% of those intervals would contain the true mean. More generally, 95% of all confidence intervals constructed using this procedure should contain the true population mean. This idea is illustrated in Figure 8.12, which shows 50 confidence intervals constructed for a "measure 10 IQ scores" experiment (top panel) and another 50 confidence intervals for a "measure 25 IQ scores" experiment (bottom panel). We'd expect that around 95 of our confidence intervals would contain the true population mean, and that's what we found in Figure 8.12. The critical difference here is that the Bayesian claim makes a probability statement about the population mean (i.e., it refers to our uncertainty about the population mean), which is not allowed under the frequentist interpretation of probability because you can't "replicate" a population! In the frequentist claim, the population mean is fixed and no probabilistic claims can be made about it. Confidence intervals, however, are repeatable so we can replicate experiments. Therefore a *frequentist* is allowed to talk about the probability that the *confidence interval* (a random variable) contains the true mean, but is not allowed to talk about the probability that the *true population mean* (not a repeatable event) falls within the confidence interval I know that this seems a little pedantic, but it does matter. It matters because the difference in interpretation leads to a difference in the mathematics. There is a Bayesian alternative to confidence intervals, known as *credible intervals*. In most situations credible intervals are quite similar to confidence intervals, but in other cases they are drastically different. As promised, though, I'll talk more about the Bayesian perspective in Chapter 16.

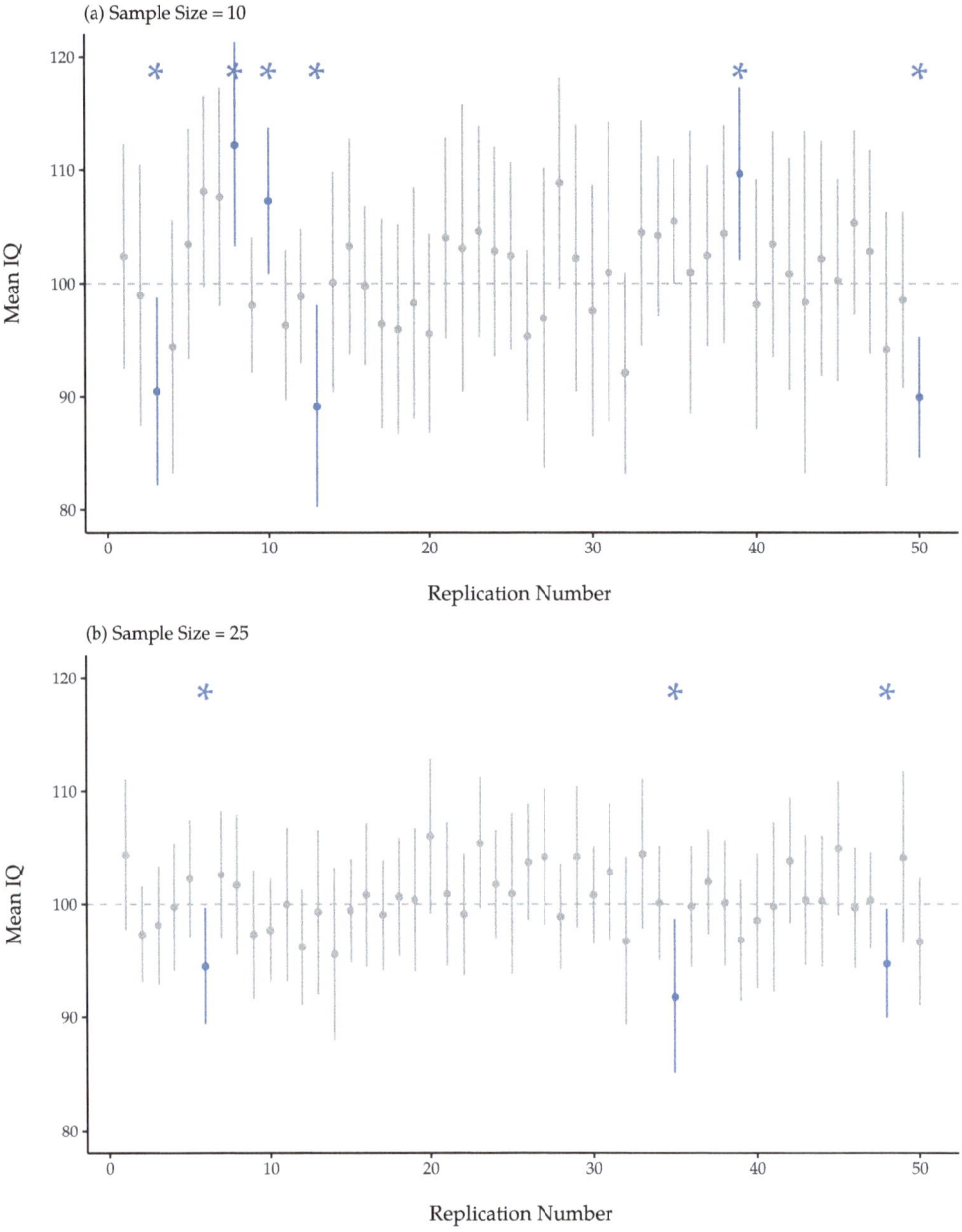

Figure 8.12: 95% confidence intervals. The top panel (a) shows 50 simulated replications of an experiment in which we measure the IQs of 10 people. The dot marks the location of the sample mean and the line shows the 95% confidence interval. Most of the 50 confidence intervals do contain the true mean (i.e., 100), but a few – in blue and marked with asterisks – do not. The lower graph (panel b) shows a similar simulation, but this time we simulate replications of an experiment that measures the IQs of 25 people

8.5.2 Calculating confidence intervals in jamovi

jamovi includes a simple way to calculate confidence intervals for the mean as part of the 'Descriptives' functionality. Under 'Descriptives' – 'Statistics' there is a check box for both the 'Std. error of Mean' and 'Confidence interval for the mean', so you can use this to find out the 95% confidence interval (which is the default). So, for example, if I load the *IQsim.omv* file, check 'Confidence interval for the mean', I can see the confidence interval associated with the simulated mean IQ: Lower 95% CI = 99.39 and Upper 95% CI = 99.97 So, in our simulated large sample data with $N = 10,000$, the mean IQ score is 99.68 with a 95% CI from 99.39 to 99.97.

When it comes to plotting confidence intervals in jamovi, you can specify that the mean is included as an option in a box plot. Moreover, when we get onto learning about specific statistical tests, for example in Chapter 13, we will see that we can also plot confidence intervals as part of the data analysis. That's pretty cool, so we'll show you how to do that later on.

8.6 Summary

In this chapter I've covered two main topics. The first half of the chapter talks about sampling theory, and the second half talks about how we can use sampling theory to construct estimates of the population parameters. The section breakdown looks like this:

- Basic ideas about Samples, populations and sampling.
- Statistical theory of sampling: The law of large numbers and Sampling distributions and the central limit theorem.
- Estimating population parameters. Means and standard deviations.
- Estimating a confidence interval.

As always, there's a lot of topics related to sampling and estimation that aren't covered in this chapter, but for an introductory psychology class this is fairly comprehensive I think. For most applied researchers you won't need much more theory than this. One big question that I haven't touched on in this chapter is what you do when you don't have a simple random sample. There is a lot of statistical theory you can draw on to handle this situation, but it's well beyond the scope of this book.

Chapter 9

Hypothesis testing

The process of induction is the process of assuming the simplest law that can be made to harmonize with our experience. This process, however, has no logical foundation but only a psychological one. It is clear that there are no grounds for believing that the simplest course of events will really happen. It is an hypothesis that the sun will rise tomorrow: and this means that we do not know whether it will rise.
– Ludwig Wittgenstein[65]

In the last chapter I discussed the ideas behind estimation, which is one of the two "big ideas" in inferential statistics. It's now time to turn our attention to the other big idea, which is *hypothesis testing*. In its most abstract form, hypothesis testing is really a very simple idea. The researcher has some theory about the world and wants to determine whether or not the data actually support that theory. However, the details are messy and most people find the theory of hypothesis testing to be the most frustrating part of statistics. The structure of the chapter is as follows. First, I'll describe how hypothesis testing works in a fair amount of detail, using a simple running example to show you how a hypothesis test is "built". I'll try to avoid being too dogmatic while doing so, and focus instead on the underlying logic of the testing procedure.[66] Afterwards, I'll spend a bit of time talking about the various dogmas, rules and heresies that surround the theory of hypothesis testing.

9.1 A menagerie of hypotheses

Eventually we all succumb to madness. For me, that day will arrive once I'm finally promoted to full professor. Safely ensconced in my ivory tower, happily protected by tenure, I will finally be able to take leave of my senses (so to speak) and indulge in that most thoroughly unproductive line of psychological research, the search for extrasensory perception (ESP).[67]

Let's suppose that this glorious day has come. My first study is a simple one in which I seek to test whether clairvoyance exists. Each participant sits down at a table and is shown a card by an experimenter. The card is black on one side and white on the other. The experimenter takes the card away and places it on a table in an adjacent room. The card is placed black side up or white side up completely at random, with the randomisation occurring only after the experimenter has left the room with the participant. A second experimenter comes in and asks the participant which side of the card is now facing upwards. It's purely a one-shot experiment. Each person sees only one card and gives only one answer, and at no stage is the participant actually in contact with someone who knows the right answer. My data set, therefore, is very simple. I have asked the question of N people and some number X of these people have given the correct response. To make things concrete, let's suppose that I have tested $N = 100$ people and $X = 62$ of these got the answer right. A surprisingly large number, sure, but is it large enough for me to feel safe in claiming I've found evidence for ESP? This is the situation where hypothesis testing comes in useful. However, before we talk about how to test hypotheses, we need to be clear about what we mean by hypotheses.

9.1.1 Research hypotheses versus statistical hypotheses

The first distinction that you need to keep clear in your mind is between research hypotheses and statistical hypotheses. In my ESP study my overall scientific goal is to demonstrate that clairvoyance exists. In this situation I have a clear research goal: I am hoping to discover evidence for ESP. In other situations I might actually be a lot more neutral than that, so I might say that my research goal is to determine whether or not clairvoyance exists. Regardless of how I want to portray myself, the basic point that I'm trying to convey here is that a research hypothesis involves making a substantive, testable scientific claim. If you are a psychologist then your research hypotheses are fundamentally about psychological constructs. Any of the following would count as **research hypotheses**:

- *Listening to music reduces your ability to pay attention to other things.* This is a claim about the causal relationship between two psychologically meaningful concepts (listening to music and paying attention to things), so it's a perfectly reasonable research hypothesis.
- *Intelligence is related to personality.* Like the last one, this is a relational claim about two psychological constructs (intelligence and personality), but the claim is weaker: correlational not causal.
- *Intelligence is speed of information processing.* This hypothesis has a quite different character. It's not actually a relational claim at all. It's an ontological claim about the fundamental character of intelligence. It's usually easier to think about how to construct experiments to test research hypotheses of the form "does X affect Y?" than it is to address claims like "what is X?" And in practice what usually happens is that you find ways of testing relational claims that follow from your ontological ones. For instance, if I believe that intelligence is speed of information processing in the brain, my experiments will often involve looking for relationships between measures of intelligence and measures of speed. As a consequence most everyday research questions do tend to be relational in nature, but they're almost always motivated by deeper ontological questions about the state of nature.

Notice that in practice, my research hypotheses could overlap a lot. My ultimate goal in the ESP experiment might be to test an ontological claim like "ESP exists", but I might operationally restrict myself to a narrower hypothesis like "Some people can 'see' objects in a clairvoyant fashion". That said, there are some things that really don't count as proper research hypotheses in any meaningful sense:

- *Love is a battlefield.* This is too vague to be testable. Whilst it's okay for a research hypothesis to have a degree of vagueness to it, it has to be possible to operationalise your theoretical ideas. Maybe I'm just not creative enough to see it, but I can't see how this can be converted into any concrete research design. If that's true then this isn't a scientific research hypothesis, it's a pop song. That doesn't mean it's not interesting. A lot of deep questions that humans have fall into this category. Maybe one day science will be able to construct testable theories of love, or to test to see if God exists, and so on. But right now we can't, and I wouldn't bet on ever seeing a satisfying scientific approach to either.
- *The first rule of tautology club is the first rule of tautology club.* This is not a substantive claim of any kind. It's true by definition. No conceivable state of nature could possibly be inconsistent with this claim. We say that this is an unfalsifiable hypothesis, and as such it is outside the domain of science. Whatever else you do in science your claims must have the possibility of being wrong.
- *More people in my experiment will say "yes" than "no".* This one fails as a research hypothesis because it's a claim about the data set, not about the psychology (unless of course your actual research question is whether people have some kind of "yes" bias!). Actually, this hypothesis is starting to sound more like a statistical hypothesis than a research hypothesis.

As you can see, research hypotheses can be somewhat messy at times and ultimately they are scientific claims. **Statistical hypotheses** are neither of these two things. Statistical hypotheses must be mathematically precise and they must correspond to specific claims about the characteristics of the data generating mechanism (i.e., the "population"). Even so, the intent is that statistical hypotheses bear a clear relationship to the substantive research hypotheses that you care about! For instance, in my ESP study my research hypothesis is that some people are able to see through walls or whatever. What I want to do is to "map" this onto a statement about how the data were generated. So let's think about what that statement would be. The quantity that I'm interested in within the experiment is $P(correct)$, the true-but-unknown probability with which the participants in my experiment answer the question correctly. Let's use the Greek letter θ (*theta*) to refer to this probability. Here are four different statistical hypotheses:

- If ESP doesn't exist and if my experiment is well designed then my participants are just guessing. So I should expect them to get it right half of the time and so my statistical hypothesis is that the true probability of choosing correctly is $\theta = 0.5$.
- Alternatively, suppose ESP does exist and participants can see the card. If that's true people will perform better than chance and the statistical hypothesis is that $\theta > 0.5$.
- A third possibility is that ESP does exist, but the colours are all reversed and people don't realise it (okay, that's wacky, but you never know). If that's how it works then you'd expect people's performance to be below chance. This would correspond to a statistical hypothesis that $\theta < 0.5$.
- Finally, suppose ESP exists but I have no idea whether people are seeing the right colour or the wrong one. In that case the only claim I could make about the data

would be that the probability of making the correct answer is not equal to 0.5. This corresponds to the statistical hypothesis that $\theta \neq 0.5$.

All of these are legitimate examples of a statistical hypothesis because they are statements about a population parameter and are meaningfully related to my experiment.

What this discussion makes clear, I hope, is that when attempting to construct a statistical hypothesis test the researcher actually has two quite distinct hypotheses to consider. First, he or she has a research hypothesis (a claim about psychology), and this then corresponds to a statistical hypothesis (a claim about the data generating population). In my ESP example these might be as shown in Table 9.1.

Table 9.1: Research and statistical hypotheses

Dani's research hypothesis:	"ESP exists"
Dani's statistical hypothesis:	$\theta \neq 0.5$

And a key thing to recognise is this. A statistical hypothesis test is a test of the statistical hypothesis, not the research hypothesis. If your study is badly designed then the link between your research hypothesis and your statistical hypothesis is broken. To give a silly example, suppose that my ESP study was conducted in a situation where the participant can actually see the card reflected in a window. If that happens I would be able to find very strong evidence that $\theta \neq 0.5$, but this would tell us nothing about whether "ESP exists".

9.1.2 Null hypotheses and alternative hypotheses

So far, so good. I have a research hypothesis that corresponds to what I want to believe about the world, and I can map it onto a statistical hypothesis that corresponds to what I want to believe about how the data were generated. It's at this point that things get somewhat counter-intuitive for a lot of people. Because what I'm about to do is invent a new statistical hypothesis (the "null" hypothesis, H_0) that corresponds to the exact opposite of what I want to believe, and then focus exclusively on that almost to the neglect of the thing I'm actually interested in (which is now called the "alternative" hypothesis, H1). In our ESP example, the null hypothesis is that $\theta = 0.5$, since that's what we'd expect if ESP didn't exist. My hope, of course, is that ESP is totally real and so the alternative to this null hypothesis is $\theta \neq 0.5$. In essence, what we're doing here is dividing up the possible values of θ into two groups: those values that I really hope aren't true (the null), and those values that I'd be happy with if they turn out to be right (the alternative). Having done so, the important thing to recognise is that the goal of a hypothesis test is not to show that the alternative hypothesis is (probably) true. The goal is to show that the null hypothesis is (probably) false. Most people find this pretty weird.

The best way to think about it, in my experience, is to imagine that a hypothesis test is a criminal trial,[68] **the trial of the null hypothesis**. The null hypothesis is the defendant, the researcher is the prosecutor, and the statistical test itself is the judge. Just like a criminal trial, there is a presumption of innocence. The null hypothesis is deemed to be true unless you, the researcher, can prove beyond a reasonable doubt that it is false. You are free to design your experiment however you like (within reason, obviously!) and your goal when doing so is to maximise the chance that the data will yield a conviction for the crime of being false. The catch is that the statistical test sets the rules of the trial and those rules are designed to protect the null hypothesis, specifically to ensure that if the null hypothesis is actually true the chances of a false conviction are guaranteed to be low. This is pretty important. After all, the null hypothesis doesn't get a lawyer, and given that the researcher is trying desperately to prove it to be false someone has to protect it.

9.2 Two types of errors

Before going into details about how a statistical test is constructed it's useful to understand the philosophy behind it. I hinted at it when pointing out the similarity between a null hypothesis test and a criminal trial, but I should now be explicit. Ideally, we would like to construct our test so that we never make any errors. Unfortunately, since the world is messy, this is never possible. Sometimes you're just really unlucky. For instance, suppose you flip a coin 10 times in a row and it comes up heads all 10 times. That feels like very strong evidence for a conclusion that the coin is biased, but of course there's a 1 in 1024 chance that this would happen even if the coin was totally fair. In other words, in real life we always have to accept that there's a chance that we made a mistake. As a consequence the goal behind statistical hypothesis testing is not to eliminate errors, but to minimise them.

At this point, we need to be a bit more precise about what we mean by "errors". First, let's state the obvious. It is either the case that the null hypothesis is true or that it is false, and our test will either retain the null hypothesis or reject it.[69] So, as Table 9.2 illustrates, after we run the test and make our choice one of four things might have happened:

Table 9.2: Null hypothesis statistical testing (NHST)

	retain H_0	reject H_0
H_0 is true	correct decision	error (type I)
H_0 is false	error (type II)	correct decision

As a consequence there are actually two different types of error here. If we reject a null hypothesis that is actually true then we have made a **type I error**. On the other hand, if we retain the null hypothesis when it is in fact false then we have made a **type II error**.

Remember how I said that statistical testing was kind of like a criminal trial? Well, I meant it. A criminal trial requires that you establish "beyond a reasonable doubt" that the defendant did it. All of the evidential rules are (in theory, at least) designed to ensure that there's (almost) no chance of wrongfully convicting an innocent defendant. The trial is designed to protect the rights of a defendant, as the English jurist William Blackstone famously said, it is "better that ten guilty persons escape than that one innocent suffer." In other words, a criminal trial doesn't treat the two types of error in the same way. Punishing the innocent is deemed to be much worse than letting the guilty go free. A statistical test is pretty much the same. The single most important design principle of the test is to control the probability of a type I error, to keep it below some fixed probability. This probability, which is denoted α, is called the **significance level** of the test. And I'll say it again, because it is so central to the whole set-up, a hypothesis test is said to have significance level α if the type I error rate is no larger than α.

So, what about the type II error rate? Well, we'd also like to keep those under control too, and we denote this probability by β. However, it's much more common to refer to the **power** of the test, that is the probability with which we reject a null hypothesis when it really is false, which is $1 - \beta$. To help keep this straight, here's the same table again but with the relevant numbers added (Table 9.3):

Table 9.3: Null hypothesis statistical testing (NHST) – additional detail

	retain H_0	**reject H_0**
H_0 is true	$1 - \alpha$ (probability of correct retention)	α (type I error rate)
H_0 is false	β (type II error rate)	$1 - \beta$ (power of the test)

A "powerful" hypothesis test is one that has a small value of β, while still keeping α fixed at some (small) desired level. By convention, scientists make use of three different α levels: .05, .01 and .001. Notice the asymmetry here; the tests are designed to ensure that the α level is kept small but there's no corresponding guarantee regarding β. We'd certainly like the type II error rate to be small and we try to design tests that keep it small, but this is typically secondary to the overwhelming need to control the type I error rate. As Blackstone might have said if he were a statistician, it is "better to retain 10 false null hypotheses than to reject a single true one". To be honest, I don't know that I agree with this philosophy. There are situations where I think it makes sense, and situations where I think it doesn't, but that's neither here nor there. It's how the tests are built.

9.3 Test statistics and sampling distributions

At this point we need to start talking specifics about how a hypothesis test is constructed. To that end, let's return to the ESP example. Let's ignore the actual data that we obtained, for the moment, and think about the structure of the experiment. Regardless of what the actual numbers are, the form of the data is that X out of N people correctly identified the colour of the hidden card. Moreover, let's suppose for the moment that the null hypothesis really is true, that ESP doesn't exist and the true probability that anyone picks the correct colour is exactly $\theta = 0.5$. What would we expect the data to look like? Well, obviously we'd expect the proportion of people who make the correct response to be pretty close to 50%. Or, to phrase this in more mathematical terms, we'd say that $\frac{X}{N}$ is approximately 0.5. Of course, we wouldn't expect this fraction to be exactly 0.5. If, for example, we tested $N = 100$ people and $X = 53$ of them got the question right, we'd probably be forced to concede that the data are quite consistent with the null hypothesis. On the other hand, if $X = 99$ of our participants got the question right then we'd feel pretty confident that the null hypothesis is wrong. Similarly, if only $X = 3$ people got the answer right we'd be similarly confident that the null was wrong. Let's be a little more technical about this. We have a quantity X that we can calculate by looking at our data. After looking at the value of X we make a decision about whether to believe that the null hypothesis is correct, or to reject the null hypothesis in favour of the alternative. The name for this thing that we calculate to guide our choices is a **test statistic**.

Having chosen a test statistic, the next step is to state precisely which values of the test statistic would cause is to reject the null hypothesis, and which values would cause us to keep it. In order to do so we need to determine what the **sampling distribution of the test statistic** would be if the null hypothesis were actually true (we talked about sampling distributions earlier in Section 8.3.1. Why do we need this? Because this distribution tells us exactly what values of X our null hypothesis would lead us to expect. And, therefore, we can use this distribution as a tool for assessing how closely the null hypothesis agrees with our data.

How do we actually determine the sampling distribution of the test statistic? For a lot of hypothesis tests this step is actually quite complicated, and later on in the book you'll see me being slightly evasive about it for some of the tests (some of them I don't even understand myself). However, sometimes it's very easy. And, fortunately for us, our ESP example provides us with one of the easiest cases. Our population parameter θ is just the overall probability that people respond correctly when asked the question, and our test statistic X is the count of the number of people who did so out of a sample size of N. We've seen a distribution like this before, in Section 7.4, and that's exactly what the binomial distribution describes! So, to use the notation and terminology that I introduced in that section, we would say that the null hypothesis predicts that X is binomially distributed, which is written:

$$X \sim Binomial(\theta, N)$$

Since the null hypothesis states that $\theta = 0.5$ and our experiment has $N = 100$ people, we have the sampling distribution we need. This sampling distribution is plotted in Figure 9.1. No surprises really, the null hypothesis says that $X = 50$ is the most likely outcome, and it says that we're almost certain to see somewhere between 40 and 60 correct responses.

9.4 Making decisions

Okay, we're very close to being finished. We've constructed a test statistic (X) and we chose this test statistic in such a way that we're pretty confident that if X is close to $\frac{N}{2}$ then we should retain the null, and if not we should reject it. The question that remains is this. Exactly which values of the test statistic should we associate with the null hypothesis, and exactly which values go with the alternative hypothesis? In my ESP study, for example, I've observed a value of $X = 62$. What decision should I make? Should I choose to believe the null hypothesis or the alternative hypothesis?

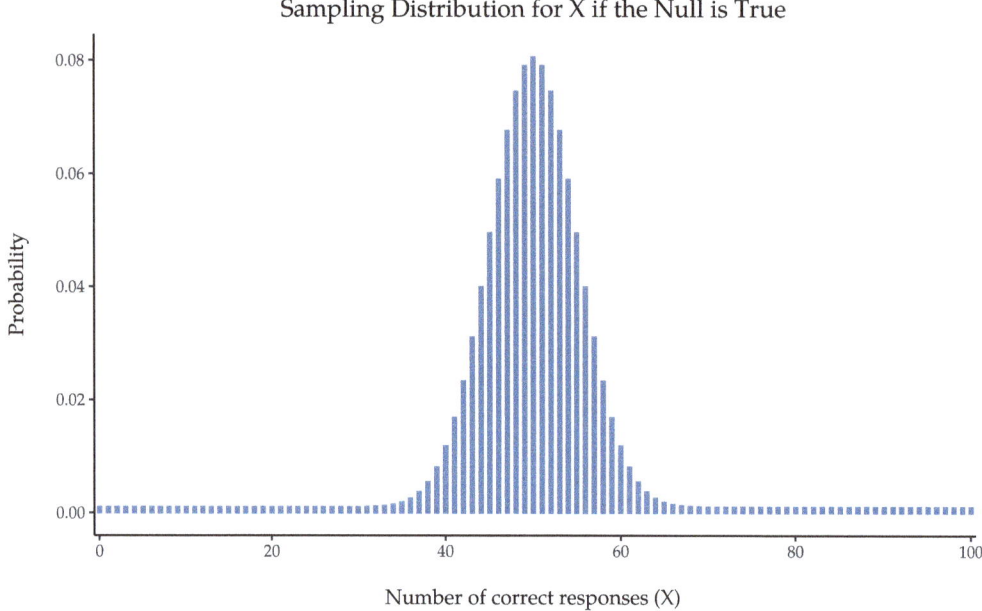

Figure 9.1: The sampling distribution for our test statistic X when the null hypothesis is true. For our ESP scenario this is a binomial distribution. Not surprisingly, since the null hypothesis says that the probability of a correct response is $\theta = .5$, the sampling distribution says that the most likely value is 50 (out of 100) correct responses. Most of the probability mass lies between 40 and 60

9.4.1 Critical regions and critical values

To answer this question we need to introduce the concept of a **critical region** for the test statistic X. The critical region of the test corresponds to those values of X that would lead us to reject null hypothesis (which is why the critical region is also sometimes called the rejection region). How do we find this critical region? Well, let's consider what we know:

- X should be very big or very small in order to reject the null hypothesis.
- If the null hypothesis is true, the sampling distribution of X is $Binomial(0.5, N)$.
- If $\alpha = .05$, the critical region must cover 5% of this sampling distribution.

It's important to make sure you understand this last point. The critical region corresponds to those values of X for which we would reject the null hypothesis, and the sampling distribution in question describes the probability that we would obtain a particular value of X if the null hypothesis were actually true. Now, let's suppose that we chose a critical region that covers 20% of the sampling distribution, and suppose that the null hypothesis is actually true. What would be the probability of incorrectly rejecting the null? The answer is of course 20%. And, therefore, we would have built a test that had an α level of 0.2. If we want $\alpha = .05$, the critical region is only allowed to cover 5% of the sampling distribution of our test statistic.

As it turns out those three things uniquely solve the problem. Our critical region consists of the most extreme values, known as the **tails** of the distribution. This is illustrated in Figure 9.2. If we want $\alpha = .05$ then our critical regions correspond to $X \leq 40$ and $X \geq 60$.[70] That is, if the number of people saying "true" is between 41 and 59, then we should retain the null hypothesis. If the number is between 0 to 40, or between 60 to 100, then we should reject the null hypothesis. The numbers 40 and 60 are often referred to as the **critical values** since they define the edges of the critical region.

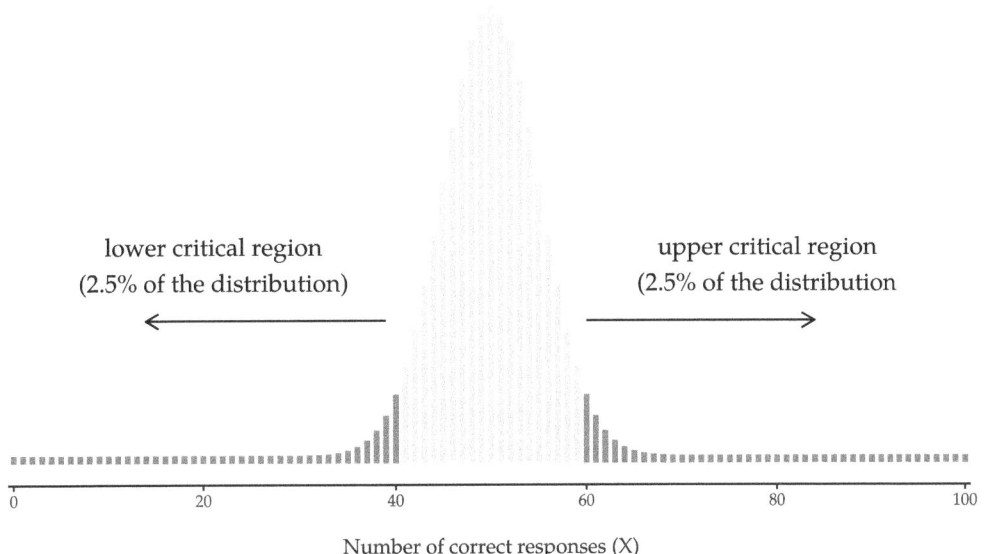

Figure 9.2: The critical region associated with the hypothesis test for the ESP study, for a hypothesis test with a significance level of $\alpha = .05$. The plot shows the sampling distribution of X under the null hypothesis (i.e., same as Figure 9.1). The grey bars correspond to those values of X for which we would retain the null hypothesis. The blue (darker shaded) bars show the critical region, those values of X for which we would reject the null. Because the alternative hypothesis is two-sided (i.e., allows both $\theta < .5$ and $\theta > .5$), the critical region covers both tails of the distribution. To ensure an α level of .05, we need to ensure that each of the two regions encompasses 2.5% of the sampling distribution

At this point, our hypothesis test is essentially complete:

1. We choose an α level (e.g., $\alpha = .05$);
2. Come up with some test statistic (e.g., X) that does a good job (in some meaningful sense) of comparing H_0 to H_1;
3. Figure out the sampling distribution of the test statistic on the assumption that the null hypothesis is true (in this case, binomial); and then
4. Calculate the critical region that produces an appropriate α level (0-40 and 60-100).

All that we have to do now is calculate the value of the test statistic for the real data (e.g., $X = 62$) and then compare it to the critical values to make our decision. Since 62 is greater than the critical value of 60 we would reject the null hypothesis. Or, to phrase it slightly differently, we say that the test has produced a statistically **significant** result.

9.4.2 A note on statistical "significance"

> *Like other occult techniques of divination, the statistical method has a private jargon deliberately contrived to obscure its methods from non-practitioners.*
> – Attributed to G. O. Ashley[71]

A very brief digression is in order at this point, regarding the word "significant". The concept of statistical significance is actually a very simple one, but has a very unfortunate name. If the data allow us to reject the null hypothesis, we say that "the result is statistically significant", which is often shortened to "the result is significant". This terminology is rather old and dates back to a time when "significant" just meant something like "indicated", rather than its modern meaning which is much closer to "important". As a result, a lot of modern readers get very confused when they start learning statistics because they think that a "significant result" must be an important one. It doesn't mean that at all. All that "statistically significant" means is that the data allowed us to reject a null hypothesis. Whether or not the result is actually important in the real world is a very different question, and depends on all sorts of other things.

9.4.3 The difference between one-sided and two-sided tests

There's one more thing I want to point out about the hypothesis test that I've just constructed. If we take a moment to think about the statistical hypotheses I've been using:

$$H_0 : \theta = 0.5$$
$$H_1 : \theta \neq 0.5$$

we notice that the alternative hypothesis covers both the possibility that $\theta < .5$ and the possibility that θ .5. This makes sense if I really think that ESP could produce either better-than chance performance or worse-than-chance performance (and there are some people who think that). In statistical language this is an example of a **two-sided test**. It's called this because the alternative hypothesis covers the area on both "sides" of the null hypothesis, and as a consequence the critical region of the test covers both tails of the sampling distribution (2.5% on either side if $\alpha = .05$), as illustrated earlier in Figure 9.2. However, that's not the only possibility. I might only be willing to believe in ESP if it produces better than chance performance. If so, then my alternative hypothesis would only covers the possibility that $\theta > .5$, and as a consequence the null hypothesis now becomes:

$$H_0 : \theta \leq 0.5$$
$$H_1 : \theta > 0.5$$

When this happens, we have what's called a **one-sided test** and the critical region only covers one tail of the sampling distribution. This is illustrated in Figure 9.3.

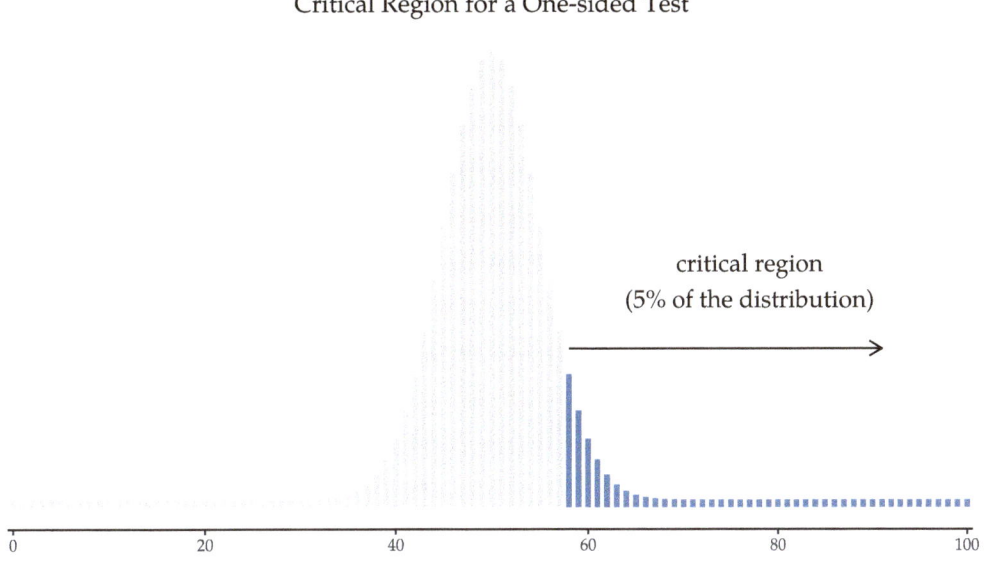

Figure 9.3: The critical region for a one-sided test. In this case, the alternative hypothesis is that $\theta \geq .5$ so we would only reject the null hypothesis for large values of X. As a consequence, the critical region only covers the upper tail of the sampling distribution, specifically the upper 5% of the distribution. Contrast this to the two-sided version in Figure 9.2

9.5 The p-value of a test

In one sense, our hypothesis test is complete. We've constructed a test statistic, figured out its sampling distribution if the null hypothesis is true, and then constructed the critical region for the test. Nevertheless, I've actually omitted the most important number of all, **the p-value**. It is to this topic that we now turn. There are two somewhat different ways of interpreting a p-value, one proposed by Sir Ronald Fisher and the other by Jerzy Neyman. Both versions are legitimate, though they reflect very different ways of thinking about hypothesis tests. Most introductory textbooks tend to give Fisher's version only, but I think that's a bit of a shame. To my mind, Neyman's version is cleaner and actually better reflects the logic of the null hypothesis test. You might disagree though, so I've included both. I'll start with Neyman's version.

9.5.1 A softer view of decision making

One problem with the hypothesis testing procedure that I've described is that it makes no distinction at all between a result that is "barely significant" and those that are "highly significant". For instance, in my ESP study the data I obtained only just fell inside the critical region, so I did get a significant effect but it was a pretty near thing. In contrast, suppose that I'd run a study in which $X = 97$ out of my $N = 100$ participants got the answer right. This would obviously be significant too but my a much larger margin, such that there's really no ambiguity about this at all. The procedure that I have already described makes no distinction between the two. If I adopt the standard convention of allowing $\alpha = .05$ as my acceptable type I error rate, then both of these are significant results.

This is where the p-value comes in handy. To understand how it works, let's suppose that we ran lots of hypothesis tests on the same data set, but with a different value of α in each case. When we do that for my original ESP data what we'd get is something like Table 9.4.

Table 9.4: Rejecting the null hypothesis at different levels of alpha

Value of α	0.05	0.04	0.03	0.02	0.01
Reject the null?	Yes	Yes	Yes	No	No

When we test the ESP data ($X = 62$ successes out of $N = 100$ observations), using α levels of .03 and above, we'd always find ourselves rejecting the null hypothesis. For α levels of .02 and below we always end up retaining the null hypothesis. Therefore, somewhere between .02 and .03 there must be a smallest value of α that would allow us to reject the null hypothesis for this data. This is the p-value. As it turns out the ESP data has $p = .021$. In short, p is defined to be the smallest type I error rate (α) that you have to be willing to tolerate if you want to reject the null hypothesis.

If it turns out that p describes an error rate that you find intolerable, then you must retain the null. If you're comfortable with an error rate equal to p, then it's okay to reject the null hypothesis in favour of your preferred alternative.

In effect, p is a summary of all the possible hypothesis tests that you could have run, taken across all possible α values. And as a consequence it has the effect of "softening" our decision process. For those tests in which p d' α you would have rejected the null hypothesis, whereas for those tests in which p ą α you would have retained the null. In my ESP study I obtained $X = 62$ and as a consequence I've ended up with $p = .021$. So the error rate I have to tolerate is 2.1%. In contrast, suppose my experiment had yielded $X = 97$. What happens to my p-value now? This time it's shrunk to $p = 1.36 \times 10^{-25}$, which is a tiny, tiny[72] type I error rate. For this second case I would be able to reject the null hypothesis with a lot more confidence, because I only have to be "willing" to tolerate a type I error rate of about 1 in 10 trillion trillion in order to justify my decision to reject.

9.5.2 The probability of extreme data

The second definition of the p-value comes from Sir Ronald Fisher, and it's actually this one that you tend to see in most introductory statistics textbooks. Notice how, when I constructed the critical region, it corresponded to the tails (i.e., extreme values) of the sampling distribution? That's not a coincidence, almost all "good" tests have this characteristic (good in the sense of minimising our type II error rate, β). The reason for that is that a good critical region almost always corresponds to those values of the test statistic that are least likely to be observed if the null hypothesis is true. If this rule is true, then we can define the *p*-value as the probability that we would have observed a test statistic that is at least as extreme as the one we actually did get. In other words, if the data are extremely implausible according to the null hypothesis, then the null hypothesis is probably wrong.

9.5.3 A common mistake

Okay, so you can see that there are two rather different but legitimate ways to interpret the *p*-value, one based on Neyman's approach to hypothesis testing and the other based on Fisher's. Unfortunately, there is a third explanation that people sometimes give, especially when they're first learning statistics, and it is *absolutely and completely wrong*. This mistaken approach is to refer to the *p*-value as "the probability that the null hypothesis is true". It's an intuitively appealing way to think, but it's wrong in two key respects. First, null hypothesis testing is a frequentist tool and the frequentist approach to probability does not allow you to assign probabilities to the null hypothesis. According to this view of probability, the null hypothesis is either true or it is not, it cannot have a "5% chance" of being true. Second, even within the Bayesian approach, which does let you assign probabilities to hypotheses, the *p*-value would not correspond to the probability that the null is true. This interpretation is entirely inconsistent with the mathematics of how the *p*-value is calculated. Put bluntly, despite the intuitive appeal of thinking this way, there is no justification for interpreting a *p*-value this way. Never do it.

9.6 Reporting the results of a hypothesis test

When writing up the results of a hypothesis test there's usually several pieces of information that you need to report, but it varies a fair bit from test to test. Throughout the rest of the book I'll spend a little time talking about how to report the results of different tests (see Section 10.1.9 for a particularly detailed example), so that you can get a feel for how it's usually done. However, regardless of what test you're doing, the one thing that you always have to do is say something about the *p*-value and whether or not the outcome was significant.

The fact that you have to do this is unsurprising, it's the whole point of doing the test. What might be surprising is the fact that there is some contention over exactly how you're supposed to do it. Leaving aside those people who completely disagree with the entire framework underpinning null hypothesis testing, there's a certain amount of tension that exists regarding whether or not to report the exact *p*-value that you obtained, or if you should state only that $p < \alpha$ for a significance level that you chose in advance (e.g., $p < .05$).

9.6.1 The issue

To see why this is an issue, the key thing to recognise is that *p*-values are terribly convenient. In practice, the fact that we can compute a *p*-value means that we don't actually have to specify any α level at all in order to run the test. Instead, what you can do is calculate your *p*-value and interpret it directly. If you get $p = .062$, then it means that you'd have to be willing to tolerate a type I error rate of 6.2% to justify rejecting the null. If you personally find 6.2% intolerable then you retain the null. Therefore, the argument goes, why don't we just report the actual *p*-value and let the reader make up their own minds about what an acceptable type I error rate is? This approach has the big advantage of "softening" the decision making process. In fact, if you accept the Neyman definition of the *p*-value, that's the whole point of the *p*-value. We no longer have a fixed significance level of $\alpha = .05$ as a bright line separating "accept" from "reject" decisions, and this removes the rather pathological problem of being forced to treat $p = .051$ in a fundamentally different way to $p = .049$.

This flexibility is both the advantage and the disadvantage to the *p*-value. The reason why a lot of people don't like the idea of reporting an exact *p*-value is that it gives the researcher a bit too much freedom. In particular, it lets you change your mind about what error tolerance you're willing to put up with after you look at the data. For instance, consider my ESP experiment. Suppose I ran my test and ended up with a *p*-value of .09. Should I accept or reject? Now, to be honest, I haven't yet bothered to think about what level of type I error I'm "really" willing to accept. I don't have an opinion on that topic. But I *do* have an opinion about whether or not ESP exists, and I *definitely* have an opinion about whether my research should be published in a reputable scientific journal. And amazingly, now that I've looked at the data I'm starting to think that a 9% error rate isn't so bad, especially when compared to how annoying it would be to have to admit to the world that my experiment has failed. So, to avoid looking like I just made it up after the fact, I now say that my α is .1, with the argument that a 10% type I error rate isn't too bad and at that level my test is significant! I win.

In other words, the worry here is that I might have the best of intentions, and be the most honest of people, but the temptation to just "shade" things a little bit here and there is really, really strong. As anyone who has ever run an experiment can attest, it's a long and difficult process and you often get very attached to your hypotheses. It's hard to let go and admit the experiment didn't find what you wanted it to find. And that's the danger here. If we use the "raw" *p*-value, people will start interpreting the data in terms of what they want to believe, not what the data are actually saying and, if we allow that, why are we even bothering to do science at all? Why not let everyone believe whatever they like about anything, regardless of what the facts are? Okay, that's a bit extreme, but that's where the worry comes from. According to this view, you really must specify your α value in advance and then only report whether the test was significant or not. It's the only way to keep ourselves honest.

9.6.2 Two proposed solutions

In practice, it's pretty rare for a researcher to specify a single α level ahead of time. Instead, the convention is that scientists rely on three standard significance levels: .05, .01 and .001. When reporting your results, you indicate which (if any) of these significance levels allow you to reject the null hypothesis. This is summarised in Table 9.5. This allows us to soften the decision rule a little bit, since $p < .01$ implies that the data meet a stronger evidential standard than $p < .05$ would. Nevertheless, since these levels are fixed in advance by convention, it does prevent people choosing their α level after looking at the data.

Table 9.5: Typical translations of p-value levels

Usual notation	Signif. stars	English translation	The null is...
$p > .05$		The test wasn't significant	Retained
$p < .05$	*	The test was significant at $\alpha = .05$ but not at $\alpha = .01$ or $\alpha = .001$.	Rejected
$p < .01$	**	The test was significant at $\alpha = .05$ and $\alpha = .01$ but not at $\alpha = .001$.	Rejected
$p < .001$	***	The test was significant at all levels	Rejected

Nevertheless, quite a lot of people still prefer to report exact p-values. To many people, the advantage of allowing the reader to make up their own mind about how to interpret $p = .06$ outweighs any disadvantages. In practice, however, even among those researchers who prefer exact p-values it is quite common to just write $p < .001$ instead of reporting an exact value for small p. This is in part because a lot of software doesn't actually print out the p-value when it's that small (e.g., SPSS just writes $p = .000$ whenever $p < .001$), and in part because a very small p-value can be kind of misleading. The human mind sees a number like .0000000001 and it's hard to suppress the gut feeling that the evidence in favour of the alternative hypothesis is a near certainty. In practice however, this is usually wrong. Life is a big, messy, complicated thing, and every statistical test ever invented relies on simplifications, approximations and assumptions. As a consequence, it's probably not reasonable to walk away from any statistical analysis with a feeling of confidence stronger than $p < .001$ implies. In other words, $p < .001$ is really code for "as far as this test is concerned, the evidence is overwhelming."

In light of all this, you might be wondering exactly what you should do. There's a fair bit of contradictory advice on the topic, with some people arguing that you should report the exact p-value, and other people arguing that you should use the tiered approach illustrated in Table 9.1. As a result, the best advice I can give is to suggest that you look at papers/reports written in your field and see what the convention seems to be. If there doesn't seem to be any consistent pattern, then use whichever method you prefer.

9.7 Running the hypothesis test in practice

At this point some of you might be wondering if this is a "real" hypothesis test, or just a toy example that I made up. It's real. In the previous discussion I built the test from first principles, thinking that it was the simplest possible problem that you might ever encounter in real life. However, this test already exists. It's called the *binomial test*, and it's implemented by jamovi as one of the statistical analyses available when you hit the 'Frequencies' button. To test the null hypothesis that the response probability is one-half $p = .5$,[73] and using data in which $x = 62$ of $N = 100$ people made the correct response, available in the *binomialtest.omv* data file, we get the results shown in Figure 9.4.

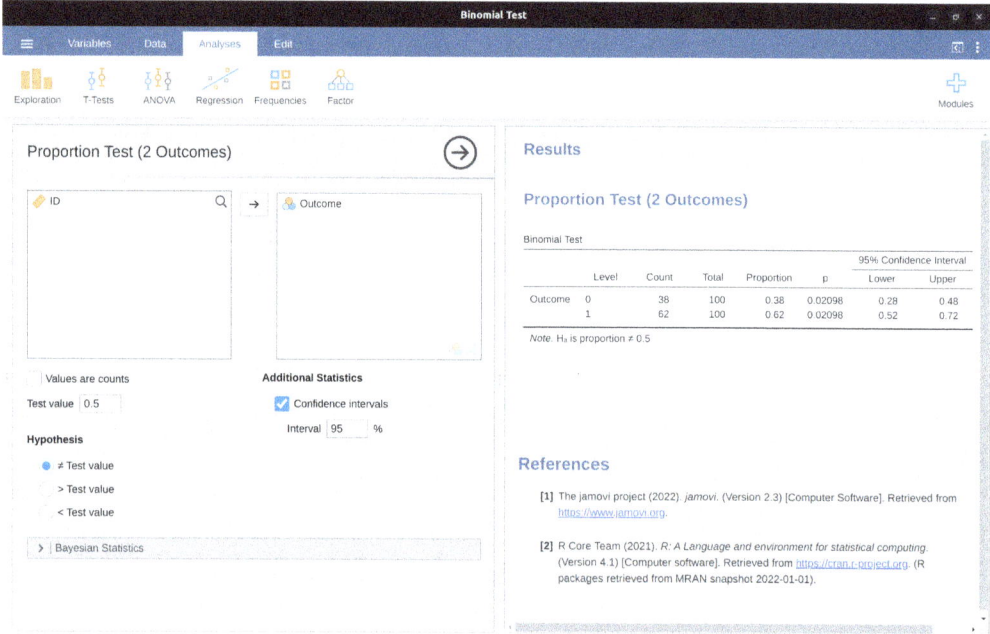

Figure 9.4: Binomial test analysis and results in jamovi

Right now, this output looks pretty unfamiliar to you, but you can see that it's telling you more or less the right things. Specifically, the p-value of 0.02 is less than the usual choice of $\alpha = .05$, so you can reject the null. We'll talk a lot more about how to read this sort of output as we go along, and after a while you'll hopefully find it quite easy to read and understand.

9.8 Effect size, sample size and power

In previous sections I've emphasised the fact that the major design principle behind statistical hypothesis testing is that we try to control our type I error rate. When we fix $\alpha = .05$ we are attempting to ensure that only 5% of true null hypotheses are incorrectly rejected. However, this doesn't mean that we don't care about type II errors. In fact, from the researcher's perspective, the error of failing to reject the null when it is actually false is an extremely annoying one. With that in mind, a secondary goal of hypothesis testing is to try to minimise β, the type II error rate, although we don't usually talk in terms of minimising type II errors. Instead, we talk about maximising the power of the test. Since power is defined as $1 - \beta$, this is the same thing.

9.8.1 The power function

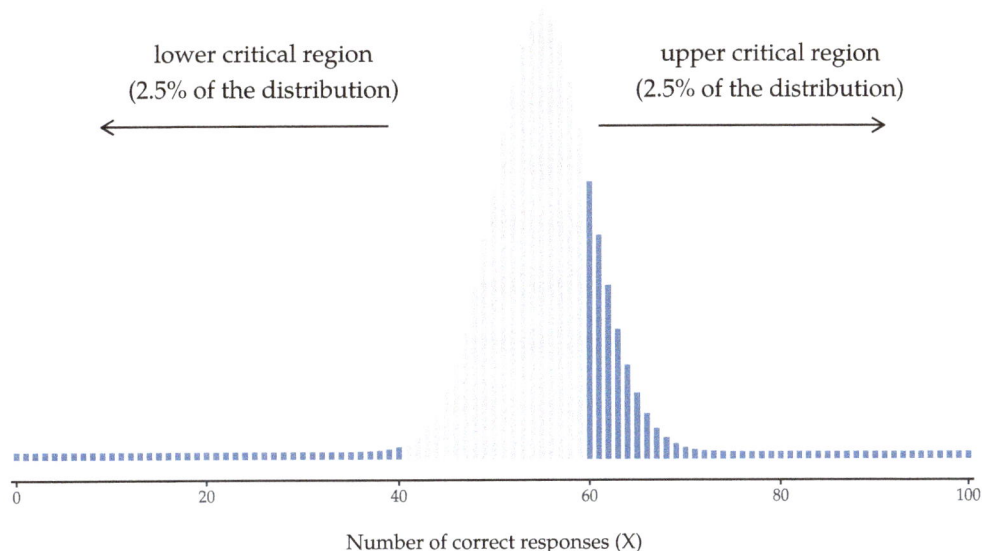

Figure 9.5: Sampling distribution under the alternative hypothesis for a population parameter value of $\theta = 0.55$. A reasonable proportion of the distribution lies in the rejection region

Let's take a moment to think about what a type II error actually is. A type II error occurs when the alternative hypothesis is true, but we are nevertheless unable to reject the null hypothesis. Ideally, we'd be able to calculate a single number β that tells us the type II error rate, in the same way that we can set $\alpha = .05$ for the type I error rate. Unfortunately, this is a lot trickier to do. To see this, notice that in my ESP study the alternative hypothesis actually corresponds to lots of possible values of θ. In fact, the alternative hypothesis corresponds to every value of θ except 0.5. Let's suppose that the true probability of someone choosing the correct response is 55% (i.e., $\theta = .55$). If so, then the true sampling distribution for X is not the same one that the null hypothesis predicts, as the most likely value for X is now 55 out of 100. Not only that, the whole sampling distribution has now shifted, as shown in Figure 9.5. The critical regions, of course, do not change. By definition the critical regions are based on what the null hypothesis predicts, but when the null hypothesis is wrong, a much larger proportion of the sampling distribution distribution falls in the critical region. The probability of rejecting the null hypothesis is larger when the null hypothesis is actually false! However $\theta = .55$ is not the only possibility consistent with the alternative hypothesis. Let's instead suppose that the true value of θ is actually 0.7. What happens to the sampling distribution when this occurs? The answer, shown in Figure 9.6, is that almost the entirety of the sampling distribution has now moved into the critical region. Therefore, if $\theta = 0.7$, the probability of us correctly rejecting the null hypothesis (i.e., the power of the test) is much larger than if $\theta = 0.55$. In short, while $\theta = .55$ and $\theta = .70$ are both part of the alternative hypothesis, the type II error rate is different.

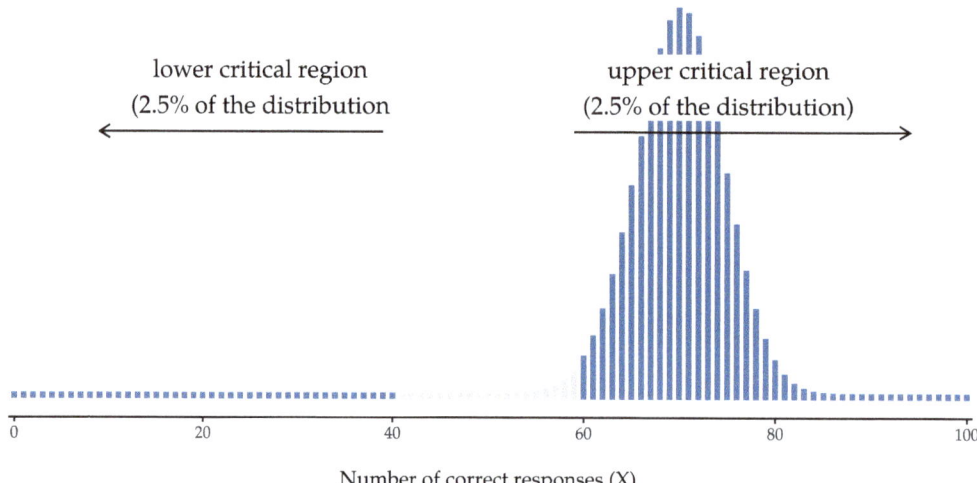

Figure 9.6: Sampling distribution under the *alternative* hypothesis for a population parameter value of $\theta = 0.70$. Almost all of the distribution lies in the rejection region

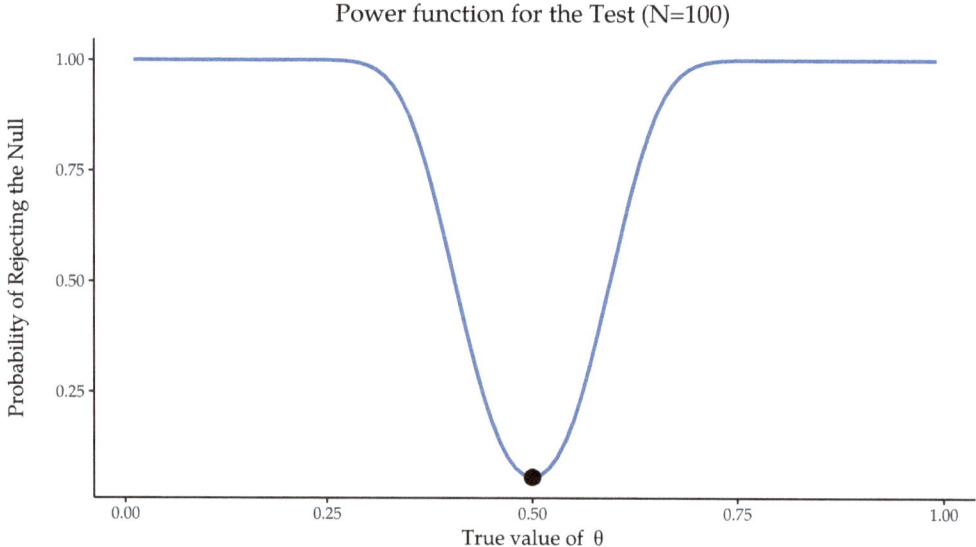

Figure 9.7: The probability that we will reject the null hypothesis, plotted as a function of the true value of θ. Obviously, the test is more powerful (greater chance of correct rejection) if the true value of θ is very different from the value that the null hypothesis specifies (i.e., $\theta = .5$). Notice that when θ actually is equal to .5 (plotted as a black dot), the null hypothesis is in fact true and rejecting the null hypothesis in this instance would be a type I error

What all this means is that the power of a test (i.e., $1-\beta$) depends on the true value of θ. To illustrate this, I've calculated the expected probability of rejecting the null hypothesis for all values of θ, and plotted it in Figure 9.7. This plot describes what is usually called the power function of the test. It's a nice summary of how good the test is, because it actually tells you the power $(1-\beta)$ for all possible values of θ. As you can see, when the true value of θ is very close to 0.5, the power of the test drops very sharply, but when it is further away, the power is large.

9.8.2 The power function

> *Since all models are wrong the scientist must be alert to what is importantly wrong. It is inappropriate to be concerned with mice when there are tigers abroad.*
> – George Box (Box, 1976, p. 792).

The plot shown in Figure 9.7 captures a fairly basic point about hypothesis testing. If the true state of the world is very different from what the null hypothesis predicts then your power will be very high, but if the true state of the world is similar to the null (but not identical) then the power of the test is going to be very low. Therefore, it's useful to be able to have some way of quantifying how "similar" the true state of the world is to the null hypothesis. A statistic that does this is called a measure of **effect size** (e.g., Cohen (1988); Ellis (2010)). Effect size is defined slightly differently in different contexts (and so this section just talks in general terms) but the qualitative idea that it tries to capture is always the same (see e.g. Table 9.6). How big is the difference between the *true* population parameters and the parameter values that are assumed by the null hypothesis? In our ESP example, if we let $\theta_0 = 0.5$ denote the value assumed by the null hypothesis and let θ denote the true value, then a simple measure of effect size could be something like the difference between the true value and null (i.e., $\theta - \theta_0$), or possibly just the magnitude of this difference, $abs(\theta - \theta_0)$.

Table 9.6: A crude guide to understanding the relationship between statistical significance and effect sizes. Basically, if you don't have a significant result then the effect size is pretty meaningless because you don't have any evidence that it's even real. On the other hand, if you do have a significant effect but your effect size is small then there's a pretty good chance that your result (although real) isn't all that interesting. It does depend a lot on what exactly you're studying; small effects can be of massive practical importance in some situations. So don't take this table too seriously – it's a rough guide at best

	big effect size	small effect size
significant result	difference is real, and of practical importance	difference is real, but might not be interesting
non-significant result	no effect observed	no effect observed

Why calculate effect size? Let's assume that you've run your experiment, collected the data, and gotten a significant effect when you ran your hypothesis test. Isn't it enough just to say that you've gotten a significant effect? Surely that's the point of hypothesis testing? Well, sort of. Yes, the point of doing a hypothesis test is to try to demonstrate that the null hypothesis is wrong, but that's hardly the only thing we're interested in. If the null hypothesis claimed that $\theta = .5$ and we show that it's wrong, we've only really told half of the story. Rejecting the null hypothesis implies that we believe that $\theta \neq .5$, but there's a big difference between $\theta = .51$ and $\theta = .8$. If we find that $\theta = .8$, then not only have we found that the null hypothesis is wrong, it appears to be very wrong. On the other hand, suppose we've successfully rejected the null hypothesis, but it looks like the true value of θ is only .51 (this would only be possible with a very large study). Sure, the null hypothesis is wrong but it's not at all clear that we actually care because the effect size is so small. In the context of my ESP study we might still care since any demonstration of real psychic powers would actually be pretty cool,[74] but in other contexts a 1% difference usually isn't very interesting, even if it is a real difference. For instance, suppose we're looking at differences in high school exam scores between males and females and it turns out that the female scores are 1% higher on average than the males. If I've got data from thousands of students then this difference will almost certainly be statistically significant, but regardless of how small the p-value is it's just not very interesting. You'd hardly want to go around proclaiming a crisis in boys education on the basis of such a tiny difference would you? It's for this reason that it is becoming more standard (slowly, but surely) to report some kind of standard measure of effect size along with the the results of the hypothesis test. The hypothesis test itself tells you whether you should believe that the effect you have observed is real (i.e., not just due to chance), whereas the effect size tells you whether or not you should care.

9.8.3 Increasing the power of your study

Not surprisingly, scientists are fairly obsessed with maximising the power of their experiments. We want our experiments to work and so we want to maximise the chance of rejecting the null hypothesis if it is false. As we've seen, one factor that influences power is the effect size. So the first thing you can do to increase your power is to increase the effect size. In practice, what this means is that you want to design your study in such a way that the effect size gets magnified. For instance, in my ESP study I might believe that psychic powers work best in a quiet, darkened room with fewer distractions to cloud the mind. Therefore I would try to conduct my experiments in just such an environment. If I can strengthen people's ESP abilities somehow then the true value of θ will go up[75] and therefore my effect size will be larger. In short, clever experimental design is one way to boost power, because it can alter the effect size.

Unfortunately, it's often the case that even with the best of experimental designs you may have only a small effect. Perhaps, for example, ESP really does exist but even under the best of conditions it's very, very weak. Under those circumstances your best bet for increasing power is to increase the sample size. In general, the more observations that you have, the more likely it is that you can discriminate between two hypotheses. If I ran my ESP experiment with 10 participants and 7 of them correctly guessed the colour of the hidden card you wouldn't be terribly impressed. But if I ran it with 10,000 participants, and 7,000 of them got the answer right, you would be much more likely to think I had discovered something. In other words, power increases with the sample size. This is illustrated in Figure 9.8, which shows the power of the test for a true parameter of $\theta = 0.7$ for all sample sizes N from 1 to 100, where I'm assuming that the null hypothesis predicts that $\theta_0 = 0.5$.

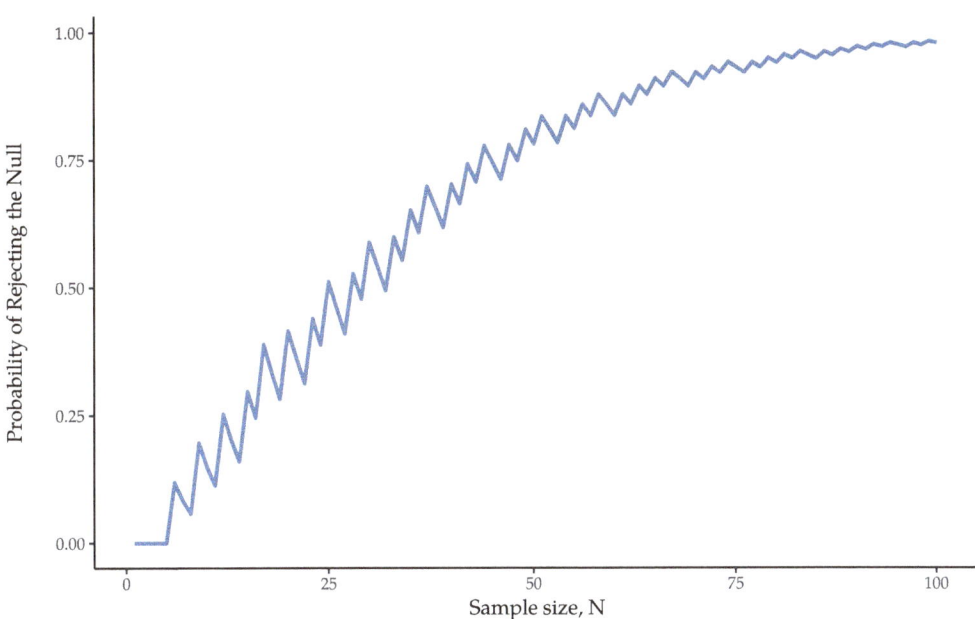

Figure 9.8: The power of our test plotted as a function of the sample size N. In this case, the true value of θ is 0.7 but the null hypothesis is that $\theta = 0.5$. Overall, larger N means greater power. (The small zig-zags in this function occur because of some odd interactions between θ, α and the fact that the binomial distribution is discrete, it does not matter for any serious purpose)

Because power is important, whenever you're contemplating running an experiment it would be pretty useful to know how much power you're likely to have. It's never possible to know for sure since you can't possibly know what your real effect size is. However, it's often (well, sometimes) possible to guess how big it should be. If so, you can guess what sample size you need! This idea is called **power analysis**, and if it's feasible to do it then it's very helpful. It can tell you something about whether you have enough time or money to be able to run the experiment successfully. It's increasingly common to see people arguing that power analysis should be a required part of experimental design, so it's worth knowing about. I don't discuss power analysis in this book, however. This is partly for a boring reason and partly for a substantive one. The boring reason is that I haven't had time to write about power analysis yet. The substantive one is that I'm still a little suspicious of power analysis. Speaking as a researcher, I have very rarely found myself in a position to be able to do one. It's either the case that (a) my experiment is a bit non-standard and I don't know how to define effect size properly, or (b) I literally have so little idea about what the effect size will be that I wouldn't know how to interpret the answers. Not only that, after extensive conversations with someone who does stats consulting for a living (my wife, as it happens), I can't help but notice that in practice the only time anyone ever asks her for a power analysis is when she's helping someone write a grant application. In other words, the only time any scientist ever seems to want a power analysis in real life is when they're being forced to do it by bureaucratic process. It's not part of anyone's day to day work. In short, I've always been of the view that whilst power is an important concept, power analysis is not as useful as people make it sound, except in the rare cases where (a) someone has figured out how to calculate power for your actual experimental design and (b) you have a pretty good idea what the effect size is likely to be.[76] Maybe other people have had better experiences than me, but I've personally never been in a situation where both (a) and (b) were true. Maybe I'll be convinced otherwise in the future, and probably a future version of this book would include a more detailed discussion of power analysis, but for now this is about as much as I'm comfortable saying about the topic.

9.9 Some issues to consider

What I've described to you in this chapter is the orthodox framework for null hypothesis significance testing (NHST). Understanding how NHST works is an absolute necessity because it has been the dominant approach to inferential statistics ever since it came to prominence in the early 20th century. It's what the vast majority of working scientists rely on for their data analysis, so even if you hate it you need to know it. However, the approach is not without problems. There are a number of quirks in the framework, historical oddities in how it came to be, theoretical disputes over whether or not the framework is right, and a lot of practical traps for the unwary. I'm not going to go into a lot of detail on this topic, but I think it's worth briefly discussing a few of these issues.

9.9.1 Neyman versus Fisher

The first thing you should be aware of is that orthodox NHST is actually a mash-up of two rather different approaches to hypothesis testing, one proposed by Sir Ronald Fisher and the other proposed by Jerzy Neyman (see Lehmann (2011) for a historical summary). The history is messy because Fisher and Neyman were real people whose opinions changed over time, and at no point did either of them offer "the definitive statement" of how we should interpret their work many decades later. That said, here's a quick summary of what I take these two approaches to be.

First, let's talk about Fisher's approach. As far as I can tell, Fisher assumed that you only had the one hypothesis (the null) and that what you want to do is find out if the null hypothesis is inconsistent with the data. From his perspective, what you should do is check to see if the data are "sufficiently unlikely" according to the null. In fact, if you remember back to our earlier discussion, that's how Fisher defines the p-value. According to Fisher, if the null hypothesis provided a very poor account of the data then you could safely reject it. But, since you don't have any other hypotheses to compare it to, there's no way of "accepting the alternative" because you don't necessarily have an explicitly stated alternative. That's more or less all there is to it.

In contrast, Neyman thought that the point of hypothesis testing was as a guide to action and his approach was somewhat more formal than Fisher's. His view was that there are multiple things that you could do (accept the null or accept the alternative) and the point of the test was to tell you which one the data support. From this perspective, it is critical to specify your alternative hypothesis properly. If you don't know what the alternative hypothesis is, then you don't know how powerful the test is, or even which action makes sense. His framework genuinely requires a competition between different hypotheses. For Neyman, the *p*-value didn't directly measure the probability of the data (or data more extreme) under the null, it was more of an abstract description about which "possible tests" were telling you to accept the null, and which "possible tests" were telling you to accept the alternative.

As you can see, what we have today is an odd mishmash of the two. We talk about having both a null hypothesis and an alternative (Neyman), but usually[77] define the *p*-value in terms of extreme data (Fisher), but we still have α values (Neyman). Some of the statistical tests have explicitly specified alternatives (Neyman) but others are quite vague about it (Fisher). And, according to some people at least, we're not allowed to talk about accepting the alternative (Fisher). It's a mess, but I hope this at least explains why it's a mess.

9.9.2 Bayesians versus frequentists

Earlier on in this chapter I was quite emphatic about the fact that you *cannot* interpret the *p*-value as the probability that the null hypothesis is true. NHST is fundamentally a frequentist tool (see Chapter 7) and as such it does not allow you to assign probabilities to hypotheses. The null hypothesis is either true or it is not. The Bayesian approach to statistics interprets probability as a degree of belief, so it's totally okay to say that there is a 10% chance that the null hypothesis is true. That's just a reflection of the degree of confidence that you have in this hypothesis. You aren't allowed to do this within the frequentist approach. Remember, if you're a frequentist, a probability can only be defined in terms of what happens after a large number of independent replications (i.e., a long run frequency). If this is your interpretation of probability, talking about the "probability" that the null hypothesis is true is complete gibberish: a null hypothesis is either true or it is false. There's no way you can talk about a long run frequency for this statement. To talk about "the probability of the null hypothesis" is as meaningless as "the colour of freedom". It doesn't have one!

Most importantly, this isn't a purely ideological matter. If you decide that you are a Bayesian and that you're okay with making probability statements about hypotheses, you have to follow the Bayesian rules for calculating those probabilities. I'll talk more about this in Chapter 16, but for now what I want to point out to you is the *p*-value is a terrible approximation to the probability that H_0 is true. If what you want to know is the probability of the null, then the *p*-value is not what you're looking for!

9.9.3 Traps

As you can see, the theory behind hypothesis testing is a mess, and even now there are arguments in statistics about how it "should" work. However, disagreements among statisticians are not our real concern here. Our real concern is practical data analysis. And while the "orthodox" approach to null hypothesis significance testing has many drawbacks, even an unrepentant Bayesian like myself would agree that they can be useful if used responsibly. Most of the time they give sensible answers and you can use them to learn interesting things. Setting aside the various ideologies and historical confusions that we've discussed, the fact remains that the biggest danger in all of statistics is *thoughtlessness*. I don't mean stupidity, I literally mean thoughtlessness. The rush to interpret a result without spending time thinking through what each test actually says about the data, and checking whether that's consistent with how you've interpreted it. That's where the biggest trap lies.

To give an example of this, consider the following example (see Gelman & Stern (2006)). Suppose I'm running my ESP study and I've decided to analyse the data separately for the male participants and the female participants. Of the male participants, 33 out of 50 guessed the colour of the card correctly. This is a significant effect ($p = .03$). Of the female participants, 29 out of 50 guessed correctly. This is not a significant effect ($p = .32$). Upon observing this, it is extremely tempting for people to start wondering why there is a difference between males and females in terms of their psychic abilities. However, this is wrong. If you think about it, we haven't actually run a test that explicitly compares males to females. All we have done is compare males to chance (binomial test was significant) and compared females to chance (binomial test was non significant). If we want to argue that there is a real difference between the males and the females, we should probably run a test of the null hypothesis that there is no difference! We can do that using a different hypothesis test,[78] but when we do that it turns out that we have no evidence that males and females are significantly different ($p = .54$). Now do you think that there's anything fundamentally different between the two groups? Of course not. What's happened here is that the data from both groups (male and female) are pretty borderline. By pure chance one of them happened to end up on the magic side of the $p = .05$ line, and the other one didn't. That doesn't actually imply that males and females are different. This mistake is so common that you should always be wary of it. The difference between significant and not-significant is not evidence of a real difference. If you want to say that there's a difference between two groups, then you have to test for that difference!

The example above is just that, an example. I've singled it out because it's such a common one, but the bigger picture is that data analysis can be tricky to get right. Think about what it is you want to test, why you want to test it, and whether or not the answers that your test gives could possibly make any sense in the real world.

9.10 Summary

Null hypothesis testing is one of the most ubiquitous elements to statistical theory. The vast majority of scientific papers report the results of some hypothesis test or another. As a consequence it is almost impossible to get by in science without having at least a cursory understanding of what a p-value means, making this one of the most important chapters in the book. As usual, I'll end the chapter with a quick recap of the key ideas that we've talked about:

- A menagerie of hypotheses. Research hypotheses and statistical hypotheses. Null and alternative hypotheses.
- Two types of errors. Type I and type II.
- Test statistics and sampling distributions.
- Hypothesis testing for Making decisions.
- The p-value of a test. p-values as "soft" decisions.
- Reporting the results of a hypothesis test.
- Running the hypothesis test in practice.
- Effect size, sample size and power.

- Some issues to consider regarding hypothesis testing.

Later in the book, in Chapter 16, I'll revisit the theory of null hypothesis tests from a Bayesian perspective and introduce a number of new tools that you can use if you aren't particularly fond of the orthodox approach. But for now, though, we're done with the abstract statistical theory, and we can start discussing specific data analysis tools.

Part V

Statistical tools

Chapter 10

Categorical data analysis

Now that we've covered the basic theory behind hypothesis testing it's time to start looking at specific tests that are commonly used in psychology. So where should we start? Not every textbook agrees on where to start, but I'm going to start with χ^2 tests (this chapter, pronounced "chi-square"[79] and t-tests in Chapter 11). Both of these tools are very frequently used in scientific practice, and whilst they're not as powerful as "regression" and "analysis of variance" which we cover in later chapters, they're much easier to understand.

The term "categorical data" is just another name for "nominal scale data". It's nothing that we haven't already discussed, it's just that in the context of data analysis people tend to use the term "categorical data" rather than "nominal scale data". I don't know why. In any case, **categorical data analysis** refers to a collection of tools that you can use when your data are nominal scale. However, there are a lot of different tools that can be used for categorical data analysis, and this chapter covers only a few of the more common ones.

10.1 The χ^2 (chi-square) goodness-of-fit test

The χ^2 goodness-of-fit test is one of the oldest hypothesis tests around. It was invented by Karl Pearson around the turn of the century (Pearson, 1900), with some corrections made later by Sir Ronald Fisher (Fisher, 1922a). It tests whether an observed frequency distribution of a nominal variable matches an expected frequency distribution. For example, suppose a group of patients has been undergoing an experimental treatment and have had their health assessed to see whether their condition has improved, stayed the same or worsened. A goodness-of-fit test could be used to determine whether the numbers in each category – improved, no change, worsened – match the numbers that would be expected given the standard treatment option. Let's think about this some more, with some psychology.

10.1.1 The cards data

Over the years there have been many studies showing that humans find it difficult to simulate randomness. Try as we might to "act" random, we think in terms of patterns and structure and so, when asked to "do something at random", what people actually do is anything but random. As a consequence, the study of human randomness (or non-randomness, as the case may be) opens up a lot of deep psychological questions about how we think about the world.

With this in mind, let's consider a very simple study. Suppose I asked people to imagine a shuffled deck of cards, and mentally pick one card from this imaginary deck "at random". After they've chosen one card I ask them to mentally select a second one. For both choices what we're going to look at is the suit (hearts, clubs, spades or diamonds) that people chose. After asking, say, $N = 200$ people to do this, I'd like to look at the data and figure out whether or not the cards that people pretended to select were really random. The data are contained in the *randomness.csv* file in which, when you open it up in jamovi and take a look at the spreadsheet view, you will see three variables. These are: an id variable that assigns a unique identifier to each participant, and the two variables choice_1 and choice_2 that indicate the card suits that people chose.

For the moment, let's just focus on the first choice that people made. We'll use the Frequency tables option under 'Exploration' – 'Descriptives' to count the number of times that we observed people choosing each suit. This is what we get: (Table 10.1).

Table 10.1: Number of times each suit was chosen

clubs	diamonds	hearts	spades
35	51	64	50

That little frequency table is quite helpful. Looking at it, there's a bit of a hint that people might be more likely to select hearts than clubs, but it's not completely obvious just from looking at it whether that's really true, or if this is just due to chance. So we'll probably have to do some kind of statistical analysis to find out, which is what I'm going to talk about in the next section.

Excellent. From this point on, we'll treat this table as the data that we're looking to analyse. However, since I'm going to have to talk about this data in mathematical terms (sorry!) it might be a good idea to be clear about what the notation is. In mathematical notation, we shorten the human-readable word "observed" to the letter O, and we use subscripts to denote the position of the observation. So the second observation in our table is written as O_2 in maths. The relationship between the English descriptions and the mathematical symbols are illustrated in Table 10.2.

Hopefully that's pretty clear. It's also worth noting that mathematicians prefer to talk about general rather than specific things, so you'll also see the notation O_i, which refers to the number of observations that fall within the i-th category (where i could be 1, 2, 3 or 4). Finally, if we want to refer to the set of all observed frequencies, statisticians group all observed values into a vector,[80] which I'll refer to as O.

$$O = (O_1, O_2, O_3, O_4)$$

Table 10.2: Relationship between English descriptions and mathematical symbols

label	index, i	math. symbol	the value
clubs, ♣	1	O_1	35
diamonds, ♢	2	O_2	51
hearts, ♡	3	O_3	64
spades, ♠	4	O_4	50

Again, this is nothing new or interesting. It's just notation. If I say that $O = (35, 51, 64, 50)$ all I'm doing is describing the table of observed frequencies (i.e., observed), but I'm referring to it using mathematical notation.

10.1.2 The null hypothesis and the alternative hypothesis

As the last section indicated, our research hypothesis is that "people don't choose cards randomly". What we're going to want to do now is translate this into some statistical hypotheses and then construct a statistical test of those hypotheses. The test that I'm going to describe to you is **Pearson's** χ^2 (chi-square) goodness-of-fit test, and as is so often the case we have to begin by carefully constructing our null hypothesis. In this case, it's pretty easy. First, let's state the null hypothesis in words:

$$H_0 : \text{All four suits are chosen with equal probability}$$

Now, let's say this in a mathematical way. To do this, let's use the notation P_j to refer to the true probability that the j-th suit is chosen. If the null hypothesis is true, then each of the four suits has a 25% chance of being selected. In other words, our null hypothesis claims that $P_1 = .25$, $P_2 = .25$, $P_3 = .25$ and finally that $P_4 = .25$. However, in the same way that we can group our observed frequencies into a vector O that summarises the entire data set, we can use P to refer to the probabilities that correspond to our null hypothesis. So if I let the vector $P = (P_1, P_2, P_3, P_4)$ refer to the collection of probabilities that describe our null hypothesis, then we have:

$$H_0 : P = (.25, .25, .25, .25)$$

In this instance, our null hypothesis corresponds to a vector of probabilities P in which all of the probabilities are equal to one another. But this doesn't have to be the case: if the experimental task was for people to imagine they were drawing from a deck that had twice as many clubs as any other suit, then the null hypothesis would be $P = (.4, .2, .2, .2)$. As long as the probabilities are all positive numbers, and they all sum to 1, then it's a perfectly legitimate choice for the null hypothesis. However, the typical use of the goodness-of-fit test is with a null hypothesis that all the categories are equally likely, so we'll stick to that for our example.

What about our alternative hypothesis, H_1? All we're really interested in is demonstrating that the probabilities involved aren't all identical (that is, people's choices weren't completely random). As a consequence, the "human friendly" versions of our hypotheses look like this:

$$H_0 : \text{All four suits are chosen with equal probability}$$
$$H_1 : \text{At least one of the suit-choice probabilities isn't 0.25}$$

...and the "mathematician friendly" version is:

$$H_0 : P = (.25, .25, .25, .25)$$
$$H_1 : P \neq (.25, .25, .25, .25)$$

10.1.3 The "goodness-of-fit" test statistic

At this point, we have our observed frequencies O and a collection of probabilities P corresponding to the null hypothesis that we want to test. What we now want to do is construct a test of the null hypothesis. As always, if we want to test H_0 against H_1, we're going to need a test statistic. The basic trick that a goodness-of-fit test uses is to construct a test statistic that measures how "close" the data are to the null hypothesis. If the data don't resemble what you'd "expect" to see if the null hypothesis were true, then it probably isn't true. Okay, if the null hypothesis were true, what would we expect to see? Or, to use the correct terminology, what are the **expected frequencies**. There are $N = 200$ observations, and (if the null is true) the probability of any one of them choosing a heart is $P_3 = .25$, so I guess we're expecting $200 \times .25 = 50$ hearts, right? Or, more specifically, if we let E_i refer to "the number of category i responses that we're expecting if the null is true", then:

$$E_i = N \times P_i$$

This is pretty easy to calculate. If there are 200 observations that can fall into four categories, and we think that all four categories are equally likely, then on average we'd expect to see 50 observations in each category, right?

Now, how do we translate this into a test statistic? Clearly, what we want to do is compare the expected number of observations in each category (E_i) with the observed number of observations in that category (O_i). And on the basis of this comparison we ought to be able to come up with a good test statistic. To start with, let's calculate the difference between what the null hypothesis expected us to find and what we actually did find. That is, we calculate the "observed minus expected" difference score, $O_i - E_i$. This is illustrated in Table 10.3.

So, based on our calculations, it's clear that people chose more hearts and fewer clubs than the null hypothesis predicted. However, a moment's thought suggests that these raw differences aren't quite what we're looking for. Intuitively, it feels like it's just as bad when the null hypothesis predicts too few observations (which is what happened with hearts) as it is when it predicts too many (which is what happened with clubs). So it's a bit weird that we have a negative number for clubs and a positive number for hearts. One easy way to fix this is to square everything, so that we now calculate the squared differences, $(E_i - O_i)^2$. As before, we can do this by hand (Table 10.4).

Table 10.3: Expected and observed frequencies

	♣	♦	♡	♠
expected frequency E_i	50	50	50	50
observed frequency O_i	35	51	64	50
difference score $O_i - E_i$	-15	1	14	0

Table 10.4: Squaring the difference scores

♣	♦	♡	♠
225	1	196	0

Now we're making progress. What we've got now is a collection of numbers that are big whenever the null hypothesis makes a bad prediction (clubs and hearts), but are small whenever it makes a good one (diamonds and spades). Next, for some technical reasons that I'll explain in a moment, let's also divide all these numbers by the expected frequency E_i, so we're actually calculating $\frac{(E_i - O_i)^2}{E_i}$. Since $E_i = 50$ for all categories in our example, it's not a very interesting calculation, but let's do it anyway (Table 10.5).

Table 10.5: Dividing the squared difference scores by the expected frequency to provide an "error" score

♣	♦	♡	♠
4.50	0.02	3.92	0.00

In effect, what we've got here are four different "error" scores, each one telling us how big a "mistake" the null hypothesis made when we tried to use it to predict our observed frequencies. So, in order to convert this into a useful test statistic, one thing we could do is just add these numbers up. The result is called the **goodness-of-fit** statistic, conventionally referred to either as χ^2 (chi-square) or GOF. We can calculate it as:

$$\sum ((observed - expected)^2 / expected)$$

This gives us a value of 8.44.

[Additional technical detail[81]]

As we've seen from our calculations, in our cards data set we've got a value of $\chi^2 = 8.44$. So now the question becomes is this a big enough value to reject the null?

10.1.4 The sampling distribution of the GOF statistic

To determine whether or not a particular value of χ^2 is large enough to justify rejecting the null hypothesis, we're going to need to figure out what the sampling distribution for χ^2 would be if the null hypothesis were true. So that's what I'm going to do in this section. I'll show you in a fair amount of detail how this sampling distribution is constructed, and then, in the next section, use it to build up a hypothesis test. If you want to cut to the chase and are willing to take it on faith that the sampling distribution is a χ^2 (chi-square) distribution with $k-1$ degrees of freedom, you can skip the rest of this section. However, if you want to understand *why* the goodness-of-fit test works the way it does, read on.

Okay, let's suppose that the null hypothesis is actually true. If so, then the true probability that an observation falls in the i-th category is P_i. After all, that's pretty much the definition of our null hypothesis. Let's think about what this actually means. This is kind of like saying that "nature" makes the decision about whether or not the observation ends up in category i by flipping a weighted coin (i.e., one where the probability of getting a head is P_j). And therefore we can think of our observed frequency O_i by imagining that nature flipped N of these coins (one for each observation in the data set), and exactly O_i of them came up heads. Obviously, this is a pretty weird way to think about the experiment. But what it does (I hope) is remind you that we've actually seen this scenario before. It's exactly the same set up that gave rise to Section 7.4 in Chapter 7. In other words, if the null hypothesis is true, then it follows that our observed frequencies were generated by sampling from a binomial distribution:

$$O_i \sim Binomial(P_i, N)$$

Now, if you remember from our discussion of Section 8.3.3 the binomial distribution starts to look pretty much identical to the normal distribution, especially when N is large and when P_i isn't too close to 0 or 1. In other words as long as N_i^P is large enough. Or, to put it another way, when the expected frequency E_i is large enough then the theoretical distribution of O_i is approximately normal. Better yet, if O_i is normally distributed, then so is $(O_i - E_i)/\sqrt{(E_i)}$. Since E_i is a fixed value, subtracting off E_i and dividing by $\sqrt{(E_i)}$ changes the mean and standard deviation of the normal distribution but that's all it does. Okay, so now let's have a look at what our goodness-of-fit statistic actually is. What we're doing is taking a bunch of things that are normally-distributed, squaring them, and adding them up. Wait. We've seen that before too! As we discussed in the section on Section 7.6, when you take a bunch of things that have a standard normal distribution (i.e., mean 0 and standard deviation 1), square them and then add them up, the resulting quantity has a chi-square distribution. So now we know that the null hypothesis predicts that the sampling distribution of the goodness-of-fit statistic is a chi-square distribution. Cool.

There's one last detail to talk about, namely the degrees of freedom. If you remember back to Section 7.6, I said that if the number of things you're adding up is k, then the degrees of freedom for the resulting chi-square distribution is k. Yet, what I said at the start of this section is that the actual degrees of freedom for the chi-square goodness-of-fit test is $k-1$. What's up with that? The answer here is that what we're supposed to be looking at is the number of genuinely independent things that are getting added together. And, as I'll go on to talk about in the next section, even though there are k

things that we're adding only $k-1$ of them are truly independent, and so the degrees of freedom is actually only $k-1$. That's the topic of the next section.[82]

10.1.5 Degrees of freedom

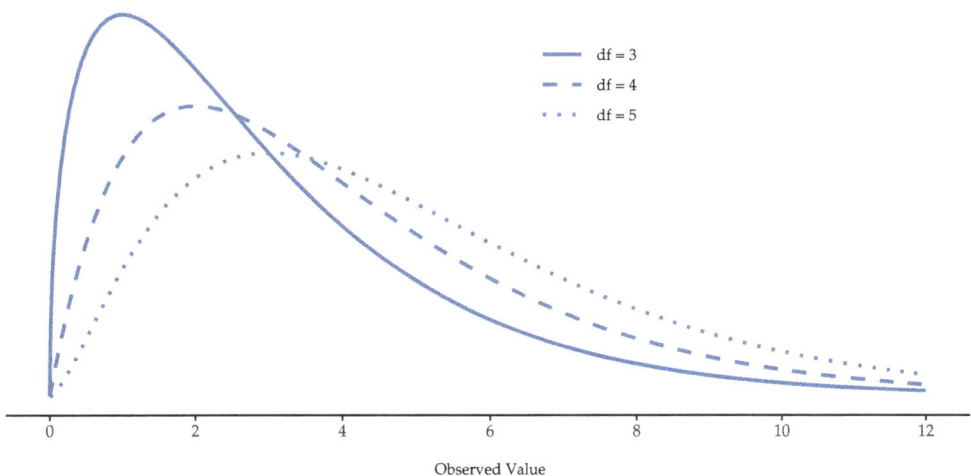

Figure 10.1: χ^2 (chi-square) distributions with different values for the "degrees of freedom"

When I introduced the chi-square distribution in Section 7.6, I was a bit vague about what **"degrees of freedom"** actually means. Obviously, it matters. Looking at Figure 10.1, you can see that if we change the degrees of freedom then the chi-square distribution changes shape quite substantially. But what exactly is it? Again, when I introduced the distribution and explained its relationship to the normal distribution, I did offer an answer: it's the number of "normally distributed variables" that I'm squaring and adding together. But, for most people, that's kind of abstract and not entirely helpful. What we really need to do is try to understand degrees of freedom in terms of our data. So here goes.

The basic idea behind degrees of freedom is quite simple. You calculate it by counting up the number of distinct "quantities" that are used to describe your data and then subtracting off all of the "constraints" that those data must satisfy.[83] This is a bit vague, so let's use our cards data as a concrete example. We describe our data using four numbers, O_1, O_2, O_3 and O_4 corresponding to the observed frequencies of the four different categories (hearts, clubs, diamonds, spades). These four numbers are the random outcomes of our experiment. But my experiment actually has a fixed constraint built into it: the sample size N.[84] That is, if we know.

How many people chose hearts, how many chose diamonds and how many chose clubs, then we'd be able to figure out exactly how many chose spades. In other words, although our data are described using four numbers, they only actually correspond to

$4 - 1 = 3$ degrees of freedom. A slightly different way of thinking about it is to notice that there are four probabilities that we're interested in (again, corresponding to the four different categories), but these probabilities must sum to one, which imposes a constraint. Therefore the degrees of freedom is $4 - 1 = 3$. Regardless of whether you want to think about it in terms of the observed frequencies or in terms of the probabilities, the answer is the same. In general, when running the χ^2(chi-square) goodness-of-fit test for an experiment involving k groups, then the degrees of freedom will be $k - 1$.

10.1.6 Testing the null hypothesis

The final step in the process of constructing our hypothesis test is to figure out what the rejection region is. That is, what values of χ^2 would lead us to reject the null hypothesis. As we saw earlier, large values of χ^2 imply that the null hypothesis has done a poor job of predicting the data from our experiment, whereas small values of χ^2 imply that it's actually done pretty well. Therefore, a pretty sensible strategy would be to say there is some critical value such that if χ^2 is bigger than the critical value we reject the null, but if χ^2 is smaller than this value we retain the null. In other words, to use the language we introduced in Chapter 9 the chi-square goodness-of-fit test is always a **one-sided test**. Right, so all we have to do is figure out what this critical value is. And it's pretty straightforward. If we want our test to have significance level of $\alpha = .05$ (that is, we are willing to tolerate a type I error rate of 5), then we have to choose our critical value so that there is only a 5% chance that χ^2 could get to be that big if the null hypothesis is true. This is illustrated in Figure 10.2.

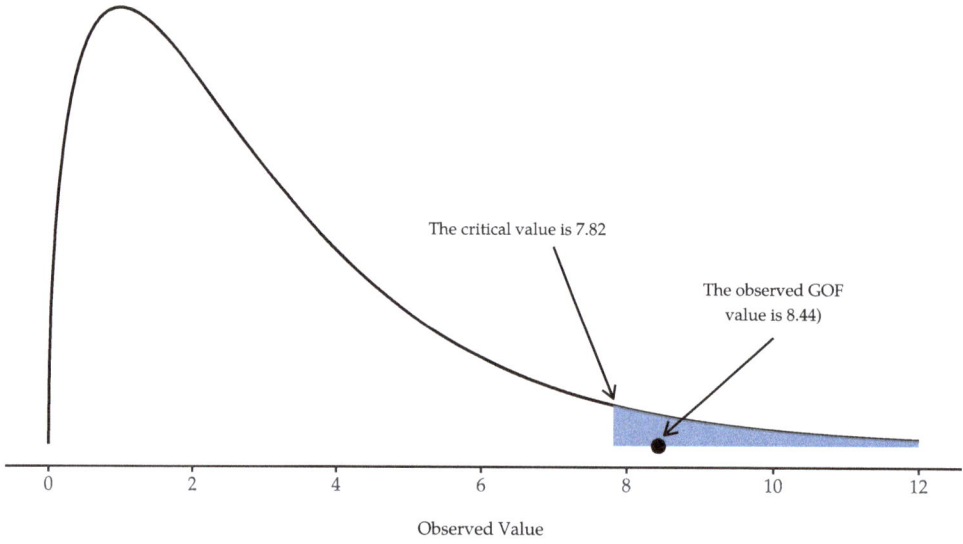

Figure 10.2: Illustration of how the hypothesis testing works for the χ^2 (chi-square) goodness of-fit test

Ah but, I hear you ask, how do I find the critical value of a chi-square distribution with $k-1$ degrees of freedom? Many many years ago when I first took a psychology statistics class we used to look up these critical values in a book of critical value tables, like the one in Table 10.6. Looking at this Figure, we can see that the critical value for a χ^2 distribution with 3 degrees of freedom and $p = 0.05$ is 7.815.

Table 10.6: Table of critical values for the chi-square distribution

Degrees of freedom	Probability							
	0.95	0.9	0.7	0.5	0.1	0.05	0.01	0.001
1	0.004	0.016	0.148	0.455	2.706	3.841	6.635	10.828
2	0.103	0.211	0.713	1.386	4.605	5.991	9.210	13.816
3	0.352	0.584	1.424	2.366	6.251	7.815	11.345	16.266
4	0.711	1.064	2.195	3.357	7.779	9.488	13.277	18.467
5	1.145	1.610	3.000	4.351	9.236	11.070	15.086	20.515
6	1.635	2.204	3.828	5.348	10.645	12.592	16.812	22.458
7	2.167	2.833	4.671	6.346	12.017	14.067	18.475	24.322
8	2.733	3.490	5.527	7.344	13.362	15.507	20.090	26.124
9	3.325	4.168	6.393	8.343	14.684	16.919	21.666	27.877
10	3.940	4.865	7.267	9.342	15.987	18.307	23.209	29.588
	Non-significant					Significant		

So, if our calculated χ^2 statistic is bigger than the critical value of 7.815, then we can reject the null hypothesis (remember that the null hypothesis, H_0, is that all four suits are chosen with equal probability). Since we actually already calculated that before (i.e., $\chi^2 = 8.44$) we can reject the null hypothesis. And that's it, basically. You now know "Pearson's χ^2 test for the goodness-of-fit". Lucky you.

10.1.7 Doing the test in jamovi

Not surprisingly, jamovi provides an analysis that will do these calculations for you. Let's use the *Randomness.omv* file. From the main 'Analyses' toolbar select 'Frequencies' – 'One Sample Proportion Tests' – 'N Outcomes'. Then in the analysis window that appears move the variable you want to analyse (choice 1 across into the 'Variable' box. Also, click on the 'Expected counts' check box so that these are shown on the results table. When you have done all this, you should see the analysis results in jamovi as in Figure 10.3. No surprise then that jamovi provides the same expected counts and statistics that we calculated by hand above, with a χ^2 value of (8.44 with 3 df and $p = 0.038$. Note that we don't need to look up a critical p-value threshold value any more, as jamovi gives us the actual p-value of the calculated χ^2 for 3 df.

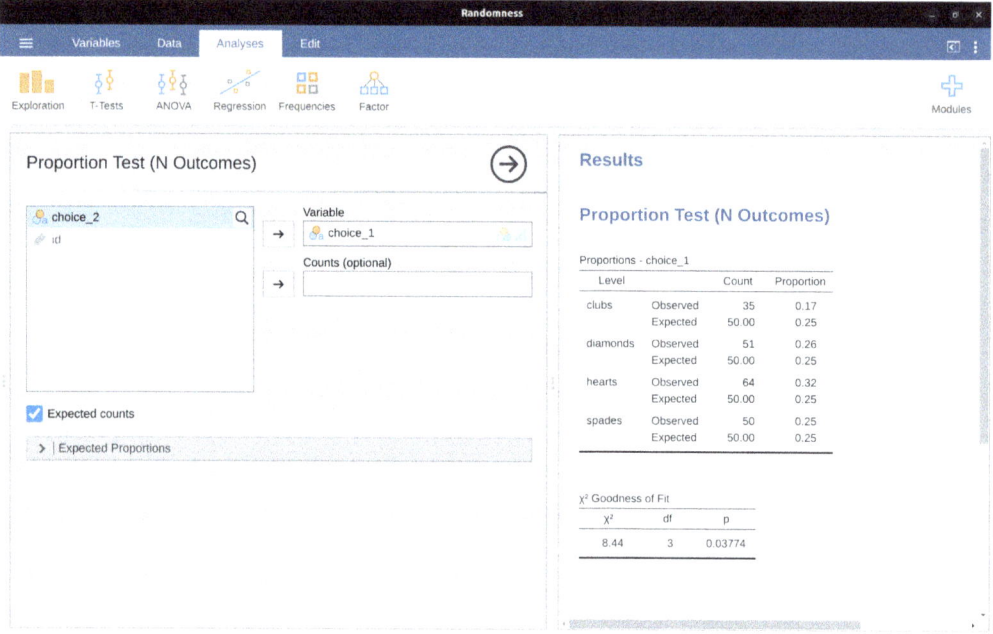

Figure 10.3: A χ^2 One Sample Proportion Test in jamovi, with table showing both observed and expected frequencies and proportions

10.1.8 Specifying a different null hypothesis

At this point you might be wondering what to do if you want to run a goodness-of-fit test but your null hypothesis is not that all categories are equally likely. For instance, let's suppose that someone had made the theoretical prediction that people should choose red cards 60% of the time, and black cards 40% of the time (I've no idea why you'd predict that), but had no other preferences. If that were the case, the null hypothesis would be to expect 30% of the choices to be hearts, 30% to be diamonds, 20% to be spades and 20% to be clubs. In other words we would expect hearts and diamonds to appear 1.5 times more often than spades and clubs (the ratio 30% : 20% is the same as 1.5 : 1). The expected counts are now shown in Table 10.7.

Table 10.7: Expected counts for a different null hypothesis

	♣	♦	♡	♠
expected frequency E_i	40	60	60	40

This seems like a silly theory to me, and it's pretty easy to test this explicitly specified null hypothesis with the data in our jamovi analysis. In the analysis window (labelled 'Proportion Test (N Outcomes)' in Figure 10.3 you can expand the options for 'Expected Proportions'. When you do this, there are options for entering different ratio values for the variable you have selected, in our case this is choice 1. Change the ratio to reflect the new null hypothesis, as in Figure 10.4, and check the results.

Expected Proportions		
Level	Ratio	Proportion
clubs	1	0.200
diamonds	1.5	0.300
hearts	1.5	0.300
spades	1	0.200

Figure 10.4: Changing the expected proportions in the χ^2 One Sample Proportion Test in jamovi

You can see that the χ^2 statistic is 4.74, 3 df, $p = 0.192$. Now, the results of our updated hypotheses and the expected frequencies are different from what they were last time. As a consequence our χ^2 test statistic is different, and our p-value is different too. Annoyingly, the p-value is .192, so we can't reject the null hypothesis (look back at Section 9.5 to remind yourself why). Sadly, despite the fact that the null hypothesis corresponds to a very silly theory, these data don't provide enough evidence against it.

10.1.9 How to report the results of the test

So now you know how the test works, and you know how to do the test using a wonderful jamovi flavoured magic computing box. The next thing you need to know is how to write up the results. After all, there's no point in designing and running an experiment and then analysing the data if you don't tell anyone about it! So let's now talk about what you need to do when reporting your analysis. Let's stick with our card-suits example. If I wanted to write this result up for a paper or something, then the conventional way to report this would be to write something like this:

> Of the 200 participants in the experiment, 64 selected hearts for their first choice, 51 selected diamonds, 50 selected spades, and 35 selected clubs. A chi-square goodness-of-fit test was conducted to test whether the choice probabilities were identical for all four suits. The results were significant ($\chi^2(3) = 8.44, p < .05$), suggesting that people did not select suits purely at random.

This is pretty straightforward and hopefully it seems pretty unremarkable. That said, there's a few things that you should note about this description:

- *The statistical test is preceded by the descriptive statistics.* That is, I told the reader something about what the data look like before going on to do the test. In general, this is good practice. Always remember that your reader doesn't know your data anywhere near as well as you do. So, unless you describe it to them properly, the statistical tests won't make any sense to them and they'll get frustrated and cry.

- *The description tells you what the null hypothesis being tested is*. To be honest, writers don't always do this but it's often a good idea in those situations where some ambiguity exists, or when you can't rely on your readership being intimately familiar with the statistical tools that you're using. Quite often the reader might not know (or remember) all the details of the test that your using, so it's a kind of politeness to "remind" them! As far as the goodness-of-fit test goes, you can usually rely on a scientific audience knowing how it works (since it's covered in most intro stats classes). However, it's still a good idea to be explicit about stating the null hypothesis (briefly!) because the null hypothesis can be different depending on what you're using the test for. For instance, in the cards example my null hypothesis was that all the four suit probabilities were identical (i.e., $P_1 = P_2 = P_3 = P_4 = 0.25$), but there's nothing special about that hypothesis. I could just as easily have tested the null hypothesis that $P_1 = 0.7$ and $P_2 = P_3 = P_4 = 0.1$ using a goodness-of-fit test. So it's helpful to the reader if you explain to them what your null hypothesis was. Also, notice that I described the null hypothesis in words, not in maths. That's perfectly acceptable. You can describe it in maths if you like, but since most readers find words easier to read than symbols, most writers tend to describe the null using words if they can.

- *A "stat block" is included.* When reporting the results of the test itself, I didn't just say that the result was significant, I included a "stat block" (i.e., the dense mathematical looking part in the parentheses) which reports all the "key" statistical information. For the chi-square goodness-of-fit test, the information that gets reported is the test statistic (that the goodness-of-fit statistic was 8.44), the information about the distribution used in the test (χ^2 with 3 degrees of freedom which is usually shortened to $\chi^2(3)$), and then the information about whether the result was significant (in this case $p < .05$). The particular information that needs to go into the stat block is different for every test, and so each time I introduce a new test I'll show you what the stat block should look like.[85] However the general principle is that you should always provide enough information so that the reader could check the test results themselves if they really wanted to.

- *The results are interpreted*. In addition to indicating that the result was significant, I provided an interpretation of the result (i.e., that people didn't choose randomly). This is also a kindness to the reader, because it tells them something about what they should believe about what's going on in your data. If you don't include something like this, it's really hard for your reader to understand what's going on.[86]

As with everything else, your overriding concern should be that you explain things to your reader. Always remember that the point of reporting your results is to communicate to another human being. I cannot tell you just how many times I've seen the results section of a report or a thesis or even a scientific article that is just gibberish, because the writer has focused solely on making sure they've included all the numbers and forgotten to actually communicate with the human reader.

Satan delights equally in statistics and in quoting scripture[87]
– H.G. Wells

10.2 The χ^2 test of independence (or association)

> GUARDBOT 1: Halt!
> GUARDBOT 2: Be you robot or human?
> LEELA: Robot...we be.
> FRY: Uh, yup! Just two robots out roboting it up! Eh?
> GUARDBOT 1: Administer the test.
> GUARDBOT 2: Which of the following would you most prefer? A: A puppy, B: A pretty flower from your sweetie, or C: A large properly-formatted data file?
> GUARDBOT 1: Choose!
> – Futurama, *Fear of a Bot Planet*

The other day I was watching an animated documentary examining the quaint customs of the natives of the planet *Chapek 9*. Apparently, in order to gain access to their capital city a visitor must prove that they're a robot, not a human. In order to determine whether or not a visitor is human, the natives ask whether the visitor prefers puppies, flowers, or large, properly formatted data files. "Pretty clever," I thought to myself "but what if humans and robots have the same preferences? That probably wouldn't be a very good test then, would it?" As it happens, I got my hands on the testing data that the civil authorities of *Chapek 9* used to check this. It turns out that what they did was very simple. They found a bunch of robots and a bunch of humans and asked them what they preferred. I saved their data in a file called *chapek9.omv*, which we can now load into jamovi. As well as the ID variable that identifies individual people, there are two nominal text variables, species and choice. In total there are 180 entries in the data set, one for each person (counting both robots and humans as "people") who was asked to make a choice. Specifically, there are 93 humans and 87 robots, and overwhelmingly the preferred choice is the data file. You can check this yourself by asking jamovi for Frequency Tables, under the 'Exploration' – 'Descriptives' button. However, this summary does not address the question we're interested in. To do that, we need a more detailed description of the data. What we want to do is look at the choices broken down *by species*. That is, we need to cross-tabulate the data (see Section 6.1). In jamovi we do this using the 'Frequencies' – 'Contingency Tables' – 'Independent Samples' analysis, and we should get a table something like Table 10.8.

Table 10.8: Cross-tabulating the data

	Robot	Human	Total
Puppy	13	15	28
Flower	30	13	43
Data	44	65	109
Total	87	93	180

From this, it's quite clear that the vast majority of the humans chose the data file, whereas the robots tended to be a lot more even in their preferences. Leaving aside the question of why the humans might be more likely to choose the data file for the moment (which does seem quite odd, admittedly), our first order of business is to determine if the discrepancy between human choices and robot choices in the data set is statistically significant.

10.2.1 Constructing our hypothesis test

How do we analyse this data? Specifically, since my research hypothesis is that "humans and robots answer the question in different ways", how can I construct a test of the null hypothesis that "humans and robots answer the question the same way"? As before, we begin by establishing some notation to describe the data (Table 10.9).

Table 10.9: Notation to describe the data

	Robot	Human	Total
Puppy	O_{11}	O_{12}	R_1
Flower	O_{21}	O_{22}	R_2
Data	O_{31}	O_{32}	R_3
Total	C_1	C_2	N

In this notation we say that O_{ij} is a count (observed frequency) of the number of respondents that are of species j (robots or human) who gave answer i (puppy, flower or data) when asked to make a choice. The total number of observations is written N, as usual. Finally, I've used R_i to denote the row totals (e.g., R_1 is the total number of people who chose the flower), and C_j to denote the column totals (e.g., C_1 is the total number of robots).[88]

So now let's think about what the null hypothesis says. If robots and humans are responding in the same way to the question, it means that the probability that "a robot says puppy" is the same as the probability that "a human says puppy", and so on for the other two possibilities. So, if we use P_{ij} to denote "the probability that a member of species j gives response i" then our null hypothesis is that:

H_0 : All of the following are true:
$P_{11} = P_{12}$ (same probability of saying "puppy"),
$P_{21} = P_{22}$ (same probability of saying "flower"), and
$P_{31} = P_{32}$ (same probability of saying "data").

And actually, since the null hypothesis is claiming that the true choice probabilities don't depend on the species of the person making the choice, we can let Pi refer to this probability, e.g., P_1 is the true probability of choosing the puppy.

Next, in much the same way that we did with the goodness-of-fit test, what we need to do is calculate the expected frequencies. That is, for each of the observed counts O_{ij} we need to figure out what the null hypothesis would tell us to expect. Let's denote this expected frequency by E_{ij}. This time, it's a little bit trickier. If there are a total of C_j people that belong to species j, and the true probability of anyone (regardless of species) choosing option i is P_i, then the expected frequency is just:

$$E_{ij} = C_j \times P_i$$

Now, this is all very well and good, but we have a problem. Unlike the situation we had with the goodness-of-fit test, the null hypothesis doesn't actually specify a particular value for Pi .

It's something we have to estimate (see Chapter 8) from the data! Fortunately, this is pretty easy to do. If 28 out of 180 people selected the flowers, then a natural estimate for the probability of choosing flowers is $\frac{28}{180}$, which is approximately .16. If we phrase this in mathematical terms, what we're saying is that our estimate for the probability of choosing option i is just the row total divided by the total sample size:

$$\hat{P}_i = \frac{R_i}{N}$$

Therefore, our expected frequency can be written as the product (i.e. multiplication) of the row total and the column total, divided by the total number of observations:

$$E_{ij} = \frac{R_i \times C_j}{N}$$

[Additional technical detail[89]]

As before, large values of X^2 indicate that the null hypothesis provides a poor description of the data, whereas small values of X^2 suggest that it does a good job of accounting for the data. Therefore, just like last time, we want to reject the null hypothesis if X^2 is too large.

Not surprisingly, this statistic is χ^2 distributed. All we need to do is figure out how many degrees of freedom are involved, which actually isn't too hard. As I mentioned before, you can (usually) think of the degrees of freedom as being equal to the number of data points that you're analysing, minus the number of constraints. A contingency table with r rows and c columns contains a total of r^c observed frequencies, so that's the total number of observations. What about the constraints? Here, it's slightly trickier. The answer is always the same:

$$df = (r-1)(c-1)$$

But the explanation for why the degrees of freedom takes this value is different depending on the experimental design. For the sake of argument, let's suppose that we had honestly intended to survey exactly 87 robots and 93 humans (column totals fixed by the experimenter), but left the row totals free to vary (row totals are random variables). Let's think about the constraints that apply here. Well, since we deliberately fixed the column totals by Act of Experimenter, we have c constraints right there. But, there's actually more to it than that.

Remember how our null hypothesis had some free parameters (i.e., we had to estimate the Pi values)? Those matter too. I won't explain why in this book, but every free parameter in the null hypothesis is rather like an additional constraint. So, how many of those are there? Well, since these probabilities have to sum to 1, there's only $r-1$ of these.

So our total degrees of freedom is:

$$df = \text{(number of observations) - (number of constraints)}$$

$$= (r \times c) - (c + (r-1))$$

$$= rc - c - r + 1$$

$$= (r-1)(c-1)$$

Alternatively, suppose that the only thing that the experimenter fixed was the total sample size N. That is, we quizzed the first 180 people that we saw and it just turned out that 87 were robots and 93 were humans. This time around our reasoning would be slightly different, but would still lead us to the same answer. Our null hypothesis still has $r-1$ free parameters corresponding to the choice probabilities, but it now also has $c-1$ free parameters corresponding to the species probabilities, because we'd also have to estimate the probability that a randomly sampled person turns out to be a robot.[90]

Finally, since we did actually fix the total number of observations N, that's one more constraint. So, now we have rc observations, and $(c-1) + (r-1) + 1$ constraints. And that gives:

$$df = \text{(number of observations) - (number of constraints)}$$

$$= (r \times c) - ((c-1) + (r-1) + 1)$$

$$= (r-1)(c-1)$$

Amazing.

10.2.2 Doing the test in jamovi

Okay, now that we know how the test works let's have a look at how it's done in jamovi. As tempting as it is to lead you through the tedious calculations so that you're forced to learn it the long way, I figure there's no point. I already showed you how to do it the long way for the goodness-of-fit test in the last section, and since the test of independence isn't conceptually any different, you won't learn anything new by doing it the long way. So instead I'll go straight to showing you the easy way.

After you have run the test in jamovi ('Frequencies' – 'Contingency Tables' – 'Independent Samples'), all you have to do is look underneath the contingency table in the jamovi results window and there is the χ^2 statistic for you. This shows a χ^2 statistic value of 10.72, with 2 df and p-value = 0.005.

That was easy, wasn't it! You can also ask jamovi to show you the expected counts - just click on the check box for 'Counts' – 'Expected' in the 'Cells' options and the expected counts will appear in the contingency table. And whilst you are doing that, an effect size measure would be helpful. We'll choose Cramér's V, and you can specify this from a check box in the 'Statistics' options, and it gives a value for Cramér's V of 0.24. See Figure 10.5. We will talk about this some more in just a moment.

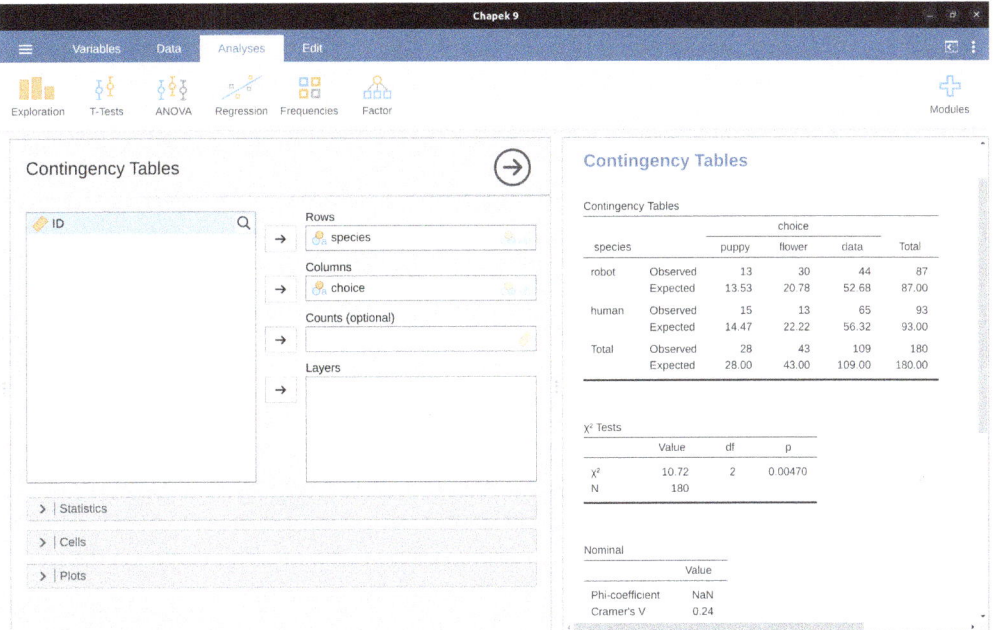

Figure 10.5: Independent samples χ^2 test in jamovi using the *Chapek 9* data

This output gives us enough information to write up the result:

> *Pearson's* χ^2 revealed a significant association between species and choice ($\chi^2(2) = 10.7, p < .01$). Robots appeared to be more likely to say that they prefer flowers, but the humans were more likely to say they prefer data.

Notice that, once again, I provided a little bit of interpretation to help the human reader understand what's going on with the data. Later on in my discussion section I'd provide a bit more context. To illustrate the difference, here's what I'd probably say later on:

> The fact that humans appeared to have a stronger preference for raw data files than robots is somewhat counter-intuitive. However, in context it makes some sense, as the civil authority on Chapek 9 has an unfortunate tendency to kill and dissect humans when they are identified. As such it seems most likely that the human participants did not respond honestly to the question, so as to avoid potentially undesirable consequences. This should be considered to be a substantial methodological weakness.

This could be classified as a rather extreme example of a reactivity effect, I suppose. Obviously, in this case the problem is severe enough that the study is more or less worthless as a tool for understanding the difference preferences among humans and robots. However, I hope this illustrates the difference between getting a statistically significant result (our null hypothesis is rejected in favour of the alternative), and finding something of scientific value (the data tell us nothing of interest about our research hypothesis due to a big methodological flaw).

10.3 The continuity correction

Okay, time for a little bit of a digression. I've been lying to you a little bit so far. There's a tiny change that you need to make to your calculations whenever you only have 1 degree of freedom. It's called the "continuity correction", or sometimes the **Yates correction**. Remember what I pointed out earlier: the χ^2 test is based on an approximation, specifically on the assumption that the binomial distribution starts to look like a normal distribution for large N. One problem with this is that it often doesn't quite work, especially when you've only got 1 degree of freedom (e.g., when you're doing a test of independence on a 2×2 contingency table). The main reason for this is that the true sampling distribution for the X^2 statistic is actually discrete (because you're dealing with categorical data!) but the χ^2 distribution is continuous. This can introduce systematic problems. Specifically, when N is small and when $df = 1$, the goodness-of-fit statistic tends to be "too big", meaning that you actually have a bigger α value than you think (or, equivalently, the p-values are a bit too small).

As far as I can tell from reading Yates' paper,[91] the correction is basically a hack. It's not derived from any principled theory. Rather, it's based on an examination of the behaviour of the test, and observing that the corrected version seems to work better. You can specify this correction in jamovi from a check box in the 'Statistics' options, where it is called 'χ^2 continuity correction'.

10.4 Effect size

As we discussed earlier in Section 9.8, it's becoming commonplace to ask researchers to report some measure of effect size. So, let's suppose that you've run your chi-square test, which turns out to be significant. So you now know that there is some association between your variables (independence test) or some deviation from the specified probabilities (goodness-of-fit test). Now you want to report a measure of effect size. That is, given that there is an association or deviation, how strong is it?

There are several different measures that you can choose to report, and several different tools that you can use to calculate them. I won't discuss all of them but will instead focus on the most commonly reported measures of effect size.

By default, the two measures that people tend to report most frequently are the ϕ statistic and the somewhat superior version, known as Cramér's V.

[Additional technical detail[92]]

And you're done. This seems to be a fairly popular measure, presumably because it's easy to calculate, and it gives answers that aren't completely silly. With Cramér's V, you know that the value really does range from 0 (no association at all) to 1 (perfect association).

10.5 Assumptions of the test(s)

All statistical tests make assumptions, and it's usually a good idea to check that those assumptions are met. For the chi-square tests discussed so far in this chapter, the assumptions are:

- *Expected frequencies are sufficiently large*. Remember how in the previous section we saw that the χ^2 sampling distribution emerges because the binomial distribution is pretty similar to a normal distribution? Well, like we discussed in Chapter 7 this is only true when the number of observations is sufficiently large. What that means in practice is that all of the expected frequencies need to be reasonably big. How big is reasonably big? Opinions differ, but the default assumption seems to be that you generally would like to see all your expected frequencies larger than about 5, though for larger tables you would probably be okay if at least 80% of the the expected frequencies are above 5 and none of them are below 1. However, from what I've been able to discover (e.g., Cochran, 1954) these seem to have been proposed as rough guidelines, not hard and fast rules, and they seem to be somewhat conservative (Larntz, 1978).
- *Data are independent of one another*. One somewhat hidden assumption of the chi-square test is that you have to genuinely believe that the observations are independent. Here's what I mean. Suppose I'm interested in proportion of babies born at a particular hospital that are boys. I walk around the maternity wards and observe 20 girls and only 10 boys. Seems like a pretty convincing difference, right? But later on, it turns out that I'd actually walked into the same ward 10 times and in fact I'd only seen 2 girls and 1 boy. Not as convincing, is it? My original 30 observations were massively non-independent, and were only in fact equivalent to 3 independent observations. Obviously this is an extreme (and extremely silly) example, but it illustrates the basic issue. Non-independence "stuffs things up". Sometimes it causes you to falsely reject the null, as the silly hospital example illustrates, but it can go the other way too. To give a slightly less stupid example, let's consider what would happen if I'd done the cards experiment slightly differently Instead of asking 200 people to try to imagine sampling one card at random, suppose I asked 50 people to select 4 cards. One possibility would be that *everyone* selects one heart, one club, one diamond and one spade (in keeping with the "representativeness heuristic" (Tversky & Kahneman, 1974)). This is highly non-random behaviour from people, but in this case I would get an observed frequency of 50 for all four suits. For this example the fact that the observations are non-independent (because the four cards that you pick will be related to each other) actually leads to the opposite effect, falsely retaining the null.

If you happen to find yourself in a situation where independence is violated, it may be possible to use the McNemar test (which we'll discuss) or the Cochran test (which we won't). Similarly, if your expected cell counts are too small, check out the Fisher exact test. It is to these topics that we now turn.

10.6 The Fisher exact test

What should you do if your cell counts are too small, but you'd still like to test the null hypothesis that the two variables are independent? One answer would be "collect more data", but that's far too glib There are a lot of situations in which it would be either infeasible or unethical do that. If so, statisticians have a kind of moral obligation to provide scientists with better tests. In this instance, Fisher (1922a) kindly provided the right answer to the question. To illustrate the basic idea let's suppose that we're analysing data from a field experiment looking at the emotional status of people who have been accused of Witchcraft, some of whom are currently being burned at the stake.[93] Unfortunately for the scientist (but rather fortunately for the general populace), it's actually quite hard to find people in the process of being set on fire, so the cell counts are awfully small in some cases. A contingency table of the *salem.csv* data illustrates the point (Table 10.10).

Table 10.10: Contingency table of the *salem.csv* data

	happy	FALSE	TRUE
on.fire	FALSE	3	10
	TRUE	3	0

Looking at this data, you'd be hard pressed not to suspect that people not on fire are more likely to be happy than people on fire. However, the chi-square test makes this very hard to test because of the small sample size. So, speaking as someone who doesn't want to be set on fire, I'd *really* like to be able to get a better answer than this. This is where **Fisher's exact test** (Fisher, 1922a) comes in very handy.

The Fisher exact test works somewhat differently to the chi-square test (or in fact any of the other hypothesis tests that I talk about in this book) insofar as it doesn't have a test statistic, but it calculates the *p*-value "directly". I'll explain the basics of how the test works for a 2 × 2 contingency table. As before, let's have some notation (Table 10.11).

Table 10.11: Notation for the Fisher exact test

	Happy	Sad	Total
Set on fire	O_{11}	O_{12}	R_1
Not set on fire	O_{21}	O_{22}	R_2
Total	C_1	C_2	N

In order to construct the test Fisher treats both the row and column totals (R_1, R_2, C_1 and C_2) as known, fixed quantities and then calculates the probability that we would have obtained the observed frequencies that we did (O_{11}, O_{12}, O_{21} and O_{22}) given those totals. In the notation that we developed in Chapter 7 this is written:

$$P(O_{11}, O_{12}, O_{21}, O_{22} \mid R_1, R_2, C_1, C_2)$$

and as you might imagine, it's a slightly tricky exercise to figure out what this probability is. But it turns out that this probability is described by a distribution known as the hypergeometric distribution. What we have to do to calculate our *p*-value is calculate the probability of observing this particular table or a table that is *"more extreme"*.[94] Back in the 1920s, computing this sum was daunting even in the simplest of situations, but these days it's pretty easy as long as the tables aren't too big and the sample size isn't too large. The conceptually tricky issue is to figure out what it means to say that one contingency table is more "extreme" than another. The easiest solution is to say that the table with the lowest probability is the most extreme. This then gives us the *p*-value.

You can specify this test in jamovi from a check box in the 'Statistics' options of the 'Contingency Tables' analysis. When you do this with the data from the *salem.csv* file, the Fisher exact test statistic is shown in the results. The main thing we're interested in here is the *p*-value, which in this case is small enough (p = .036) to justify rejecting the null hypothesis that people on fire are just as happy as people not on fire. See Figure 10.6.

Contingency Tables

happy	on.fire FALSE	on.fire TRUE	Total
FALSE	3	3	6
TRUE	10	0	10
Total	13	3	16

χ^2 Tests

	Value	df	p
χ^2	6.15	1	0.01311
Fisher's exact test			0.03571
N	16		

Figure 10.6: Fisher exact test analysis in jamovi

10.7 The McNemar test

Suppose you've been hired to work for the *Australian Generic Political Party* (AGPP), and part of your job is to find out how effective the AGPP political advertisements are. So you decide to put together a sample of $N = 100$ people and ask them to watch the AGPP ads. Before they see anything, you ask them if they intend to vote for the AGPP, and then after showing the ads you ask them again to see if anyone has changed their minds. Obviously, if you're any good at your job, you'd also do a whole lot of other things too, but let's consider just this one simple experiment. One way to describe your data is via the contingency table shown in Table 10.12.

Table 10.12: Contingency table with data on AGPP political advertisements

	Before	After	Total
Yes	30	10	40
No	70	90	160
Total	100	100	200

At first pass, you might think that this situation lends itself to the Pearson χ^2 test of independence (as per The χ^2 test of independence (or association)). However, a little bit of thought reveals that we've got a problem. We have 100 participants but 200 observations. This is because each person has provided us with an answer in both the before column and the after column. What this means is that the 200 observations aren't independent of each other. If voter A says "yes" the first time and voter B says "no", then you'd expect that voter A is more likely to say "yes" the second time than voter B! The consequence of this is that the usual χ^2 test won't give trustworthy answers due to the violation of the independence assumption. Now, if this were a really uncommon situation, I wouldn't be bothering to waste your time talking about it. But it's not uncommon at all. This is a standard repeated measures design, and none of the tests we've considered so far can handle it.

The solution to the problem was published by McNemar (1947). The trick is to start by tabulating your data in a slightly different way (Table 10.13).

Table 10.13: Tabulate the data in a different way when you have repeated measures data

	After: No	After: Yes	Total
Before: No	65	5	70
Before: Yes	25	5	30
Total	90	10	100

Next, let's think about what our null hypothesis is: it's that the "before" test and the "after" test have the same proportion of people saying "Yes, I will vote for AGPP". Because of the way that we have rewritten the data, it means that we're now testing the hypothesis that the row totals and column totals come from the same distribution. Thus, the null hypothesis in McNemar's test is that we have "marginal homogeneity". That is, the row totals and column totals have the same distribution: $P_a + P_b = P_a + P_c$, and similarly that $P_c + P_d = P_b + P_d$. Notice that this means that the null hypothesis actually simplifies to $Pb = Pc$. In other words, as far as the McNemar test is concerned, it's only the off-diagonal entries in this table (i.e., b and c) that matter! After noticing this, the **McNemar test of marginal homogeneity** is no different to a usual χ^2 test.

After applying the Yates correction, our test statistic becomes:

$$\chi^2 = \frac{(|b - c| - 0.5)^2}{b + c}$$

or, to revert to the notation that we used earlier in this chapter:

$$\chi^2 = \frac{(|O_{12} - O_{21}| - 0.5)^2}{O_{12} + O_{21}}$$

and this statistic has a χ^2 distribution (approximately) with $df = 1$. However, remember that just like the other χ^2 tests it's only an approximation, so you need to have reasonably large expected cell counts for it to work.

10.7.1 Doing the McNemar test in jamovi

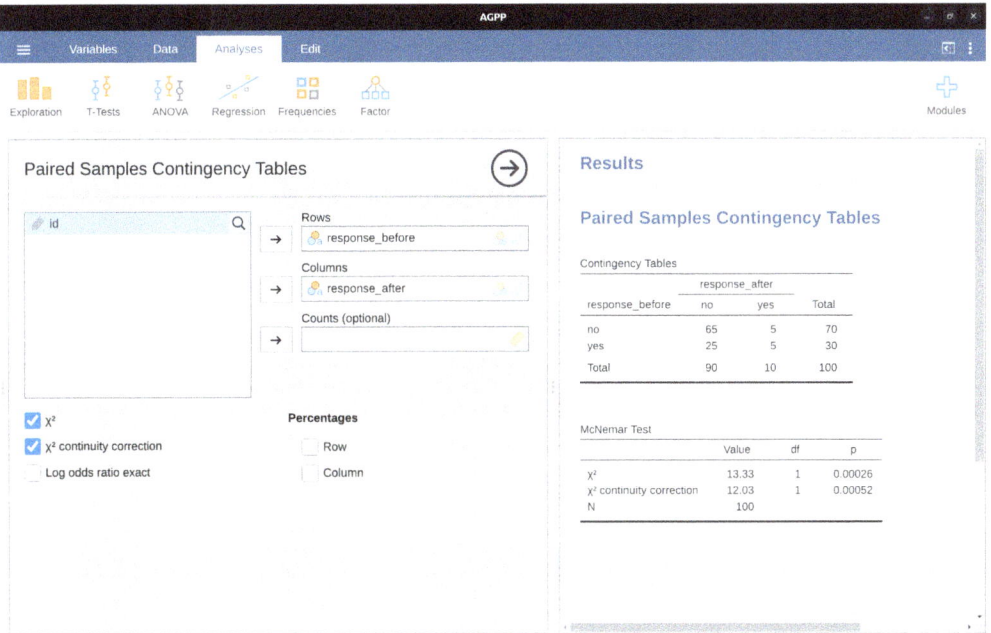

Figure 10.7: McNemar test output in jamovi

Now that you know what the McNemar test is all about, lets actually run one. The *agpp.csv* file contains the raw data that I discussed previously. The *agpp* data set contains three variables, an id variable that labels each participant in the data set (we'll see why that's useful in a moment), a **response_before** variable that records the person's answer when they were asked the question the first time, and a response_after variable that shows the answer that they gave when asked the same question a second time. Notice that each participant appears only once in this data set. Go to the 'Analyses' – 'Frequencies' – 'Contingency Tables' – 'Paired Samples' analysis in jamovi, and move **response_before** into the 'Rows' box, and **response_after** into the 'Columns' box. You will then get a contingency table in the results window, with the statistic for the McNemar test just below it, see Figure 10.7.

And we're done. We've just run a McNemar's test to determine if people were just as likely to vote AGPP after the ads as they were beforehand. The test was significant

($\chi^2(1) = 12.03, p < .001$), suggesting that they were not. And, in fact it looks like the ads had a negative effect: people were less likely to vote AGPP after seeing the ads. Which makes a lot of sense when you consider the quality of a typical political advertisement.

10.8 What's the difference between McNemar and independence?

Let's go all the way back to the beginning of the chapter and look at the cards data set again. If you recall, the actual experimental design that I described involved people making two choices. Because we have information about the first choice and the second choice that everyone made, we can construct the following contingency table that cross-tabulates the first choice against the second choice (Table 10.14).

Table 10.14: Cross-tabulating first against second choice with the *Randomness.omv* (cards) data

	Before: Yes	Before: No	Total
After: Yes	a	b	$a + b$
After: No	c	d	$c + d$
Total	$a + c$	$b + d$	n

Suppose I wanted to know whether the choice you make the second time is dependent on the choice you made the first time. This is where a test of independence is useful, and what we're trying to do is see if there's some relationship between the rows and columns of this table. Alternatively, suppose I wanted to know if on average, the frequencies of suit choices were different the second time than the first time. In that situation, what I'm really trying to see is if the row totals are different from the column totals. That's when you use the McNemar test.

The different statistics produced by these different analyses are shown in Figure 10.8. Notice that the results are different! These aren't the same test.

Contingency Tables

Contingency Tables

	choice_2				
choice_1	clubs	diamonds	hearts	spades	Total
clubs	10	9	10	6	35
diamonds	20	4	13	14	51
hearts	20	18	3	23	64
spades	18	13	15	4	50
Total	68	44	41	47	200

χ^2 Tests

	Value	df	p
χ^2	29.24	9	0.00059
N	200		

Paired Samples Contingency Tables

Contingency Tables

	choice_2				
choice_1	clubs	diamonds	hearts	spades	Total
clubs	10	9	10	6	35
diamonds	20	4	13	14	51
hearts	20	18	3	23	64
spades	18	13	15	4	50
Total	68	44	41	47	200

McNemar Test

	Value	df	p
χ^2	16.03	6	0.01358
N	200		

Figure 10.8: Independent vs. Paired (McNemar) with the *Randomness.omv* (cards) data

10.9 Summary

The key ideas discussed in this chapter are:

- The χ^2 (chi-square) goodness-of-fit test is used when you have a table of observed frequencies of different categories, and the null hypothesis gives you a set of "known" probabilities to compare them to.
- The χ^2 test of independence (or association) is used when you have a contingency table (cross-tabulation) of two categorical variables. The null hypothesis is that there is no relationship or association between the variables.
- Effect size for a contingency table can be measured in several ways. In particular we noted the Cramér's V statistic.
- Both versions of the Pearson test rely on two assumptions: that the expected frequencies are sufficiently large, and that the observations are independent (Assumptions of the test(s). The Fisher exact test can be used when the expected frequencies are small. The McNemar test can be used for some kinds of violations of independence.

If you're interested in learning more about categorical data analysis a good first choice would be Agresti (1996) which, as the title suggests, provides an *Introduction to Categorical Data Analysis*. If the introductory book isn't enough for you (or can't solve the problem you're working on) you could consider Agresti (2002), *Categorical Data Analysis*. The latter is a more advanced text, so it's probably not wise to jump straight from this book to that one.

Chapter 11

Comparing two means

In Chapter 10 we covered the situation when your outcome variable is nominal scale and your predictor variable is also nominal scale. Lots of real-world situations have that character, and so you'll find that chi-square tests in particular are quite widely used. However, you're much more likely to find yourself in a situation where your outcome variable is interval scale or higher, and what you're interested in is whether the average value of the outcome variable is higher in one group or another. For instance, a psychologist might want to know if anxiety levels are higher among parents than non-parents, or if working memory capacity is reduced by listening to music (relative to not listening to music). In a medical context we might want to know if a new drug increases or decreases blood pressure. An agricultural scientist might want to know whether adding phosphorus to Australian native plants will kill them.[95] In all these situations our outcome variable is a fairly continuous, interval or ratio scale variable, and our predictor is a binary "grouping" variable. In other words, we want to compare the means of the two groups.

The standard answer to the problem of comparing means is to use a *t*-test, of which there are several varieties depending on exactly what question you want to solve. As a consequence, the majority of this chapter focuses on different types of *t*-test: one sample *t*-tests, independent samples *t*-tests and paired samples *t*-tests. We'll then talk about one-sided tests and, after that, we'll talk a bit about Cohen's *d*, which is the standard measure of effect size for a *t*-test. The later sections of the chapter focus on the assumptions of the *t*-tests, and possible remedies if they are violated. However, before discussing any of these useful things, we'll start with a discussion of the *z*-test.

11.1 The one-sample *z*-test

In this section I'll describe one of the most useless tests in all of statistics: the *z*-test. Seriously – this test is almost never used in real life. Its only real purpose is that, when teaching statistics, it's a very convenient stepping stone along the way towards the *t*-test, which is probably the most (over)used tool in all statistics.

11.1.1 The inference problem that the test addresses

To introduce the idea behind the z-test, let's use a simple example. A friend of mine, Dr Zeppo, grades his introductory statistics class on a curve. Let's suppose that the average grade in his class is 67.5, and the standard deviation is 9.5. Of his many hundreds of students, it turns out that 20 of them also take psychology classes. Out of curiosity, I find myself wondering if the psychology students tend to get the same grades as everyone else (i.e., mean 67.5) or do they tend to score higher or lower? He emails me the *zeppo.csv* file, which I use to look at the grades of those students, in the jamovi spreadsheet view, and then calculate the mean in 'Exploration' – 'Descriptives'.[96] The mean value is 72.3.

> 50 60 60 64 66 66 67 69 70 74 76 76 77 79 79 79 81 82 82 89

Hmm. It might be that the psychology students are scoring a bit higher than normal. That sample mean of $\bar{X} = 72.3$ is a fair bit higher than the hypothesised population mean of $\mu = 67.5$ but, on the other hand, a sample size of $N = 20$ isn't all that big. Maybe it's pure chance.

To answer the question, it helps to be able to write down what it is that I think I know. Firstly, I know that the sample mean is $\bar{X} = 72.3$. If I'm willing to assume that the psychology students have the same standard deviation as the rest of the class then I can say that the population standard deviation is $\sigma = 9.5$. I'll also assume that since Dr Zeppo is grading to a curve, the psychology student grades are normally distributed.

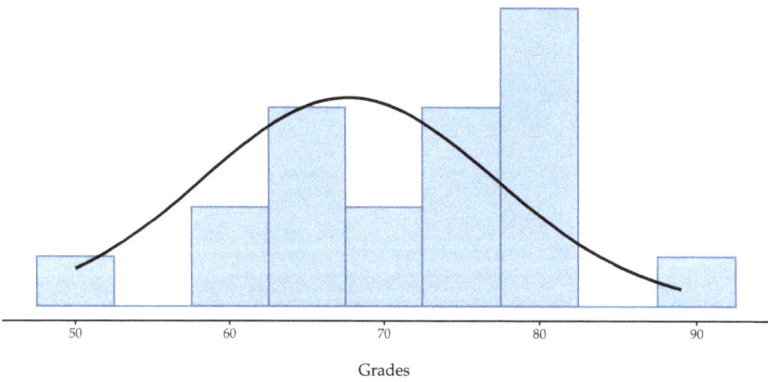

Figure 11.1: The theoretical distribution (solid line) from which the psychology student grades (bars) are supposed to have been generated

Next, it helps to be clear about what I want to learn from the data. In this case my research hypothesis relates to the population mean μ for the psychology student grades, which is unknown. Specifically, I want to know if $\mu = 67.5$ or not. Given that this is what I know, can we devise a hypothesis test to solve our problem? The data, along with the hypothesised distribution from which they are thought to arise, are shown in Figure 11.1. Not entirely obvious what the right answer is, is it? For this, we are going to need some statistics.

11.1.2 Constructing the hypothesis test

The first step in constructing a hypothesis test is to be clear about what the null and alternative hypotheses are. This isn't too hard to do. Our null hypothesis, H_0, is that the true population mean μ for psychology student grades is 67.5%, and our alternative hypothesis is that the population mean isn't 67.5%. If we write this in mathematical notation, these hypotheses become:

$$H_0 : \mu = 67.5$$
$$H_1 : \mu \neq 67.5$$

though to be honest this notation doesn't add much to our understanding of the problem, it's just a compact way of writing down what we're trying to learn from the data. The null hypotheses H_0 and the alternative hypothesis H_1 for our test are both illustrated in Figure 11.2. In addition to providing us with these hypotheses, the scenario outlined above provides us with a fair amount of background knowledge that might be useful. Specifically, there are two special pieces of information that we can add:

1. The psychology grades are normally distributed.
2. The true standard deviation of these scores σ is known to be 9.5.

For the moment, we'll act as if these are absolutely trustworthy facts. In real life, this kind of absolutely trustworthy background knowledge doesn't exist, and so if we want to rely on these facts we'll just have make the *assumption* that these things are true. However, since these assumptions may or may not be warranted, we might need to check them. For now though, we'll keep things simple.

The next step is to figure out what would be a good choice for a diagnostic test statistic, something that would help us discriminate between H_0 and H_1. Given that the hypotheses all refer to the population mean μ, you'd feel pretty confident that the sample mean $\bar{X}$ would be a pretty useful place to start. What we could do is look at the difference between the sample mean $\bar{X}$ and the value that the null hypothesis predicts for the population mean. In our example that would mean we calculate $\bar{X} - 67.5$. More generally, if we let μ_0 refer to the value that the null hypothesis claims is our population mean, then we'd want to calculate:

$$\bar{X} - \mu_0$$

If this quantity equals or is very close to 0, things are looking good for the null hypothesis. If this quantity is a long way away from 0, then it's looking less likely that the null hypothesis is worth retaining. But how far away from zero should it be for us to reject H_0?

To figure that out we need to be a bit more sneaky, and we'll need to rely on those two pieces of background knowledge that I wrote down previously; namely that the raw data are normally distributed and that we know the value of the population standard deviation σ. If the null hypothesis is actually true, and the true mean is μ_0, then these facts together mean that we know the complete population distribution of the data: a normal distribution with mean μ_0 and standard deviation σ.[97]

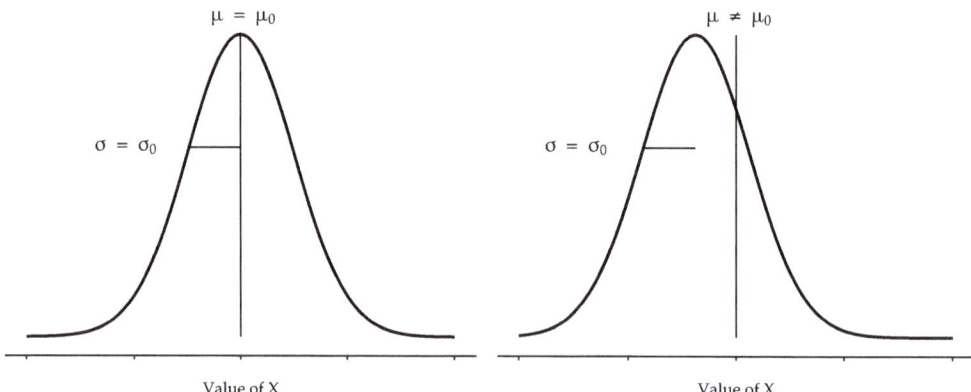

Figure 11.2: Graphical illustration of the null and alternate hypotheses assumed by the one sample z-test (the two-sided version, that is). The null and alternate hypotheses both assume that the population distribution is normal, and additionally assumes that the population standard deviation is known (fixed at some value σ_0). The null hypothesis (left) is that the population mean μ is equal to some specified value μ_0. The alternative hypothesis (right) is that the population mean differs from this value, $\mu \neq \mu_0$

Okay, if that's true, then what can we say about the distribution of $\bar{X}$? Well, as we discussed earlier (see Section 8.3.3), the sampling distribution of the mean $\bar{X}$ is also normal, and has mean μ. But the standard deviation of this sampling distribution $se\bar{X}$, which is called the standard error of the mean, is:[98]

$$se(\bar{X} = \frac{\sigma}{\sqrt{N}})$$

Now comes the trick. What we can do is convert the sample mean $\bar{X}$ into a standard score (see Section 4.5). This is conventionally written as z, but for now I'm going to refer to it as $z_{\bar{X}}$. The reason for using this expanded notation is to help you remember that we're calculating a standardised version of a sample mean, not a standardised version of a single observation, which is what a z-score usually refers to. When we do so the z-score for our sample mean is:

$$z_{\bar{X}} = \frac{\bar{X} - \mu_0}{SE(\bar{X})}$$

or, equivalently:

$$z_{\bar{X}} = \frac{\bar{X} - \mu_0}{\frac{\sigma}{\sqrt{N}}}$$

This z-score is our test statistic. The nice thing about using this as our test statistic is that like all z-scores, it has a standard normal distribution:[99]

In other words, regardless of what scale the original data are on, the z-statistic itself always has the same interpretation: it's equal to the number of standard errors that separate the observed sample mean $\bar{X}$ from the population mean μ_0 predicted by the null hypothesis. Better yet, regardless of what the population parameters for the raw scores actually are, the 5% critical regions for the z-test are always the same, as illustrated in Figure 11.3. And what this meant, way back in the days where people did all their statistics by hand, is that someone could publish a table like Table 11.1. This, in turn, meant that researchers could calculate their z-statistic by hand and then look up the critical value in a textbook.

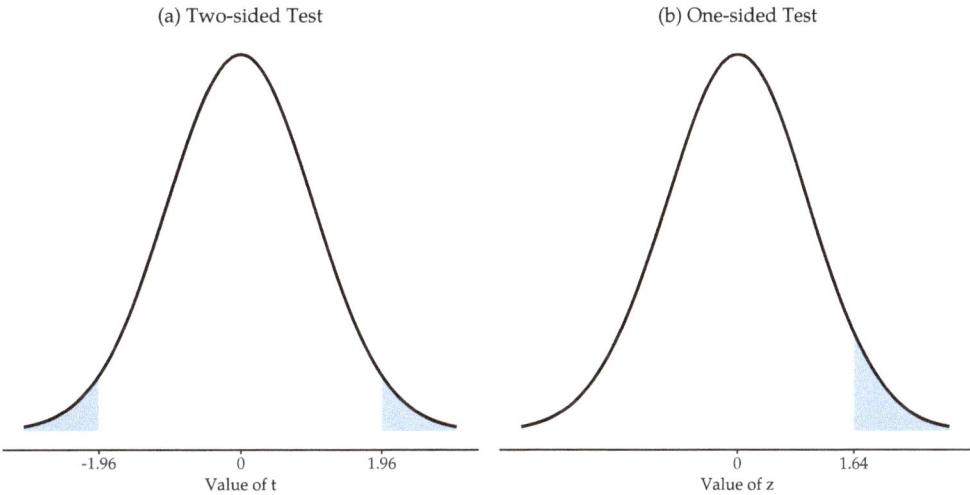

Figure 11.3: Rejection regions for the two-sided z-test (a) and the one-sided z-test (b)

Table 11.1: Critical values for different alpha levels

	critical z-value	
desired α level	two-sided test	one-sided test
.1	1.644854	1.281552
.05	1.959964	1.644854
.01	2.575829	2.326348
.001	3.290527	3.090232

11.1.3 A worked example, by hand

Now, as I mentioned earlier, the z-test is almost never used in practice. It's so rarely used in real life that the basic installation of jamovi doesn't have a built in function for it. However, the test is so incredibly simple that it's really easy to do one manually. Let's go back to the data from Dr Zeppo's class. Having loaded the grades data, the first thing I need to do is calculate the sample mean, which I've already done (72.3). We already have the known population standard deviation ($\sigma = 9.5$), and the value of the population mean that the null hypothesis specifies ($\mu_0 = 67.5$), and we know the sample size ($N = 20$).

Next, let's calculate the (true) standard error of the mean (easily done with a calculator):

$$sem.true = \frac{sd.true}{\sqrt{N}}$$

$$= \frac{9.5}{\sqrt{20}}$$

$$= 2.124265$$

And finally, we calculate our z-score:

$$z.score = \frac{sample.mean - mu.null}{sem.true}$$

$$= \frac{(72.3 - 67.5)}{2.124265}$$

$$= 2.259606$$

At this point, we would traditionally look up the value 2.26 in our table of critical values. Our original hypothesis was two-sided (we didn't really have any theory about whether psych students would be better or worse at statistics than other students) so our hypothesis test is two-sided (or two-tailed) also. Looking at the little table that I showed earlier, we can see that 2.26 is bigger than the critical value of 1.96 that would be required to be significant at $\alpha = .05$, but smaller than the value of 2.58 that would be required to be significant at a level of $\alpha = .01$. Therefore, we can conclude that we have a significant effect, which we might write up by saying something like this:

> *With a mean grade of* 73.2 in the sample of psychology students, and assuming a true population standard deviation of 9.5, we can conclude that the psychology students have significantly different statistics scores to the class average ($z = 2.26, N = 20, p < .05$).

11.1.4 Assumptions of the z-test

As I've said before, all statistical tests make assumptions. Some tests make reasonable assumptions, while other tests do not. The test I've just described, the one sample z-test, makes three basic assumptions. These are:

- *Normality*. As usually described, the z-test assumes that the true population distribution is normal.[100] This is often a pretty reasonable assumption, and it's also an assumption that we can check if we feel worried about it (see Section on Checking the normality of a sample).
- *Independence*. The second assumption of the test is that the observations in your data set are not correlated with each other, or related to each other in some funny way. This isn't as easy to check statistically, it relies a bit on good experimental design. An obvious (and stupid) example of something that violates this assumption is a data set where you "copy" the same observation over and over again in your data file so that you end up with a massive "sample size", which consists of only one genuine observation. More realistically, you have to ask yourself if it's really plausible to imagine that each observation is a completely random sample from the population that you're interested in. In practice this assumption is never met, but we try our best to design studies that minimise the problems of correlated data.
- *Known standard deviation*. The third assumption of the z-test is that the true standard deviation of the population is known to the researcher. This is just stupid. In no real world data analysis problem do you know the standard deviation σ of some population but are completely ignorant about the mean μ. In other words, this assumption is always wrong.

In view of the stupidity of assuming that α is known, let's see if we can live without it. This takes us out of the dreary domain of the z-test, and into the magical kingdom of the *t*-test, with unicorns and fairies and leprechauns!

11.2 The one-sample *t*-test

After some thought, I decided that it might not be safe to assume that the psychology student grades necessarily have the same standard deviation as the other students in Dr Zeppo's class (Figure 11.4). After all, if I'm entertaining the hypothesis that they don't have the same mean, then why should I believe that they absolutely have the same standard deviation? In view of this, I should really stop assuming that I know the true value of σ. This violates the assumptions of my z-test, so in one sense I'm back to square one. However, it's not like I'm completely bereft of options. After all, I've still got my raw data, and those raw data give me an estimate of the population standard deviation, which is 9.52. In other words, while I can't say that I know that $\sigma = 9.5$, I can say that $\hat{\sigma} = 9.52$.

Okay, cool. The obvious thing that you might think to do is run a z-test, but using the estimated standard deviation of 9.52 instead of relying on my assumption that the true standard deviation is 9.5. And you probably wouldn't be surprised to hear that this would still give us a significant result. This approach is close, but it's not quite correct.

Because we are now relying on an estimate of the population standard deviation we need to make some adjustment for the fact that we have some uncertainty about what the true population standard deviation actually is. Maybe our data are just a fluke... maybe the true population standard deviation is 11, for instance. But if that were actually true, and we ran the z-test assuming $\sigma = 11$, then the result would end up being non-significant. That's a problem, and it's one we're going to have to address.

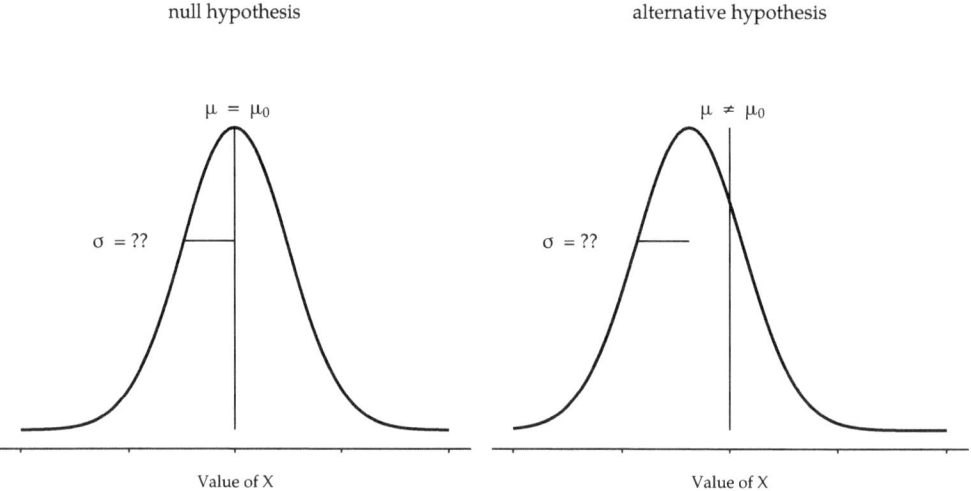

Figure 11.4: Graphical illustration of the null and alternative hypotheses assumed by the (two-sided) one sample t-test. Note the similarity to the z-test (Figure 11.2). The null hypothesis is that the population mean μ is equal to some specified value μ_0, and the alternative hypothesis is that it is not. Like the z-test, we assume that the data are normally distributed, but we do not assume that the population standard deviation σ is known in advance

11.2.1 Introducing the t-test

This ambiguity is annoying, and it was resolved in 1908 by a guy called William Sealy Gosset (Student, 1908), who was working as a chemist for the Guinness brewery at the time (see J. F. Box (1987)). Because Guinness took a dim view of its employees publishing statistical analysis (apparently they felt it was a trade secret), he published the work under the pseudonym "A Student" and, to this day, the full name of the t-test is actually **Student's t-test**. The key thing that Gosset figured out is how we should accommodate the fact that we aren't completely sure what the true standard deviation is.[101] The answer is that it subtly changes the sampling distribution. In the t-test our test statistic, now called a t-statistic, is calculated in exactly the same way I mentioned above. If our null hypothesis is that the true mean is μ, but our sample has mean $\bar{X}$ and our estimate of the population standard deviation is $\hat{\sigma}$, then our t-statistic is:

$$t = \frac{\bar{X} - \mu}{\frac{\hat{\sigma}}{\sqrt{N}}}$$

The only thing that has changed in the equation is that instead of using the known true value σ, we use the estimate $\hat{\sigma}$. And if this estimate has been constructed from N observations, then the sampling distribution turns into a t-distribution with $N-1$ **degrees of freedom** (df). The t-distribution is very similar to the normal distribution, but has "heavier" tails, as discussed earlier in Section 7.6 and illustrated in Figure 11.5. Notice, though, that as df gets larger, the t-distribution starts to look identical to the standard normal distribution. This is as it should be: if you have a sample size of $N = 70,000,000$ then your "estimate" of the standard deviation would be pretty much perfect, right? So, you should expect that for large N, the t-test would behave exactly the same way as a z-test. And that's exactly what happens!

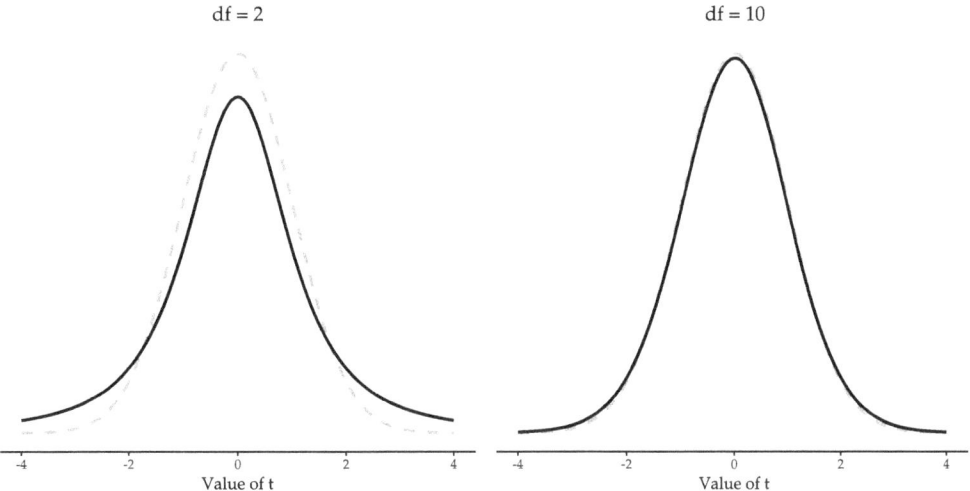

Figure 11.5: The t-distribution with 2 degrees of freedom (left) and 10 degrees of freedom (right), with a standard normal distribution (i.e., mean 0 and std dev 1) plotted as dotted lines for comparison purposes. Notice that the t-distribution has heavier tails (leptokurtic, higher kurtosis) than the normal distribution; this effect is quite exaggerated when the degrees of freedom are very small, but negligible for larger values. In other words, for large df the t-distribution is essentially identical to a normal distribution

11.2.2 Doing the test in jamovi

As you might expect, the mechanics of the t-test are almost identical to the mechanics of the z-test. So there's not much point in going through the tedious exercise of showing you how to do the calculations using low level commands. It's pretty much identical to the calculations that we did earlier, except that we use the estimated standard deviation and then we test our hypothesis using the t-distribution rather than the normal distribution. And so instead of going through the calculations in tedious detail for a second time, I'll jump straight to showing you how t-tests are actually done. jamovi comes with a dedicated analysis for t-tests that is very flexible (it can run lots of different kinds of t-tests). It's pretty straightforward to use; all you need to do is specify 'Analyses' – 'T-Tests' – 'One Sample T-Test', move the variable you are interested in (X) across into the 'Variables' box, and type in the mean value for the null hypothesis ('67.5') in the 'Hypothesis' – 'Test value' box. Easy enough. See Figure 11.6, which, amongst other things that we will get to in a moment, gives you a t-test statistic = 2.25, with 19 degrees of freedom and an associated p-value of 0.036.

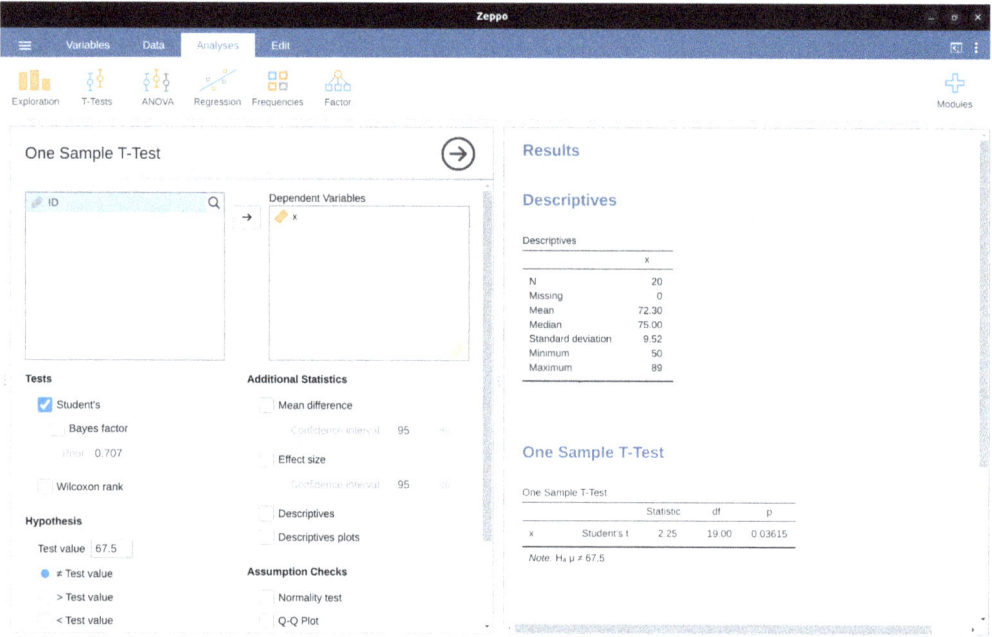

Figure 11.6: jamovi does the one-sample t-test

Also reported are two other things you might care about: the 95% confidence interval and a measure of effect size (we'll talk more about effect sizes later). So that seems straightforward enough. Now what do we do with this output? Well, since we're pretending that we actually care about my toy example, we're overjoyed to discover that the result is statistically significant (i.e. p-value below .05). We could report the result by saying something like this:

> With a mean grade of 72.3, the psychology students scored slightly higher than the average grade of 67.5 ($t(19) = 2.25$, $p = .036$); the mean difference was 4.80 and the 95% confidence interval was from 0.34 to 9.26.

...where $t(19)$ is shorthand notation for a t statistic that has 19 degrees of freedom. That said, it's often the case that people don't report the confidence interval, or do so using a much more compressed form than I've done here. For instance, it's not uncommon to see the confidence interval included as part of the stat block after reporting the mean difference, like this:

$$t(19) = 2.25, p = .036, CI_{95} = [0.34, 9.26]$$

With that much jargon crammed into half a line, you know it must be really smart.[102]

11.2.3 Assumptions of the one sample t-test

Okay, so what assumptions does the one-sample t-test make? Well, since the t-test is basically a z-test with the assumption of known standard deviation removed, you shouldn't be surprised to see that it makes the same assumptions as the z-test, minus the one about the known standard deviation. That is:

- Normality. We're still assuming that the population distribution is normal,[103] and as noted earlier, there are standard tools that you can use to check to see if this assumption is met (Checking the normality of a sample), and other tests you can do in its place if this assumption is violated (Testing non-normal data).
- Independence. Once again, we have to assume that the observations in our sample are generated independently of one another. See the earlier discussion about the z-test for specifics (Assumptions of the z-test).

Overall, these two assumptions aren't terribly unreasonable, and as a consequence the one sample t-test is pretty widely used in practice as a way of comparing a sample mean against a hypothesised population mean.

11.3 The independent samples t-test (Student test)

Although the one sample t-test has its uses, it's not the most typical example of a t-test.[104] A much more common situation arises when you've got two different groups of observations. In psychology, this tends to correspond to two different groups of participants, where each group corresponds to a different condition in your study. For each person in the study you measure some outcome variable of interest, and the research question that you're asking is whether or not the two groups have the same population mean. This is the situation that the independent samples t-test is designed for.

11.3.1 The data

Suppose we have 33 students taking Dr Harpo's statistics lectures, and Dr Harpo doesn't grade to a curve. Actually, Dr Harpo's grading is a bit of a mystery, so we don't really know anything about what the average grade is for the class as a whole. There are two tutors for the class, Anastasia and Bernadette. There are $N_1 = 15$ students in Anastasia's tutorials, and $N_2 = 18$ in Bernadette's tutorials. The research question I'm interested in is whether Anastasia or Bernadette is a better tutor, or if it doesn't make much of a difference. Dr Harpo emails me the course grades, in the *harpo.csv* file. As usual, I'll load the file into jamovi and have a look at what variables it contains - there are three variables, ID, grade and tutor. The grade variable contains each student's grade, but it is not imported into jamovi with the correct measurement level attribute, so I need to change this so it is regarded as a continuous variable (see Section 3.6). The tutor variable is a factor that indicates who each student's tutor was - either Anastasia or Bernadette.

We can calculate means and standard deviations, using the 'Exploration' – 'descriptives' analysis, and here's a nice little summary table (Table 11.2).

Table 11.2: Descriptives summary table

	mean	std dev	N
Anastasia's students	74.53	9.00	15
Bernadette's students	69.06	5.77	18

To give you a more detailed sense of what's going on here, I've plotted box and violin plots in jamovi, with mean scores added to the plot with a small solid square. These plots show the distribution of grades for both tutors (Figure 11.7),

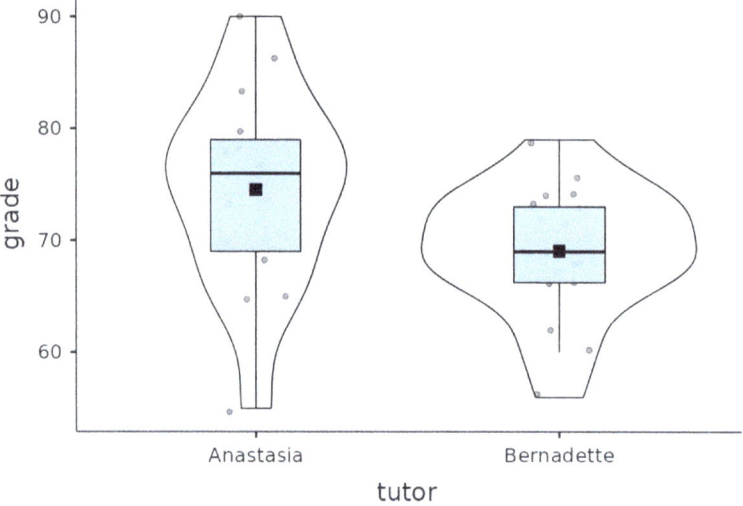

Figure 11.7: Box and violin plots from jamovi showing the distribution of grades for students in the classes of Anastasia and Bernadette. Visually, these suggest that students in the class of Anastasia may be getting slightly better grades on average, though they also seem a bit more variable

11.3.2 Introducing the test

The **independent samples t-test** comes in two different forms, Student's and Welch's. The original Student t-test, which is the one I'll describe in this section, is the simpler of the two but relies on much more restrictive assumptions than the Welch t-test. Assuming for the moment that you want to run a two-sided test, the goal is to determine whether two "independent samples" of data are drawn from populations with the same mean (the null hypothesis) or different means (the alternative hypothesis). When we say "independent" samples, what we really mean here is that there's no special relationship between observations in the two samples. This probably doesn't make a lot of sense right now, but it will be clearer when we come to talk about the paired samples t-test later on. For now, let's just point out that if we have an experimental design where participants are randomly allocated to one of two groups, and we want to compare the two groups' mean performance on some outcome measure, then an independent samples t-test (rather than a paired samples t-test) is what we're after.

Okay, so let's let μ_1 denote the true population mean for group 1 (e.g., Anastasia's students), and μ_2 will be the true population mean for group 2 (e.g., Bernadette's students),[105] and as usual we'll let $\bar{X}_1$ and $\bar{X}_2$ denote the observed sample means for both of these groups. Our null hypothesis states that the two population means are identical ($\mu_1 = \mu_2$) and the alternative to this is that they are not ($\mu_1 \neq \mu_2$) (Figure 11.8). Written in mathematical-ese, this is:

$$H_0 : \mu_1 = \mu_2$$

$$H_0 : \mu_1 \neq \mu_2$$

To construct a hypothesis test that handles this scenario we start by noting that if the null hypothesis is true, then the difference between the population means is *exactly* zero, $\mu_1 - \mu_2 = 0$. As a consequence, a diagnostic test statistic will be based on the difference between the two sample means. Because if the null hypothesis is true, then we'd expect $\bar{X}_1 - \bar{X}_2$ to be pretty close to zero. However, just like we saw with our one-sample tests (i.e., the one-sample z-test and the one-sample t-test) we have to be precise about exactly how close to zero this difference should be. And the solution to the problem is more or less the same one. We calculate a standard error estimate (SE), just like last time, and then divide the difference between means by this estimate. So our **t-statistic** will be of the form:

$$t = \frac{\bar{X}_1 - \bar{X}_2}{SE}$$

We just need to figure out what this standard error estimate actually is. This is a bit trickier than was the case for either of the two tests we've looked at so far, so we need to go through it a lot more carefully to understand how it works.

11.3.3 A "pooled estimate" of the standard deviation

In the original "Student t-test", we make the assumption that the two groups have the same population standard deviation. That is, regardless of whether the population means are the same, we assume that the population standard deviations are identical, $\sigma_1 = \sigma_2$. Since we're assuming that the two standard deviations are the same, we drop the subscripts and refer to both of them as σ. How should we estimate this? How should we construct a single estimate of a standard deviation when we have two samples? The answer is, basically, we average them. Well, sort of. Actually, what we do is take a *weighted* average of the *variance* estimates, which we use as our **pooled estimate of the variance**. The weight assigned to each sample is equal to the number of observations in that sample, minus 1.

[Additional technical detail[106]]

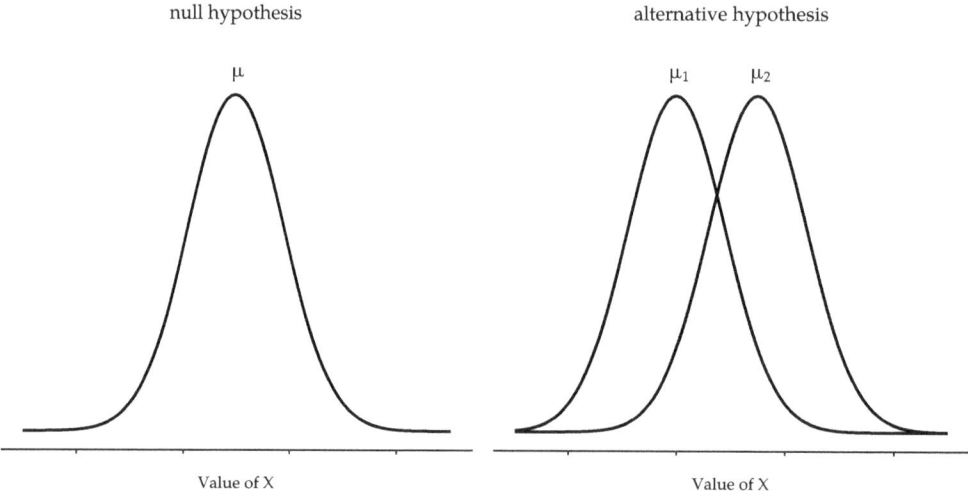

Figure 11.8: Graphical illustration of the null and alternative hypotheses assumed by the Student t-test. The null hypothesis assumes that both groups have the same mean μ, whereas the alternative assumes that they have different means μ_1 and μ_2. Notice that it is assumed that the population distributions are normal, and that, although the alternative hypothesis allows the group to have different means, it assumes they have the same standard deviation

11.4 Completing the test

Regardless of which way you want to think about it, we now have our pooled estimate of the standard deviation. From now on, I'll drop the silly p subscript, and just refer to this estimate as $\hat{\sigma}$. Great. Let's now go back to thinking about the bloody hypothesis test, shall we? Our whole reason for calculating this pooled estimate was that we knew it would be helpful when calculating our standard error estimate. But standard error of what? In the one-sample t-test it was the standard error of the sample mean, $se(\bar{X})$, and since $se(\bar{X}) = \frac{\sigma}{\sqrt{N}}$ that's what the denominator of our t-statistic looked like. This time around, however, we have two sample means. And what we're interested in, specifically, is the the difference between the two $\bar{X}_1 - \bar{X}_2$ As a consequence, the standard error that we need to divide by is in fact the **standard error of the difference** between means.

[Additional technical detail[107]]

Just as we saw with our one-sample test, the sampling distribution of this t-statistic is a t-distribution (shocking, isn't it?) as long as the null hypothesis is true and all of the assumptions of the test are met. The degrees of freedom, however, is slightly different. As usual, we can think of the degrees of freedom to be equal to the number of data points minus the number of constraints. In this case, we have N observations (N_1 in sample 1, and N_2 in sample 2), and 2 constraints (the sample means). So the total degrees of freedom for this test are $N - 2$.

11.4.1 Doing the test in jamovi

Not surprisingly, you can run an independent samples t-test easily in jamovi. The outcome variable for our test is the student grade, and the groups are defined in terms of the tutor for each class. So you probably won't be too surprised that all you have to do in jamovi is go to the relevant analysis ('Analyses' – 'T-Tests' – 'Independent Samples T-Test') and move the grade variable across to the 'Dependent Variables' box, and the tutor variable across into the 'Grouping Variable' box, as shown in Figure 11.9.

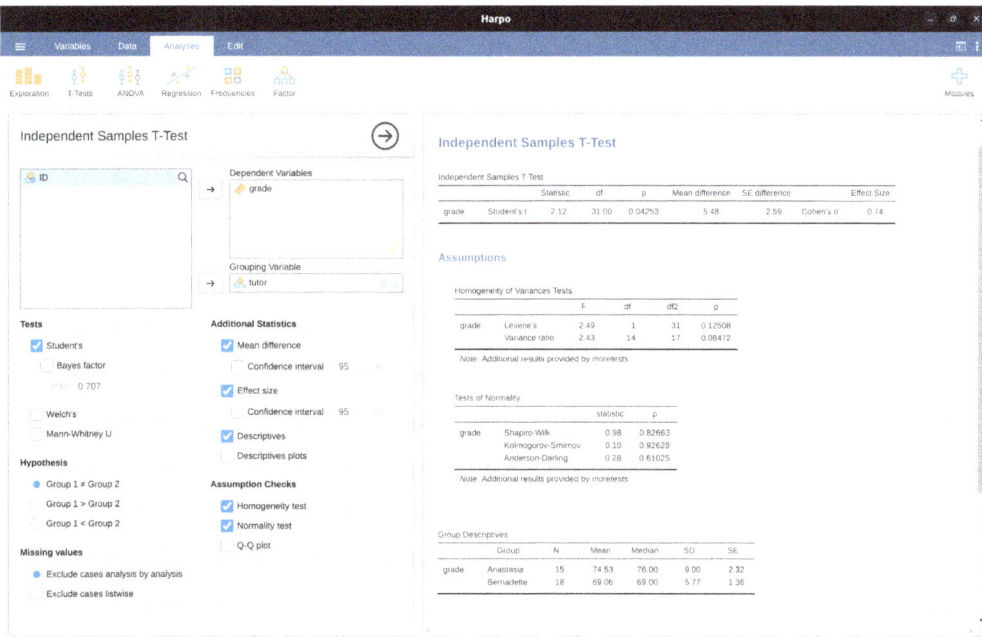

Figure 11.9: Independent t-test in jamovi, with options checked for useful results

The output has a very familiar form. First, it tells you what test was run, and it tells you the name of the dependent variable that you used. It then reports the test results. Just like last time the test results consist of a t-statistic, the degrees of freedom, and the p-value. The final section reports two things: it gives you a confidence interval and an effect size. I'll talk about effect sizes later. The confidence interval, however, I should talk about now.

It's pretty important to be clear on what this confidence interval actually refers to. It is a confidence interval for the *difference* between the group means. In our example, Anastasia's students had an average grade of 74.53, and Bernadette's students had an average grade of 69.06, so the difference between the two sample means is 5.48. But of course the difference between population means might be bigger or smaller than this. The confidence interval reported in Figure 11.10 tells you that if we replicated this study again and again, then 95% of the time the true difference in means would lie between 0.20 and 10.76. Look back at Section 8.5 for a reminder about what confidence intervals mean.

In any case, the difference between the two groups is significant (just barely), so we might write up the result using text like this:

> The mean grade in Anastasia's class was 74.5% (std dev = 9.0), whereas the mean in Bernadette's class was 69.1% (std dev = 5.8). A Student's independent samples t-test showed that this 5.4% difference was significant ($t(31) = 2.1, p < .05, CI_{95} = [0.2, 10.8], d = .74$), suggesting that a genuine difference in learning outcomes has occurred.

Notice that I've included the confidence interval and the effect size in the stat block. People don't always do this. At a bare minimum, you'd expect to see the t-statistic, the degrees of freedom and the p-value. So you should include something like this at a minimum: $t(31) = 2.1, p < .05$. If statisticians had their way, everyone would also report the confidence interval and probably the effect size measure too, because they are useful things to know. But real life doesn't always work the way statisticians want it to so you should make a judgement based on whether you think it will help your readers and, if you're writing a scientific paper, the editorial standard for the journal in question. Some journals expect you to report effect sizes, others don't. Within some scientific communities it is standard practice to report confidence intervals, in others it is not. You'll need to figure out what your audience expects. But, just for the sake of clarity, if you're taking my class, my default position is that it's usually worth including both the effect size and the confidence interval.

11.4.2 Positive and negative t-values

Before moving on to talk about the assumptions of the t-test, there's one additional point I want to make about the use of t-tests in practice. The first one relates to the sign of the t-statistic (that is, whether it is a positive number or a negative one). One very common worry that students have when they start running their first t-test is that they often end up with negative values for the t-statistic and don't know how to interpret it. In fact, it's not at all uncommon for two people working independently to end up with results that are almost identical, except that one person has a negative t-values and the other one has a positive t value. Assuming that you're running a two-sided test then the p-values will be identical. On closer inspection, the students will notice that the confidence intervals also have the opposite signs. This is perfectly okay. Whenever this happens, what you'll find is that the two versions of the results arise from slightly different ways of running the t-test. What's happening here is very simple. The t-statistic that we calculate here is always of the form:

$$t = \frac{\text{mean 1-mean 2}}{SE}$$

If "mean 1" is larger than "mean 2" the t-statistic will be positive, whereas if "mean 2" is larger then the t-statistic will be negative. Similarly, the confidence interval that jamovi reports is the confidence interval for the difference "(mean 1) minus (mean 2)", which will be the reverse of what you'd get if you were calculating the confidence interval for the difference "(mean 2) minus (mean 1)".

Okay, that's pretty straightforward when you think about it, but now consider our t-test comparing Anastasia's class to Bernadette's class. Which one should we call "mean 1" and which one should we call "mean 2". It's arbitrary. However, you really do need to designate one of them as "mean 1" and the other one as "mean 2". Not surprisingly, the way that jamovi handles this is also pretty arbitrary. In earlier versions of the book I used to try to explain it, but after a while I gave up, because it's not really all that important and to be honest I can never remember myself. Whenever I get a significant t-test result, and I want to figure out which mean is the larger one, I don't try to figure it out by looking at the t-statistic. Why would I bother doing that? It's foolish. It's easier just to look at the actual group means since the jamovi output actually shows them!

Here's the important thing. Because it really doesn't matter what jamovi shows you, I usually try to report the t-statistic in such a way that the numbers match up with the text. Suppose that what I want to write in my report is: *Anastasia's class had higher grades than Bernadette's class*. The phrasing here implies that Anastasia's group comes first, so it makes sense to report the t-statistic as if Anastasia's class corresponded to group 1. If so, I would write *Anastasia's class had higher grades than Bernadette's class* $(t(31) = 2.1, p = .04)$.

(I wouldn't actually underline the word "higher" in real life, I'm just doing it to emphasise the point that "higher" corresponds to positive t-values). On the other hand, suppose the phrasing I wanted to use has Bernadette's class listed first. If so, it makes more sense to treat her class as group 1, and if so, the write up looks like this: *Bernadette's class had lower grades than Anastasia's class* $(t(31) = -2.1, p = .04)$.

Because I'm talking about one group having "lower" scores this time around, it is more sensible to use the negative form of the t-statistic. It just makes it read more cleanly.

One last thing: please note that you can't do this for other types of test statistics. It works for t-tests, but it wouldn't be meaningful for chi-square tests, F-tests or indeed for most of the tests I talk about in this book. So don't over-generalise this advice! I'm really just talking about t-tests here and nothing else!

11.4.3 Assumptions of the Student t-test

As always, our hypothesis test relies on some assumptions. So what are they? For the Student t-test there are three assumptions, some of which we saw previously in the context of the one sample t-test (see Assumptions of the one sample t-test):

- *Normality*. Like the one-sample t-test, it is assumed that the data are normally distributed. Specifically, we assume that both groups are normally distributed.[108] In the section on Checking the normality of a sample we'll discuss how to test for normality, and in Testing non-normal data we'll discuss possible solutions.
- *Independence*. Once again, it is assumed that the observations are independently sampled. In the context of the Student test this has two aspects to it. Firstly, we assume that the observations within each sample are independent of one another (exactly the same as for the one-sample test). However, we also assume that there are no cross-sample dependencies. If, for instance, it turns out that you included some participants in both experimental conditions of your study (e.g., by accidentally allowing the same person to sign up to different conditions), then there are some cross sample dependencies that you'd need to take into account.
- *Homogeneity of variance* (also called "homoscedasticity"). The third assumption is that the population standard deviation is the same in both groups. You can test

this assumption using the Levene test, which I'll talk about later on in the book (in Section 13.6.1). However, there's a very simple remedy for this assumption if you are worried, which I'll talk about in the next section.

11.5 The independent samples t-test (Welch test)

The biggest problem with using the Student test in practice is the third assumption listed in the previous section. It assumes that both groups have the same standard deviation. This is rarely true in real life. If two samples don't have the same means, why should we expect them to have the same standard deviation? There's really no reason to expect this assumption to be true. We'll talk a little bit about how you can check this assumption later on because it does crop up in a few different places, not just the t-test. But right now I'll talk about a different form of the t-test (Welch, 1947) that does not rely on this assumption. A graphical illustration of what the Welch t-test assumes about the data is shown in Figure 11.10, to provide a contrast with the Student test version in Figure 11.8. I'll admit it's a bit odd to talk about the cure before talking about the diagnosis, but as it happens the Welch test can be specified as one of the 'Independent Samples T-Test' options in jamovi, so this is probably the best place to discuss it.

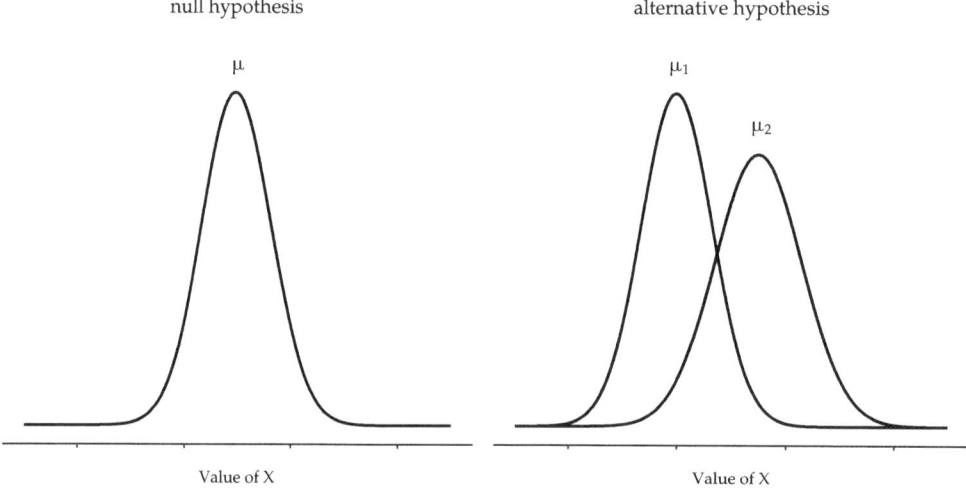

Figure 11.10: Graphical illustration of the null and alternative hypotheses assumed by the Welch t-test. Like the Student test (Figure 11.9) we assume that both samples are drawn from a normal population; but the alternative hypothesis no longer requires the two populations to have equal variance

The Welch test is very similar to the Student test. For example, the t-statistic that we use in the Welch test is calculated in much the same way as it is for the Student test. That is, we take the difference between the sample means and then divide it by some estimate of the standard error of that difference:

$$t = \frac{\bar{X}_1 - \bar{X}_2}{SE(\bar{X}_1 - \bar{X}_2)}$$

The main difference is that the standard error calculations are different. If the two populations have different standard deviations, then it's complete nonsense to try to calculate a pooled standard deviation estimate, because you're averaging apples and oranges.[109]

[Additional technical detail[110]]

The second difference between Welch and Student is that the degrees of freedom are calculated in a very different way. In the Welch test, the "degrees of freedom" doesn't have to be a whole number any more, and it doesn't correspond all that closely to the "number of data points minus the number of constraints" heuristic that I've been using up to this point.

11.5.1 Doing the Welch test in jamovi

If you tick the check box for the Welch test in the analysis we did above, then this is what it gives you (Figure 11.11).

Independent Samples T-Test

Independent Samples T-Test

		Statistic	df	p	Mean difference	SE difference		Effect Size
grade	Student's t	2.12	31.00	0.04253	5.48	2.59	Cohen's d	0.74
	Welch's t	2.03	23.02	0.05361	5.48	2.69	Cohen's d	0.72

Figure 11.11: Results showing the Welch test alongside the default Students t-test in jamovi

The interpretation of this output should be fairly obvious. You read the output for the Welch's test in the same way that you would for the Student's test. You've got your descriptive statistics, the test results and some other information. So that's all pretty easy.

Except, except...our result isn't significant anymore. When we ran the Student test we did get a significant effect, but the Welch test on the same data set is not ($t(23.02) = 2.03, p = .054$). What does this mean? Should we panic? Is the sky burning? Probably not. The fact that one test is significant and the other isn't doesn't itself mean very much, especially since I kind of rigged the data so that this would happen. As a general rule, it's not a good idea to go out of your way to try to interpret or explain the difference between a p-value of .049 and a p-value of .051. If this sort of thing happens in real life, the *difference* in these p-values is almost certainly due to chance. What does matter is that you take a little bit of care in thinking about what test you use. The Student test and the Welch test have different strengths and weaknesses. If the two populations really do have equal variances, then the Student test is slightly more powerful (lower type II error rate) than the Welch test. However, if they *don't* have the same variances, then the assumptions of the Student test are violated and you may not be able to trust it; you might end up with a higher type I error rate. So it's a trade off. However, in real life I tend to prefer the Welch test, because almost no-one actually believes that the population variances are identical.

11.5.2 Assumptions of the Welch test

The assumptions of the Welch test are very similar to those made by the Student t-test (see Assumptions of the Student t-test, except that the Welch test does not assume homogeneity of variance. This leaves only the assumption of normality and the assumption of independence. The specifics of these assumptions are the same for the Welch test as for the Student test.

11.6 The paired-samples t-test

Regardless of whether we're talking about the Student test or the Welch test, an independent samples t-test is intended to be used in a situation where you have two samples that are, well, independent of one another. This situation arises naturally when participants are assigned randomly to one of two experimental conditions, but it provides a very poor approximation to other sorts of research designs. In particular, a repeated measures design, in which each participant is measured (with respect to the same outcome variable) in both experimental conditions, is not suited for analysis using independent samples t-tests. For example, we might be interested in whether listening to music reduces people's working memory capacity. To that end, we could measure each person's working memory capacity in two conditions: with music, and without music. In an experimental design such as this one, [111] each participant appears in *both* groups. This requires us to approach the problem in a different way, by using the **paired samples t-test**.

11.6.1 The data

The data set that we'll use this time comes from Dr Chico's class.[112] In her class students take two major tests, one early in the semester and one later in the semester. To hear her tell it, she runs a very hard class, one that most students find very challenging. But she argues that by setting hard assessments students are encouraged to work harder. Her theory is that the first test is a bit of a "wake up call" for students. When they realise how hard her class really is, they'll work harder for the second test and get a better mark. Is she right? To test this, let's import the *chico.csv* file into jamovi. This time jamovi does a good job during the import of attributing measurement levels correctly. The *chico* data set contains three variables: an id variable that identifies each student in the class, the grade_test1 variable that records the student grade for the first test, and the grade_test2 variable that has the grades for the second test.

If we look at the jamovi spreadsheet it does seem like the class is a hard one (most grades are between 50% and 60%), but it does look like there's an improvement from the first test to the second one.

If we take a quick look at the descriptive statistics, in Figure 11.12, we see that this impression seems to be supported. Across all 20 students the mean grade for the first test is 57%, but this rises to 58% for the second test. Although, given that the standard deviations are 6.6% and 6.4% respectively, it's starting to feel like maybe the improvement is just illusory; maybe just random variation. This impression is reinforced when you see the means and confidence intervals plotted in Figure 11.13a. If we were to rely on this plot alone, looking at how wide those confidence intervals are, we'd be tempted to think that the apparent improvement in student performance is pure chance.

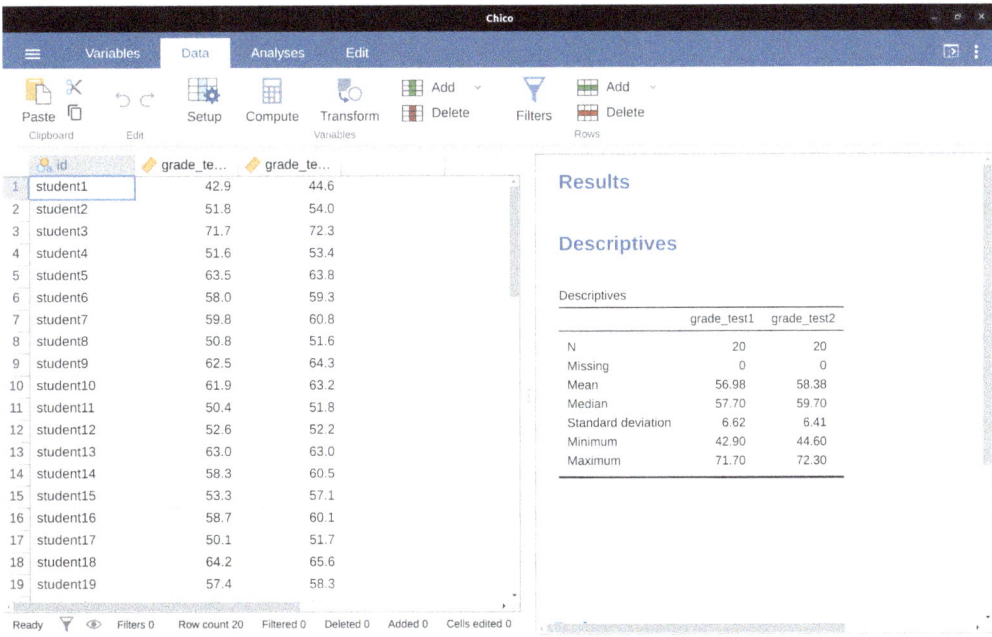

Figure 11.12: Descriptives for the two grade test variables in the *chico* data set

Nevertheless, this impression is wrong. To see why, take a look at the scatterplot of the grades for test 1 against the grades for test 2, shown in Figure 11.13b. In this plot each dot corresponds to the two grades for a given student. If their grade for test 1 (x co-ordinate) equals their grade for test 2 (y co-ordinate), then the dot falls on the line. Points falling above the line are the students that performed better on the second test. Critically, almost all of the data points fall above the diagonal line: almost all of the students do seem to have improved their grade, if only by a small amount. This suggests that we should be looking at the improvement made by each student from one test to the next and treating that as our raw data. To do this, we'll need to create a new variable for the improvement that each student makes, and add it to the *chico* data set. The easiest way to do this is to compute a new variable, with the expression grade_test2 - grade_test1.

Once we have computed this new improvement variable we can draw a histogram showing the distribution of these improvement scores, shown in Figure 11.14. When we look at the histogram, it's very clear that there is a real improvement here. The vast majority of the students scored higher on test 2 than on test 1, reflected in the fact that almost the entire histogram is above zero.

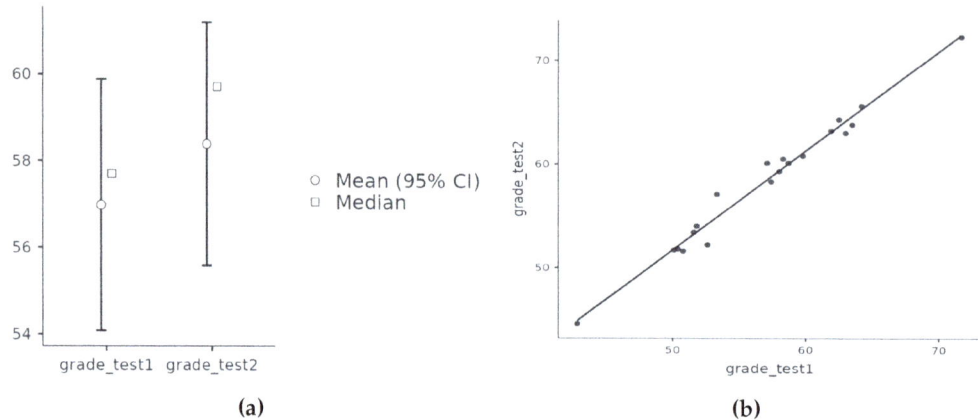

Figure 11.13: Mean grade for test 1 and test 2, with associated 95% confidence intervals (a). Scatterplot showing the individual grades for test 1 and test 2 (b)

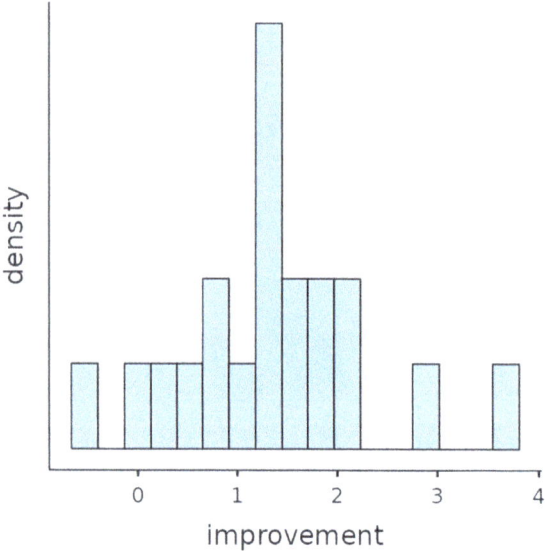

Figure 11.14: Histogram from jamovi showing the improvement made by each student in Dr Chico's class. Notice that almost the entire distribution is above zero – the vast majority of students did improve their performance from the first test to the second one

11.6.2 What is the paired samples t-test?

In light of the previous exploration, let's think about how to construct an appropriate t-test. One possibility would be to try to run an independent samples t-test using grade_test1 and grade_test2 as the variables of interest. However, this is clearly the wrong thing to do as the independent samples t-test assumes that there is no particular relationship between the two samples. Yet clearly that's not true in this case because of the repeated measures structure in the data. To use the language that I introduced in the last section, if we were to try to do an independent samples t-test, we would be conflating the **within subject** differences (which is what we're interested in testing) with the **between subject** variability (which we are not).

The solution to the problem is obvious, I hope, since we already did all the hard work in the previous section. Instead of running an independent samples t-test on grade_test1 and grade_test2, we run a one-sample t-test on the within-subject difference variable, improvement. To formalise this slightly, if X_{i1} is the score that the i-th participant obtained on the first variable, and X_{i2} is the score that the same person obtained on the second one, then the difference score is:

$$D_i = X_{i1} - X_{i2}$$

Notice that the difference scores is variable 1 minus variable 2 and not the other way around, so if we want improvement to correspond to a positive valued difference, we actually want "test 2" to be our "variable 1". Equally, we would say that $\mu_D = \mu_1 - \mu_2$ is the population mean for this difference variable. So, to convert this to a hypothesis test, our null hypothesis is that this mean difference is zero and the alternative hypothesis is that it is not:

$$H_0 : \mu_D = 0$$
$$H_1 : \mu_D \neq 0$$

This is assuming we're talking about a two-sided test here. This is more or less identical to the way we described the hypotheses for the one-sample t-test. The only difference is that the specific value that the null hypothesis predicts is 0. And so our t-statistic is defined in more or less the same way too. If we let $\bar{D}$ denote the mean of the difference scores, then:

$$t = \frac{\bar{D}}{SE(\bar{D})}$$

which is:

$$t = \frac{\bar{D}}{\frac{\hat{\sigma}_D}{\sqrt{N}}}$$

where $\hat{\sigma}_D$ is the standard deviation of the difference scores. Since this is just an ordinary, one-sample t-test, with nothing special about it, the degrees of freedom are still $N - 1$. And that's it. The paired samples t-test really isn't a new test at all. It's a one-sample t-test, but applied to the difference between two variables. It's actually very simple. The only reason it merits a discussion as long as the one we've just gone through is that you need to be able to recognise *when* a paired samples test is appropriate, and to understand *why* it's better than an independent samples t-test.

11.6.3 Doing the test in jamovi

How do you do a paired samples t-test in jamovi? One possibility is to follow the process I outlined above. That is, create a "difference" variable and then run a one sample t-test on that. Since we've already created a variable called "improvement", let's do that and see what we get, Figure 11.15.

One Sample T-Test

One Sample T-Test

		Statistic	df	p	Mean difference	95% Confidence Interval		Effect Size	
						Lower	Upper		
improvement	Student's t	6.48	19.00	< .00001	1.40	0.95	1.86	Cohen's d	1.45

Figure 11.15: Results showing a one sample t-test on paired difference scores

The output shown in Figure 11.15 is (obviously) formatted exactly the same was as it was the last time we used the one-sample t-test analysis (Section 11.2), and it confirms our intuition. There's an average improvement of 1.4% from test 1 to test 2, and this is significantly different from 0 ($t(19) = 6.48, p < .001$).

However, suppose you're lazy and you don't want to go to all the effort of creating a new variable. Or perhaps you just want to keep the difference between one-sample and paired samples tests clear in your head. If so, you can use the jamovi 'Paired Samples T-Test' analysis, getting the results shown in Figure 11.16.

Paired Samples T-Test

Paired Samples T-Test

			statistic	df	p	Mean difference	SE difference	95% Confidence Interval		Effect Size	
								Lower	Upper		
grade_test2	grade_test1	Student's t	6.48	19.00	< .00001	1.40	0.22	0.95	1.86	Cohen's d	1.45

Figure 11.16: Results showing a paired sample t-test. Compare with Figure 11.15

The numbers are identical to those that come from the one sample test, which of course they have to be given that the paired samples t-test is just a one sample test under the hood.

11.7 One-sided tests

When introducing the theory of null hypothesis tests, I mentioned that there are some situations when it's appropriate to specify a one-sided test (see Section 9.4.3). So far all of the t-tests have been two-sided tests. For instance, when we specified a one sample t-test for the grades in Dr Zeppo's class the null hypothesis was that the true mean was 67.5%. The alternative hypothesis was that the true mean was greater than or less than 67.5%. Suppose we were only interested in finding out if the true mean is greater than 67.5%, and have no interest whatsoever in testing to find out if the true mean is lower than 67.5%. If so, our null hypothesis would be that the true mean is 67.5% or less, and the alternative hypothesis would be that the true mean is greater than 67.5%. In jamovi, for the 'One Sample T-Test' analysis, you can specify this by clicking on the '> Test Value' option, under 'Hypothesis'. When you have done this, you will get the results as shown in Figure 11.17.

One Sample T-Test

One Sample T-Test

						95% Confidence Interval			
		Statistic	df	p	Mean difference	Lower	Upper		Effect Size
x	Student's t	2.25	19.00	0.01807	4.80	1.12	Inf	Cohen's d	0.50

Note. H_a μ > 67.5

Figure 11.17: jamovi results showing a 'One Sample T-Test' where the actual hypothesis is one-sided, i.e. that the true mean is greater than 67.5%

Notice that there are a few changes from the output that we saw last time. Most important is the fact that the actual hypothesis has changed, to reflect the different test. The second thing to note is that although the t-statistic and degrees of freedom have not changed, the p-value has. This is because the one-sided test has a different rejection region from the two-sided test. If you've forgotten why this is and what it means, you may find it helpful to read back over Chapter 9, and Section 9.4.3 in particular. The third thing to note is that the confidence interval is different too: it now reports a one-sided confidence interval rather than a two-sided one. In a two-sided confidence interval we're trying to find numbers a and b such that we're confident that, if we were to repeat the study many times, then 95% of the time the mean would lie between a and b. In a one-sided confidence interval, we're trying to find a single number a such that we're confident that 95% of the time the true mean would be greater than a (or less than a if you selected Measure 1 < Measure 2 in the 'Hypothesis' section).

So that's how to do a one-sided one sample t-test. However, all versions of the t-test can be one-sided. For an independent samples t-test, you could have a one-sided test if you're only interested in testing to see if group A has higher scores than group B, but have no interest in finding out if group B has higher scores than group A. Let's suppose that, for Dr Harpo's class, you wanted to see if Anastasia's students had higher grades than Bernadette's. For this analysis, in the 'Hypothesis' options, specify that 'Group 1 > Group2'. You should get the results shown in Figure 11.18.

Again, the output changes in a predictable way. The definition of the alternative hypothesis has changed, the p-value has changed, and it now reports a one-sided confidence interval rather than a two-sided one.

Independent Samples T-Test

Independent Samples T-Test

		Statistic	df	p	Mean difference	SE difference		Effect Size
grade	Student's t	2.12	31.00	0.02126	5.48	2.59	Cohen's d	0.74
	Welch's t	2.03	23.02	0.02680	5.48	2.69	Cohen's d	0.72

Note. H_a $\mu_{Anastasia} > \mu_{Bernadette}$

Figure 11.18: jamovi results showing an 'Independent Samples T-Test' where the actual hypothesis is one-sided, i.e. that Anastasia's students had higher grades than Bernadette's

What about the paired samples t-test? Suppose we wanted to test the hypothesis that grades go up from test 1 to test 2 in Dr Zeppo's class, and are not prepared to consider the idea that the grades go down. In jamovi you would do this by specifying, under the 'Hypotheses' option, that grade_test2 ('Measure 1' in jamovi, because we copied this first into the paired variables box) > grade_test1 ('Measure 2' in jamovi). You should get the results shown in Figure 11.19.

Paired Samples T-Test

Paired Samples T-Test

			statistic	df	p	Mean difference	SE difference	95% Confidence Interval			Effect Size
								Lower	Upper		
grade_test2	grade_test1	Student's t	6.48	19.00	<.00001	1.40	0.22	1.03	Inf	Cohen's d	1.45

Note. H_a $\mu_{Measure\ 1\ -\ Measure\ 2} > 0$

Figure 11.19: jamovi results showing a 'Paired Samples T-Test' where the actual hypothesis is one-sided, i.e. that grade_test2 ('Measure 1') > grade_test1 ('Measure 2')

Yet again, the output changes in a predictable way. The hypothesis has changed, the p-value has changed, and the confidence interval is now one-sided.

11.8 Effect size

The most commonly used measure of effect size for a t-test is *Cohen's d* (Cohen, 1988). It's a very simple measure in principle, with quite a few wrinkles when you start digging into the details. Cohen himself defined it primarily in the context of an independent samples t-test, specifically the Student test. In that context, a natural way of defining the effect size is to divide the difference between the means by an estimate of the standard deviation. In other words, we're looking to calculate something along the lines of this:

$$d = \frac{(\text{mean 1}) - (\text{mean 2})}{\text{std dev}}$$

and he suggested a rough guide for interpreting d in Table 11.3.

Table 11.3: A (very) rough guide to interpreting Cohen's d. My personal recommendation is to not use these blindly. The d statistic has a natural interpretation in and of itself. It re-describes the difference in means as the number of standard deviations that separates those means. So it's generally a good idea to think about what that means in practical terms. In some contexts a "small" effect could be of big practical importance. In other situations a "large" effect may not be all that interesting

d-value	rough interpretation
about 0.2	"small" effect
about 0.5	"moderate" effect
about 0.8	"large" effect

You'd think that this would be pretty unambiguous, but it's not. This is largely because Cohen wasn't too specific on what he thought should be used as the measure of the standard deviation (in his defence he was trying to make a broader point in his book, not nitpick about tiny details). As discussed by McGrath & Meyer (2006), there are several different versions in common usage, and each author tends to adopt slightly different notation. For the sake of simplicity (as opposed to accuracy), I'll use d to refer to any statistic that you calculate from the sample, and use δ to refer to a theoretical population effect. Obviously, that does mean that there are several different things all called d.

My suspicion is that the only time that you would want Cohen's d is when you're running a t-test, and jamovi has an option to calculate the effect size for all the different flavours of t-test it provides.

11.8.1 Cohen's d from one sample

The simplest situation to consider is the one corresponding to a one-sample t-test. In this case, this is the one sample mean $\bar{X}$ and one (hypothesised) population mean μ_0 to compare it to. Not only that, there's really only one sensible way to estimate the population standard deviation. We just use our usual estimate $\hat{\sigma}$. Therefore, we end up with the following as the only way to calculate d:

$$d = \frac{\bar{X} - \mu_0}{\hat{\sigma}}$$

When we look back at the results in Figure 11.6, the effect size value is Cohen's $d = 0.50$. Overall, then, the psychology students in Dr Zeppo's class are achieving grades ($mean = 72.3\%$) that are about .5 standard deviations higher than the level that you'd expect (67.5%) if they were performing at the same level as other students. Judged against Cohen's rough guide, this is a moderate effect size.

11.8.2 Cohen's d from a Student's t-test

The majority of discussions of Cohen's d focus on a situation that is analogous to Student's independent samples t-test, and it's in this context that the story becomes messier, since there are several different versions of d that you might want to use in this situation. To understand why there are multiple versions of d, it helps to take the time to write down a formula that corresponds to the true population effect size δ. It's pretty straightforward:

$$\delta = \frac{\mu_1 - \mu_2}{\sigma}$$

where, as usual, μ_1 and μ_2 are the population means corresponding to group 1 and group 2 respectively, and σ is the standard deviation (the same for both populations). The obvious way to estimate δ is to do exactly the same thing that we did in the t-test itself, i.e., use the sample means as the top line and a pooled standard deviation estimate for the bottom line:

$$d = \frac{\bar{X}_1 - \bar{X}_2}{\hat{\sigma}_p}$$

where $\hat{\sigma}_p$ is the exact same pooled standard deviation measure that appears in the t-test. This is the most commonly used version of Cohen's d when applied to the outcome of a Student t-test, and is the one provided in jamovi. It is sometimes referred to as Hedges' g statistic (Hedges, 1981).

However, there are other possibilities which I'll briefly describe. Firstly, you may have reason to want to use only one of the two groups as the basis for calculating the standard deviation. This approach (often called Glass' Δ, pronounced delta) only makes most sense when you have good reason to treat one of the two groups as a purer reflection of "natural variation" than the other. This can happen if, for instance, one of the two groups is a control group. Secondly, recall that in the usual calculation of the pooled standard deviation we divide by $N - 2$ to correct for the bias in the sample variance. In one version of Cohen's d this correction is omitted, and instead we divide by N. This version makes sense primarily when you're trying to calculate the effect size in the sample rather than estimating an effect size in the population. Finally, there is a version called Hedge's g, based on Hedges & Olkin (1985), who point out there is a small bias in the usual (pooled) estimation for Cohen's d.[113]

In any case, ignoring all those variations that you could make use of if you wanted, let's have a look at the default version in jamovi. In Figure 11.10 Cohen's $d = 0.74$, indicating that the grade scores for students in Anastasia's class are, on average, 0.74 standard deviations higher than the grade scores for students in Bernadette's class. For a Welch test, the estimated effect size is the same (Figure 11.12).

11.8.3 Cohen's d from a paired-samples test

Finally, what should we do for a paired samples t-test? In this case, the answer depends on what it is you're trying to do. jamovi assumes that you want to measure your effect sizes relative to the distribution of difference scores, and the measure of d that you calculate is:

$$d = \frac{\bar{D}}{\hat{\sigma}_D}$$

where $\hat{\sigma}_D$ is the estimate of the standard deviation of the differences. In Figure 11.16 Cohen's $d = 1.45$, indicating that the time 2 grade scores are, on average, 1.45 standard deviations higher than the time 1 grade scores.

This is the version of Cohen's d that gets reported by the jamovi 'Paired Samples T-Test' analysis. The only wrinkle is figuring out whether this is the measure you want or not. To the extent that you care about the practical consequences of your research, you often want to measure the effect size relative to the *original* variables, not the *difference scores* (e.g., the 1% improvement in Dr Chico's class over time is pretty small when measured against the amount of between-student variation in grades), in which case you use the same versions of Cohen's d that you would use for a Student or Welch test. It's not so straightforward to do this in jamovi; essentially you have to change the structure of the data in the spreadsheet view so I won't go into that here,[114] but the Cohen's d for this perspective is quite different: it is 0.22 which is quite small when assessed on the scale of the original variables.

11.9 Checking the normality of a sample

All of the tests that we have discussed so far in this chapter have assumed that the data are normally distributed. This assumption is often quite reasonable, because the central limit theorem (see Section 8.3.3) does tend to ensure that many real world quantities are normally distributed. Any time that you suspect that your variable is *actually* an average of lots of different things, there's a pretty good chance that it will be normally distributed, or at least close enough to normal that you can get away with using t-tests. However, life doesn't come with guarantees, and besides there are lots of ways in which you can end up with variables that are highly non-normal. For example, any time you think that your variable is actually the minimum of lots of different things, there's a very good chance it will end up quite skewed. In psychology, response time (RT) data is a good example of this. If you suppose that there are lots of things that could trigger a response from a human participant, then the actual response will occur the first time one of these trigger events occurs.[115] This means that RT data are systematically non-normal. Okay, so if normality is assumed by all the tests, and is mostly but not always satisfied (at least approximately) by real world data, how can we check the normality of a sample? In this section I discuss two methods: QQ plots and the Shapiro-Wilk test.

11.9.1 QQ plots

One way to check whether a sample violates the normality assumption is to draw a "QQ plot" (Quantile-Quantile plot). This allows you to visually check whether you're seeing any systematic violations. In a QQ plot, each observation is plotted as a single dot. The x co-ordinate is the theoretical quantile that the observation should fall in if the data were normally distributed (with mean and variance estimated from the sample), and on the y co-ordinate is the actual quantile of the data within the sample. If the data are normal, the dots should form a straight line. For instance, lets see what happens if we generate data by sampling from a normal distribution, and then drawing a QQ plot. The results are shown in Figure 11.20.

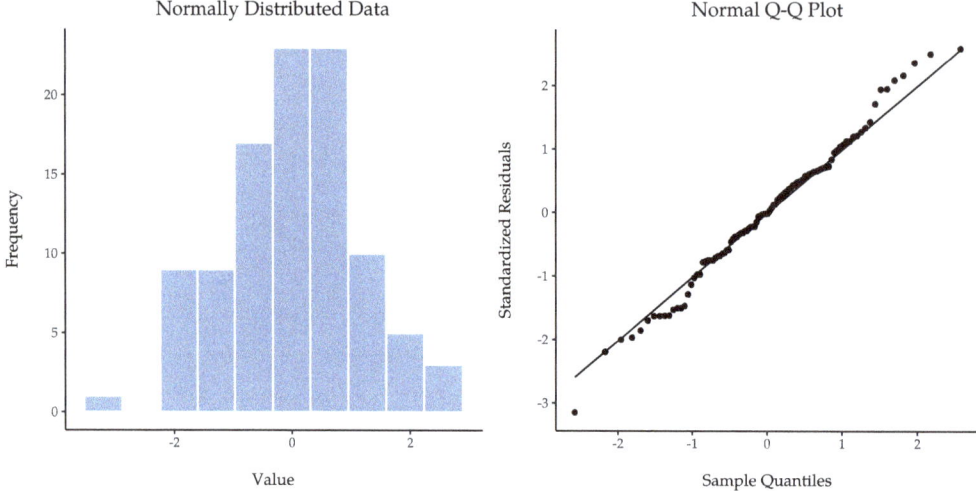

Figure 11.20: Histogram (panel (a)) and normal QQ plot (panel (b)) of normal.data, a normally distributed sample with 100 observations. The Shapiro-Wilk statistic associated with these data is $W = .99$, indicating that no significant departures from normality were detected ($p = .54$)

As you can see, these data form a pretty straight line; which is no surprise given that we sampled them from a normal distribution! In contrast, have a look at the two data sets shown in Figure 11.21. The top panels show the histogram and a QQ plot for a data set that is highly skewed: the QQ plot curves upwards. The lower panels show the same plots for a heavy tailed (i.e., high kurtosis) data set: in this case the QQ plot flattens in the middle and curves sharply at either end.

11.9.2 QQ plots for independent and paired t-tests

In our previous analyses we showed how to conduct in jamovi an independent t-test (Figure 11.10) and a paired samples t-test (Figure 11.16). And for these analyses jamovi provides an option to show a QQ plot for the difference scores (which jamovi calls the 'residuals'), which is a better way of checking the normality assumption. When we select this option for these analyses, we get the QQ plots shown in Figure 11.22 and Figure 11.23, respectively. My interpretation is that these plots both show that the difference scores are reasonably normally distributed, so we are good to go!

11.9.3 Shapiro-Wilk tests

QQ plots provide a nice way to informally check the normality of your data, but sometimes you'll want to do something a bit more formal and the **Shapiro-Wilk test** (Shapiro & Wilk, 1965) is probably what you're looking for.[116] As you'd expect, the null hypothesis being tested is that a set of N observations is normally distributed.

[Additional technical detail[117]]

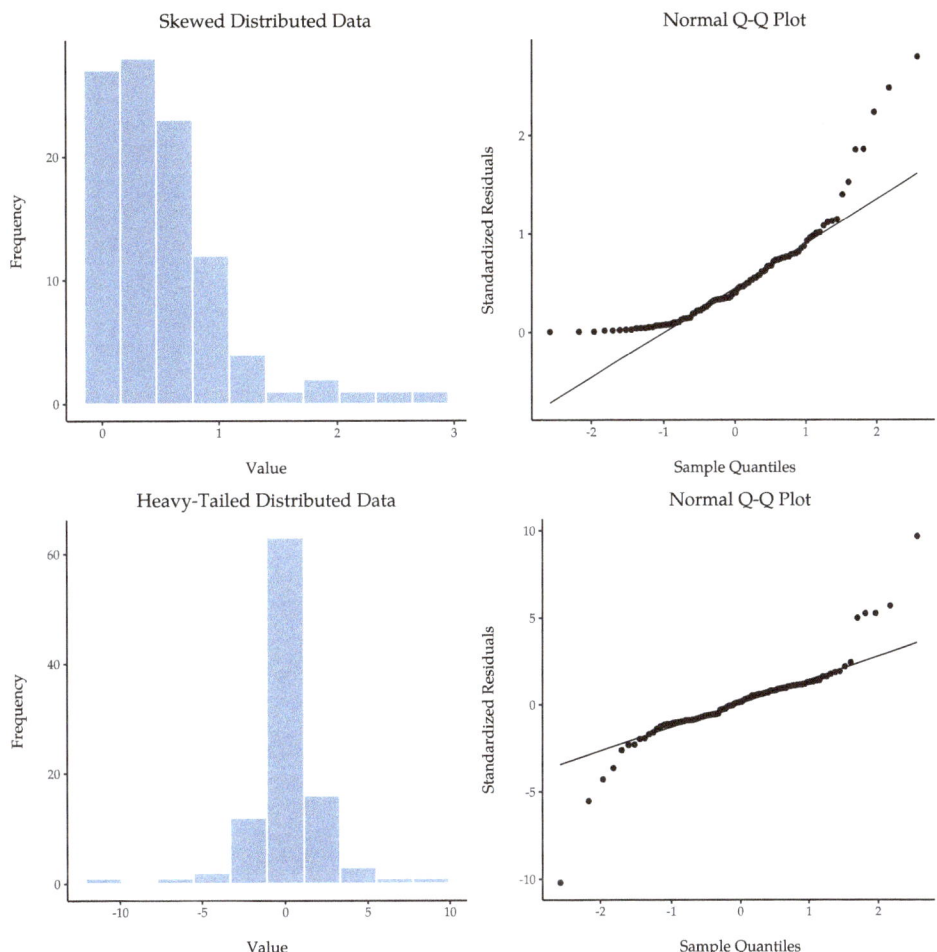

Figure 11.21: In the top row, a histogram and a normal QQ plot of the 100 observations in a skewed data set. The skewness of the data here is 1.88, and is reflected in a QQ plot that curves upwards. As a consequence, the Shapiro-Wilk statistic is $W = .80$, reflecting a significant departure from normality ($p < .001$). The bottom row shows the same plots for a heavy tailed data set, again consisting of 100 observations. In this case the heavy tails in the data produce a high kurtosis (6.57), and cause the QQ plot to flatten in the middle, and curve away sharply on either side. The resulting Shapiro-Wilk statistic is $W = .75$, again reflecting significant non-normality ($p < .001$)

To get the Shapiro-Wilk statistic in jamovi t-tests, check the option for 'Normality' listed under 'Assumptions'. In the randomly sampled data ($N = 100$) we used for the QQ plot, the value for the Shapiro-Wilk normality test statistic was $W = 0.99$ with a p-value of 0.54. So, not surprisingly, we have no evidence that these data depart from normality. When reporting the results for a Shapiro-Wilk test, you should (as usual) make sure to include the test statistic W and the p-value, though given that the sampling distribution depends so heavily on N it would probably be a politeness to include N as well.

grade

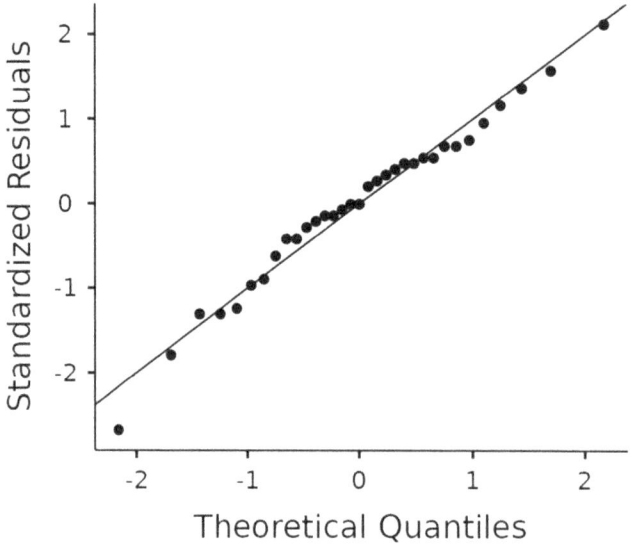

Figure 11.22: jamovi QQ plot for the independent t-test analysis shown in Figure 11.10

grade_test2 - grade_test1

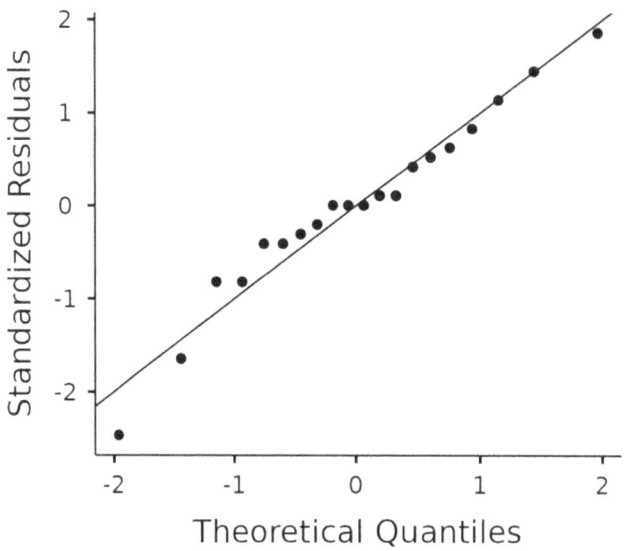

Figure 11.23: jamovi QQ plot for the paired samples t-test analysis shown in Figure 11.16

11.9.4 Example

In the meantime, it's probably worth showing you an example of what happens to the QQ plot and the Shapiro-Wilk test when the data turn out to be non-normal. For that, let's look at the distribution of our AFL winning margins data, which if you remember back to Chapter 4 it didn't look like they came from a normal distribution at all. Here's what happens to the QQ plot (Figure 11.24).

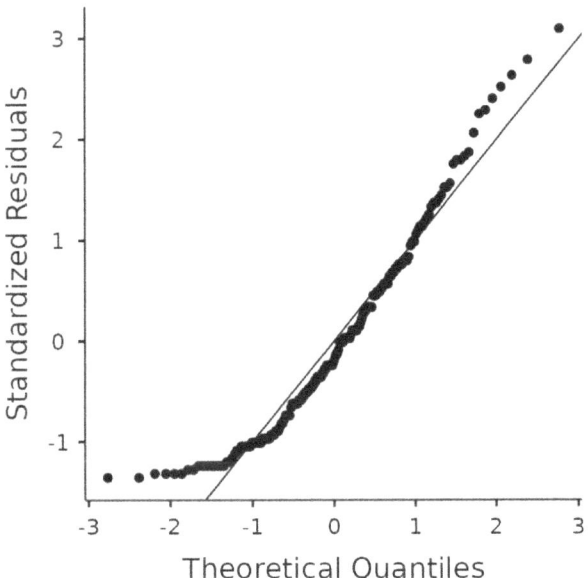

Figure 11.24: jamovi QQ plot showing non-normality from the AFL winning margins data

And when we run the Shapiro-Wilk test on the *afl.margins* data, we get a value for the Shapiro-Wilk normality test statistic of $W = 0.94$, and $p-value = 9.481x10^{-07}$. Clearly a significant effect!

11.10 Testing non-normal data

Okay, suppose your data turn out to be pretty substantially non-normal, but you still want to run something like a t-test? This situation occurs a lot in real life. For the AFL winning margins data, for instance, the Shapiro-Wilk test made it very clear that the normality assumption is violated. This is the situation where you want to use Wilcoxon tests.

Like the t-test, the Wilcoxon test comes in two forms, one-sample and two-sample, and they're used in more or less the exact same situations as the corresponding t-tests. Unlike the t-test, the Wilcoxon test doesn't assume normality, which is nice. In fact, they don't make any assumptions about what kind of distribution is involved. In statistical jargon, this makes them **nonparametric tests**. While avoiding the normality assumption is nice, there's a drawback: the Wilcoxon test is usually less powerful than the t-test (i.e., higher type II error rate). I won't discuss the Wilcoxon tests in as much detail as the t-tests, but I'll give you a brief overview.

11.10.1 Two sample Mann-Whitney U test

I'll start by describing the Mann-Whitney U test, since it's actually simpler than the one sample version. Suppose we're looking at the scores of 10 people on some test. Since my imagination has now failed me completely, let's pretend it's a "test of awesomeness" and there are two groups of people, "A" and "B". I'm curious to know which group is more awesome. The data are included in the file *awesome.csv*, and there are two variables apart from the usual ID variable: scores and group.

As long as there are no ties (i.e., people with the exact same awesomeness score) then the test that we want to do is surprisingly simple. All we have to do is construct a table that compares every observation in group A against every observation in group B. Whenever the group A datum is larger, we place a check mark in the table (Table 11.4).

Table 11.4: Comparing observations by group for a two-sample Mann-Whitney U test

		group B				
		14.5	10.4	12.4	11.7	13.0
group A	6.4	.	.	.	.	.
	10.7	.	✓	.	.	.
	11.9	.	✓	.	✓	.
	7.3	.	.	.	.	.
	10	.	.	.	.	.

We then count up the number of checkmarks. This is our test statistic, W.[118] The actual sampling distribution for W is somewhat complicated, and I'll skip the details. For our purposes, it's sufficient to note that the interpretation of W is qualitatively the same as the interpretation of t or z. That is, if we want a two-sided test then we reject the null hypothesis when W is very large or very small, but if we have a directional (i.e., one-sided) hypothesis then we only use one or the other.

In jamovi, if we run an 'Independent Samples T-Test' with scores as the dependent variable. and group as the grouping variable, and then under the options for 'tests' check the option for 'Mann-Whitney U', we will get results showing that $U = 3$ (i.e., the same number of checkmarks as shown above), and a p-value = 0.05556. See Figure 11.25. Counting up the tick marks this time we get a test statistic of $W = 7$. As before, if our test is two-sided, then we reject the null hypothesis when W is very large or very small.

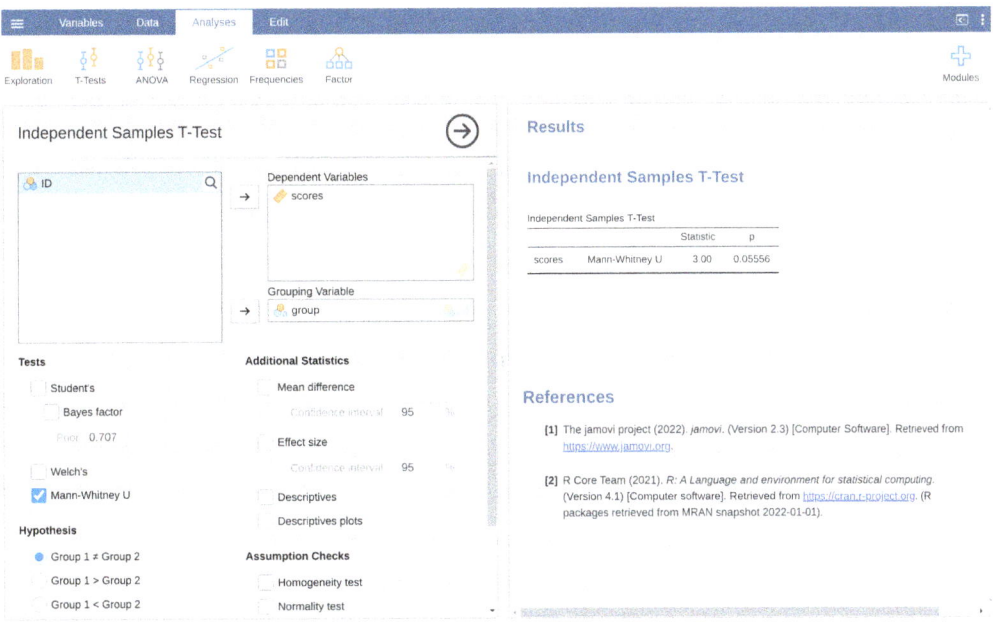

Figure 11.25: jamovi screen showing results for the Mann-Whitney U test

11.10.2 One sample Wilcoxon test

What about the **one sample Wilcoxon test** (or equivalently, the paired samples Wilcoxon test)? Suppose I'm interested in finding out whether taking a statistics class has any effect on the happiness of students. My data is in the *happiness.csv* file. What I've measured here is the happiness of each student before taking the class and after taking the class, and the change score is the difference between the two. Just like we saw with the t-test, there's no fundamental difference between doing a paired-samples test using before and after, versus doing a one-sample test using the change scores. As before, the simplest way to think about the test is to construct a tabulation. The way to do it this time is to take those change scores that are positive differences, and tabulate them against all the complete sample. What you end up with is a table that looks like Table 11.5.

Table 11.5: Comparing observations by group for a one-sample Wilcoxon U test

		all differences									
positive differ- ences		−24	−14	−10	7	−6	−38	2	−35	−30	5
	7	.	.	.	✓	✓	.	✓	.	.	✓
	2	.	.	.	.	.	.	✓	.	.	.
	5	.	.	.	.	.	.	✓	.	.	✓

As far as running it in jamovi goes, it's pretty much what you'd expect. For the one sample version, you specify the 'Wilcoxon rank' option under 'Tests' in the 'One Sample

T-Test' analysis window. This gives you Wilcoxon $W = 7$, p-value = 0.03711. As this shows, we have a significant effect. Evidently, taking a statistics class does have an effect on your happiness. Switching to a paired samples version of the test won't give us a different answer, of course; see Figure 11.26.

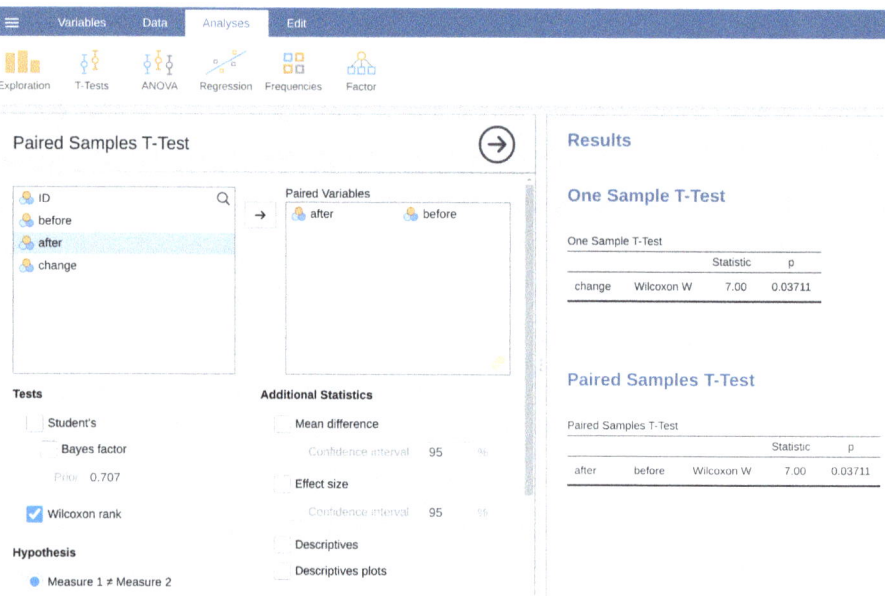

Figure 11.26: jamovi screen showing results for one-sample and paired sample Wilcoxon nonparametric tests

11.11 Summary

- The one-sample t-test is used to compare a single sample mean against a hypothesised value for the population mean.
- An independent samples t-test is used to compare the means of two groups, and tests the null hypothesis that they have the same mean. It comes in two forms: The independent samples t-test (Student test) assumes that the groups have the same standard deviation, The independent samples t-test (Welch test) does not.
- The paired-samples t-test is used when you have two scores from each person, and you want to test the null hypothesis that the two scores have the same mean. It is equivalent to taking the difference between the two scores for each person, and then running a one sample t-test on the difference scores.
- One-sided tests are perfectly legitimate as long as they are pre-planned (like all tests!).
- Effect size calculations for the difference between means can be calculated via the Cohen's d statistic.
- Checking the normality of a sample using QQ plots and the Shapiro-Wilk test.
- If your data are non-normal, you can use Mann-Whitney or Wilcoxon tests instead of t-tests for Testing non-normal data.

Chapter 12

Correlation and linear regression

The goal in this chapter is to introduce **correlation** and **linear regression**. These are the standard tools that statisticians rely on when analysing the relationship between continuous predictors and continuous outcomes.

12.1 Correlations

In this section we'll talk about how to describe the relationships between variables in the data. To do that, we want to talk mostly about the **correlation** between variables. But first, we need some data (Table 12.1).

12.1.1 The data

Table 12.1: Data for correlation analysis – descriptive statistics for the *parenthood* data

variable	min	max	mean	median	std. dev	IQR
Dani's grumpiness	41	91	63.71	62	10.05	14
Dani's hours slept	4.84	9.00	6.97	7.03	1.02	1.45
Dani's son's hours slept	3.25	12.07	8.05	7.95	2.07	3.21

Let's turn to a topic close to every parent's heart: sleep. The data set we'll use is fictitious, but based on real events. Suppose I'm curious to find out how much my infant son's sleeping habits affect my mood. Let's say that I can rate my grumpiness very precisely, on a scale from 0 (not at all grumpy) to 100 (grumpy as a very, very grumpy old man or woman). And lets also assume that I've been measuring my grumpiness, my sleeping patterns and my son's sleeping patterns for quite some time now. Let's say, for 100 days. And, being a nerd, I've saved the data as a file called *parenthood.csv*. If we load the data we can see that the file contains four variables dani.sleep, baby.sleep, dani.grump and day. Note that when you first load this data set jamovi may not have guessed the data type for each variable correctly, in which case you should fix it: dani.sleep, baby.sleep, dani.grump and day can be specified as continuous variables, and ID is a nominal(integer) variable.[119]

Next, I'll take a look at some basic descriptive statistics and, to give a graphical depiction of what each of the three interesting variables looks like, Figure 12.1 plots histograms. One thing to note: just because jamovi can calculate dozens of different statistics doesn't mean you should report all of them. If I were writing this up for a report, I'd probably pick out those statistics that are of most interest to me (and to my readership), and then put them into a nice, simple table like the one in Table 12.1.[120] Notice that when I put it into a table, I gave everything "human readable" names. This is always good practice. Notice also that I'm not getting enough sleep. This isn't good practice, but other parents tell me that it's pretty standard.

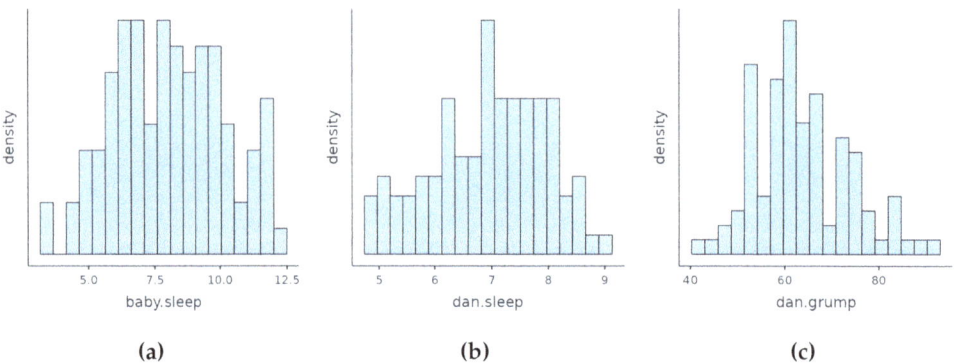

Figure 12.1: Histograms from jamovi for the three interesting variables in the *parenthood* data set

12.1.2 The strength and direction of a relationship

We can draw scatterplots to give us a general sense of how closely related two variables are. Ideally though, we might want to say a bit more about it than that. For instance, let's compare the relationship between baby.sleep and dani.grump (Figure 12.2a), left, with that between dani.sleep and dani.grump (Figure 12.2b), right. When looking at these two plots side by side, it's clear that the relationship is qualitatively the same in both cases: more sleep equals less grump! However, it's also pretty obvious that the relationship between dani.sleep and dani.grump is stronger than the relationship between baby.sleep and dani.grump. The plot on the right is "neater" than the one on the left. What it feels like is that if you want to predict what my mood is, it'd help you a little bit to know how many hours my son slept, but it'd be more helpful to know how many hours I slept.

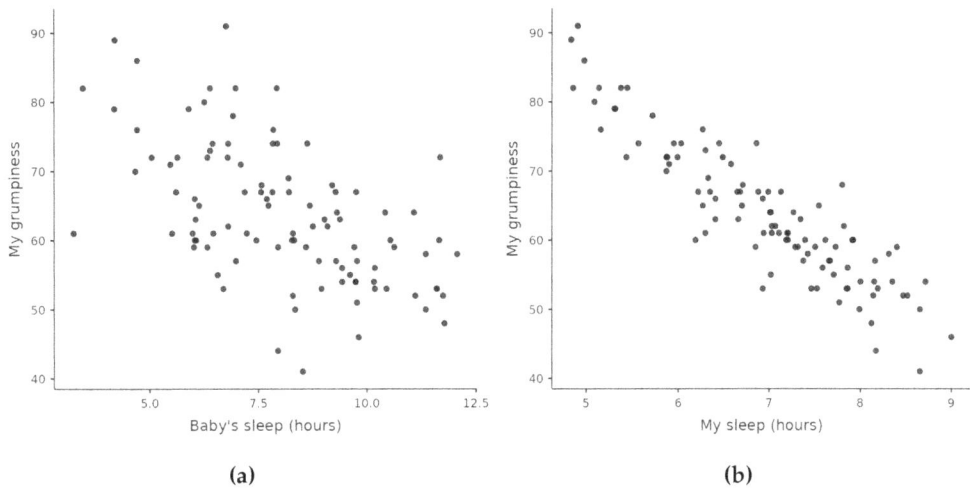

Figure 12.2: Scatterplots from jamovi showing the relationship between baby.sleep and dani.grump (left) and the relationship between dani.sleep and dani.grump (right)

In contrast, let's consider the two scatterplots shown in Figure 12.3. If we compare the scatterplot of "baby.sleep v dani.grump" (left) to the scatterplot of "baby.sleep v dani.sleep" (right), the overall strength of the relationship is the same, but the direction is different. That is, if my son sleeps more, I get more sleep (positive relationship, right-hand side), but if he sleeps more then I get less grumpy (negative relationship, left-hand side).

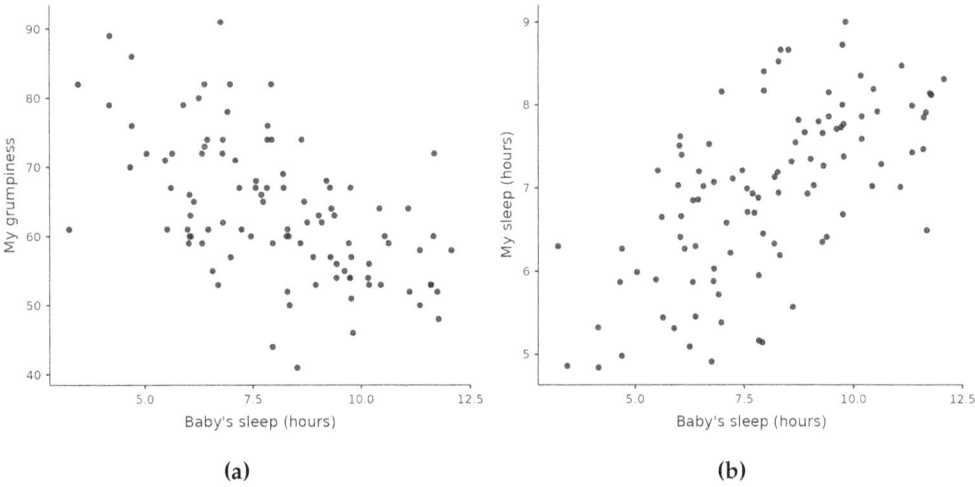

Figure 12.3: Scatterplots from jamovi showing the relationship between baby.sleep and dani.grump (left), as compared to the relationship between baby.sleep and dani.sleep (right)

12.1.3 The correlation coefficient

We can make these ideas a bit more explicit by introducing the idea of a **correlation coefficient** (or, more specifically, Pearson's correlation coefficient), which is traditionally denoted as r. The correlation coefficient between two variables X and Y (sometimes denoted r_{XY}), which we'll define more precisely in the next section, is a measure that varies from -1 to 1. When $r = -1$ it means that we have a perfect negative relationship, and when $r = 1$ it means we have a perfect positive relationship. When $r = 0$, there's no relationship at all. If you look at Figure 12.4, you can see several plots showing what different correlations look like.

[Additional technical detail[121]]

By standardising the covariance, not only do we keep all of the nice properties of the covariance discussed earlier, but the actual values of r are on a meaningful scale: r = 1 implies a perfect positive relationship and $r = -1$ implies a perfect negative relationship. I'll expand a little more on this point later, in the section on Interpreting a correlation. But before I do, let's look at how to calculate correlations in jamovi.

12.1.4 Calculating correlations in jamovi

Calculating correlations in jamovi can be done by clicking on the 'Regression' – 'Correlation Matrix' button. Transfer all four continuous variables across into the box on the right to get the output in Figure 12.5.

12.1.5 Interpreting a correlation

Naturally, in real life you don't see many correlations of 1. So how should you interpret a correlation of, say, r = .4? The honest answer is that it really depends on what you want to use the data for, and on how strong the correlations in your field tend to be. A friend of mine in engineering once argued that any correlation less than .95 is completely useless (I think he was exaggerating, even for engineering). On the other hand, there are real cases, even in psychology, where you should really expect correlations that strong. For instance, one of the benchmark data sets used to test theories of how people judge similarities is so clean that any theory that can't achieve a correlation of at least .9 really isn't deemed to be successful. However, when looking for (say) elementary correlates of intelligence (e.g., inspection time, response time), if you get a correlation above .3 you're doing very very well. In short, the interpretation of a correlation depends a lot on the context. That said, the rough guide in Table 12.2 is pretty typical.

However, something that can never be stressed enough is that you should always look at the scatterplot before attaching any interpretation to the data. A correlation might not mean what you think it means. The classic illustration of this is "Anscombe's Quartet" (Anscombe, 1973), a collection of four data sets. Each data set has two variables, an X and a Y. For all four data sets the mean value for X is 9 and the mean for Y is 7.5. The standard deviations for all X variables are almost identical, as are those for the Y variables. And in each case the correlation between X and Y is $r = 0.816$. You can verify this yourself, since I happen to have saved it in a file called *anscombe.csv*.

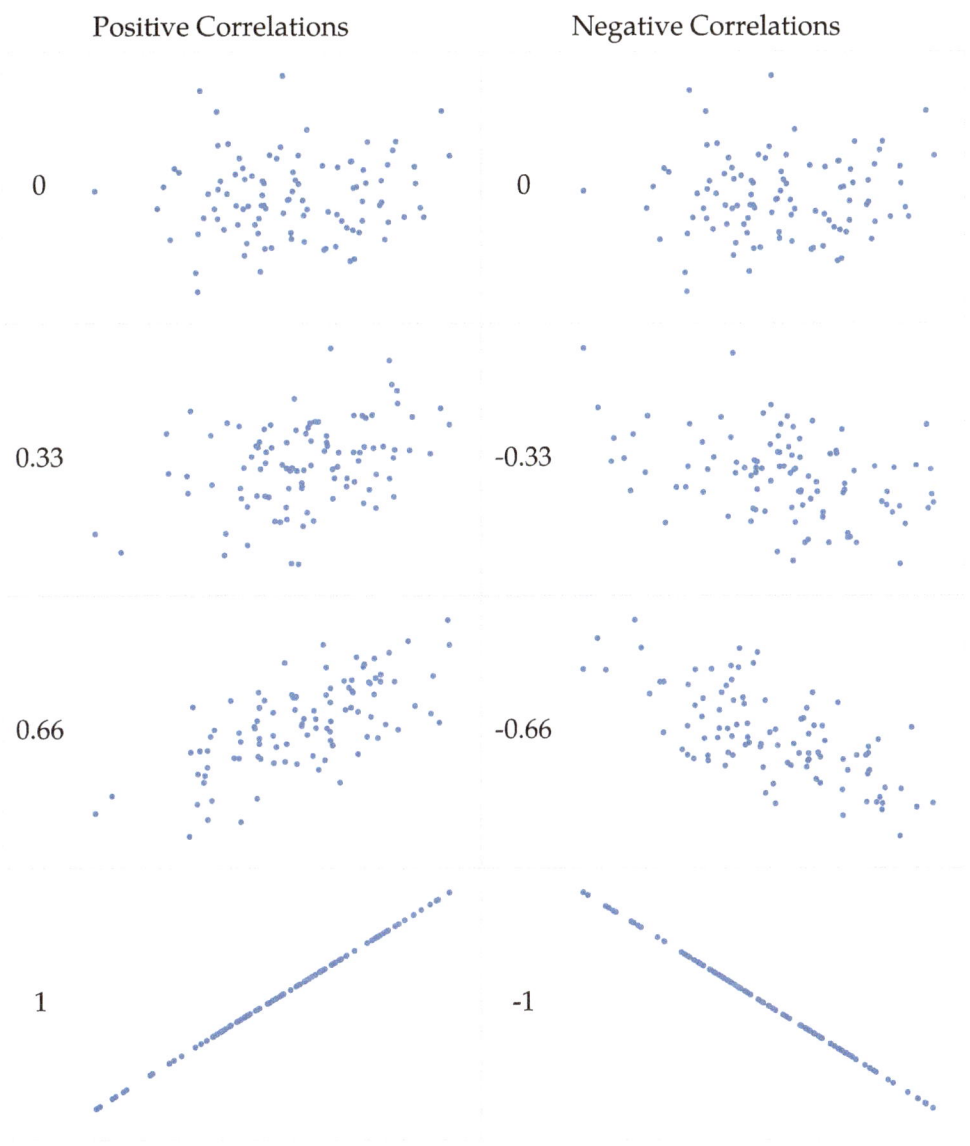

Figure 12.4: Illustration of the effect of varying the strength and direction of a correlation. In the left-hand column, the correlations are $0, .33, .66$ and 1. In the right-hand column, the correlations are $0, -.33, -.66$ and -1

You'd think that these four data sets would look pretty similar to one another. They do not. If we draw scatterplots of X against Y for all four variables, as shown in Figure 12.6, we see that all four of these are spectacularly different to each other. The lesson here, which so very many people seem to forget in real life, is "always graph your raw data" (see Chapter 5).

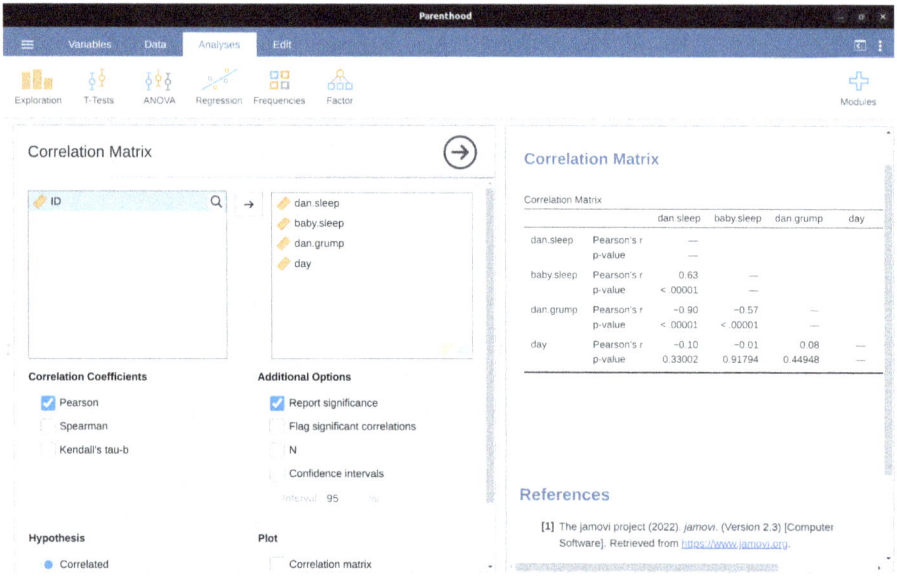

Figure 12.5: Correlations between variables in the *parenthood.csv* file

Table 12.2: A rough guide to interpreting correlations

Correlation	Strength	Direction
-1.00 to -0.90	Very strong	Negative
-0.90 to -0.70	Strong	Negative
-0.70 to -0.40	Moderate	Negative
-0.40 to -0.20	Weak	Negative
-0.20 to 0.00	Negligible	Negative
0.00 to 0.20	Negligible	Positive
0.20 to 0.40	Weak	Positive
0.40 to 0.70	Moderate	Positive
0.70 to 0.90	Strong	Positive
0.90 to 1.00	Very strong	Positive

Note that I say a rough guide. There aren't hard and fast rules for what counts as strong or weak relationships. It depends on the context

12.1.6 Spearman's rank correlations

The Pearson correlation coefficient is pretty useful, but it does have shortcomings. One issue stands out: what it actually measures is the strength of the linear relationship between two variables. In other words, what it gives you is a measure of the extent to which the data all tend to fall on a single, perfectly straight line. Often, this is a pretty good approximation to what we mean when we say "relationship", and so the Pearson correlation is a good thing to calculate. Sometimes though, it isn't.

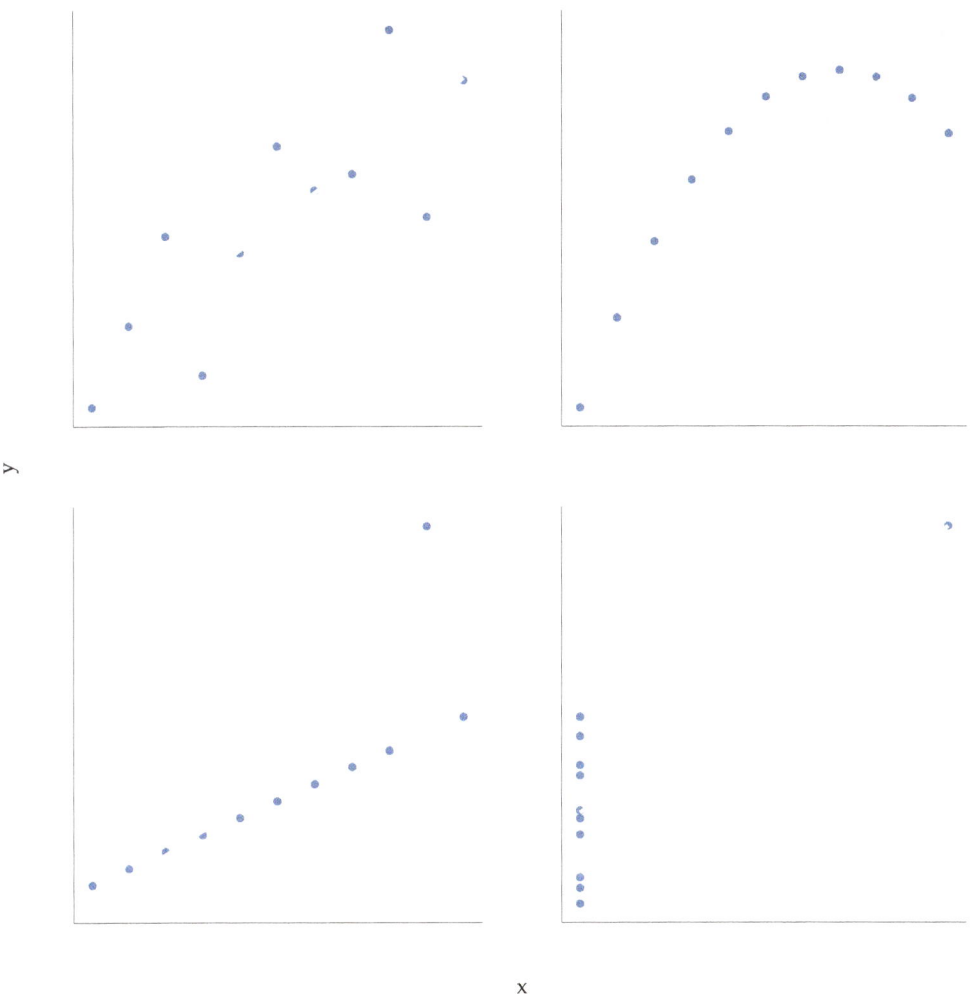

Figure 12.6: Anscombe's quartet scatterplots. All four of these data sets have a Pearson correlation of $r = .816$, but they are qualitatively different from one another

One very common situation where the Pearson correlation isn't quite the right thing to use arises when an increase in one variable X really is reflected in an increase in another variable Y, but the nature of the relationship isn't necessarily linear. An example of this might be the relationship between effort and reward when studying for an exam. If you put zero effort (X) into learning a subject then you should expect a grade of 0% (Y). However, a little bit of effort will cause a massive improvement. Just turning up to lectures means that you learn a fair bit, and if you just turn up to classes and scribble a few things down your grade might rise to 35%, all without a lot of effort. However, you just don't get the same effect at the other end of the scale. As everyone knows, it takes a lot more effort to get a grade of 90% than it takes to get a grade of 55%. What this means is that, if I've got data looking at study effort and grades, there's a pretty good chance that Pearson correlations will be misleading.

To illustrate, consider the data plotted in Figure 12.7, showing the relationship between hours worked and grade received for 10 students taking some class. The curious thing about this (highly fictitious) data set is that increasing your effort always increases your grade. It might be by a lot or it might be by a little, but increasing effort will never decrease your grade. If we run a standard Pearson correlation, it shows a strong relationship between hours worked and grade received, with a correlation coefficient of 0.91. However, this doesn't actually capture the observation that increasing hours worked always increases the grade. There's a sense here in which we want to be able to say that the correlation is perfect but for a somewhat different notion of what a "relationship" is. What we're looking for is something that captures the fact that there is a perfect **ordinal relationship** here. That is, if student 1 works more hours than student 2, then we can guarantee that student 1 will get the better grade. That's not what a correlation of $r = .91$ says at all.

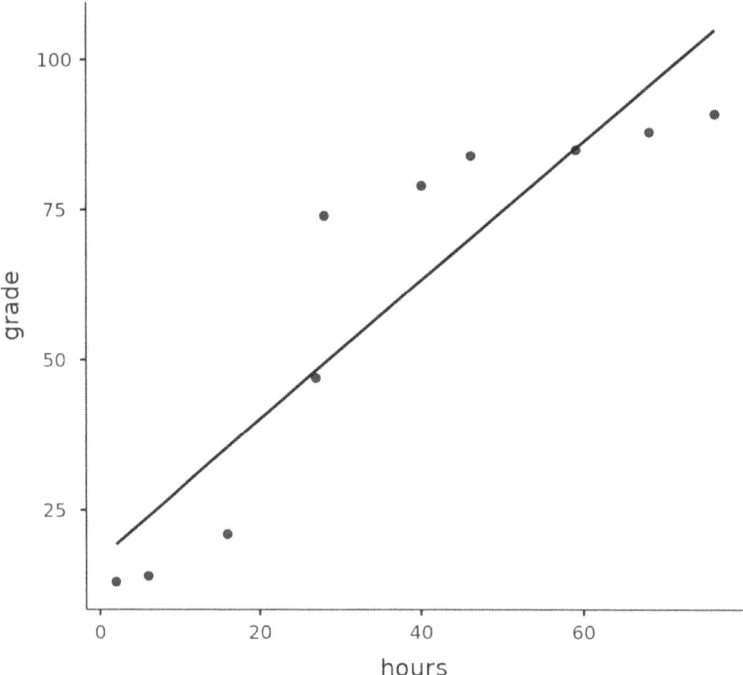

Figure 12.7: jamovi plot showing the relationship between hours worked and grade received for a toy data set consisting of only 10 students (each dot corresponds to one student). The line through the middle shows the linear relationship between the two variables. This produces a strong Pearson correlation of $r = .91$. However, the interesting thing to note here is that there's actually a perfect monotonic relationship between the two variables. In this toy example, increasing the hours worked always increases the grade received, as illustrated by the solid line. This is reflected in a Spearman correlation of $\rho = 1$. With such a small data set, however, it's an open question as to which version better describes the actual relationship involved

How should we address this? Actually, it's really easy. If we're looking for ordinal relationships all we have to do is treat the data as if it were ordinal scale! So, instead of measuring effort in terms of "hours worked", lets rank all 10 of our students in order of hours worked. That is, student 1 did the least work out of anyone (2 hours) so they get the lowest rank (rank = 1). Student 4 was the next laziest, putting in only 6 hours of work over the whole semester, so they get the next lowest rank (rank = 2). Notice that I'm using "rank =1" to mean "low rank". Sometimes in everyday language we talk about "rank = 1" to mean "top rank" rather than "bottom rank". So be careful, you can rank "from smallest value to largest value" (i.e., small equals rank 1) or you can rank "from largest value to smallest value" (i.e., large equals rank 1). In this case, I'm ranking from smallest to largest, but as it's really easy to forget which way you set things up you have to put a bit of effort into remembering!

Okay, so let's have a look at our students when we rank them from worst to best in terms of effort and reward Table 12.3.

Table 12.3: Students ranked in terms of effort and reward

	rank (hours worked)	rank (grade received)
student 1	1	1
student 2	10	10
student 3	6	6
student 4	2	2
student 5	3	3
student 6	5	5
student 7	4	4
student 8	8	8
student 9	7	7
student 10	9	9

Hmm. These are identical. The student who put in the most effort got the best grade, the student with the least effort got the worst grade, etc. As the table above shows, these two rankings are identical, so if we now correlate them we get a perfect relationship, with a correlation of 1.0.

What we've just re-invented is **Spearman's rank order correlation**, usually denoted ρ to distinguish it from the Pearson correlation r. We can calculate Spearman's ρ using jamovi simply by clicking the 'Spearman' check box in the 'Correlation Matrix' screen.

12.2 Scatterplots

Scatterplots are a simple but effective tool for visualising the relationship between two variables, like we saw with the figures in the section on Correlations. It's this latter application that we usually have in mind when we use the term "scatterplot". In this kind of plot each observation corresponds to one dot. The horizontal location of the dot plots the value of the observation on one variable, and the vertical location displays its value on the other variable. In many situations you don't really have a clear opinion about what the causal relationship is (e.g., does A cause B, or does B cause A, or does some other variable C control both A and B). If that's the case, it doesn't really matter which variable you plot on the x-axis and which one you plot on the y-axis. However, in many situations you do have a pretty strong idea which variable you think is most likely to be causal, or at least you have some suspicions in that direction. If so, then it's conventional to plot the cause variable on the x-axis, and the effect variable on the y-axis. With that in mind, let's look at how to draw scatterplots in jamovi, using the same *parenthood* data set (i.e. *parenthood.csv*) that I used when introducing correlations.

Suppose my goal is to draw a scatterplot displaying the relationship between the amount of sleep that I get (dani.sleep) and how grumpy I am the next day (dani.grump). There are two different ways in which we can use jamovi to get the plot that we're after. The first way is to use the 'Plot' option under the 'Regression' – 'Correlation Matrix' button, giving us the output shown in Figure 12.8. Note that jamovi draws a line through the points, we'll come onto this a bit later in the section on What is a linear regression model?. Plotting a scatterplot in this way also allows you to specify 'Densities for variables' and this option adds a density curve showing how the data in each variable is distributed.

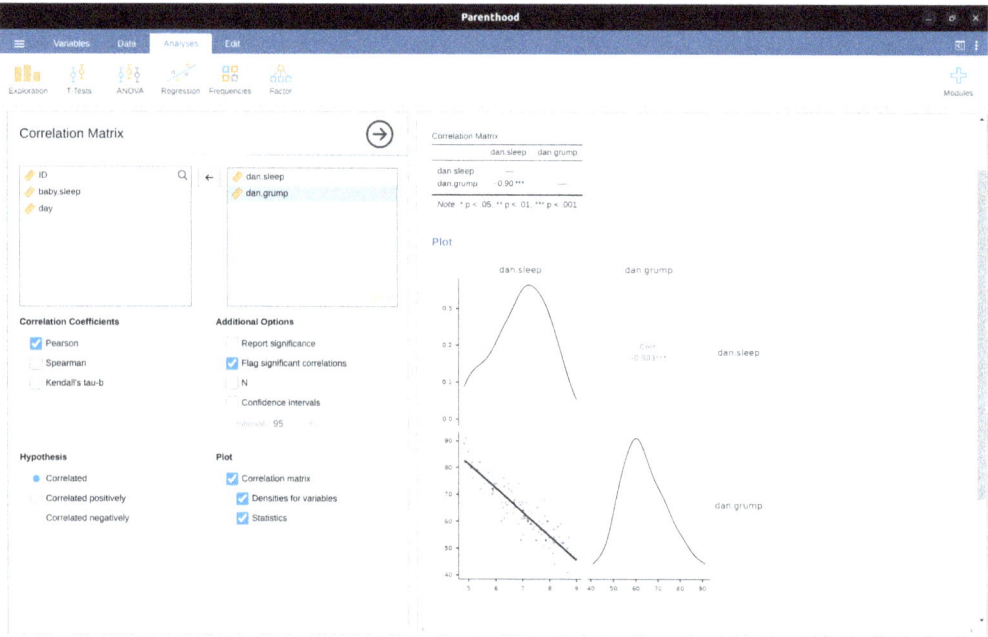

Figure 12.8: Scatterplot via the 'Correlation Matrix' command in jamovi

The second way do to it is to use one of the jamovi add-on modules. This module is called 'scatr' and you can install it by clicking on the large '+' icon in the top right of the jamovi screen, opening the jamovi library, scrolling down until you find 'scatr' and clicking 'install'. When you have done this, you will find a new 'Scatterplot' command available under the 'Exploration' button. This plot is a bit different than the first way, see Figure 12.9, but the important information is the same.

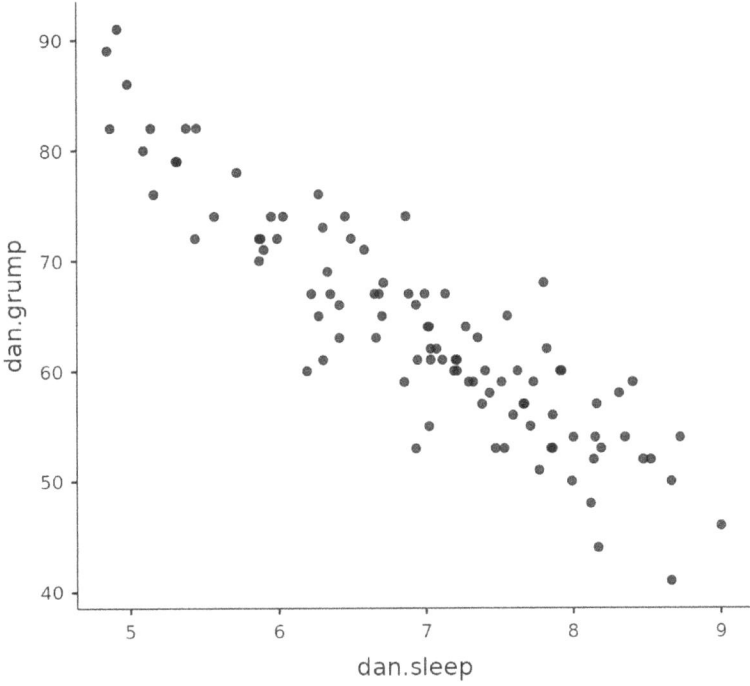

Figure 12.9: Scatterplot via the 'scatr' add-on module in jamovi

12.2.1 More elaborate options

Often you will want to look at the relationships between several variables at once, using a **scatterplot matrix** (in jamovi via the 'Correlation Matrix' – 'Plot' command). Just add another variable, for example baby.sleep to the list of variables to be correlated, and jamovi will create a scatterplot matrix for you, just like the one in Figure 12.10.

12.3 What is a linear regression model?

Stripped to its bare essentials, linear regression models are basically a slightly fancier version of the Pearson correlation (see Correlations), but they are actually much more powerful tools. We'll return to the *parenthood.csv* file that we were using to illustrate how correlations work. Recall that, in this data set we were trying to find out why Dani is so very grumpy all the time and our working hypothesis was that I'm not getting enough sleep. We drew a scatterplots to help us examine the relationship between the amount of sleep I get and my grumpiness the following day, as in Figure 12.9, and as we saw that this corresponded to a correlation of $r = -.90$, but what we find ourselves secretly imagining is something that looks closer to Figure 12.11(a). That is, we mentally draw a straight line through the middle of the data. In statistics, this line that we're drawing is called a **regression line**. Notice that, since we're not idiots, the regression line goes through the middle of the data. We don't find ourselves imagining anything like the rather silly plot shown in Figure 12.11(b).

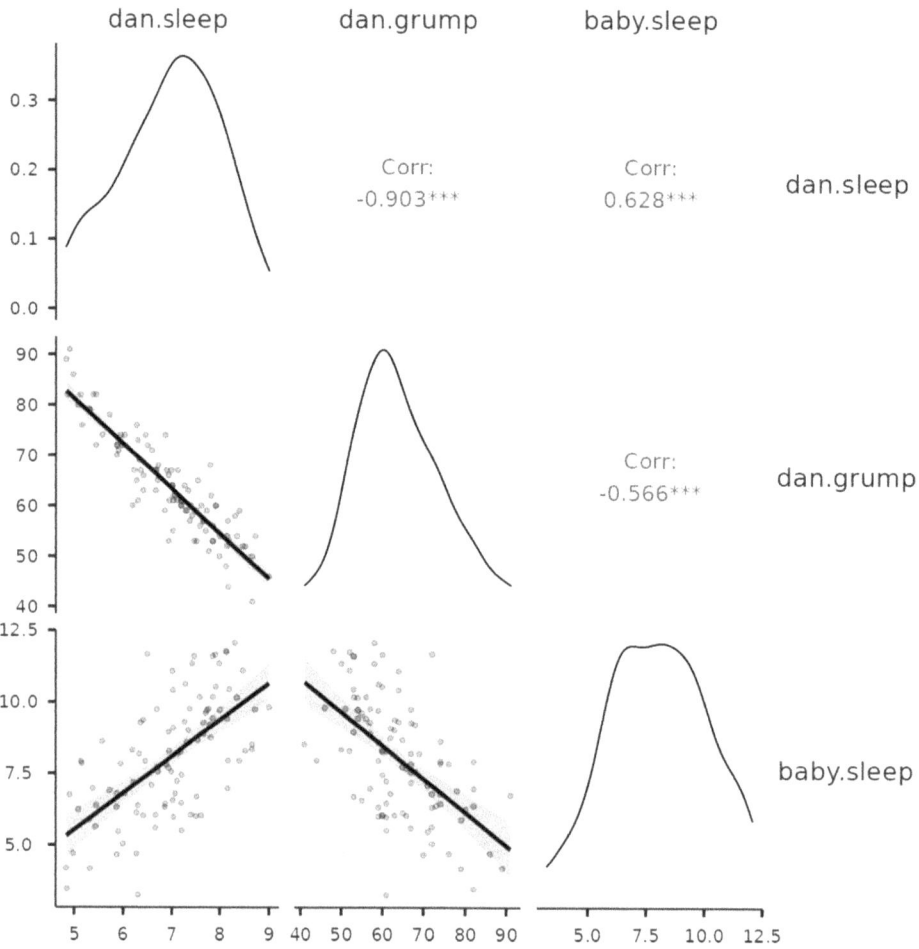

Figure 12.10: A matrix of scatterplots produced using jamovi

This is not highly surprising. The line that I've drawn in Figure 12.11(b) doesn't "fit" the data very well, so it doesn't make a lot of sense to propose it as a way of summarising the data, right? This is a very simple observation to make, but it turns out to be very powerful when we start trying to wrap just a little bit of maths around it. To do so, let's start with a refresher of some high school maths. The formula for a straight line is usually written like this:

$$y = a + bx$$

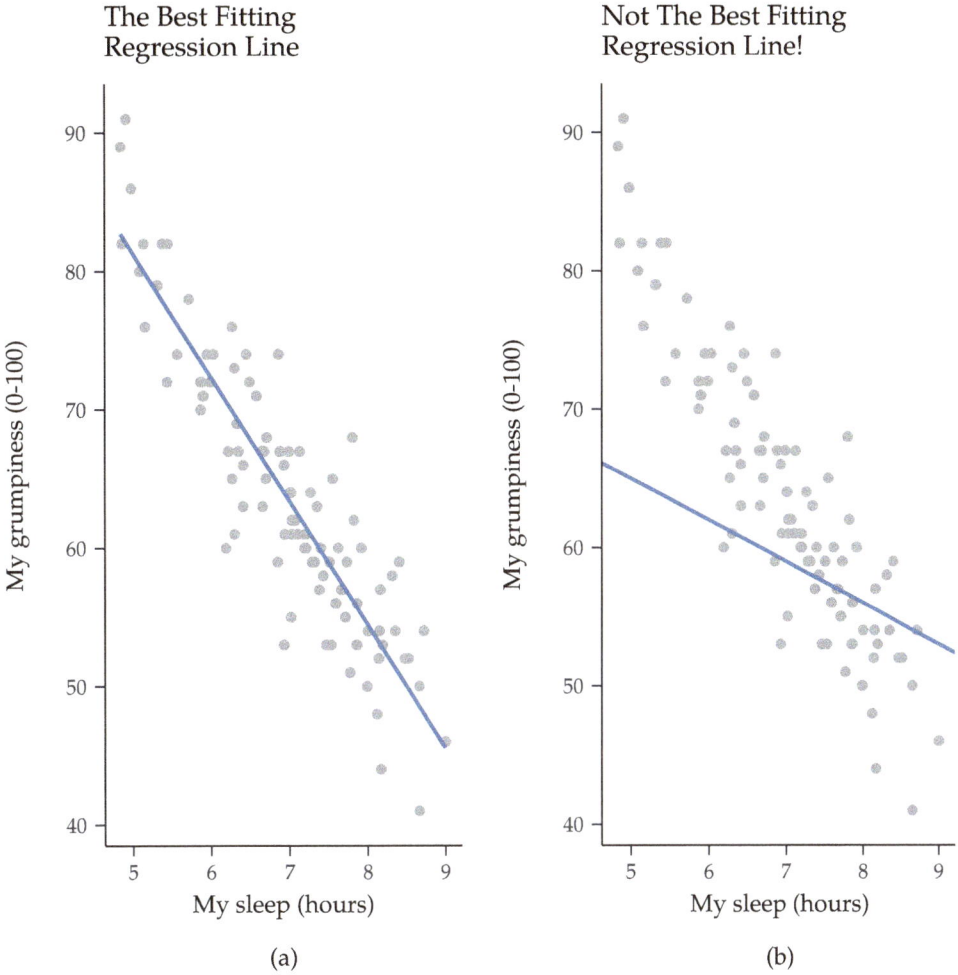

Figure 12.11: Panel (a) shows the sleep-grumpiness scatterplot from Figure 12.9 with the best fitting regression line drawn over the top. Not surprisingly, the line goes through the middle of the data. In contrast, panel (b) shows the same data, but with a very poor choice of regression line drawn over the top

The two variables are x and y, and we have two coefficients, a and b.[122] The coefficient a represents the y-intercept of the line, and coefficient b represents the slope of the line. The intercept is interpreted as "the value of y that you get when $x = 0$". Similarly, a slope of b means that if you increase the x-value by 1 unit, then the y-value goes up by b units, and a negative slope means that the y-value would go down rather than up. We use the exact same formula for a regression line. If Y is the outcome variable (the DV) and X is the predictor variable (the IV), then the formula that describes our regression is written like this:

$$\hat{Y}_i = b_0 + b_1 X_i$$

Hmm. Looks like the same formula, but there's some extra frilly bits in this version. Let's make sure we understand them. Firstly, notice that I've written X_i and Y_i rather than just plain old X and Y. This is because we want to remember that we're dealing with actual data. In this equation, X_i is the value of predictor variable for the ith observation (i.e., the number of hours of sleep that I got on day i of my little study), and Y_i is the corresponding value of the outcome variable (i.e., my grumpiness on that day). And although I haven't said so explicitly in the equation, what we're assuming is that this formula works for all observations in the data set (i.e., for all i). Secondly, notice that I wrote $\hat{Y}_i$ and not Y_i. This is because we want to make the distinction between the actual data Y_i, and the estimate $\hat{Y}_i$ (i.e., the prediction that our regression line is making). Thirdly, I changed the letters used to describe the coefficients from a and b to b_0 and b_1. That's just the way that statisticians like to refer to the coefficients in a regression model. I've no idea why they chose b, but that's what they did. In any case b_0 always refers to the intercept term, and b_1 refers to the slope.

Excellent, excellent. Next, I can't help but notice that, regardless of whether we're talking about the good regression line or the bad one, the data don't fall perfectly on the line. Or, to say it another way, the data Y_i are not identical to the predictions of the regression model $\hat{Y}_i$. Since statisticians love to attach letters, names and numbers to everything, let's refer to the difference between the model prediction and that actual data point as a residual, and we'll refer to it as ϵ_i.[123] Written using mathematics, the residuals are defined as

12.4 Estimating a linear regression model

Okay, now let's redraw our pictures but this time I'll add some lines to show the size of the residual for all observations. When the regression line is good, our residuals (the lengths of the solid black lines) all look pretty small, as shown in Figure 12.12(a), but when the regression line is a bad one the residuals are a lot larger, as you can see from looking at Figure 12.12(b). Hmm. Maybe what we "want" in a regression model is *small* residuals. Yes, that does seem to make sense. In fact, I think I'll go so far as to say that the "best fitting" regression line is the one that has the smallest residuals. Or, better yet, since statisticians seem to like to take squares of everything why not say that:

> The estimated regression coefficients, $\hat{b}_0$ and $\hat{b}_1$, are those that minimise the sum of the squared residuals, which we could either write as $\sum_i (Y_i - \hat{Y}_i)^2$ or as $\sum_i \epsilon_i^2$.

Yes, yes that sounds even better. And since I've indented it like that, it probably means that this is the right answer. And since this is the right answer, it's probably worth making a note of the fact that our regression coefficients are estimates (we're trying to guess the parameters that describe a population!), which is why I've added the little hats, so that we get $\hat{b}_0$ and $\hat{b}_1$ rather than b_0 and b_1. Finally, I should also note that, since there's actually more than one way to estimate a regression model, the more technical name for this estimation process is **ordinary least squares (OLS) regression**.

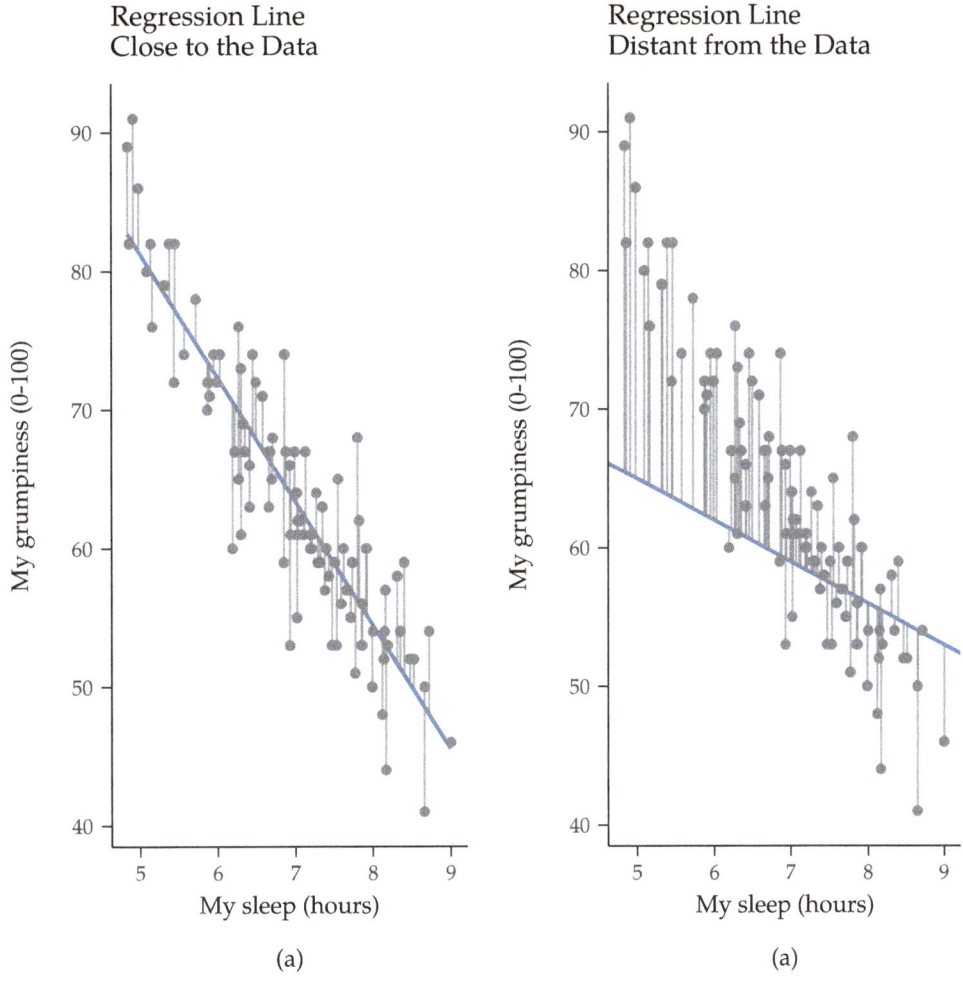

Figure 12.12: A depiction of the residuals associated with the best fitting regression line (panel a), and the residuals associated with a poor regression line (panel b). The residuals are much smaller for the good regression line. Again, this is no surprise given that the good line is the one that goes right through the middle of the data

At this point, we now have a concrete definition for what counts as our "best" choice of regression coefficients, $\hat{b}_0$ and $\hat{b}_1$. The natural question to ask next is, if our optimal regression coefficients are those that minimise the sum squared residuals, how do we find these wonderful numbers? The actual answer to this question is complicated and doesn't help you understand the logic of regression.[124] This time I'm going to let you off the hook. Instead of showing you the long and tedious way first and then "revealing" the wonderful shortcut that jamovi provides, let's cut straight to the chase and just use jamovi to do all the heavy lifting.

12.4.1 Linear regression in jamovi

To run my linear regression, open up the 'Regression' – 'Linear Regression' analysis in jamovi, using the *parenthood.csv* data file. Then specify dani.grump as the 'Dependent Variable' and dani.sleep as the variable entered in the 'Covariates' box. This gives the results shown in Figure 12.13, showing an intercept $\hat{b}_0 = 125.96$ and the slope $\hat{b}_1 = -8.94$. In other words, the best fitting regression line that I plotted in Figure 12.12 has this formula:

$$\hat{Y}_i = 125.96 + (-8.94 X_i)$$

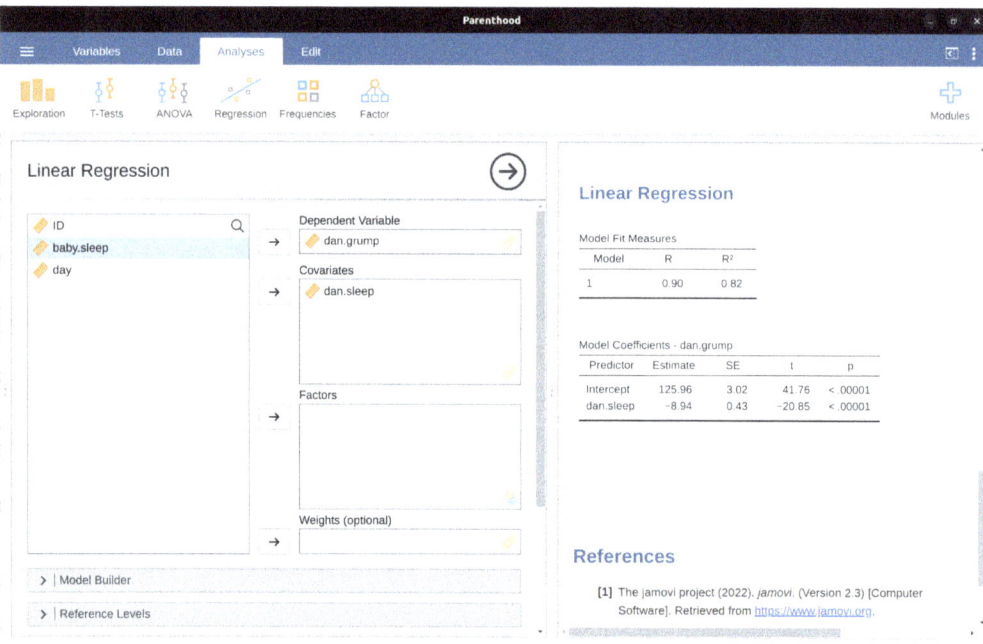

Figure 12.13: A jamovi screenshot showing a simple linear regression analysis

12.4.2 Interpreting the estimated model

The most important thing to be able to understand is how to interpret these coefficients. Let's start with $\hat{b}_1$, the slope. If we remember the definition of the slope, a regression coefficient of $\hat{b}_1 = -8.94$ means that if I increase X_i by 1, then I'm decreasing Y_i by 8.94. That is, each additional hour of sleep that I gain will improve my mood, reducing my grumpiness by 8.94 grumpiness points. What about the intercept? Well, since $\hat{b}_0$ corresponds to "the expected value of Y_i when X_i equals 0", it's pretty straightforward. It implies that if I get zero hours of sleep ($X_i = 0$) then my grumpiness will go off the scale, to an insane value of ($Y_i = 125.96$). Best to be avoided, I think.

12.5 Multiple linear regression

The simple linear regression model that we've discussed up to this point assumes that there's a single predictor variable that you're interested in, in this case dani.sleep. In fact, up to this point every statistical tool that we've talked about has assumed that your analysis uses one predictor variable and one outcome variable. However, in many (perhaps most) research projects you actually have multiple predictors that you want to examine. If so, it would be nice to be able to extend the linear regression framework to be able to include multiple predictors. Perhaps some kind of **multiple regression** model would be in order?

Multiple regression is conceptually very simple. All we do is add more terms to our regression equation. Let's suppose that we've got two variables that we're interested in; perhaps we want to use both dani.sleep and baby.sleep to predict the dani.grump variable. As before, we let Y_i refer to my grumpiness on the i-th day. But now we have two X variables: the first corresponding to the amount of sleep I got and the second corresponding to the amount of sleep my son got. So we'll let X_{i1} refer to the hours I slept on the i-th day and X_{i2} refers to the hours that the baby slept on that day. If so, then we can write our regression model like this:

$$Y_i = b_0 + b_1 X_{i1} + b_2 X_{i2} + \epsilon_i$$

As before, ϵ_i is the residual associated with the i-th observation, $\epsilon_i = Y_i - \hat{Y}_i$. In this model, we now have three coefficients that need to be estimated: b_0 is the intercept, b_1 is the coefficient associated with my sleep, and b_2 is the coefficient associated with my son's sleep. However, although the number of coefficients that need to be estimated has changed, the basic idea of how the estimation works is unchanged: our estimated coefficients $\hat{b}_0$, $\hat{b}_1$ and $\hat{b}_2$ are those that minimise the sum squared residuals.

12.5.1 Doing it in jamovi

Multiple regression in jamovi is no different to simple regression. All we have to do is add additional variables to the 'Covariates' box in jamovi. For example, if we want to use both dani.sleep and baby.sleep as predictors in our attempt to explain why I'm so grumpy, then move baby.sleep across into the 'Covariates' box alongside dani.sleep. By default, jamovi assumes that the model should include an intercept. The coefficients we get this time are shown in Table 12.4.

Table 12.4: Adding multiple variables as predictors in a regression

(Intercept)	dani.sleep	baby.sleep
125.97	-8.95	0.01

The coefficient associated with dani.sleep is quite large, suggesting that every hour of sleep I lose makes me a lot grumpier. However, the coefficient for baby.sleep is very small, suggesting that it doesn't really matter how much sleep my son gets. What matters as far as my grumpiness goes is how much sleep I get. To get a sense of what this multiple regression model looks like, Figure 12.14 shows a 3D plot that plots all three variables, along with the regression model itself.

[Additional technical detail[125]]

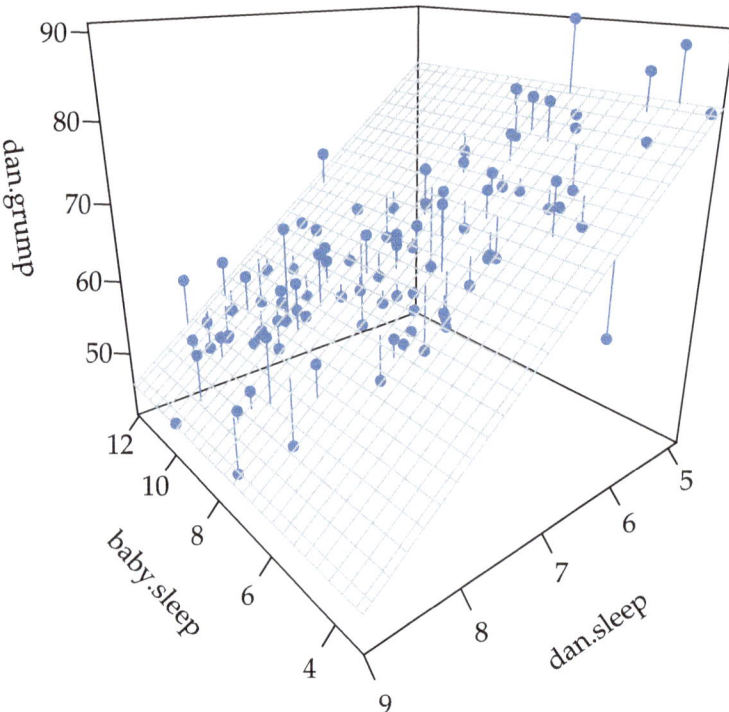

Figure 12.14: A 3D visualisation of a multiple regression model. There are two predictors in the model, dani.sleep and baby.sleep and the outcome variable is dani.grump. Together, these three variables form a 3D space. Each observation (dot) is a point in this space. In much the same way that a simple linear regression model forms a line in 2D space, this multiple regression model forms a plane in 3D space. When we estimate the regression coefficients what we're trying to do is find a plane that is as close to all the blue dots as possible

12.6 Quantifying the fit of the regression model

So we now know how to estimate the coefficients of a linear regression model. The problem is, we don't yet know if this regression model is any good. For example, the regression.1 model claims that every hour of sleep will improve my mood by quite a lot, but it might just be rubbish. Remember, the regression model only produces a prediction $\hat{Y}_i$ about what my mood is like, but my actual mood is Y_i. If these two are very close, then the regression model has done a good job. If they are very different, then it has done a bad job.

12.6.1 The R^2 value

Once again, let's wrap a little bit of mathematics around this. Firstly, we've got the sum of the squared residuals:

$$SS_{res} = \sum_i (Y_i - \hat{Y}_i)^2$$

which we would hope to be pretty small. Specifically, what we'd like is for it to be very small in comparison to the total variability in the outcome variable:

$$SS_{tot} = \sum_i (Y_i - \bar{Y})^2$$

While we're here, let's calculate these values ourselves, not by hand though. Let's use something like Excel or another standard spreadsheet programme. I have done this by opening up the *parenthood.csv* file in Excel and saving it as parenthood rsquared.xls so that I can work on it. The first thing to do is calculate the $\hat{Y}$ values, and for the simple model that uses only a single predictor we would do the following:

1. Create a new column called ' Y.pred ' using the formula ' = 125.97 + (-8.94 × dani.sleep) '.
2. Calculate the SS_{resid} by creating a new column called ' (Y-Y.pred)^2 ' using the formula ' = (dani.grump - Y.pred)^2 '.
3. Then, at the bottom of this column calculate the sum of these values, i.e. ' sum((Y-Y.pred)^2) '.
4. At the bottom of the dani.grump column, calculate the mean value for dani.grump (NB Excel uses the word ' AVERAGE ' rather than ' mean ' in its function).
5. Then create a new column, called ' (Y - mean(Y))^2) ' using the formula ' = (dani.grump - AVERAGE(dani.grump))^2 '.
6. Then, at the bottom of this column calculate the sum of these values, i.e. 'sum((Y - mean(Y))^2)'.
7. Calculate R^2 by typing into a blank cell the following: '= 1 - (SS(resid) / SS(tot))'.

This gives a value for R^2 of '0.8161018'. The R^2 value, sometimes called the **coefficient of determination**[126] has a simple interpretation: it is the proportion of the variance in the outcome variable that can be accounted for by the predictor. So, in this case the fact that we have obtained $R^2 = .816$ means that the predictor (my.sleep) explains 81.6% of the variance in the outcome (my.grump).

Naturally, you don't actually need to type all these commands into Excel yourself if you want to obtain the R^2 value for your regression model. As we'll see later on in the section on Running the hypothesis tests in jamovi, all you need to do is specify this as an option in jamovi. However, let's put that to one side for the moment. There's another property of R^2 that I want to point out.

12.6.2 The relationship between regression and correlation

At this point we can revisit my earlier claim that regression, in this very simple form that I've discussed so far, is basically the same thing as a correlation. Previously, we used the symbol r to denote a Pearson correlation. Might there be some relationship between the value of the correlation coefficient r and the R^2 value from linear regression? Of course there is: the squared correlation r^2 is identical to the R^2 value for a linear regression with only a single predictor. In other words, running a Pearson correlation is more or less equivalent to running a linear regression model that uses only one predictor variable.

12.6.3 The adjusted R^2 value

One final thing to point out before moving on. It's quite common for people to report a slightly different measure of model performance, known as "adjusted R^2". The motivation behind calculating the adjusted R^2 value is the observation that adding more predictors into the model will always cause the R^2 value to increase (or at least not decrease).

[Additional technical detail[127]]

This adjustment is an attempt to take the degrees of freedom into account. The big advantage of the adjusted R^2 value is that when you add more predictors to the model, the adjusted R^2 value will only increase if the new variables improve the model performance more than you'd expect by chance. The big disadvantage is that the adjusted R^2 value can't be interpreted in the elegant way that R^2 can. R^2 has a simple interpretation as the proportion of variance in the outcome variable that is explained by the regression model. To my knowledge, no equivalent interpretation exists for adjusted R^2.

An obvious question then is whether you should report R^2 or adjusted R^2. This is probably a matter of personal preference. If you care more about interpretability, then R^2 is better. If you care more about correcting for bias, then adjusted R^2 is probably better. Speaking just for myself, I prefer R^2. My feeling is that it's more important to be able to interpret your measure of model performance. Besides, as we'll see in Hypothesis tests for regression models, if you're worried that the improvement in R^2 that you get by adding a predictor is just due to chance and not because it's a better model, well we've got hypothesis tests for that.

12.7 Hypothesis tests for regression models

So far we've talked about what a regression model is, how the coefficients of a regression model are estimated, and how we quantify the performance of the model (the last of these, incidentally, is basically our measure of effect size). The next thing we need to talk about is hypothesis tests. There are two different (but related) kinds of hypothesis tests that we need to talk about: those in which we test whether the regression model as a whole is performing significantly better than a null model, and those in which we test whether a particular regression coefficient is significantly different from zero.

12.7.1 Testing the model as a whole

Okay, suppose you've estimated your regression model. The first hypothesis test you might try is the null hypothesis that there is no relationship between the predictors and the outcome, and the alternative hypothesis that the data are distributed in exactly the way that the regression model predicts.

[Additional technical detail[128]]

We'll see much more of the F-statistic in Chapter 13, but for now just know that we can interpret large F-values as indicating that the null hypothesis is performing poorly in comparison to the alternative hypothesis. In a moment I'll show you how to do the test in jamovi the easy way, but first let's have a look at the tests for the individual regression coefficients.

12.7.2 Tests for individual coefficients

The F-test that we've just introduced is useful for checking that the model as a whole is performing better than chance. If your regression model doesn't produce a significant result for the F-test then you probably don't have a very good regression model (or, quite possibly, you don't have very good data). However, while failing this test is a pretty strong indicator that the model has problems, passing the test (i.e., rejecting the null) doesn't imply that the model is good! Why is that, you might be wondering? The answer to that can be found by looking at the coefficients for the Multiple linear regression model we have already looked at (Table 12.4)

I can't help but notice that the estimated regression coefficient for the baby.sleep variable is tiny (0.01), relative to the value that we get for dani.sleep (−8.95). Given that these two variables are absolutely on the same scale (they're both measured in "hours slept"), I find this illuminating. In fact, I'm beginning to suspect that it's really only the amount of sleep that I get that matters in order to predict my grumpiness. We can re-use a hypothesis test that we discussed earlier, the t-test. The test that we're interested in has a null hypothesis that the true regression coefficient is zero ($b = 0$), which is to be tested against the alternative hypothesis that it isn't ($b \neq 0$). That is:

$$H_0 : b = 0$$
$$H_1 : b \neq 0$$

How can we test this? Well, if the central limit theorem is kind to us we might be able to guess that the sampling distribution of $\hat{b}$, the estimated regression coefficient, is a normal distribution with mean centred on b. What that would mean is that if the null hypothesis were true, then the sampling distribution of $\hat{b}$ has mean zero and unknown standard deviation. Assuming that we can come up with a good estimate for the standard error of the regression coefficient, $se(\hat{b})$, then we're in luck. That's exactly the situation for which we introduced the one-sample t-test back in Chapter 11. So let's define a t-statistic like this:

$$t = \frac{\hat{b}}{SE(\hat{b})}$$

I'll skip over the reasons why, but our degrees of freedom in this case are $df = N - K - 1$. Irritatingly, the estimate of the standard error of the regression coefficient, $se(\hat{b})$, is not as easy to calculate as the standard error of the mean that we used for the simpler t-tests in Chapter 11. In fact, the formula is somewhat ugly, and not terribly helpful to look at.[129] For our purposes it's sufficient to point out that the standard error of the estimated regression coefficient depends on both the predictor and outcome variables, and it is somewhat sensitive to violations of the homogeneity of variance assumption (discussed shortly).

In any case, this t-statistic can be interpreted in the same way as the t-statistics that we discussed in Chapter 11. Assuming that you have a two-sided alternative (i.e., you don't really care if b > 0 or b < 0), then it's the extreme values of t (i.e., a lot less than zero or a lot greater than zero) that suggest that you should reject the null hypothesis.

12.7.3 Running the hypothesis tests in jamovi

To compute all of the statistics that we have talked about so far, all you need to do is make sure the relevant options are checked in jamovi and then run the regression. If we do that, as in Figure 12.15, we get a whole bunch of useful output.

The 'Model Coefficients' at the bottom of the jamovi analysis results shown in Figure 12.15 provides the coefficients of the regression model. Each row in this table refers to one of the coefficients in the regression model. The first row is the intercept term, and the later ones look at each of the predictors. The columns give you all of the relevant information. The first column is the actual estimate of b (e.g., 125.97 for the intercept, and -8.95 for the dani.sleep predictor). The second column is the standard error estimate $\hat{\sigma}_b$. The third and fourth columns provide the lower and upper values for the 95% confidence interval around the b estimate (more on this later). The fifth column gives you the t-statistic, and it's worth noticing that in this table $t = \frac{\hat{b}}{se(\hat{b})}$ every time. Finally, the last column gives you the actual p-value for each of these tests.[130]

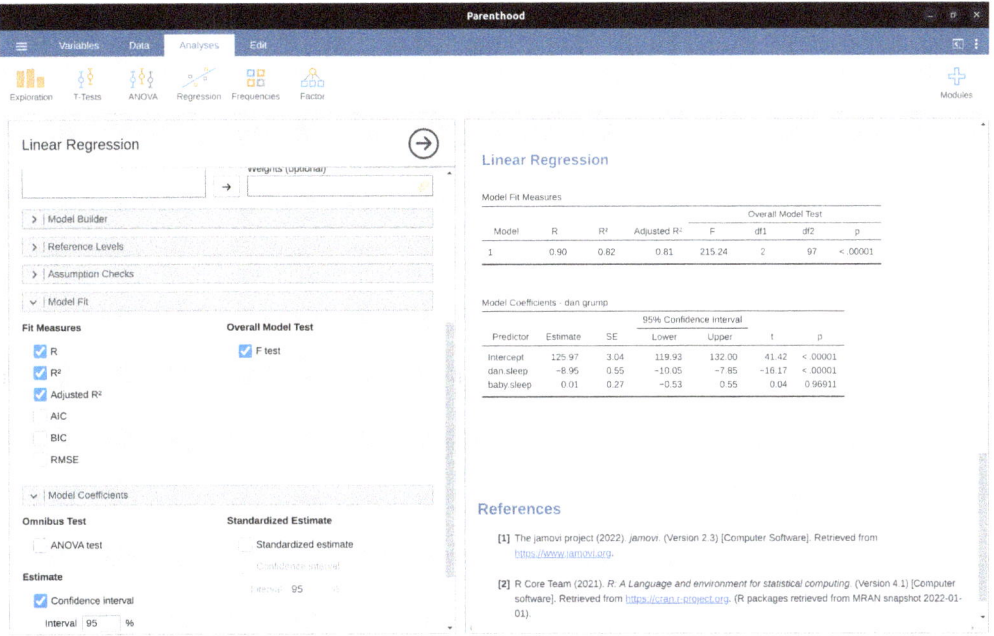

Figure 12.15: A jamovi screenshot showing a multiple linear regression analysis, with some useful options checked

The only thing that the coefficients table itself doesn't list is the degrees of freedom used in the t-test, which is always $N - K - 1$ and is listed in the table at the top of the output, labelled 'Model Fit Measures'. We can see from this table that the model performs significantly better than you'd expect by chance ($F(2, 97) = 215.24, p < .001$), which isn't all that surprising: the $R^2 = .81$ value indicate that the regression model accounts for 81% of the variability in the outcome measure (and 82% for the adjusted R^2). However, when we look back up at the t-tests for each of the individual coefficients, we have pretty strong evidence that the baby.sleep variable has no significant effect. All the work in this model is being done by the dani.sleep variable. Taken together, these results suggest that this regression model is actually the wrong model for the data. You'd probably be better off dropping the baby.sleep predictor entirely. In other words, the simple regression model that we started with is the better model.

12.8 Regarding regression coefficients

Before moving on to discuss the assumptions underlying linear regression and what you can do to check if they're being met, there's two more topics I want to briefly discuss, both of which relate to the regression coefficients. The first thing to talk about is calculating confidence intervals for the coefficients. After that, I'll discuss the somewhat murky question of how to determine which predictor is most important.

12.8.1 Confidence intervals for the coefficients

Like any population parameter, the regression coefficients b cannot be estimated with complete precision from a sample of data; that's part of why we need hypothesis tests. Given this, it's quite useful to be able to report confidence intervals that capture our uncertainty about the true value of b. This is especially useful when the research question focuses heavily on an attempt to find out how strongly variable X is related to variable Y, since in those situations the interest is primarily in the regression weight b.

[Additional technical detail[131]]

In jamovi we had already specified the '95% Confidence interval' as shown in Figure 12.15, although we could easily have chosen another value, say a '99% Confidence interval' if that is what we decided on.

12.8.2 Calculating standardised regression coefficients

One more thing that you might want to do is to calculate "standardised" regression coefficients, often denoted β. The rationale behind standardised coefficients goes like this. In a lot of situations, your variables are on fundamentally different scales. Suppose, for example, my regression model aims to predict people's IQ scores using their educational attainment (number of years of education) and their income as predictors. Obviously, educational attainment and income are not on the same scales. The number of years of schooling might only vary by 10s of years, whereas income can vary by 10,000s of dollars (or more). The units of measurement have a big influence on the regression coefficients. The b coefficients only make sense when interpreted in light of the units, both of the predictor variables and the outcome variable. This makes it very difficult to compare the coefficients of different predictors. Yet there are situations where you really do want to make comparisons between different coefficients. Specifically, you might want some kind of standard measure of which predictors have the strongest relationship to the outcome. This is what **standardised coefficients** aim to do.

The basic idea is quite simple; the standardised coefficients are the coefficients that you would have obtained if you'd converted all the variables to z-scores before running the regression.[132] The idea here is that, by converting all the predictors to z-scores, they all go into the regression on the same scale, thereby removing the problem of having variables on different scales. Regardless of what the original variables were, a β value of 1 means that an increase in the predictor of 1 standard deviation will produce a corresponding 1 standard deviation increase in the outcome variable. Therefore, if variable A has a larger absolute value of β than variable B, it is deemed to have a stronger relationship with the outcome. Or at least that's the idea. It's worth being a little cautious here, since this does rely very heavily on the assumption that "a 1 standard deviation change" is fundamentally the same kind of thing for all variables. It's not always obvious that this is true.

[Additional technical detail[133]]

To make things even simpler, jamovi has an option that computes the β coefficients for you using the 'Standardized estimate' checkbox in the 'Model Coefficients' options, see results in Figure 12.16.

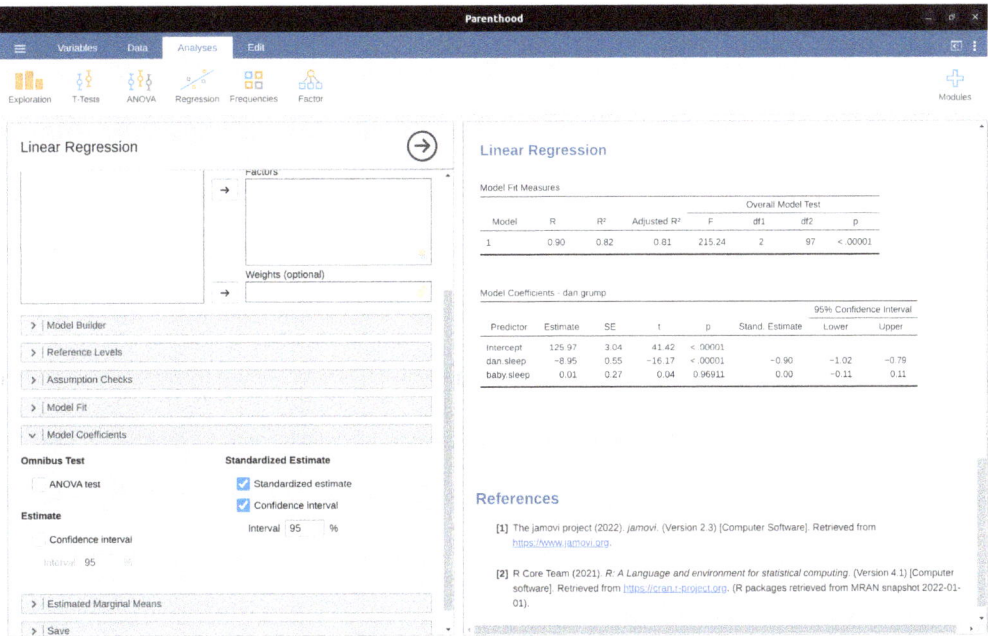

Figure 12.16: Standardised coefficients, with 95% confidence intervals, for multiple linear regression

These results clearly show that the dani.sleep variable has a much stronger effect than the baby.sleep variable. However, this is a perfect example of a situation where it would probably make sense to use the original coefficients b rather than the standardised coefficients β. After all, my sleep and the baby's sleep are already on the same scale: number of hours slept. Why complicate matters by converting these to z-scores?

12.9 Assumptions of regression

The linear regression model that I've been discussing relies on several assumptions. In Model checking we'll talk a lot more about how to check that these assumptions are being met, but first let's have a look at each of them.

- **Linearity.** A pretty fundamental assumption of the linear regression model is that the relationship between X and Y actually is linear! Regardless of whether it's a simple regression or a multiple regression, we assume that the relationships involved are linear.
- **Independence:** residuals are independent of each other. This is really just a "catch all" assumption, to the effect that "there's nothing else funny going on in the residuals". If there is something weird (e.g., the residuals all depend heavily on some other unmeasured variable) going on, it might screw things up. Independence isn't something that you can check directly and specifically with diagnostic tools, but if your regression diagnostics are messed up then think carefully about the independence of your observations and residuals.

- **N**ormality. Like many of the models in statistics, basic simple or multiple linear regression relies on an assumption of normality. Specifically, it assumes that the residuals are normally distributed. It's actually okay if the predictors X and the outcome Y variables are non-normal, so long as the residuals ϵ are normal. See the Checking the normality of the residuals section.
- **E**quality (or "homogeneity") of variance. Strictly speaking, the regression model assumes that each residual ϵ_i is generated from a normal distribution with mean 0, and (more importantly for the current purposes) with a standard deviation σ that is the same for every single residual. In practice, it's impossible to test the assumption that every residual is identically distributed. Instead, what we care about is that the standard deviation of the residual is the same for all values of $\hat{Y}$, and (if we're being especially diligent) all values of every predictor X in the model.

So, we have four main assumptions for linear regression (that neatly form the acronym **LINE**). And there are also a couple of other things we should also check for:

- Uncorrelated predictors. The idea here is that, in a multiple regression model, you don't want your predictors to be too strongly correlated with each other. This isn't "technically" an assumption of the regression model, but in practice it's required. Predictors that are too strongly correlated with each other (referred to as "collinearity") can cause problems when evaluating the model. See Checking for collinearity section.
- No "bad" outliers. Again, not actually a technical assumption of the model (or rather, it's sort of implied by all the others), but there is an implicit assumption that your regression model isn't being too strongly influenced by one or two anomalous data points because this raises questions about the adequacy of the model and the trustworthiness of the data in some cases. See the section on Outliers and anomalous data.

12.10 Model checking

The main focus of this section is **regression diagnostics**, a term that refers to the art of checking that the assumptions of your regression model have been met, figuring out how to fix the model if the assumptions are violated, and generally to check that nothing "funny" is going on. I refer to this as the "art" of model checking with good reason. It's not easy, and while there are a lot of easily available tools that you can use to diagnose and maybe even cure the problems that affect your model (if there are any, that is!), you really do need to exercise a certain amount of judgement when doing this.

In this section I describe several different things you can do to check that your regression model is doing what it's supposed to. It doesn't cover the full space of things you could do, but it's still much more detailed than what is often done in practice – unfortunately! But it's important that you get a sense of what tools are at your disposal, so I'll try to introduce a bunch of them here. Finally, I should note that this section draws quite heavily from Fox & Weisberg (2011), the book associated with the "car" package that is used to conduct regression analysis in R. The "car" package is notable for providing some excellent tools for regression diagnostics, and the book itself talks about them in an admirably clear fashion. I don't want to sound too gushy about it, but I do think that Fox & Weisberg (2011) is well worth reading, even if some of the advanced diagnostic techniques are only available in "R" and not jamovi.

12.10.1 Three kinds of residuals

The majority of regression diagnostics revolve around looking at the residuals, and there are several different kinds of residual that we might consider. In particular, the following three kinds of residuals are referred to in this section: "ordinary residuals", "standardised residuals", and "Studentised residuals". There is a fourth kind that you'll see referred to in some of the Figures, and that's the "Pearson residual". However, for the models that we're talking about in this chapter the Pearson residual is identical to the ordinary residual.

The first and simplest kind of residuals that we care about are **ordinary residuals**. These are the actual raw residuals that I've been talking about throughout this chapter so far. The ordinary residual is just the difference between the predicted value $\hat{Y}_i$ and the observed value Y_i. I've been using the notation ϵ_i to refer to the i-th ordinary residual and so, with this in mind, we have the very simple equation:

$$\epsilon_i = Y_i - \hat{Y}_i$$

This is of course what we saw earlier, and unless I specifically refer to some other kind of residual, this is the one I'm talking about. So there's nothing new here. I just wanted to repeat myself. One drawback to using ordinary residuals is that they're always on a different scale, depending on what the outcome variable is and how good the regression model is. That is, unless you've decided to run a regression model without an intercept term, the ordinary residuals will have mean 0 but the variance is different for every regression. In a lot of contexts, especially where you're only interested in the pattern of the residuals and not their actual values, it's convenient to estimate the **standardised residuals**, which are normalised in such a way as to have a standard deviation of 1.

[Additional technical detail[134]]

The third kind of residuals are **Studentised residuals** (also called "jackknifed residuals") and they're even fancier than standardised residuals. Again, the idea is to take the ordinary residual and divide it by some quantity in order to estimate some standardised notion of the residual.[135]

Before moving on, I should point out that you don't often need to obtain these residuals yourself, even though they are at the heart of almost all regression diagnostics. Most of the time the various options that provide the diagnostics, or assumption checks, will take care of these calculations for you. Even so, it's always nice to know how to actually get hold of these things yourself in case you ever need to do something non-standard.

12.10.2 Checking the linearity of the relationship

We should check for the linearity of the relationships between the predictors and the outcomes. There's a few different things that you might want to do in order to check this. Firstly, it never hurts to just plot the relationship between the predicted values $\hat{Y}_i$ and the observed values Y_i for the outcome variable, as illustrated in Figure 12.17. To draw this in jamovi we saved the predicted values to the data set, and then drew a scatterplot of the observed against the predicted (fitted) values. This gives you a kind of "big picture view" – if this plot looks approximately linear, then we're probably not doing too badly (though that's not to say that there aren't problems). However, if you can see big departures from linearity here, then it strongly suggests that you need to make some changes.

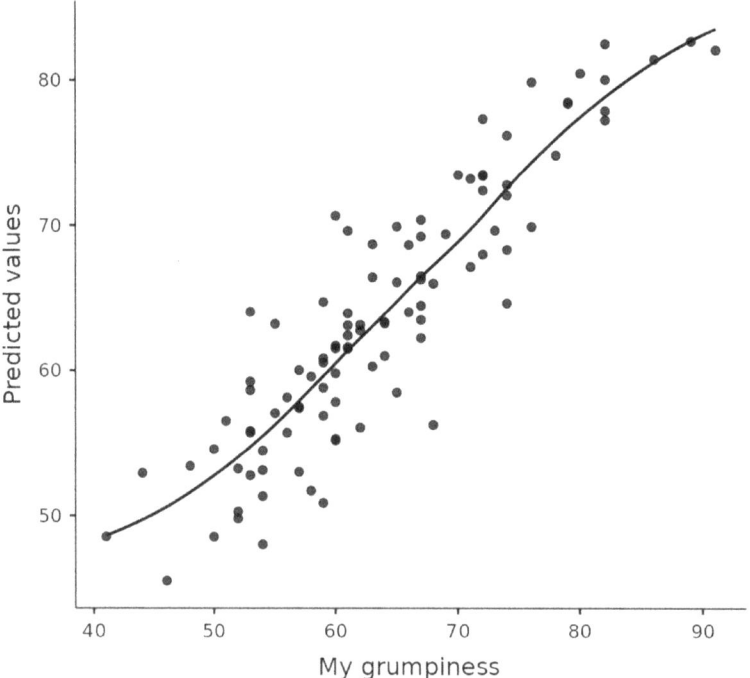

Figure 12.17: jamovi plot of the predicted values against the observed values of the outcome variable. A straight(-ish) line is what we are hoping to see here. This looks pretty good, suggesting that there is nothing grossly wrong

In any case, in order to get a more detailed picture it's often more informative to look at the relationship between the predicted values and the residuals themselves. Again, in jamovi you can save the residuals to the data set and then draw a scatterplot of the predicted values against the residual values, as in Figure 12.18. As you can see, not only does it draw the scatterplot showing the predicted value against the residuals, you can also plot a line through the data that shows the relationship between the two. Ideally, this should be a straight, perfectly horizontal line. In practice, we're looking for a reasonably straight or flat line. This is a matter of judgement.

More advanced versions of the same plot are produced by checking 'Residuals plots' in the regression analysis 'Assumption checks' options in jamovi. These are useful for checking linearity, normality and equality of variance assumptions, and we look at these in more detail in Section 12.10.3. This option not only draws plots comparing the predicted values to the residuals, it does so for each individual predictor.

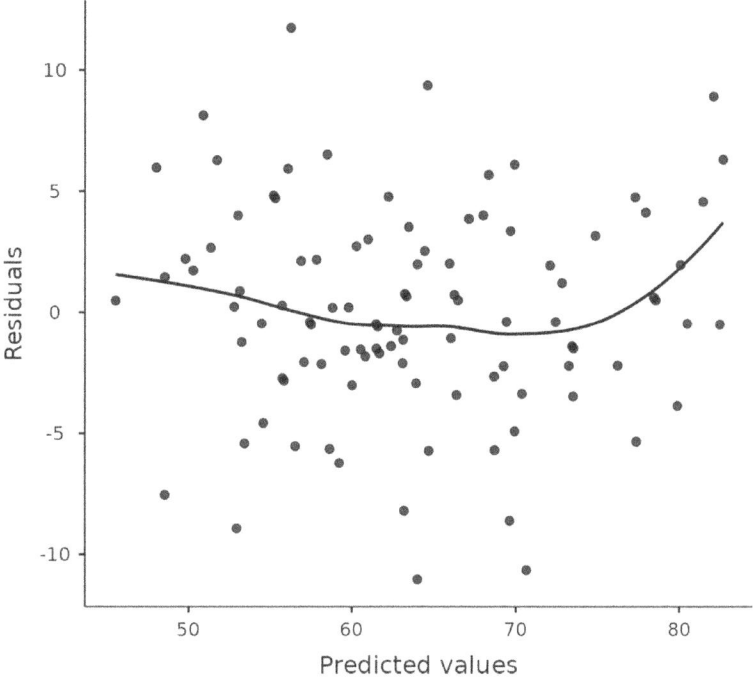

Figure 12.18: jamovi plot of the predicted values against the residuals, with a line showing the relationship between the two. If this is horizontal and straight(-ish), then we can feel reasonably confident that the "average residual" for all "predicted values" is more or less the same.

12.10.3 Checking the normality of the residuals

Like many of the statistical tools we've discussed in this book, regression models rely on a normality assumption. In this case, we assume that the residuals are normally distributed. The first thing we can do is draw a QQ-plot via the 'Assumption Checks' – 'Assumption Checks' – 'Q-Q plot of residuals' option. The output is shown in Figure 12.19, showing the standardised residuals plotted as a function of their theoretical quantiles according to the regression model.

Another thing we should check is the relationship between the predicted (fitted) values and the residuals themselves. We can get jamovi to do this using the 'Residuals Plots' option, which provides a scatterplot for each predictor variable, the outcome variable, and the predicted values against residuals, see Figure 12.20. In these plots we are looking for a fairly uniform distribution of dots, with no clear bunching or patterning of the dots. Looking at these plots, there is nothing particularly worrying as the dots are fairly evenly spread across the whole plot. There may be a little bit of non-uniformity in plot (b), but it is not a strong deviation and probably not worth worrying about.

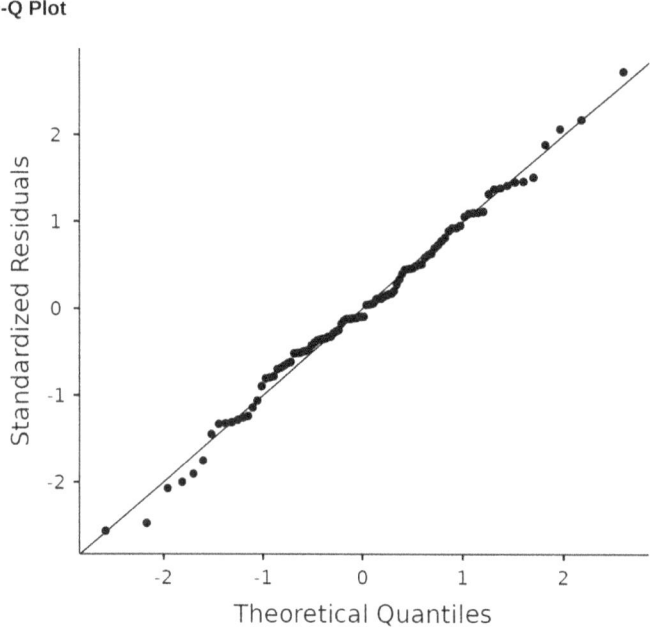

Figure 12.19: Plot of the theoretical quantiles according to the model, against the quantiles of the standardised residuals, produced in jamovi

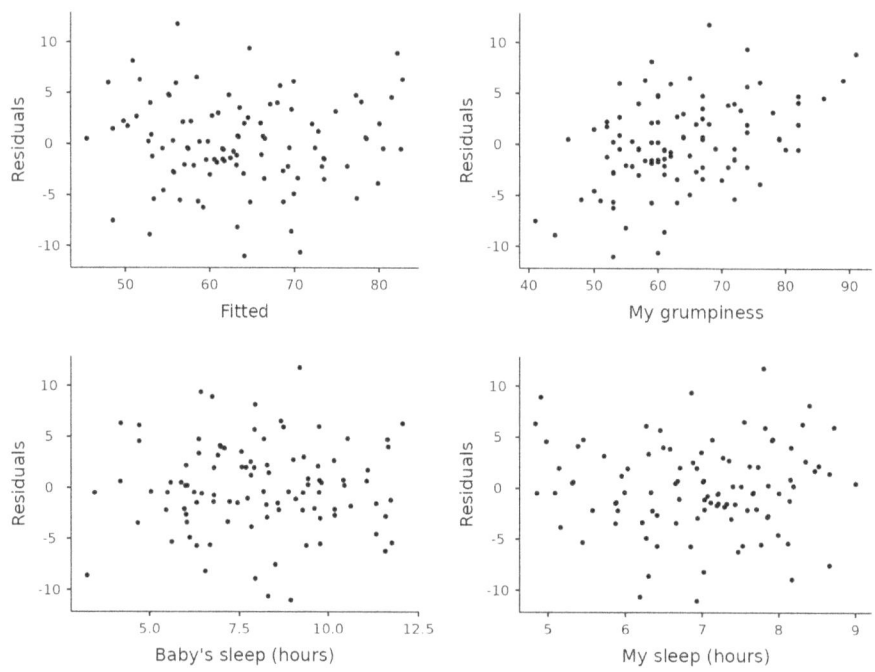

Figure 12.20: Residuals plots produced in jamovi

If we were worried, then in a lot of cases the solution to this problem (and many others) is to transform one or more of the variables. We discussed the basics of variable transformation in Section 6.3, but I do want to make special note of one additional possibility that I didn't explain fully earlier: the Box-Cox transform. The Box-Cox function is a fairly simple one and it's very widely used.[136]

You can calculate it using the BOXCOX function in the 'Compute' variables screen in jamovi.

12.10.4 Checking equality of variance

The regression models that we've talked about all make an equality (i.e.homogeneity) of variance assumption: the variance of the residuals is assumed to be constant. To plot this in jamovi first we need to calculate the square root of the (absolute) size of the residual[137] and then plot this against the predicted values, as in Figure 12.21. Note that this plot actually uses the standardised residuals rather than the raw ones, but it's immaterial from our point of view. What we're looking to see here is a straight, horizontal line running through the middle of the plot.[138]

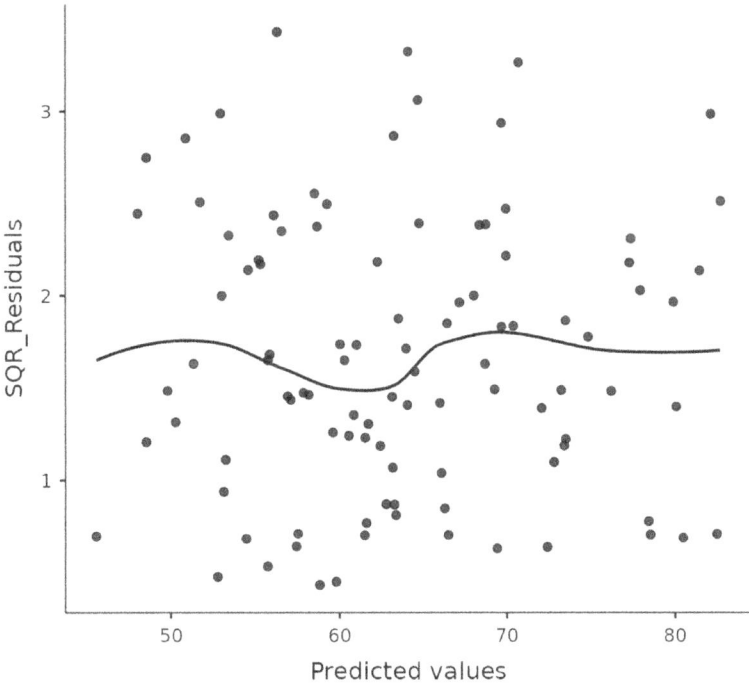

Figure 12.21: jamovi plot of the predicted values (model predictions) against the square root of the absolute standardised residuals. This plot is used to diagnose violations of homogeneity of variance. If the variance is really constant, then the line through the middle should be horizontal and flat(-ish).

12.10.5 Checking for collinearity

Another regression diagnostic is provided by **variance inflation factors** (VIFs), which are useful for determining whether or not the predictors in your regression model are too highly correlated with each other. There is a variance inflation factor associated with each predictor X_k in the model.[139]

If you've only got two predictors, the VIF values are always going to be the same, as we can see if we click on the 'Collinearity' checkbox in the 'Regression' – 'Assumptions' options in jamovi. For both dani.sleep and baby.sleep the VIF is 1.65. And since the square root of 1.65 is 1.28, we see that the correlation between our two predictors isn't causing much of a problem.

To give a sense of how we could end up with a model that has bigger collinearity problems, suppose I were to run a much less interesting regression model, in which I tried to predict the day on which the data were collected, as a function of all the other variables in the data set. To see why this would be a bit of a problem, let's have a look at the correlation matrix for all four variables (Figure 12.22).

Correlation Matrix

Correlation Matrix

	My grumpiness	Baby's sleep (hours)	My sleep (hours)	day
My grumpiness	—			
Baby's sleep (hours)	−0.57	—		
My sleep (hours)	−0.90	0.63	—	
day	0.08	−0.01	−0.10	—

Figure 12.22: Correlation matrix in jamovi for all four variables

We have some fairly large correlations between some of our predictor variables! When we run the regression model and look at the VIF values, we see that the collinearity is causing a lot of uncertainty about the coefficients. First, run the regression, as in Figure 12.23 and you can see from the VIF values that, yep, that's some mighty fine collinearity there.

12.10.6 Outliers and anomalous data

One danger that you can run into with linear regression models is that your analysis might be disproportionately sensitive to a smallish number of "unusual" or "anomalous" observations. I discussed this idea previously in Section 5.2.3 in the context of discussing the outliers that get automatically identified by the boxplot option under 'Exploration' – 'Descriptives', but this time we need to be much more precise. In the context of linear regression, there are three conceptually distinct ways in which an observation might be called "anomalous". All three are interesting, but they have rather different implications for your analysis.

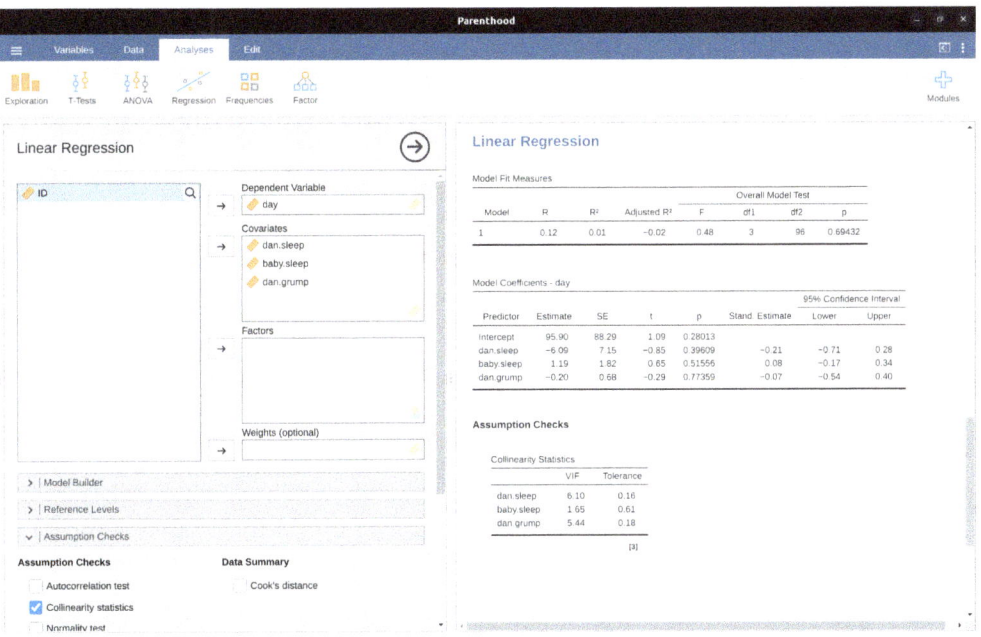

Figure 12.23: Collinearity statistics for multiple regression, produced in jamovi

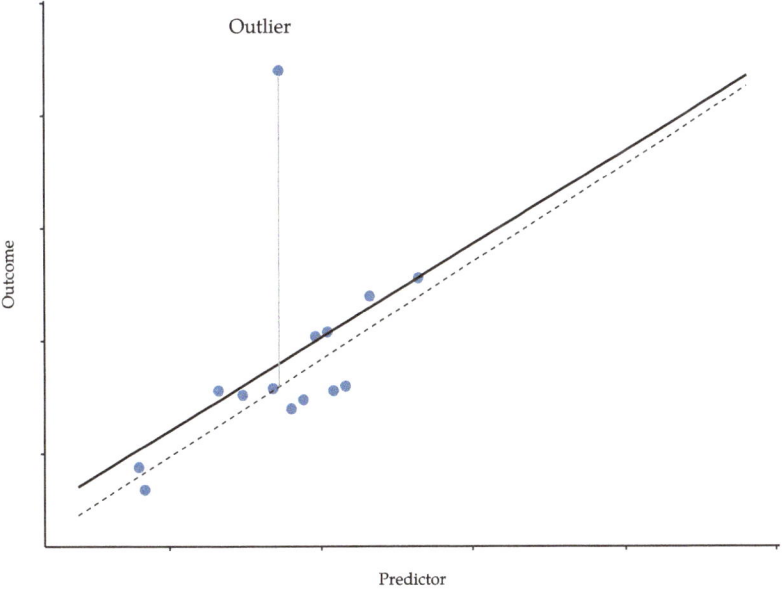

Figure 12.24: An illustration of outliers. The solid line shows the regression line with the anomalous outlier observation included. The dashed line plots the regression line estimated without the anomalous outlier observation included. The vertical line from the outlier point to the dashed regression line illustrates the large residual error for the outlier

The first kind of unusual observation is an **outlier**. The definition of an outlier (in this context) is an observation that is very different from what the regression model predicts. An example is shown in Figure 12.24, the outlier has an unusual value on the outcome (y-axis location), but not the predictor (x-axis location) and lies a long way from the regression line. In practice, we operationalise this concept by saying that an outlier is an observation that has a very large residual, ϵ_i^*. Also see the lower left plot of Anscombe's quartet, Figure 12.6.

Outliers are interesting: a big outlier might correspond to junk data, e.g., the variables might have been recorded incorrectly in the data set, or some other defect may be detectable. Note that you shouldn't throw an observation away just because it's an outlier. But the fact that it's an outlier is often a cue to look more closely at that case and try to find out why it's so different.

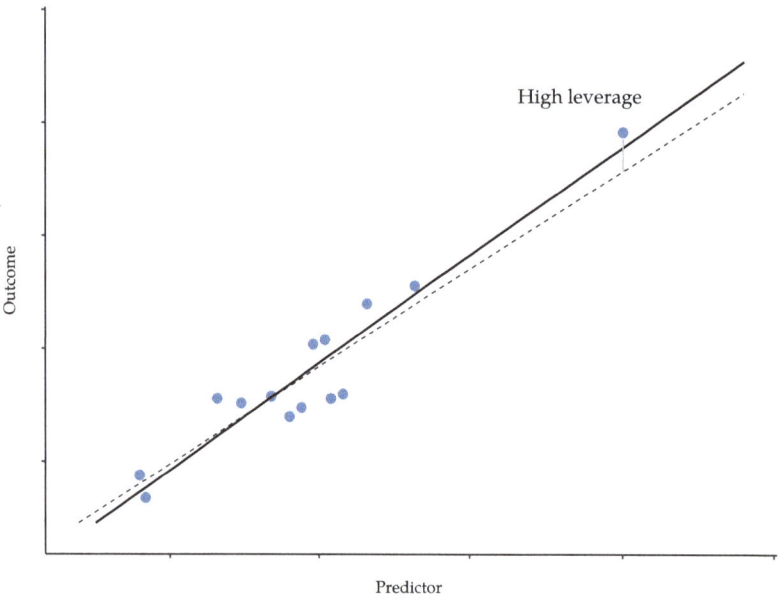

Figure 12.25: An illustration of high leverage points. The anomalous observation in this case is unusual both in terms of the predictor (x-axis) and the outcome (y-axis), but this unusualness is highly consistent with the pattern of correlations that exists among the other observations. The observation falls very close to the regression line and does not distort it by very much

The second way in which an observation can be unusual is if it has high **leverage**, which happens when the observation is very different from all the other observations. This doesn't necessarily have to correspond to a large residual. If the observation happens to be unusual on all variables in precisely the same way, it can actually lie very close to the regression line. An example of this is shown in Figure 12.25. The leverage of an observation is operationalised in terms of its hat value, usually written h_i. The formula for the hat value is rather complicated[140] but its interpretation is not: h_i is a measure of the extent to which the i-th observation is "in control" of where the regression line ends up going.

In general, if an observation lies far away from the other ones in terms of the predictor variables, it will have a large hat value (as a rough guide, high leverage is when the hat value is more than 2-3 times the average; and note that the sum of the hat values is constrained to be equal to $K + 1$). High leverage points are also worth looking at in more detail, but they're much less likely to be a cause for concern unless they are also outliers.

This brings us to our third measure of unusualness, the **influence** of an observation. A high influence observation is an outlier that has high leverage. That is, it is an observation that is very different to all the other ones in some respect, and also lies a long way from the regression line. This is illustrated in Figure 12.26. Notice the contrast to the previous two figures. Outliers don't move the regression line much and neither do high leverage points. But something that is both an outlier and has high leverage, well that has a big effect on the regression line. That's why we call these points high influence, and it's why they're the biggest worry. We operationalise influence in terms of a measure known as **Cook's distance**.[141]

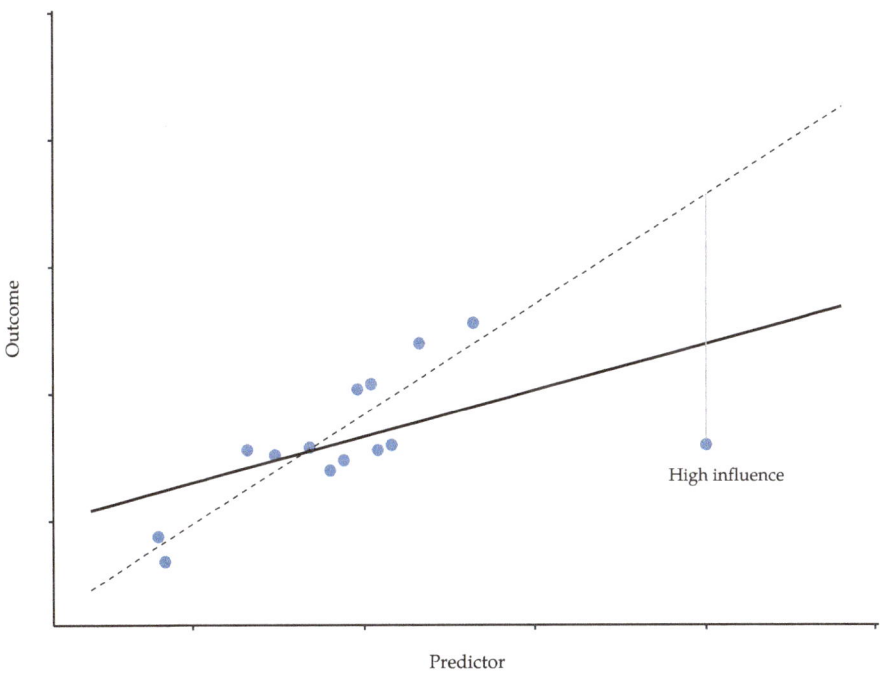

Figure 12.26: An illustration of high influence points. In this case, the anomalous observation is highly unusual on the predictor variable (x-axis), and falls a long way from the regression line. As a consequence, the regression line is highly distorted, even though (in this case) the anomalous observation is entirely typical in terms of the outcome variable (y-axis)

In order to have a large Cook's distance an observation must be a fairly substantial outlier and have high leverage. As a rough guide, Cook's distance greater than 1 is often considered large (that's what I typically use as a quick and dirty rule).

In jamovi, information about Cook's distance can be calculated by clicking on the 'Cook's Distance' checkbox in the 'Assumption Checks' – 'Data Summary' options.

When you do this, for the multiple regression model we have been using as an example in this chapter, you get the results as shown in Figure 12.27.

Data Summary

Cook's Distance

Mean	Median	SD	Range	
			Min	Max
0.01	0.00	0.02	0.00	0.11

Figure 12.27: jamovi output showing the table for the Cooks distance statistics

You can see that, in this example, the mean Cook's distance value is 0.01, and the range is from 0.00 to 0.11, so this is some way off the rule of thumb figure mentioned above that a Cook's distance greater than 1 is considered large.

An obvious question to ask next is, if you do have large values of Cook's distance what should you do? As always, there's no hard and fast rule. Probably the first thing to do is to try running the regression with the outlier with the greatest Cook's distance[142] excluded and see what happens to the model performance and to the regression coefficients. If they really are substantially different, it's time to start digging into your data set and your notes that you no doubt were scribbling as your ran your study. Try to figure out why the point is so different. If you start to become convinced that this one data point is badly distorting your results then you might consider excluding it, but that's less than ideal unless you have a solid explanation for why this particular case is qualitatively different from the others and therefore deserves to be handled separately.

12.11 Model selection

One fairly major problem that remains is the problem of "model selection". That is, if we have a data set that contains several variables, which ones should we include as predictors, and which ones should we not include? In other words, we have a problem of **variable selection**. In general, model selection is a complex business but it's made somewhat simpler if we restrict ourselves to the problem of choosing a subset of the variables that ought to be included in the model. Nevertheless, I'm not going to try covering even this reduced topic in a lot of detail. Instead, I'll talk about two broad principles that you need to think about, and then discuss one concrete tool that jamovi provides to help you select a subset of variables to include in your model. First, the two principles:

- It's nice to have an actual substantive basis for your choices. That is, in a lot of situations you the researcher have good reasons to pick out a smallish number of possible regression models that are of theoretical interest. These models will have a sensible interpretation in the context of your field. Never discount the importance of this. Statistics serves the scientific process, not the other way around.

- To the extent that your choices rely on statistical inference, there is a trade off between simplicity and goodness of fit. As you add more predictors to the model you make it more complex. Each predictor adds a new free parameter (i.e., a new regression coefficient), and each new parameter increases the model's capacity to "absorb" random variations. So the goodness of fit (e.g., R^2) continues to rise, sometimes trivially or by chance, as you add more predictors no matter what. If you want your model to be able to generalise well to new observations you need to avoid throwing in too many variables.

This latter principle is often referred to as **Ockham's razor** and is often summarised in terms of the following pithy saying: do not multiply entities beyond necessity. In this context, it means don't chuck in a bunch of largely irrelevant predictors just to boost your R^2. Hmm. Yeah, the original was better.

In any case, what we need is an actual mathematical criterion that will implement the qualitative principle behind Ockham's razor in the context of selecting a regression model. As it turns out there are several possibilities. The one that I'll talk about is the **Akaike information criterion** (AIC) (Akaike, 1974) simply because it's available as an option in jamovi.[143]

The smaller the AIC value, the better the model performance. If we ignore the low level details it's fairly obvious what the AIC does. On the left we have a term that increases as the model predictions get worse; on the right we have a term that increases as the model complexity increases. The best model is the one that fits the data well (low residuals, left-hand side) using as few predictors as possible (low K, right-hand side). In short, this is a simple implementation of Ockham's razor.

AIC can be added to the 'Model Fit Measures' output Table when the 'AIC' checkbox is clicked, and a rather clunky way of assessing different models is seeing if the AIC value is lower if you remove one or more of the predictors in the regression model. This is the only way currently implemented in jamovi, but there are alternatives in other more powerful programmes, such as R. These alternative methods can automate the process of selectively removing (or adding) predictor variables to find the best AIC. Although these methods are not implemented in jamovi, I will mention them briefly below just so you know about them.

12.11.1 Backward elimination

In backward elimination you start with the complete regression model, including all possible predictors. Then, at each "step" we try all possible ways of removing one of the variables, and whichever of these is best (in terms of lowest AIC value) is accepted. This becomes our new regression model, and we then try all possible deletions from the new model, again choosing the option with lowest AIC. This process continues until we end up with a model that has a lower AIC value than any of the other possible models that you could produce by deleting one of its predictors.

12.11.2 Forward selection

As an alternative, you can also try **forward selection**. This time around we start with the smallest possible model as our start point, and only consider the possible additions to the model. However, there's one complication. You also need to specify what the largest possible model you're willing to entertain is.

Although backward and forward selection can lead to the same conclusion, they don't always.

12.11.3 A caveat

Automated variable selection methods are seductive things, especially when they're bundled up in (fairly) simple functions in powerful statistical programmes. They provide an element of objectivity to your model selection, and that's kind of nice. Unfortunately, they're sometimes used as an excuse for thoughtlessness. No longer do you have to think carefully about which predictors to add to the model and what the theoretical basis for their inclusion might be. Everything is solved by the magic of AIC. And if we start throwing around phrases like Ockham's razor, well it sounds like everything is wrapped up in a nice neat little package that no-one can argue with.

Or, perhaps not. Firstly, there's very little agreement on what counts as an appropriate model selection criterion. When I was taught backward elimination as an undergraduate, we used F-tests to do it, because that was the default method used by the software. I've described using AIC, and since this is an introductory text that's the only method I've described, but the AIC is hardly the Word of the Gods of Statistics. It's an approximation, derived under certain assumptions, and it's guaranteed to work only for large samples when those assumptions are met. Alter those assumptions and you get a different criterion, like the Bayesian Information Criterion (BIC) for instance (also available in jamovi). Take a different approach again and you get the normalised maximum likelihood (NML) criterion. Decide that you're a Bayesian and you get model selection based on posterior odds ratios. Then there are a bunch of regression specific tools that I haven't mentioned. And so on. All of these different methods have strengths and weaknesses, and some are easier to calculate than others (AIC is probably the easiest of the lot, which might account for its popularity). Almost all of them produce the same answers when the answer is "obvious" but there's a fair amount of disagreement when the model selection problem becomes hard.

What does this mean in practice? Well, you could go and spend several years teaching yourself the theory of model selection, learning all the ins and outs of it so that you could finally decide on what you personally think the right thing to do is. Speaking as someone who actually did that, I wouldn't recommend it. You'll probably come out the other side even more confused than when you started. A better strategy is to show a bit of common sense. If you're staring at the results of an automated backwards or forwards selection procedure, and the model that makes sense is close to having the smallest AIC but is narrowly defeated by a model that doesn't make any sense, then trust your instincts. Statistical model selection is an inexact tool, and as I said at the beginning, interpretability matters.

12.11.4 Comparing two regression models

An alternative to using automated model selection procedures is for the researcher to explicitly select two or more regression models to compare to each other. You can do this in a few different ways, depending on what research question you're trying to answer. Suppose we want to know whether or not the amount of sleep that my son got has any relationship to my grumpiness, over and above what we might expect from the amount of sleep that I got. We also want to make sure that the day on which we took the measurement has no influence on the relationship. That is, we're interested in the relationship between baby.sleep and dani.grump, and from that perspective dani.sleep and day are nuisance variable or **covariates** that we want to control for. In this situation, what we would like to know is whether dani.grump ~ dani.sleep + day + baby.sleep (which I'll call Model 2, or M2) is a better regression model for these data than dani.grump ~ dani.sleep + day (which I'll call Model 1, or M1). There are two different ways we can compare these two models, one based on a model selection criterion like AIC, and the other based on an explicit hypothesis test. I'll show you the AIC based approach first because it's simpler, and follows naturally from discussion in the last section. The first thing I need to do is actually run the two regressions, note the AIC for each one, and then select the model with the smaller AIC value as it is judged to be the better model for these data. Actually, don't do this just yet. Read on because there is an easy way in jamovi to get the AIC values for different models included in one table.[144]

A somewhat different approach to the problem comes out of the hypothesis testing framework. Suppose you have two regression models, where one of them (Model 1) contains a subset of the predictors from the other one (Model 2). That is, Model 2 contains all of the predictors included in Model 1, plus one or more additional predictors. When this happens we say that Model 1 is nested within Model 2, or possibly that Model 1 is a submodel of Model 2. Regardless of the terminology, what this means is that we can think of Model 1 as a null hypothesis and Model 2 as an alternative hypothesis. And in fact we can construct an F-test for this in a fairly straightforward fashion.[145]

Okay, so that's the hypothesis test that we use to compare two regression models to one another. Now, how do we do it in jamovi? The answer is to use the 'Model Builder' option and specify the Model 1 predictors dani.sleep and day in 'Block 1' and then add the additional predictor from Model 2 (baby.sleep) in 'Block 2', as in Figure 12.28. This shows, in the 'Model Comparisons' Table, that for the comparisons between Model 1 and Model 2, $F(1,96) = 0.00$, $p = 0.954$. Since we have p > .05 we retain the null hypothesis (M1). This approach to regression, in which we add all of our covariates into a null model, then add the variables of interest into an alternative model, and then compare the two models in a hypothesis testing framework, is often referred to as **hierarchical regression**.

We can also use this 'Model Comparison' option to display a table that shows the AIC and BIC for each model, making it easy to compare and identify which model has the lowest value, as in Figure 12.28.

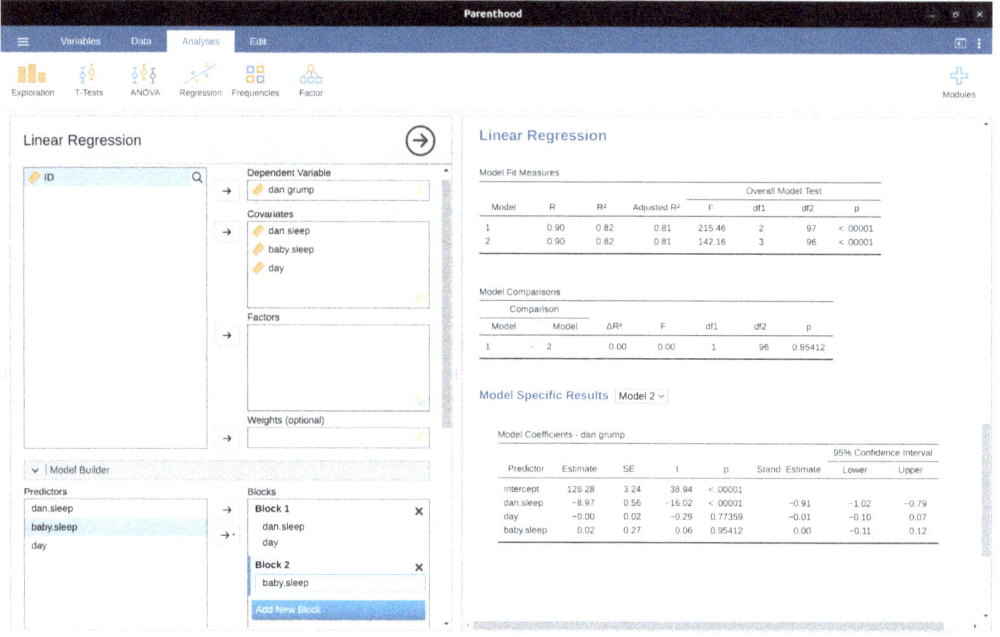

Figure 12.28: Model comparison in jamovi using the 'Model Builder' option

12.12 Summary

- Want to know how strong the relationship is between two variables? Calculate Correlations.
- Drawing Scatterplots.
- Basic ideas about What is a linear regression model? and Estimating a linear regression model.
- Multiple linear regression.
- Quantifying the fit of the regression model using R^2.
- Hypothesis tests for regression models.
- In Regarding regression coefficients we talked about calculating Confidence intervals for the coefficients and Calculating standardised regression coefficients.
- The Assumptions of regression and Model checking.
- Regression Model selection.

Chapter 13

Comparing several means (one-way ANOVA)

This chapter introduces one of the most widely used tools in psychological statistics, known as "the analysis of variance", but usually referred to as ANOVA. The basic technique was developed by Sir Ronald Fisher in the early 20th century and it is to him that we owe the rather unfortunate terminology. The term ANOVA is a little misleading, in two respects. Firstly, although the name of the technique refers to variances, ANOVA is concerned with investigating differences in means. Secondly, there are several different things out there that are all referred to as ANOVAs, some of which have only a very tenuous connection to one another. Later on in the book we'll encounter a range of different ANOVA methods that apply in quite different situations, but for the purposes of this chapter we'll only consider the simplest form of ANOVA, in which we have several different groups of observations, and we're interested in finding out whether those groups differ in terms of some outcome variable of interest. This is the question that is addressed by a one-way ANOVA.

The structure of this chapter is as follows: first I'll introduce a fictitious data set that we'll use as a running example throughout the chapter. After introducing the data, I'll describe the mechanics of how a one-way ANOVA actually works How ANOVA works and then focus on how you can run one in jamovi Running an ANOVA in jamovi. These two sections are the core of the chapter.

The remainder of the chapter discusses a range of important topics that inevitably arise when running an ANOVA, namely how to calculate effect sizes, post hoc tests and corrections for multiple comparisons and the assumptions that ANOVA relies upon. We'll also talk about how to check those assumptions and some of the things you can do if the assumptions are violated. Then we'll cover repeated measures ANOVA.

13.1 An illustrative data set

Suppose you've become involved in a clinical trial in which you are testing a new antidepressant drug called *Joyzepam*. In order to construct a fair test of the drug's effectiveness, the study involves three separate drugs to be administered. One is a placebo, and the other is an existing antidepressant / anti-anxiety drug called *Anxifree*. A collection of 18 participants with moderate to severe depression are recruited for your initial testing. Because the drugs are sometimes administered in conjunction with psychological therapy, your study includes 9 people undergoing cognitive behavioural therapy (CBT) and 9 who are not. Participants are randomly assigned (doubly blinded, of course) a treatment, such that there are 3 CBT people and 3 no-therapy people assigned to each of the 3 drugs. A psychologist assesses the mood of each person after a 3-month run with each drug, and the overall improvement in each person's mood is assessed on a scale ranging from -5 to $+5$. With that as the study design, let's now load up the data file in *clinicaltrial.csv* . We can see that this data set contains the three variables drug, therapy and mood.gain.

For the purposes of this chapter, what we're really interested in is the effect of drug on mood.gain. The first thing to do is calculate some descriptive statistics and draw some graphs. In the Chapter 4 chapter we showed you how to do this, and some of the descriptive statistics we can calculate in jamovi are shown in Figure 13.1. As the plot makes clear, there is a larger improvement in mood for participants in the Joyzepam group than for either the Anxifree group or the placebo group. The Anxifree group shows a larger mood gain than the control group, but the difference isn't as large. The question that we want to answer is are these difference "real", or are they just due to chance?

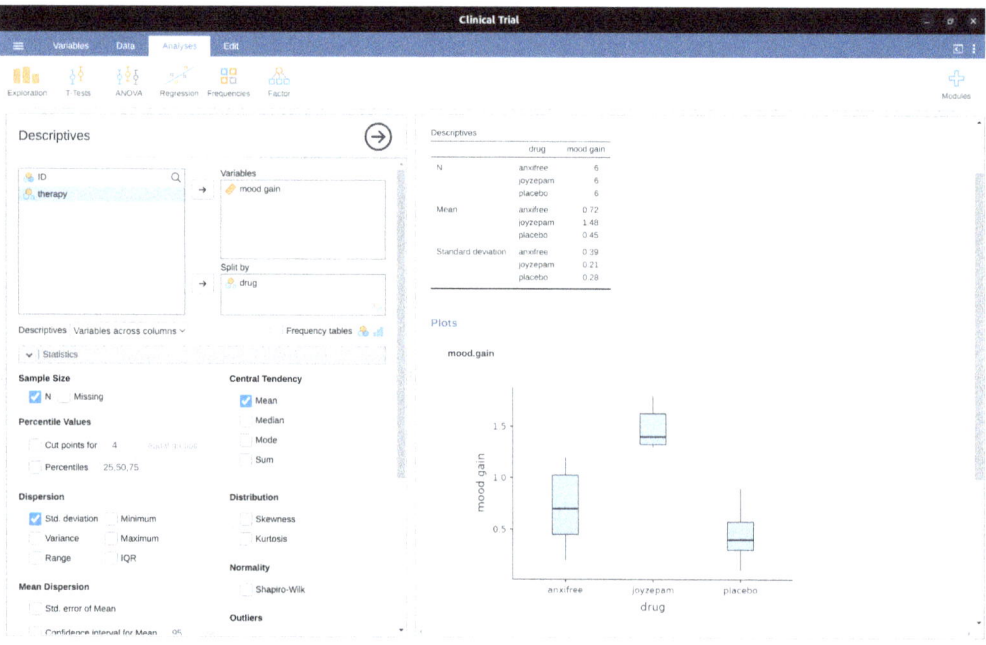

Figure 13.1: Descriptives for mood gain, and box plots by drug administered

13.2 How ANOVA works

In order to answer the question posed by our clinical trial data we're going to run a one-way ANOVA. I'm going to start by showing you how to do it the hard way, building the statistical tool from the ground up and showing you how you could do it if you didn't have access to any of the cool built-in ANOVA functions in jamovi. And I hope you'll read it carefully, try to do it the long way once or twice to make sure you really understand how ANOVA works.

The experimental design that I described in the previous section strongly suggests that we're interested in comparing the average mood change for the three different drugs. In that sense, we're talking about an analysis similar to the t-test (see Chapter 11) but involving more than two groups. If we let μ_P denote the population mean for the mood change induced by the placebo, and let μ_A and μ_J denote the corresponding means for our two drugs, Anxifree and Joyzepam, then the (somewhat pessimistic) null hypothesis that we want to test is that all three population means are identical. That is, neither of the two drugs is any more effective than a placebo. We can write out this null hypothesis as:

$$H_0 : \text{it is true that } \mu_P = \mu_A = \mu_J$$

As a consequence, our alternative hypothesis is that at least one of the three different treatments is different from the others. It's a bit tricky to write this mathematically, because (as we'll discuss) there are quite a few different ways in which the null hypothesis can be false. So for now we'll just write the alternative hypothesis like this:

$$H_1 : \text{it } \underline{\textit{is not}} \text{ true that } \mu_P = \mu_A = \mu_J$$

This null hypothesis is a lot trickier to test than any of the ones we've seen previously. Given the title of this chapter, a sensible guess for how to test this would be to "do an ANOVA", but it's not particularly clear why "analysis of variances" will help us learn anything useful about the means. In fact, this is one of the biggest conceptual difficulties that people have when first encountering ANOVA. To see how this works, let's start by talking about variances, specifically between group variability and within-group variability (Figure 13.2).

13.2.1 Two formulas for the variance of Y

First, let's start by introducing some notation. We'll use G to refer to the total number of groups. For our data set there are three drugs, so there are $G = 3$ groups. Next, we'll use N to refer to the total sample size; there are a total of $N = 18$ people in our data set. Similarly, let's use N_k to denote the number of people in the k-th group. In our fake clinical trial, the sample size is $N_k = 6$ for all three groups.[146] Finally, we'll use Y to denote the outcome variable. In our case, Y refers to mood change. Specifically, we'll use Y_{ik} to refer to the mood change experienced by the i-th member of the k-th group. Similarly, we'll use $\bar{Y}$ to be the average mood change, taken across all 18 people in the experiment, and $\bar{Y}_k$ to refer to the average mood change experienced by the 6 people in group k.

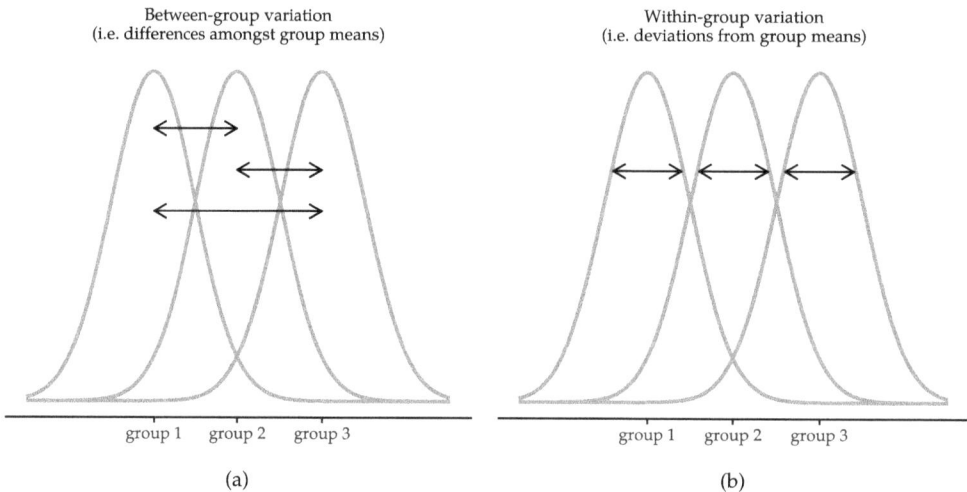

Figure 13.2: Graphical illustration of "between groups" variation (panel (a)) and "within groups" variation (panel (b)). On the left the arrows show the differences in the group means. On the right the arrows highlight the variability within each group

Now that we've got our notation sorted out we can start writing down formulas. To start with, let's recall the formula for the variance that we used in Section 4.2, way back in those kinder days when we were just doing descriptive statistics. The sample variance of Y is defined as follows:

$$Var(Y) = \frac{1}{N} \sum_{k=1}^{G} \sum_{i=1}^{N_k} (Y_{ik} - \bar{Y})^2$$

This formula looks pretty much identical to the formula for the variance in Section 4.2. The only difference is that this time around I've got two summations here: I'm summing over groups (i.e., values for k) and over the people within the groups (i.e., values for i). This is purely a cosmetic detail. If I'd instead used the notation Y_p to refer to the value of the outcome variable for person p in the sample, then I'd only have a single summation. The only reason that we have a double summation here is that I've classified people into groups, and then assigned numbers to people within groups.

A concrete example might be useful here. Let's consider Table 13.1, in which we have a total of $N = 5$ people sorted into $G = 2$ groups. Arbitrarily, let's say that the "cool" people are group 1 and the "uncool" people are group 2. It turns out that we have three cool people ($N_1 = 3$) and two uncool people ($N_2 = 2$).

Notice that I've constructed two different labelling schemes here. We have a "person" variable p so it would be perfectly sensible to refer to Y_p as the grumpiness of the p-th person in the sample. For instance, the table shows that Tim is the fourth so we'd say $p = 4$. So, when talking about the grumpiness Y of this "Tim" person, whoever he might be, we could refer to his grumpiness by saying that $Y_p = 91$, for person $p = 4$.

Table 13.1: Grumpiness for people in cool and uncool groups

name	person P	group	group num. k	index in group	grumpiness Y_{ik} or Y_p
Ann	1	cool	1	1	20
Ben	2	cool	1	2	55
Cat	3	cool	1	3	21
Tim	4	uncool	2	1	91
Egg	5	uncool	2	2	22

However, that's not the only way we could refer to Tim. As an alternative we could note that Tim belongs to the "uncool" group ($k = 2$), and is in fact the first person listed in the uncool group ($i = 1$). So it's equally valid to refer to Tim's grumpiness by saying that $Y_{ik} = 91$, where $k = 2$ and $i = 1$. In other words, each person p corresponds to a unique ik combination, and so the formula that I gave above is actually identical to our original formula for the variance, which would be:

$$Var(Y) = \frac{1}{N} \sum_{p=1}^{N} (Y_p - \bar{Y})^2$$

In both formulas, all we're doing is summing over all of the observations in the sample. Most of the time we would just use the simpler Y_p notation; the equation using Y_p is clearly the simpler of the two. However, when doing an ANOVA it's important to keep track of which participants belong in which groups, and we need to use the Y_{ik} notation to do this.

13.2.2 From variances to sums of squares

Okay, now that we've got a good grasp on how the variance is calculated, let's define something called the **total sum of squares**, which is denoted $SStot$. This is very simple. Instead of averaging the squared deviations, which is what we do when calculating the variance, we just add them up.[147]

When we talk about analysing variances in the context of ANOVA, what we're really doing is working with the total sums of squares rather than the actual variance.[148]

Next, we can define a third notion of variation which captures only the differences between groups. We do this by looking at the differences between the group means $\bar{Y}_k$ and grand mean $\bar{Y}$.[149]

It's not too difficult to show that the total variation among people in the experiment SS_{tot} is actually the sum of the differences between the groups SS_b and the variation inside the groups SS_w. That is:

$$SS_w + SS_b = SS_{tot}$$

Okay, so what have we found out? We've discovered that the total variability associated with the outcome variable (SS_{tot}) can be mathematically carved up into the sum of "the variation due to the differences in the sample means for the different groups" (SS_b) plus "all the rest of the variation" (SS_w).[150]

How does that help me find out whether the groups have different population means? Um. Wait. Hold on a second. Now that I think about it, this is exactly what we were looking for. If the null hypothesis is true then you'd expect all the sample means to be pretty similar to each other, right? And that would imply that you'd expect SS_b to be really small, or at least you'd expect it to be a lot smaller than "the variation associated with everything else", SS_w. Hmm. I detect a hypothesis test coming on.

13.2.3 From sums of squares to the F-test

As we saw in the last section, the qualitative idea behind ANOVA is to compare the two sums of squares values SS_b and SS_w to each other. If the between-group variation SS_b is large relative to the within-group variation SS_w then we have reason to suspect that the population means for the different groups aren't identical to each other. In order to convert this into a workable hypothesis test, there's a little bit of "fiddling around" needed. What I'll do is first show you what we do to calculate our test statistic, the F-**ratio**, and then try to give you a feel for why we do it this way.

In order to convert our SS values into an F-ratio the first thing we need to calculate is the **degrees of freedom** associated with the SS_b and SS_w values. As usual, the degrees of freedom corresponds to the number of unique "data points" that contribute to a particular calculation, minus the number of "constraints" that they need to satisfy. For the within groups variability what we're calculating is the variation of the individual observations (N data points) around the group means (G constraints). In contrast, for the between groups variability we're interested in the variation of the group means (G data points) around the grand mean (1 constraint). Therefore, the degrees of freedom here are:

$$df_b = G - 1$$
$$df_w = N - G$$

Okay, that seems simple enough. What we do next is convert our summed squares value into a "mean squares" value, which we do by dividing by the degrees of freedom:

$$MS_b = \frac{SS_b}{df_b}$$
$$MS_w = \frac{SS_w}{df_w}$$

Finally, we calculate the F-ratio by dividing the between groups MS by the within groups MS:

$$F = \frac{MS_b}{MS_w}$$

At a very general level, the intuition behind the F-statistic is straightforward. Bigger values of F means that the between groups variation is large relative to the within groups variation. As a consequence, the larger the value of F the more evidence we have against the null hypothesis. But how large does F have to be in order to actually reject H_0? In order to understand this, you need a slightly deeper understanding of what ANOVA is and what the mean squares values actually are.

The next section discusses that in a bit of detail, but for readers that aren't interested in the details of what the test is actually measuring I'll cut to the chase. In order to complete our hypothesis test we need to know the sampling distribution for F if the null hypothesis is true. Not surprisingly, the sampling distribution for the F-statistic under the null hypothesis is an F-distribution. If you recall our discussion of the F-distribution in Chapter 7, the F-distribution has two parameters, corresponding to the two degrees of freedom involved. The first one df_1 is the between groups degrees of freedom df_b, and the second one df_2 is the within groups degrees of freedom df_w.

A summary of all the key quantities involved in a one-way ANOVA, including the formulas showing how they are calculated, is shown in Table 13.2.

Table 13.2: All of the key quantities involved in an ANOVA organised into a "standard" ANOVA table. The formulas for all quantities (except the p-value which has a very ugly formula and would be nightmarishly hard to calculate without a computer) are shown

	between groups	within groups
df	$df_b = G - 1$	$df_w = N - G$
sum of squares	$SS_b = \sum_{k=1}^{G} N_k (\bar{Y}_k - \bar{Y})^2$	$SS_w = \sum_{k=1}^{G} \sum_{i=1}^{N_k} (Y_{ik} - \bar{Y}_k)^2$
mean squares	$MS_b = \frac{SS_b}{df_b}$	$MS_w = \frac{SS_w}{df_w}$
F-statistic	$F = \frac{MS_b}{df_b}$	-
p-value	[complicated]	-

[Additional technical detail[151]]

13.2.4 A worked example

The previous discussion was fairly abstract and a little on the technical side, so I think that at this point it might be useful to see a worked example. For that, let's go back to the clinical trial data that I introduced at the start of the chapter. The descriptive statistics that we calculated at the beginning tell us our group means: an average mood gain of 0.45 for the placebo, 0.72 for Anxifree, and 1.48 for Joyzepam. With that in mind, let's party like it's 1899[152] and start doing some pencil and paper calculations. I'll only do this for the first 5 observations because it's not bloody 1899 and I'm very lazy. Let's start by calculating SS_w, the within-group sums of squares. First, let's draw up a nice table to help us with our calculations (Table 13.3).

Table 13.3: A worked example...1

group k	outcome Y_{ik}
placebo	0.5
placebo	0.3
placebo	0.1
anxifree	0.6
anxifree	0.4

At this stage, the only thing I've included in the table is the raw data itself. That is, the grouping variable (i.e., drug) and outcome variable (i.e. mood.gain) for each person. Note that the outcome variable here corresponds to the Y_{ik} value in our equation previously. The next step in the calculation is to write down, for each person in the study, the corresponding group mean, $\bar{Y}_k$. This is slightly repetitive but not particularly difficult since we already calculated those group means when doing our descriptive statistics, see Table 13.4.

Table 13.4: A worked example...2

group k	outcome Y_{ik}	group mean $\bar{Y}_k$
placebo	0.5	0.45
placebo	0.3	0.45
placebo	0.1	0.45
anxifree	0.6	0.72
anxifree	0.4	0.72

Now that we've written those down, we need to calculate, again for every person, the deviation from the corresponding group mean. That is, we want to subtract $Y_{ik} - \bar{Y}_k$. After we've done that, we need to square everything. When we do that, here's what we get (Table 13.5).

Table 13.5: A worked example...3

group k	outcome Y_{ik}	group mean $\bar{Y}_k$	dev. from group mean $Y_{ik} - \bar{Y}_k$	squared deviation $(Y_{ik} - \bar{Y}_k)^2$
placebo	0.5	0.45	0.05	0.0025
placebo	0.3	0.45	-0.15	0.0225
placebo	0.1	0.45	-0.35	0.1225
anxifree	0.6	0.72	-0.12	0.0136
anxifree	0.4	0.72	-0.32	0.1003

The last step is equally straightforward. In order to calculate the within-group sum of squares we just add up the squared deviations across all observations:

$$SS_w = 0.0025 + 0.0225 + 0.1225 + 0.0136 + 0.1003$$
$$= 0.2614$$

Of course, if we actually wanted to get the right answer we'd need to do this for all 18 observations in the data set, not just the first five. We could continue with the pencil and paper calculations if we wanted to, but it's pretty tedious. Alternatively, it's not too hard to do this in a dedicated spreadsheet programme such as OpenOffice or Excel. Try and do it yourself. The one that I did, in Excel, is in the file clinicaltrial_anova.xls. When you do it you should end up with a within-group sum of squares value of 1.39.

Okay. Now that we've calculated the within groups variation, SS_w, it's time to turn our attention to the between-group sum of squares, SS_b. The calculations for this case are very similar. The main difference is that instead of calculating the differences between an observation Yik and a group mean $\bar{Y}_k$ for all of the observations, we calculate the differences between the group means $\bar{Y}_k$ and the grand mean $\bar{Y}$ (in this case 0.88) for all of the groups (Table 13.6).

Table 13.6: A worked example...4

group k	group mean $\bar{Y}_k$	grand mean $\bar{Y}$	deviation $\bar{Y}_k - \bar{Y}$	squared deviation $(\bar{Y}_k - \bar{Y})^2$
placebo	0.45	0.88	-0.43	0.19
anxifree	0.72	0.88	-0.16	0.03
joyzepam	1.48	0.88	0.60	0.36

However, for the between group calculations we need to multiply each of these squared deviations by N_k, the number of observations in the group. We do this because every observation in the group (all N_k of them) is associated with a between group difference. So if there are six people in the placebo group and the placebo group mean differs from the grand mean by 0.19, then the total between group variation associated with these six people is $6 \times 0.19 = 1.14$. So we have to extend our little table of calculations (Table 13.7).

Table 13.7: A worked example...5

group k	...	squared deviations $(\bar{Y}_k - \bar{Y})^2$	sample size N_k	weighted squared dev $N_k(\bar{Y}_k - \bar{Y})^2$
placebo	...	0.19	6	1.14
anxifree	...	0.03	6	0.18
joyzepam	...	0.36	6	2.16

And so now our between group sum of squares is obtained by summing these "weighted squared deviations" over all three groups in the study:

$$SS_b = 1.14 + 0.18 + 2.16$$
$$= 3.48$$

As you can see, the between group calculations are a lot shorter.[153] Now that we've calculated our sums of squares values, SS_b and SS_w, the rest of the ANOVA is pretty painless. The next step is to calculate the degrees of freedom. Since we have $G = 3$ groups and $N = 18$ observations in total our degrees of freedom can be calculated by simple subtraction:

$$df_b = G - 1 = 2$$
$$df_w = N - G = 15$$

Next, since we've now calculated the values for the sums of squares and the degrees of freedom, for both the within groups variability and the between groups variability, we can obtain the mean square values by dividing one by the other:

$$MS_b = \frac{SS_b}{df_b} = \frac{3.48}{2} = 1.74$$
$$MS_w = \frac{SS_w}{df_w} = \frac{1.39}{15} = 0.09$$

We're almost done. The mean square values can be used to calculate the F-value, which is the test statistic that we're interested in. We do this by dividing the between groups MS value by the within groups MS value:

$$F = \frac{MS_b}{MS_w} = \frac{1.74}{0.09}$$
$$= 19.3$$

Woohooo! Now that we have our test statistic, the last step is to find out whether the test itself gives us a significant result. As discussed in Chapter 9 what we'd do in the old days is open up a statistics textbook and in a huge lookup table we would find the threshold F-value corresponding to a particular value of alpha (the null hypothesis rejection region), e.g. 0.05, 0.01 or 0.001, for 2 and 15 degrees of freedom. Doing it this way would give us a threshold F-value for an alpha of 0.001 of 11.34. As this is less than our calculated F-value we say that $p < 0.001$. But nowadays fancy stats software calculates the exact p-value for you, which is 0.000071. So, unless we're being *extremely* conservative about our type I error rate, we're pretty much guaranteed to reject the null hypothesis.

At this point, we're basically done. Having completed our calculations, it's traditional to organise all these numbers into an ANOVA table like the one in Table 13.1. For our clinical trial data, the ANOVA table would look like Table 13.8.

Table 13.8: The ANOVA results table

	df	sum of squares	mean squares	F-statistic	p-value
between groups	2	3.48	1.74	19.3	0.000071
within groups	15	1.39	0.09	-	-

These days, you'll probably never have much reason to want to construct one of these tables yourself, but you will find that almost all statistical software (jamovi included) tends to organise the output of an ANOVA into a table like this, so it's a good idea to get used to reading them. However, although the software will output a full ANOVA table, there's almost never a good reason to include the whole table in your write up. A pretty standard way of reporting the stats block for this result would be to write something like this:

> One-way ANOVA showed a significant effect of drug on mood gain ($F(2,15) = 19.3, p < .001$).

Sigh. So much work for one short sentence.

13.3 Running an ANOVA in jamovi

I'm pretty sure I know what you're thinking after reading the last section, especially if you followed my advice and did all of that by pencil and paper (i.e., in a spreadsheet) yourself. Doing the ANOVA calculations yourself sucks. There's quite a lot of calculations that we needed to do along the way, and it would be tedious to have to do this over and over again every time you wanted to do an ANOVA.

13.3.1 Using jamovi to specify your ANOVA

To make life easier for you, jamovi can do ANOVA...hurrah! Go to the 'ANOVA' – 'ANOVA' analysis, and move the mood.gain variable across so it is in the 'Dependent Variable' box, and then move the drug variable across so it is in the 'Fixed Factors' box. This should give the results as shown in Figure 13.3.[154] Note I have also checked the η^2 checkbox, pronounced "eta" squared, under the 'Effect Size' option and this is also shown on the results table. We will come back to effect sizes a bit later.

The jamovi results table shows you the sums of squares values, the degrees of freedom, and a couple of other quantities that we're not really interested in right now. Notice, however, that jamovi doesn't use the names "between group" and "within group". Instead, it tries to assign more meaningful names. In our particular example, the between groups variance corresponds to the effect that the drug has on the outcome variable, and the within groups variance corresponds to the "leftover" variability so it calls that the residuals.

ANOVA

ANOVA - mood.gain

	Sum of Squares	df	Mean Square	F	p	η²
drug	3.45	2	1.73	18.61	0.00009	0.71
Residuals	1.39	15	0.09			

Figure 13.3: jamovi results table for ANOVA of mood gain by drug administered

If we compare these numbers to the numbers that I calculated by hand in A worked example, you can see that they're more or less the same, apart from rounding errors. The between groups sums of squares is $SS_b = 3.45$, the within groups sums of squares is $SS_w = 1.39$, and the degrees of freedom are 2 and 15 respectively. We also get the F-value and the p-value and, again, these are more or less the same, give or take rounding errors, to the numbers that we calculated ourselves when doing it the long and tedious way.

13.4 Effect size

There's a few different ways you could measure the effect size in an ANOVA, but the most commonly used measures are η^2 (eta squared) and partial η^2. For a one-way analysis of variance they're identical to each other, so for the moment I'll just explain η^2. The definition of η^2 is actually really simple:

$$\eta^2 = \frac{SS_b}{SS_{tot}}$$

That's all it is. So when I look at the ANOVA table in Figure 13.3, I see that $SS_b = 3.45$ and $SS_tot = 3.45 + 1.39 = 4.84$. Thus we get an η^2 value of:

$$\eta^2 = \frac{3.45}{4.84} = 0.71$$

The interpretation of η^2 is equally straightforward. It refers to the proportion of the variability in the outcome variable (mood.gain) that can be explained in terms of the predictor (drug). A value of $\eta^2 = 0$ means that there is no relationship at all between the two, whereas a value of $\eta^2 = 1$ means that the relationship is perfect. Better yet, the η^2 value is very closely related to R^2, as discussed previously in Section 12.6.1, and has an equivalent interpretation. Although many statistics textbooks suggest η^2 as the default effect size measure in ANOVA, there's an interesting blog post[155] by Daniel Lakens suggesting that eta-squared is perhaps not the best measure of effect size in real-world data analysis, because it can be a biased estimator. Usefully, there is also an option in jamovi to specify omega-squared (ω^2), which is less biased, alongside eta-squared.

13.5 Multiple comparisons and post hoc tests

Any time you run an ANOVA with more than two groups and you end up with a significant effect, the first thing you'll probably want to ask is which groups are actually different from one another. In our drugs example, our null hypothesis was that all three drugs (placebo, Anxifree and Joyzepam) have the exact same effect on mood. But if you think about it, the null hypothesis is actually claiming three different things all at once here. Specifically, it claims that:

- Your competitor's drug (Anxifree) is no better than a placebo (i.e., $\mu_A = \mu_P$)
- Your drug (Joyzepam) is no better than a placebo (i.e., $\mu_J = \mu_P$)
- Anxifree and Joyzepam are equally effective (i.e., $\mu_J = \mu_A$)

If any one of those three claims is false, then the null hypothesis is also false. So, now that we've rejected our null hypothesis, we're thinking that at least one of those things isn't true. But which ones? All three of these propositions are of interest. Since you certainly want to know if your new drug Joyzepam is better than a placebo, it would be nice to know how well it stacks up against an existing commercial alternative (i.e., Anxifree). It would even be useful to check the performance of Anxifree against the placebo. Even if Anxifree has already been extensively tested against placebos by other researchers, it can still be very useful to check that your study is producing similar results to earlier work.

When we characterise the null hypothesis in terms of these three distinct propositions, it becomes clear that there are eight possible "states of the world" that we need to distinguish between (Table 13.9).

Table 13.9: The null hypothesis and eight possible "states of the world"

possibility:	is $\mu_P = \mu_A$?	is $\mu_P = \mu_J$?	is $\mu_A = \mu_J$?	which hypothesis?
1	✓	✓	✓	null
2	✓	✓		alternative
3	✓		✓	alternative
4	✓			alternative
5		✓	✓	alternative
6		✓		alternative
7			✓	alternative
8				alternative

By rejecting the null hypothesis, we've decided that we don't believe that #1 is the true state of the world. The next question to ask is, which of the other seven possibilities *do* we think is right? When faced with this situation, it's usually helps to look at the data. For instance, if we look at the plots in Figure 13.1, it's tempting to conclude that Joyzepam is better than the placebo and better than Anxifree, but there's no real difference between Anxifree and the placebo. However, if we want to get a clearer answer about this, it might help to run some tests.

13.5.1 Running "pairwise" t-tests

How might we go about solving our problem? Given that we've got three separate pairs of means (placebo versus Anxifree, placebo versus Joyzepam, and Anxifree versus Joyzepam) to compare, what we could do is run three separate t-tests and see what happens. This is easy to do in jamovi. Go to the ANOVA 'Post Hoc Tests' options, move the 'drug' variable across into the active box on the right, and then click on the 'No correction' checkbox. This will produce a neat table showing all the pairwise t-test comparisons amongst the three levels of the drug variable, as in Figure 13.4.

Post Hoc Tests

Post Hoc Comparisons - drug

Comparison						
drug	drug	Mean Difference	SE	df	t	p
anxifree - joyzepam		−0.77	0.18	15.00	−4.36	0.00056
- placebo		0.27	0.18	15.00	1.52	0.15021
joyzepam - placebo		1.03	0.18	15.00	5.88	0.00003

Note. Comparisons are based on estimated marginal means

Figure 13.4: Uncorrected pairwise t-tests as post hoc comparisons in jamovi

13.5.2 Corrections for multiple testing

In the previous section I hinted that there's a problem with just running lots and lots of t-tests. The concern is that, when running these analyses, what we're doing is going on a "fishing expedition". We're running lots and lots of tests without much theoretical guidance in the hope that some of them come up significant. This kind of theory-free search for group differences is referred to as **post hoc analysis** ("post hoc" being Latin for "after this").[156]

It's okay to run post hoc analyses, but a lot of care is required. For instance, the analysis that I ran in the previous section should be avoided, as each individual t-test is designed to have a 5% type I error rate (i.e., $\alpha = .05$) and I ran three of these tests. Imagine what would have happened if my ANOVA involved 10 different groups, and I had decided to run 45 "post hoc" t-tests to try to find out which ones were significantly different from each other, you'd expect 2 or 3 of them to come up significant by chance alone. As we saw in Chapter 9, the central organising principle behind null hypothesis testing is that we seek to control our type I error rate, but now that I'm running lots of t-tests at once in order to determine the source of my ANOVA results, my actual type I error rate across this whole family of tests has gotten completely out of control.

The usual solution to this problem is to introduce an adjustment to the p-value, which aims to control the total error rate across the family of tests (see Shaffer (1995)). An adjustment of this form, which is usually (but not always) applied because one is doing post hoc analysis, is often referred to as a **correction for multiple comparisons**, though it is sometimes referred to as "simultaneous inference". In any case, there are quite a few different ways of doing this adjustment. I'll discuss a few of them in this section and in Section 14.8 in the next chapter, but you should be aware that there are many other methods out there (see, e.g., Hsu (1996)).

13.5.3 Bonferroni corrections

The simplest of these adjustments is called the **Bonferroni correction** (Dunn, 1961), and it's very very simple indeed. Suppose that my post hoc analysis consists of m separate tests, and I want to ensure that the total probability of making *any* type I errors at all is at most α.[157] If so, then the Bonferroni correction just says "multiply all your raw p-values by m". If we let p denote the original p-value, and let p'_j be the corrected value, then the Bonferroni correction tells that:

$$p'_j = m \times p$$

And therefore, if you're using the Bonferroni correction, you would reject the null hypothesis if $p'_j < \alpha$. The logic behind this correction is very straightforward. We're doing m different tests, so if we arrange it so that each test has a type I error rate of at most $\frac{\alpha}{m}$, then the *total* type I error rate across these tests cannot be larger than α. That's pretty simple, so much so that in the original paper, the author writes:

> *The method given here is so simple and so general that I am sure it must have been used before this. I do not find it, however, so can only conclude that perhaps its very simplicity has kept statisticians from realizing that it is a very good method in some situations* (Dunn (1961), pp. 52-53).

To use the Bonferroni correction in jamovi, just click on the 'Bonferroni' checkbox in the 'Correction' options, and you will see another column added to the ANOVA results table showing the adjusted p-values for the Bonferroni correction (Table 13.8). If we compare these three p-values to those for the uncorrected, pairwise t-tests, it is clear that the only thing that jamovi has done is multiply them by 3.

13.5.4 Holm corrections

Although the Bonferroni correction is the simplest adjustment out there, it's not usually the best one to use. One method that is often used instead is the **Holm correction** (Holm, 1979). The idea behind the Holm correction is to pretend that you're doing the tests sequentially, starting with the smallest (raw) p-value and moving onto the largest one. For the j-th largest of the p-values, the adjustment is *either*:

$$p'_j = j \times p_j$$

(i.e., the biggest p-value remains unchanged, the second biggest p-value is doubled, the third biggest p-value is tripled, and so on), *or*:

$$p'_j = p'_{j+1}$$

whichever one is larger. This might sound a little confusing, so let's go through it a little more slowly. Here's what the Holm correction does. First, you sort all of your p-values in order, from smallest to largest. For the smallest p-value all you do is multiply it by m, and you're done. However, for all the other ones it's a two-stage process. For instance, when you move to the second smallest p-value, you first multiply it by $m - 1$. If this produces a number that is bigger than the adjusted p-value that you got last time, then you keep it. But if it's smaller than the last one, then you copy the last p-value. To illustrate how this works, consider Table 13.10 which shows the calculations of a Holm correction for a collection of five p-values.

Hopefully that makes things clear.

Although it's a little harder to calculate, the Holm correction has some very nice properties. It's more powerful than Bonferroni (i.e., it has a lower type II error rate) but, counter-intuitive as it might seem, it has the same type I error rate. As a consequence, in practice there's never any reason to use the simpler Bonferroni correction since it is always outperformed by the slightly more elaborate Holm correction. Because of this, the Holm correction should be your *go to* multiple comparison correction.

Table 13.10: Holm corrected p-values

raw p	rank j	$p \times j$	Holm p
.001	5	.005	.005
.005	4	.020	.020
.019	3	.057	.057
.022	2	.044	.057
.103	1	.103	.103

Figure 13.4 also shows the Holm corrected p-values and, as you can see, the biggest p-value (corresponding to the comparison between Anxifree and the placebo) is unaltered. At a value of .15, it is exactly the same as the value we got originally when we applied no correction at all. In contrast, the smallest p-value (Joyzepam versus placebo) has been multiplied by three.

13.5.5 Writing up the post hoc test

Finally, having run the post hoc analysis to determine which groups are significantly different to one another, you might write up the result like this:

> Post hoc tests (using the Holm correction to adjust p) indicated that Joyzepam produced a significantly larger mood change than both Anxifree ($p = .001$) and the placebo (($p = 9.0 \times 10^{-5}$). We found no evidence that Anxifree performed better than the placebo ($p = .15$).

Or, if you don't like the idea of reporting exact p-values, then you'd change those numbers to $p < .01$, $p < .001$ and $p > .05$ respectively. Either way, the key thing is that you indicate that you used Holm's correction to adjust the p-values. And of course, I'm assuming that elsewhere in the write up you've included the relevant descriptive statistics (i.e., the group means and standard deviations), since these p-values on their own aren't terribly informative.

13.6 The assumptions of one-way ANOVA

Like any statistical test, analysis of variance relies on some assumptions about the data, specifically the residuals. There are three key assumptions that you need to be aware of: normality, homogeneity of variance and independence.

[Additional technical detail[158]]

So, how do we check whether the assumption about the residuals is accurate? Well, as I indicated above, there are three distinct claims buried in this one statement, and we'll consider them separately.

- **Homogeneity of variance**. Notice that we've only got the one value for the population standard deviation (i.e., σ), rather than allowing each group to have it's own value (i.e., σ_k). This is referred to as the homogeneity of variance (sometimes called homoscedasticity) assumption. ANOVA assumes that the population standard deviation is the same for all groups. We'll talk about this extensively in the Checking the homogeneity of variance assumption section.
- **Normality**. The residuals are assumed to be normally distributed. As we saw in Section 11.9, we can assess this by looking at QQ plots (or running a Shapiro-Wilk test. I'll talk about this more in an ANOVA context in the Checking the normality assumption section.
- **Independence**. The independence assumption is a little trickier. What it basically means is that, knowing one residual tells you nothing about any other residual. All of the ϵ_{ik} values are assumed to have been generated without any "regard for" or "relationship to" any of the other ones. There's not an obvious or simple way to test for this, but there are some situations that are clear violations of this. For instance, if you have a repeated measures design, where each participant in your study appears in more than one condition, then independence doesn't hold. There's a special relationship between some observations, namely those that correspond to the same person! When that happens, you need to use something like a Repeated measures one-way ANOVA.

13.6.1 Checking the homogeneity of variance assumption

To make the preliminary test on variances is rather like putting to sea in a rowing boat to find out whether conditions are sufficiently calm for an ocean liner to leave port!
– George Box (Box, 1953)

There's more than one way to skin a cat, as the saying goes, and more than one way to test the homogeneity of variance assumption, too (though for some reason no-one made a saying out of that). The most commonly used test for this that I've seen in the literature is the Levene test (Levene, 1960), and the closely related Brown-Forsythe test (Brown & Forsythe, 1974).

Regardless of whether you're doing the standard Levene test or the Brown-Forsythe test, the test statistic, which is sometimes denoted F but also sometimes written as W, is calculated in exactly the same way that the F-statistic for the regular ANOVA is calculated, just using a Z_{ik} rather than Y_{ik}. With that in mind, we can go on to look at how to run the test in jamovi.

[Additional technical detail[159]]

13.6.2 Running the Levene test in jamovi

Okay, so how do we run the Levene test? Simple really – under the ANOVA 'Assumption Checks' option, just click on the 'Homogeneity tests' checkbox. If we look at the output, shown in Figure 13.5, we see that the test is non-significant ($F_{2,15} = 1.45, p = .266$), so it looks like the homogeneity of variance assumption is fine. However, looks can be deceptive! If your sample size is pretty big, then the Levene test could show up a significant effect (i.e. p < .05) even when the homogeneity of variance assumption is not violated to an extent which troubles the robustness of ANOVA. This was the point George Box was making in the quote above. Similarly, if your sample size is quite small, then the homogeneity of variance assumption might not be satisfied and yet a Levene test could be non-significant (i.e. p > .05). What this means is that, alongside any statistical test of the assumption being met, you should always plot the standard deviation around the means for each group / category in the analysis...just to see if they look fairly similar (i.e. homogeneity of variance) or not.

Assumption Checks

Homogeneity of Variances Test (Levene's)

F	df1	df2	p
1.45	2	15	0.26569

Figure 13.5: Levene test output for one-way ANOVA in jamovi

13.6.3 Removing the homogeneity of variance assumption

In our example, the homogeneity of variance assumption turned out to be a pretty safe one: the Levene test came back non-significant (notwithstanding that we should also look at the plot of standard deviations), so we probably don't need to worry. However, in real life we aren't always that lucky. How do we save our ANOVA when the homogeneity of variance assumption is violated? If you recall from our discussion of t-tests, we've seen this problem before. The Student t-test assumes equal variances, so the solution was to use the Welch t-test, which does not. In fact, Welch (1951) also showed how we can solve this problem for ANOVA too (the **Welch one-way test**). It's implemented in jamovi using the One-Way ANOVA analysis. This is a specific analysis approach just for one-way ANOVA, and to run the Welch one-way ANOVA for our example, we would re-run the analysis as previously, but this time use the jamovi ANOVA - one-way ANOVA analysis command, and check the option for Welch's test (see Figure 13.6). To understand what's happening here, let's compare these numbers to what we got earlier when Running an ANOVA in jamovi originally. To save you the trouble of flicking back, this is what we got last time: $F(2, 15) = 18.611, p = .00009$, also shown as the Fisher's test in the One-Way ANOVA shown in Figure 13.6.

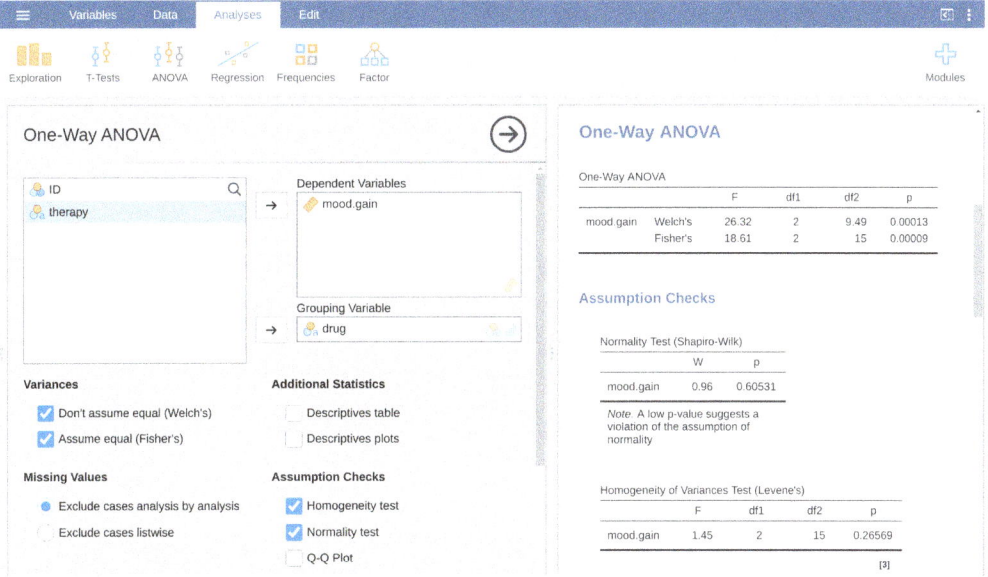

Figure 13.6: Welch test as part of the one-way ANOVA analysis in jamovi

Okay, so originally our ANOVA gave us the result $F(2, 15) = 18.6$, whereas the Welch one-way test gave us $F(2, 9.49) = 26.32$. In other words, the Welch test has reduced the within groups degrees of freedom from 15 to 9.49, and the F-value has increased from 18.6 to 26.32.

13.6.4 Checking the normality assumption

Testing the normality assumption is relatively straightforward. We covered most of what you need to know in Section 11.9. The only thing we really need to do is draw a QQ plot and, in addition if it is available, run the Shapiro-Wilk test. The QQ plot is shown in Figure 13.7 and it looks pretty normal to me.

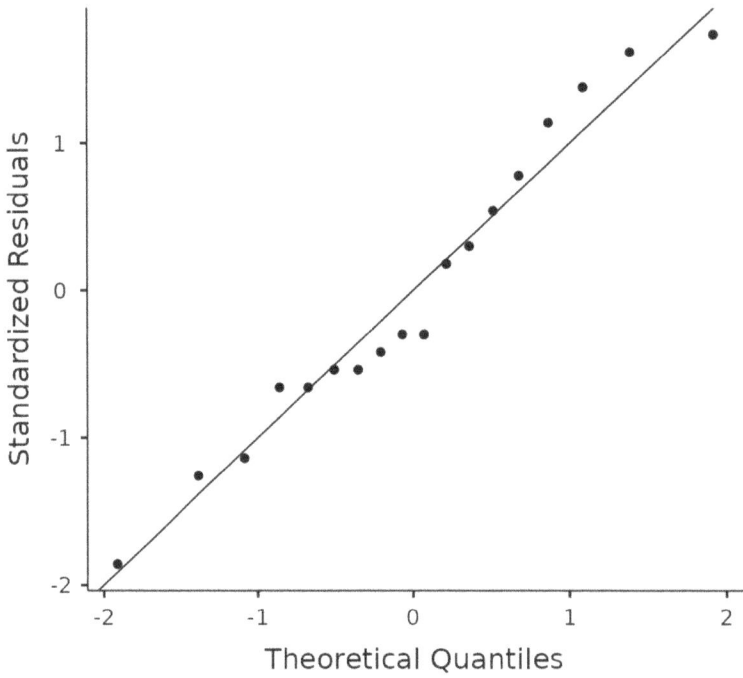

Figure 13.7: QQ plot in the one-way ANOVA analysis in jamovi

If the Shapiro-Wilk test is not significant (i.e. $p > .05$) then this indicates that the assumption of normality is not violated. However, as with Levene's test, if the sample size is large then a significant Shapiro-Wilk test may in fact be a false positive, where the assumption of normality is not violated in any substantive problematic sense for the analysis. And, similarly, a very small sample can produce false negatives. That's why a visual inspection of the QQ plot is important.

Alongside inspecting the QQ plot for any deviations from normality, the Shapiro-Wilk test for our data does show a non-significant effect, with $p = 0.6053$ (see Figure 13.6). This therefore supports the QQ plot assessment; both checks find no indication that normality is violated.

13.6.5 Removing the normality assumption

Now that we've seen how to check for normality, we are led naturally to ask what we can do to address violations of normality. In the context of a one-way ANOVA, the easiest solution is probably to switch to a non-parametric test (i.e., one that doesn't rely on any particular assumption about the kind of distribution involved). We've seen non-parametric tests before, in Chapter 11. When you only have two groups, the Mann-Whitney or the Wilcoxon test provides the non-parametric alternative that you need. When you've got three or more groups, you can use the **Kruskal-Wallis rank sum test** (Kruskal & Wallis, 1952). So that's the test we'll talk about next.

13.6.6 The logic behind the Kruskal-Wallis test

The Kruskal-Wallis test is surprisingly similar to ANOVA, in some ways. In ANOVA we started with Y_{ik}, the value of the outcome variable for the ith person in the kth group. For the Kruskal-Wallis test what we'll do is rank order all of these Y_{ik} values and conduct our analysis on the ranked data.[160]

13.6.7 Additional details

The description in the previous section illustrates the logic behind the Kruskal-Wallis test. At a conceptual level, this is the right way to think about how the test works.[161]

But wait, there's more! Dear lord, why is there always more? The story I've told so far is only actually true when there are no ties in the raw data. That is, if there are no two observations that have exactly the same value. If there are ties, then we have to introduce a correction factor to these calculations. At this point I'm assuming that even the most diligent reader has stopped caring (or at least formed the opinion that the tie-correction factor is something that doesn't require their immediate attention). So I'll very quickly tell you how it's calculated, and omit the tedious details about why it's done this way. Suppose we construct a frequency table for the raw data, and let fj be the number of observations that have the j-th unique value. This might sound a bit abstract, so here's a concrete example from the frequency table of mood.gain from the *clinicaltrials.csv* data set (Table 13.11).

Table 13.11: Frequency table of mood gain from the *clinicaltrials.csv* data

0.1	0.2	0.3	0.4	0.5	0.6	0.8	0.9	1.1	1.2	1.3	1.4	1.7	1.8
1	1	2	1	1	2	1	1	1	1	2	2	1	1

Looking at this table, notice that the third entry in the frequency table has a value of 2. Since this corresponds to a mood.gain of 0.3, this table is telling us that two people's mood increased by 0.3.[162]

And so jamovi uses a tie-correction factor to calculate the tie-corrected Kruskall-Wallis statistic. And at long last, we're actually finished with the theory of the Kruskal-Wallis test. I'm sure you're all terribly relieved that I've cured you of the existential anxiety that naturally arises when you realise that you don't know how to calculate the tie-correction factor for the Kruskal-Wallis test. Right?

13.6.8 How to run the Kruskal-Wallis test in jamovi

Despite the horror that we've gone through in trying to understand what the Kruskal-Wallis test actually does, it turns out that running the test is pretty painless, since jamovi has an analysis as part of the ANOVA analysis set called 'Non-Parametric' – 'one-way ANOVA (Kruskal-Wallis)' Most of the time you'll have data like the *clinicaltrial.csv* data set, in which you have your outcome variable mood.gain and a grouping variable drug. If so, you can just go ahead and run the analysis in jamovi. What this gives us is a Kruskal-Wallis $\chi^2 = 12.076, df = 2, p = 0.00239$, as in Figure 13.8.

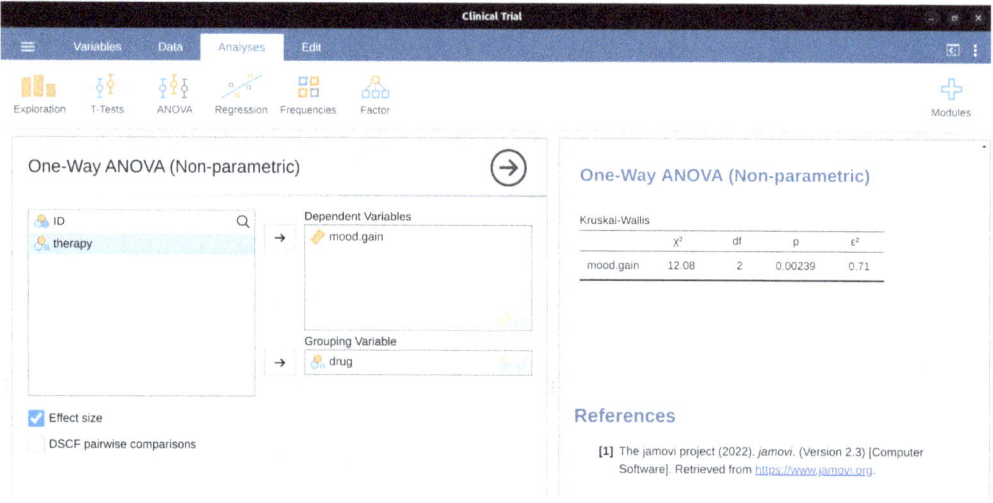

Figure 13.8: Kruskal-Wallis one-way non-parametric ANOVA in jamovi

13.7 Repeated measures one-way ANOVA

The one-way repeated measures ANOVA is a statistical test for significant differences between three or more groups where the same participants are in each group (or each participant is closely matched with participants in other experimental groups). For this reason, there should always be an equal number of scores (data points) in each experimental group. This type of design and analysis can also be called a "related ANOVA" or a "within subjects ANOVA".

The logic behind a repeated measures ANOVA is very similar to that of an independent ANOVA (sometimes called a "between subjects" ANOVA). You'll remember that earlier we showed that in a between subjects ANOVA total variability is partitioned into between groups variability (SS_b) and within groups variability (SS_w), then divided by the respective degrees of freedom to give MS_b and MS_w (see Table 13.1), whereupon the F-ratio is calculated as:

$$F = \frac{MS_b}{MS_w}$$

In a repeated measures ANOVA, the F-ratio is calculated in a similar way, but whereas in an independent ANOVA the within-group variability (SS_w) is used as the basis for the MS_w denominator, in a repeated measures ANOVA the SS_w is partioned into two parts. As we are using the same subjects in each group, we can remove the variability due to the individual differences between subjects (referred to as $SS_{subjects}$) from the within groups variability.

We won't go into too much technical detail about how this is done, but essentially each subject becomes a level of a factor called subjects. The variability in this within subjects factor is then calculated in the same way as any between subjects factor.

And then we can subtract $SS_{subjects}$ from SS_w to provide a smaller SS_{error} term:

$$\text{Independent ANOVA: } SS_{error} = SS_w$$
$$\text{Repeated Measures ANOVA: } SS_{error} = SS_w - SS_{subjects}$$

This change in SS_{error} term often leads to a more powerful statistical test, but this does depend on whether the reduction in the SS_{error} more than compensates for the reduction in degrees of freedom for the error term (as degrees of freedom go from $(n-k)$[163] to $(n-1)(k-1)$ (remembering that there are more subjects in the independent ANOVA design).

13.7.1 Repeated measures ANOVA in jamovi

First, we need some data. Geschwind (1972) has suggested that the exact nature of a patient's language deficit following a stroke can be used to diagnose the specific region of the brain that has been damaged. A researcher is concerned with identifying the specific communication difficulties experienced by six patients suffering from Broca's Aphasia (a language deficit commonly experienced following a stroke) (Table 13.12).

Table 13.12: Word recognition task scores in stroke patients

Participant	Speech	Conceptual	Syntax
1	8	7	6
2	7	8	6
3	9	5	3
4	5	4	5
5	6	6	2
6	8	7	4

The patients were required to complete three word recognition tasks. On the first (speech production) task, patients were required to repeat single words read out aloud by the researcher. On the second (conceptual) task, designed to test word comprehension, patients were required to match a series of pictures with their correct name. On the third (syntax) task, designed to test knowledge of correct word order, patients were asked to reorder syntactically incorrect sentences. Each patient completed all three tasks. The order in which patients attempted the tasks was counterbalanced between participants. Each task consisted of a series of 10 attempts. The number of attempts successfully completed by each patient are shown in Table 13.11. Enter these data into jamovi ready for analysis (or take a short-cut and load up the *broca.csv* file).

To perform a one-way related ANOVA in jamovi, open the one-way repeated measures ANOVA dialogue box, as in Figure 13.9, via 'ANOVA - Repeated Measures ANOVA'.

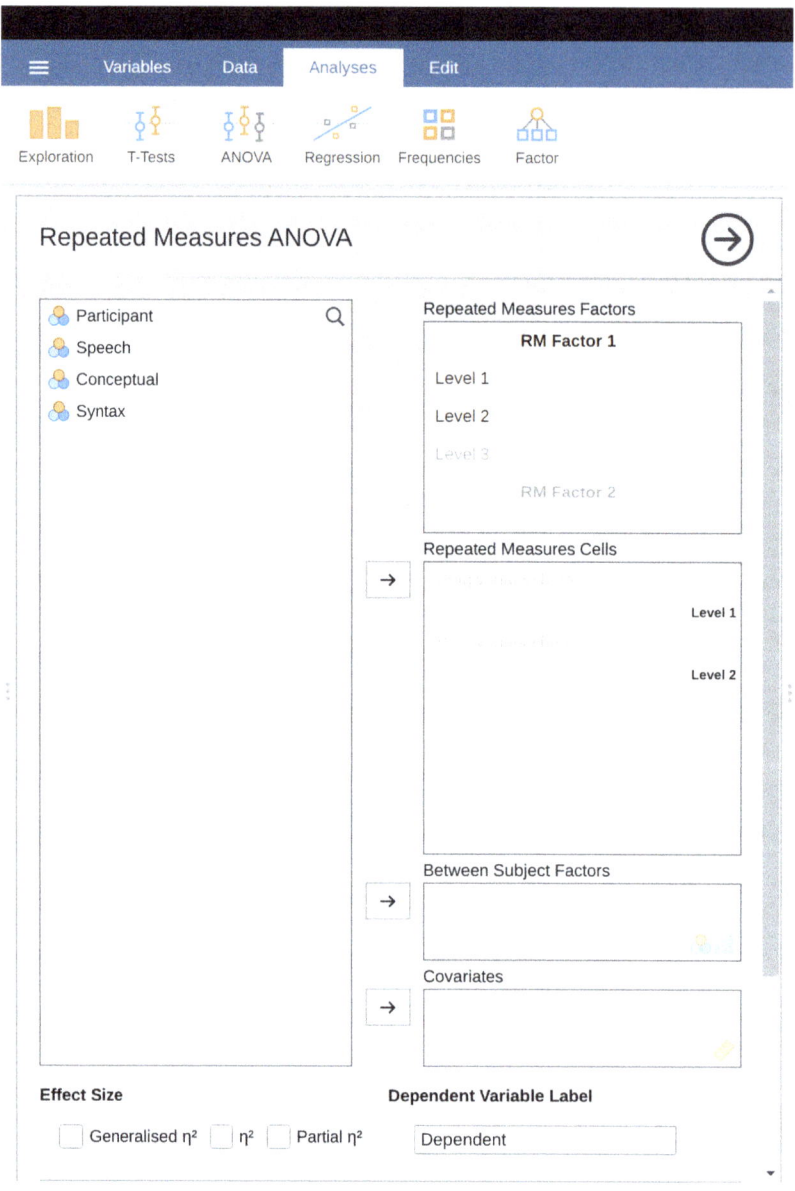

Figure 13.9: Repeated measures ANOVA dialogue box in jamovi

Then:

- Enter a 'Repeated Measures' factor name. This should be a label that you choose to describe the conditions repeated by all participants. For example, to describe the speech, conceptual and syntax tasks completed by all participants a suitable label would be 'Task'. Note that this new factor name represents the independent variable in the analysis.
- Add a third level in the 'Repeated Measures Factors' text box, as there are three levels representing the three tasks: speech, conceptual and syntax. Change the

labels of the levels accordingly.
- Then move each of the levels variables across to the 'Repeated Measures' Cells text box.
- Finally, under the 'Assumption Checks' option, tick the 'Sphericity checks' text box.

jamovi output for a one-way repeated measures ANOVA is produced as shown in Figure 13.10 to Figure 13.13. The first output we should look at is Mauchly's Test of Sphericity, which tests the hypothesis that the variances of the differences between the conditions are equal (meaning that the spread of difference scores between the study conditions is approximately the same). In Figure 13.10 the significance level in Mauchly's test is $p = .720$. If Mauchly's test is non-significant (i.e. $p > .05$, as is the case in this analysis) then it is reasonable to conclude that the variances of the differences are not significantly different (i.e. they are roughly equal and sphericity can be assumed).

Assumptions

Tests of Sphericity

	Mauchly's W	p	Greenhouse-Geisser ε	Huynh-Feldt ε
Task	0.85	0.72009	0.87	1.00

Figure 13.10: One-way repeated measures ANOVA output – Mauchly Test of Sphericity

If, on the other hand, Mauchly's test had been significant ($p < .05$) then we would conclude that there are significant differences between the variance of the differences, and the requirement of sphericity has not been met. In this case, we should apply a correction to the F-value obtained in the one-way related ANOVA analysis:

- If the Greenhouse-Geisser value in the 'Tests of Sphericity' table is > .75 then you should use the Huynh-Feldt correction
- But if the Greenhouse-Geisser value is < .75, then you should use the Greenhouse-Geisser correction.

Both these corrected F-values can be specified in the 'Sphericity Corrections' check boxes under the 'Assumption Checks' options, and the corrected F-values are then shown in the results table, as in Figure 13.11.

In our analysis, we saw that the significance of Mauchly's Test of Sphericity was $p = .720$ (i.e., $p > 0.05$). So, this means we can assume that the requirement of sphericity has been met so no correction to the F-value is needed. Therefore, we can use the 'None' Sphericity Correction output values for the repeated measure 'Task': $F = 6.93$, $df = 2$, $p = .013$, and we can conclude that the number of tests successfully completed on each language task did vary significantly depending on whether the task was speech, comprehension or syntax based ($F(2, 10) = 6.93, p = .013$).

Repeated Measures ANOVA

Within Subjects Effects

	Sphericity Correction	Sum of Squares	df	Mean Square	F	p
Task	None	24.78	2	12.39	6.93	0.01296
	Greenhouse-Geisser	24.78	1.74	14.26	6.93	0.01802
	Huynh-Feldt	24.78	2.00	12.39	6.93	0.01296
Residual	None	17.89	10	1.79		
	Greenhouse-Geisser	17.89	8.68	2.06		
	Huynh-Feldt	17.89	10.00	1.79		

Note. Type 3 Sums of Squares

Figure 13.11: One-way repeated measures ANOVA output – Tests of Within Subjects Effects

Post hoc tests can also be specified in jamovi for repeated measures ANOVA in the same way as for independent ANOVA. The results are shown in Figure 13.12. These indicate that there is a significant difference between Speech and Syntax, but not between other levels.

Post Hoc Tests

Post Hoc Comparisons - Task

Comparison						
Task	Task	Mean Difference	SE	df	t	p_{tukey}
Speech	- Conceptual	1.00	0.68	5.00	1.46	0.38130
	- Syntax	2.83	0.91	5.00	3.11	0.05814
Conceptual	- Syntax	1.83	0.70	5.00	2.61	0.10245

Figure 13.12: Post hoc tests in repeated measures ANOVA in jamovi

Descriptive statistics (marginal means) can be reviewed to help interpret the results, produced in the jamovi output as in Figure 13.13. Comparison of the mean number of trials successfully completed by participants shows that Broca's Aphasics perform reasonably well on speech production (mean = 7.17) and language comprehension (mean = 6.17) tasks. However, their performance was considerably worse on the syntax task (mean = 4.33), with a significant difference in post hoc tests between Speech and Syntax task performance.

Estimated Marginal Means - Task

Task	Mean	SE	95% Confidence Interval	
			Lower	Upper
Speech	7.17	0.60	5.62	8.71
Conceptual	6.17	0.60	4.62	7.71
Syntax	4.33	0.67	2.62	6.05

Figure 13.13: One-way repeated measures ANOVA output – Descriptive Statistics

13.8 The Friedman non-parametric repeated measures ANOVA test

The Friedman test is a non-parametric version of a repeated measures ANOVA and can be used instead of the Kruskal-Wallis test when testing for differences between three or more groups where the same participants are in each group, or each participant is closely matched with participants in other conditions. If the dependent variable is ordinal, or if the assumption of normality is not met, then the Friedman test can be used.

As with the Kruskal-Wallis test, the underlying mathematics is complicated, and won't be presented here. For the purpose of this book, it is sufficient to note that jamovi calculates the tie-corrected version of the Friedman test, and in Figure 13.14 there is an example using the Broca's Aphasia data we have already looked at.

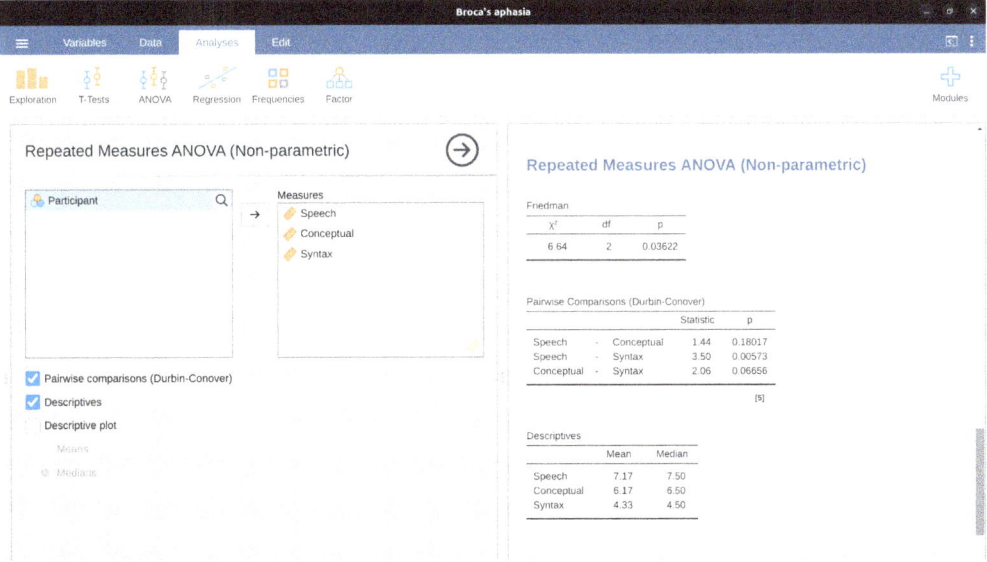

Figure 13.14: The 'Repeated Measures ANOVA (Non-parametric)' dialogue box and results in jamovi

It's pretty straightforward to run a Friedman test in jamovi. Just select 'Analyses - ANOVA - Repeated Measures ANOVA (Non-parametric)' as in Figure 13.14. Then highlight and transfer the names of the repeated measures variables you wish to compare (Speech, Conceptual, Syntax) into the 'Measures:' text box. To produce descriptive statistics (means and medians) for the three repeated measures variables, click on the 'Descriptives' button.

The jamovi results show descriptive statistics, chi-square value, degrees of freedom, and the p-value (Figure 13.14). Since the p-value is less than the level conventionally used to determine significance ($p < .05$), we can conclude that Broca's Aphasics perform reasonably well on speech production (median = 7.5) and language comprehension (median = 6.5) tasks. However, their performance was considerably worse on the syntax task (median = 4.5), with a significant difference in post hoc tests between Speech and Syntax task performance.

13.9 On the relationship between ANOVA and the Student t-test

There's one last thing I want to point out before finishing. It's something that a lot of people find kind of surprising, but it's worth knowing about. An ANOVA with two groups is identical to the Student t-test. No, really. It's not just that they are similar, but they are actually equivalent in every meaningful way. I won't try to prove that this is always true, but I will show you a single concrete demonstration. Suppose that, instead of running an ANOVA on our mood.gain ~ drug model, let's instead do it using therapy as the predictor. If we run this ANOVA we get an F-statistic of $F(1, 16) = 1.71$, and a p-value = 0.21. Since we only have two groups, I didn't actually need to resort to an ANOVA, I could have just decided to run a Student t-test. So let's see what happens when I do that: I get a t-statistic of $t(16) = -1.3068$ and a p-value = 0.21. Curiously, the p-values are identical. Once again we obtain a value of $p = .21$. But what about the test statistic? Having run a t-test instead of an ANOVA, we get a somewhat different answer, namely $t(16) = -1.3068$. However, there is a fairly straightforward relationship here. If we square the t-statistic then we get the F-statistic from before: $-1.3068^2 = 1.7077$.

13.10 Summary

There's a fair bit covered in this chapter, but there's still a lot missing.[164] Most obviously, I haven't discussed how to run an ANOVA when you are interested in more than one grouping variable, but that will be discussed in a lot of detail in Chapter 14. In terms of what we have discussed, the key topics were:

- The basic logic behind How ANOVA works and Running an ANOVA in jamovi.
- How to compute an Effect size for an ANOVA.
- Multiple comparisons and post hoc tests for multiple testing.
- The assumptions of one-way ANOVA.
- Checking the homogeneity of variance assumption and what to do if it is violated: Removing the homogeneity of variance assumption.

- Checking the normality assumption and what to do if it is violated: Removing the normality assumption.
- Repeated measures one-way ANOVA and the non-parametric equivalent, The Friedman non-parametric repeated measures ANOVA test.

324

Chapter 14

Factorial ANOVA

Over the course of the last few chapters we have done quite a lot. We have looked at statistical tests you can use when you have one nominal predictor variable with two groups (e.g., the t-test in Chapter 11) or with three or more groups (Chapter 13). Chapter 12 introduced a powerful new idea, that is building statistical models with multiple continuous predictor variables used to explain a single outcome variable. For instance, a regression model could be used to predict the number of errors a student makes in a reading comprehension test based on the number of hours they studied for the test and their score on a standardised IQ test.

The goal in this chapter is to extend the idea of using multiple predictors into the ANOVA framework. For instance, suppose we were interested in using the reading comprehension test to measure student achievements in three different schools, and we suspect that girls and boys are developing at different rates (and so would be expected to have different performance on average). Each student is classified in two different ways: on the basis of their gender and on the basis of their school. What we'd like to do is analyse the reading comprehension scores in terms of both of these grouping variables. The tool for doing so is generically referred to as **factorial ANOVA**. However, since we have two grouping variables, we sometimes refer to the analysis as a two-way ANOVA, in contrast to the one-way ANOVAs that we ran in Chapter 13.

14.1 Factorial ANOVA 1: balanced designs, focus on main effects

When we discussed analysis of variance in Chapter 13, we assumed a fairly simple experimental design. Each person is in one of several groups and we want to know whether these groups have different mean scores on some outcome variable. In this section, I'll discuss a broader class of experimental designs known as **factorial designs**, in which we have more than one grouping variable. I gave one example of how this kind of design might arise above. Another example appears in Chapter 13 in which we were looking at the effect of different drugs on the mood.gain experienced by each person. In that chapter we did find a significant effect of drug, but at the end of the chapter we also ran an analysis to see if there was an effect of therapy. We didn't find one, but there's something a bit worrying about trying to run two separate analyses trying to predict the same outcome. Maybe there actually is an effect of therapy on mood gain, but we couldn't find it because it was being "hidden" by the effect of drug? In other words, we're going to want to run a single analysis that includes both drug and therapy as predictors. For this analysis each person is cross-classified by the drug they were given (a factor with 3 levels) and what therapy they received (a factor with 2 levels). We refer to this as a 3×2 factorial design.

If we cross-tabulate drug by therapy, using the 'Frequencies' – 'Contingency Tables' analysis in jamovi (see Section 6.1), we get the table shown in Figure 14.1.

Contingency Tables

drug	therapy		Total
	CBT	no.therapy	
anxifree	3	3	6
joyzepam	3	3	6
placebo	3	3	6
Total	9	9	18

Figure 14.1: jamovi contingency table of drug by therapy

As you can see, not only do we have participants corresponding to all possible combinations of the two factors, indicating that our design is **completely crossed**, it turns out that there are an equal number of people in each group. In other words, we have a balanced design. In this section I'll talk about how to analyse data from **balanced** designs, since this is the simplest case. The story for unbalanced designs is quite tedious, so we'll put it to one side for the moment.

14.1.1 What hypotheses are we testing?

Like one-way ANOVA, factorial ANOVA is a tool for testing certain types of hypotheses about population means. So a sensible place to start would be to be explicit about what our hypotheses actually are. However, before we can even get to that point, it's really useful to have some clean and simple notation to describe the population means. Because of the fact that observations are cross-classified in terms of two different factors, there are quite a lot of different means that one might be interested in. To see this, let's start by thinking about all the different sample means that we can calculate for this kind of design. Firstly, there's the obvious idea that we might be interested in this list of group means (Table 14.1).

Now, the next Table (Table 14.2) shows a list of the group means for all possible combinations of the two factors (e.g., people who received the placebo and no therapy, people who received the placebo while getting CBT, etc.). It is helpful to organise all these numbers, plus the marginal and grand means, into a single table.

Table 14.1: Group means for drug and therapy groups in the *clinicaltrial.csv* data

drug	therapy	mood.gain
placebo	no.therapy	0.30
anxifree	no.therapy	0.40
joyzepam	no.therapy	1.47
placebo	CBT	0.60
anxifree	CBT	1.03
joyzepam	CBT	1.50

Table 14.2: Group and total means for drug and therapy groups in the *clinicaltrial.csv* data

	no therapy	CBT	total
placebo	0.30	0.60	0.45
anxifree	0.40	1.03	0.72
joyzepam	1.47	1.50	1.48
total	0.72	1.04	0.88

Now, each of these different means is of course a sample statistic. It's a quantity that pertains to the specific observations that we've made during our study. What we want to make inferences about are the corresponding population parameters. That is, the true means as they exist within some broader population. Those population means can also be organised into a similar table, but we'll need a little mathematical notation to do so (Table 14.3). As usual, I'll use the symbol μ to denote a population mean. However, because there are lots of different means, I'll need to use subscripts to distinguish between them.

Table 14.3: Notation for population means in a factorial table

	no therapy	CBT	total
placebo	μ_{11}	μ_{12}	
anxifree	μ_{21}	μ_{22}	
joyzepam	μ_{31}	μ_{32}	
total			

Here's how the notation works. Our table is defined in terms of two factors. Each row corresponds to a different level of Factor A (in this case drug), and each column corresponds to a different level of Factor B (in this case therapy). If we let R denote the number of rows in the table, and C denote the number of columns, we can refer to this as an $R \times C$ factorial ANOVA. In this case $R = 3$ and $C = 2$. We'll use lowercase letters to refer to specific rows and columns, so μ_{rc} refers to the population mean associated with the r-th level of Factor A (i.e. row number r) and the c-th level of Factor B (column number c).[165]

Okay, what about the remaining entries? For instance, how should we describe the average mood gain across the entire (hypothetical) population of people who might be given Joyzepam in an experiment like this, regardless of whether they were in CBT? We use the "dot" notation to express this. In the case of Joyzepam, notice that we're talking about the mean associated with the third row in the table. That is, we're averaging across two cell means (i.e., μ_{31} and μ_{32}). The result of this averaging is referred to as a marginal mean, and would be denoted $\mu_{3.}$ in this case. The **marginal mean** for CBT corresponds to the population mean associated with the second column in the table, so we use the notation because it is the mean obtained by averaging (marginalising)[166] over both. So our full table of population means can be written down like in Table 14.4.

Table 14.4: Notation for population and total means in a factorial table

	no therapy	CBT	total
placebo	μ_{11}	μ_{12}	$\mu_{1.}$
anxifree	μ_{21}	μ_{22}	$\mu_{2.}$
joyzepam	μ_{31}	μ_{32}	$\mu_{3.}$
total	$\mu_{.1}$	$\mu_{.2}$	$\mu_{..}$

Now that we have this notation, it is straightforward to formulate and express some hypotheses. Let's suppose that the goal is to find out two things. First, does the choice of drug have any effect on mood? And second, does CBT have any effect on mood? These aren't the only hypotheses that we could formulate of course, and we'll see a really important example of a different kind of hypothesis in the section Factorial ANOVA 2: balanced designs, interpreting interactions, but these are the two simplest hypotheses to test, and so we'll start there. Consider the first test. If the drug has no effect then we would expect all of the row means to be identical, right? So that's our null hypothesis. On the other hand, if the drug does matter then we should expect these row means to be different. Formally, we write down our null and alternative hypotheses in terms of the equality of marginal means:

Null hypothesis, H_0: row means are the same, i.e., $\mu_{1.} = \mu_{2.} = \mu_{3.}$

Alternative hypothesis, H_1: at least one row mean is different

It's worth noting that these are exactly the same statistical hypotheses that we formed when we ran a one-way ANOVA on these data in Chapter 13. Back then I used the notation $\mu \times P$ to refer to the mean mood gain for the placebo group, with μA and $\mu \times J$ corresponding to the group means for the two drugs, and the null hypothesis was $\mu P = \mu A = \mu J$. So we're actually talking about the same hypothesis, it's just that the more complicated ANOVA requires more careful notation due to the presence of multiple grouping variables, so we're now referring to this hypothesis as $\mu_{1.} = \mu_{2.} = \mu_{3.}$. However, as we'll see shortly, although the hypothesis is identical the test of that hypothesis is subtly different due to the fact that we're now acknowledging the existence of the second grouping variable.

Speaking of the other grouping variable, you won't be surprised to discover that our second hypothesis test is formulated the same way. However, since we're talking about the psychological therapy rather than drugs our null hypothesis now corresponds to the equality of the column means:

Null hypothesis, H_0: column means are the same, i.e., $\mu_{.1} = \mu_{.2}$

Alternative hypothesis, H_1: column means are different, i.e., $\mu_{.1} \neq \mu_{.2}$

14.1.2 Running the analysis in jamovi

The null and alternative hypotheses that I described in the last section should seem awfully familiar. They're basically the same as the hypotheses that we were testing in our simpler oneway ANOVAs in Chapter 13. So you're probably expecting that the hypothesis tests that are used in factorial ANOVA will be essentially the same as the F-test from Chapter 13. You're expecting to see references to sums of squares (SS), mean squares (MS), degrees of freedom (df), and finally an F-statistic that we can convert into a p-value, right? Well, you're absolutely and completely right. So much so that I'm going to depart from my usual approach. Throughout this book, I've generally taken the approach of describing the logic (and to an extent the mathematics) that underpins a particular analysis first and only then introducing the analysis in jamovi. This time I'm going to do it the other way around and show you how to do it in jamovi first. The reason for doing this is that I want to highlight the similarities between the simple one-way ANOVA tool that we discussed in Chapter 13, and the more complicated approach that we're going to use in this chapter.

If the data you're trying to analyse correspond to a balanced factorial design then running your analysis of variance is easy. To see how easy it is, let's start by reproducing the original analysis from Chapter 13. In case you've forgotten, for that analysis we were using only a single factor (i.e., drug) to predict our outcome variable (i.e., mood.gain), and we got the results shown in Figure 14.2.

ANOVA

ANOVA - mood.gain

	Sum of Squares	df	Mean Square	F	p	η²
drug	3.45	2	1.73	18.61	0.00009	0.71
Residuals	1.39	15	0.09			

Figure 14.2: jamovi one-way ANOVA of mood.gain by drug

Now, suppose I'm also curious to find out if therapy has a relationship to mood.gain. In light of what we've seen from our discussion of multiple regression in Chapter 12, you probably won't be surprised that all we have to do is add therapy as a second 'Fixed Factor' in the analysis, see Figure 14.3.

ANOVA

ANOVA - mood.gain

	Sum of Squares	df	Mean Square	F	p	η²	η²p	ω²
drug	3.45	2	1.73	31.71	<.001	0.71	0.84	0.68
therapy	0.47	1	0.47	8.58	0.013	0.10	0.42	0.08
drug ✶ therapy	0.27	2	0.14	2.49	0.125	0.06	0.29	0.03
Residuals	0.65	12	0.05					

Figure 14.3: jamovi two-way anova of mood.gain by drug and therapy

This output is pretty simple to read too. The first row of the table reports a between-group sum of squares (SS) value associated with the drug factor, along with a corresponding between-group df value. It also calculates a mean square value (MS), an F-statistic and a p-value. There is also a row corresponding to the therapy factor, a row corresponding to the interaction between the drug factor and the therapy factor (which we won't cover just yet – more on interactions later), and a row corresponding to the residuals (i.e., the within groups variation).

Not only are all of the individual quantities pretty familiar, the relationships between these different quantities have remained unchanged, just like we saw with the original one-way ANOVA. Note that the mean square value is calculated by dividing SS by the corresponding df. That is, it's still true that:

$$MS = \frac{SS}{df}$$

regardless of whether we're talking about drug, therapy or the residuals. To see this, let's not worry about how the sums of squares values are calculated. Instead, let's take it on faith that jamovi has calculated the SS values correctly, and try to verify that all the rest of the numbers make sense. First, note that for the drug factor, we divide 3.45 by 2 and end up with a mean square value of 1.73. For the therapy factor, there's only 1 degree of freedom, so our calculations are even simpler: dividing 0.47 (the SS value) by 1 gives us an answer of 0.47 (the MS value).

Turning to the F-statistics and the p-values, notice that we have one corresponding to the drug factor and one corresponding to the therapy factor. Regardless of which one we're talking about, the F-statistic is calculated by dividing the mean square value associated with the factor by the mean square value associated with the residuals. If we use "A" as shorthand notation to refer to the first factor (Factor A; in this case drug) and "R" as shorthand notation to refer to the residuals, then the F-statistic associated with Factor A is denoted F_A, and is calculated as:

$$F_A = \frac{MS_A}{MS_R}$$

and an equivalent formula exists for Factor B (i.e., therapy). Note that this use of "R" to refer to residuals is a bit awkward, since we also used the letter R to refer to the number of rows in the table, but I'm only going to use "R" to mean residuals in the context of SS_R and MS_R, so hopefully this shouldn't be confusing. Anyway, to apply this formula to the drugs factor we take the mean square of 1.73 and divide it by the residual mean square value of 0.05, which gives us an F-statistic of 31.71.[167] The corresponding calculation for the therapy variable would be to divide 0.47 by 0.05 which gives 8.58 as the F-statistic. Not surprisingly, of course, these are the same values that jamovi has reported in the ANOVA table above.

Also in the ANOVA table is the calculation of the p-values. Once again, there is nothing new here. For each of our two factors what we're trying to do is test the null hypothesis that there is no relationship between the factor and the outcome variable (I'll be a bit more precise about this later on). To that end, we've (apparently) followed a similar strategy to what we did in the one-way ANOVA and have calculated an F-statistic for each of these hypotheses. To convert these to p-values, all we need to do is note that the sampling distribution for the F-statistic under the null hypothesis (that the factor in question is irrelevant) is an F-distribution. Also note that the two degrees of freedom values are those corresponding to the factor and those corresponding to the residuals. For the drug factor we're talking about an F-distribution with 2 and 12 degrees of freedom (I'll discuss degrees of freedom in more detail later). In contrast, for the therapy factor the sampling distribution is F with 1 and 12 degrees of freedom.

At this point, I hope you can see that the ANOVA table for this more complicated factorial analysis should be read in much the same way as the ANOVA table for the simpler one way analysis. In short, it's telling us that the factorial ANOVA for our 3×2 design found a significant effect of drug ($F_{2,12} = 31.71, p < .001$) as well as a significant effect of therapy ($F_{1,12} = 8.58, p = .013$). Or, to use the more technically correct terminology, we would say that there are two **main effects** of drug and therapy. At the moment, it probably seems a bit redundant to refer to these as "main" effects, but it actually does make sense. Later on, we're going to cover "interactions" between the two factors, and so we generally make a distinction between main effects and interaction effects.

14.1.3 How are the sum of squares calculated?

In the previous section I had two goals. Firstly, to show you that the jamovi method needed to do factorial ANOVA is pretty much the same as what we used for a one-way ANOVA. The only difference is the addition of a second factor. Secondly, I wanted to show you what the ANOVA table looks like in this case, so that you can see from the outset that the basic logic and structure behind factorial ANOVA is the same as that which underpins one-way ANOVA. Try to hold onto that feeling. It's genuinely true, insofar as factorial ANOVA is built in more or less the same way as the simpler one-way ANOVA model. It's just that this feeling of familiarity starts to evaporate once you start digging into the details. Traditionally, this comforting sensation is replaced by an urge to hurl abuse at the authors of statistics textbooks.

Okay, let's start by looking at some of those details. The explanation that I gave in the last section illustrates the fact that the hypothesis tests for the main effects (of drug and therapy in this case) are F-tests, but what it doesn't do is show you how the sum of squares (SS) values are calculated. Nor does it tell you explicitly how to calculate degrees of freedom (df values) though that's a simple thing by comparison. Let's assume for now that we have only two predictor variables, Factor A and Factor B. If we use Y to refer to the outcome variable, then we would use $Yrci$ to refer to the outcome associated with the i-th member of group rc (i.e., level/row r for Factor A and level/column c for Factor B). Thus, if we use $\bar{Y}$ to refer to a sample mean, we can use the same notation as before to refer to group means, marginal means and grand means. That is, $\bar{Y}_{rc}$ is the sample mean associated with the rth level of Factor A and the cth level of Factor B, $\bar{Y}_{r.}$ would be the marginal mean for the rth level of Factor A, $\bar{Y}_{.c}$ would be the marginal mean for the cth level of Factor B, and $\bar{Y}_{..}$ is the grand mean. In other words, our sample means can be organised into the same table as the population means. For our clinical trial data, that table is shown in Table 14.5.

Table 14.5: Notation for sample means for the clinical trial data

	no therapy	CBT	total
placebo	$\bar{Y}_{11}$	$\bar{Y}_{12}$	$\bar{Y}_{1.}$
anxifree	$\bar{Y}_{21}$	$\bar{Y}_{22}$	$\bar{Y}_{2.}$
joyzepam	$\bar{Y}_{31}$	$\bar{Y}_{32}$	$\bar{Y}_{3.}$
total	$\bar{Y}_{.1}$	$\bar{Y}_{.2}$	$\bar{Y}_{..}$

And if we look at the sample means that I showed earlier, we have $\bar{Y}_{11} = 0.30$, $\bar{Y}_{12} = 0.60$ etc. In our clinical trial example, the drugs factor has 3 levels and the therapy factor has 2 levels, and so what we're trying to run is a 3×2 factorial ANOVA. However, we'll be a little more general and say that Factor A (the row factor) has r levels and Factor B (the column factor) has c levels, and so what we're running here is an $r \times c$ factorial ANOVA.

[Additional technical detail[168]]

14.1.4 What are our degrees of freedom?

The degrees of freedom are calculated in much the same way as for one-way ANOVA. For any given factor, the degrees of freedom is equal to the number of levels minus 1 (i.e., $R - 1$ for the row variable Factor A, and $C - 1$ for the column variable Factor B). So, for the drugs factor we obtain $df = 2$, and for the therapy factor we obtain $df = 1$. Later on, when we discuss the interpretation of ANOVA as a regression model (see Section 14.6), I'll give a clearer statement of how we arrive at this number. But for the moment we can use the simple definition of degrees of freedom, namely that the degrees of freedom equals the number of quantities that are observed, minus the number of constraints. So, for the drugs factor, we observe 3 separate group means, but these are constrained by 1 grand mean, and therefore the degrees of freedom is 2.

14.1.5 Factorial ANOVA versus one-way ANOVAs

Now that we've seen how a factorial ANOVA works, it's worth taking a moment to compare it to the results of the one way analyses, because this will give us a really good sense of why it's a good idea to run the factorial ANOVA. In Chapter 13 I ran a one-way ANOVA that looked to see if there are any differences between drugs, and a second one-way ANOVA to see if there were any differences between therapies. As we saw in the section Section 14.1.1, the null and alternative hypotheses tested by the one-way ANOVAs are in fact identical to the hypotheses tested by the factorial ANOVA. Looking even more carefully at the ANOVA tables, we can see that the sum of squares associated with the factors are identical in the two different analyses (3.45 for drug and 0.92 for therapy), as are the degrees of freedom (2 for drug, 1 for therapy). But they don't give the same answers! Most notably, when we ran the one-way ANOVA for therapy in Section 13.9 we didn't find a significant effect (the p-value was .21). However, when we look at the main effect of therapy within the context of the two-way ANOVA, we do get a significant effect ($p = .019$). The two analyses are clearly not the same.

Why does that happen? The answer lies in understanding how the residuals are calculated. Recall that the whole idea behind an F-test is to compare the variability that can be attributed to a particular factor with the variability that cannot be accounted for (the residuals). If you run a one-way ANOVA for therapy, and therefore ignore the effect of drug, the ANOVA will end up dumping all of the drug-induced variability into the residuals! This has the effect of making the data look more noisy than they really are, and the effect of therapy which is correctly found to be significant in the two-way ANOVA now becomes non-significant. If we ignore something that actually matters (e.g., drug) when trying to assess the contribution of something else (e.g., therapy) then our analysis will be distorted. Of course, it's perfectly okay to ignore variables that are genuinely irrelevant to the phenomenon of interest. If we had recorded the colour of the walls, and that turned out to be a non-significant factor in a three-way ANOVA, it would be perfectly okay to disregard it and just report the simpler two-way ANOVA that doesn't include this irrelevant factor. What you shouldn't do is drop variables that actually make a difference!

14.1.6 What kinds of outcomes does this analysis capture?

The ANOVA model that we've been talking about so far covers a range of different patterns that we might observe in our data. For instance, in a two-way ANOVA design there are four possibilities: (a) only Factor A matters, (b) only Factor B matters, (c) both A and B matter, and (d) neither A nor B matters. An example of each of these four possibilities is plotted in Figure 14.4.

14.2 Factorial ANOVA 2: balanced designs, interpreting interactions

The four patterns of data shown in Figure 14.4 are all quite realistic. There are many data sets that produce exactly those patterns. However, they are not the whole story and the ANOVA model that we have been talking about up to this point is not sufficient to fully account for a table of group means. Why not? Well, so far we have the ability to talk about the idea that drugs can influence mood, and therapy can influence mood, but no way of talking about the possibility of an **interaction** between the two.

An interaction between A and B is said to occur whenever the effect of Factor A is *different*, depending on which level of Factor B we're talking about. Several examples of an interaction effect with the context of a 2 × 2 ANOVA are shown in Figure 14.5.

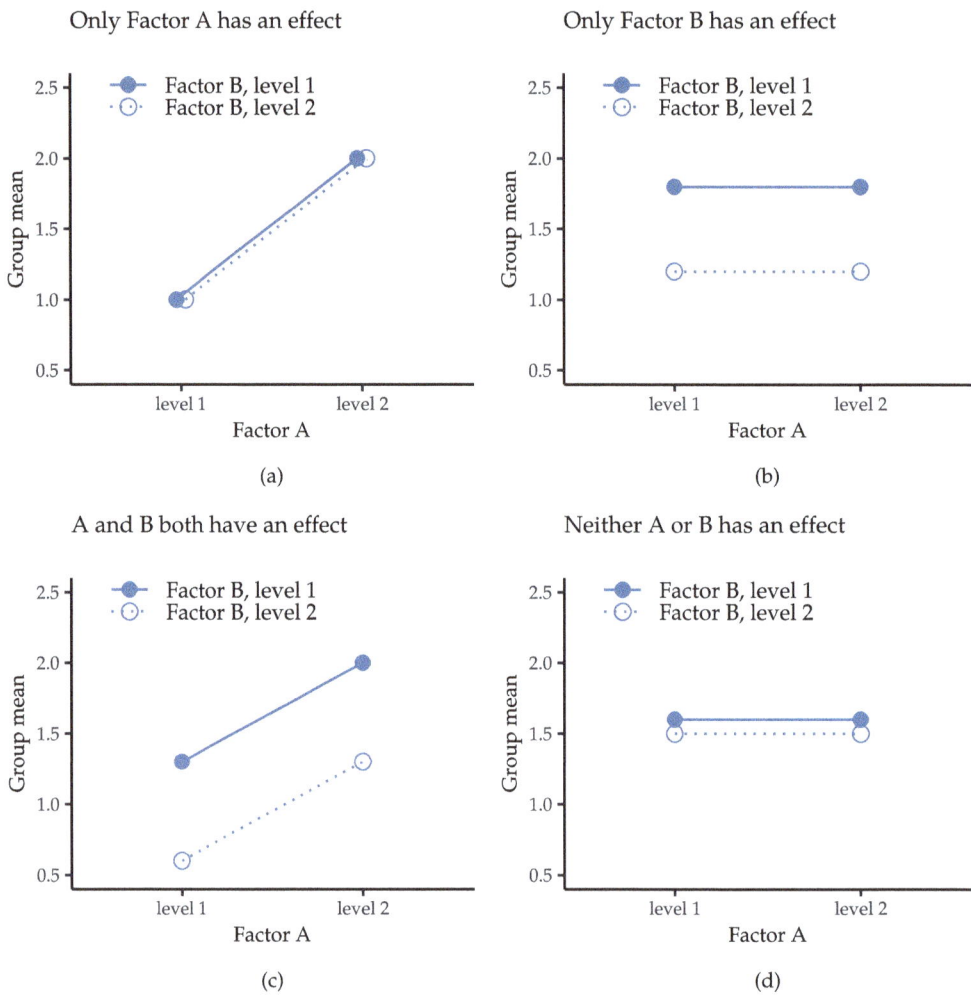

Figure 14.4: The four different outcomes for a 2 × 2 ANOVA when no interactions are present. In panel (a) we see a main effect of Factor A and no effect of Factor B. Panel (b) shows a main effect of Factor B but no effect of Factor A. Panel (c) shows main effects of both Factor A and Factor B. Finally, panel (d) shows no effect of either factor

To give a more concrete example, suppose that the operation of Anxifree and Joyzepam is governed by quite different physiological mechanisms. One consequence of this is that while Joyzepam has more or less the same effect on mood regardless of whether one is in therapy, Anxifree is actually much more effective when administered in conjunction with CBT. The ANOVA that we developed in the previous section does not capture this idea. To get some idea of whether an interaction is actually happening it helps to plot the various group means. In jamovi this is done via the ANOVA 'Estimated Marginal Means' option – just move drug and therapy across into the 'Marginal Means' box under 'Term 1'. This should look something like Figure 14.6.

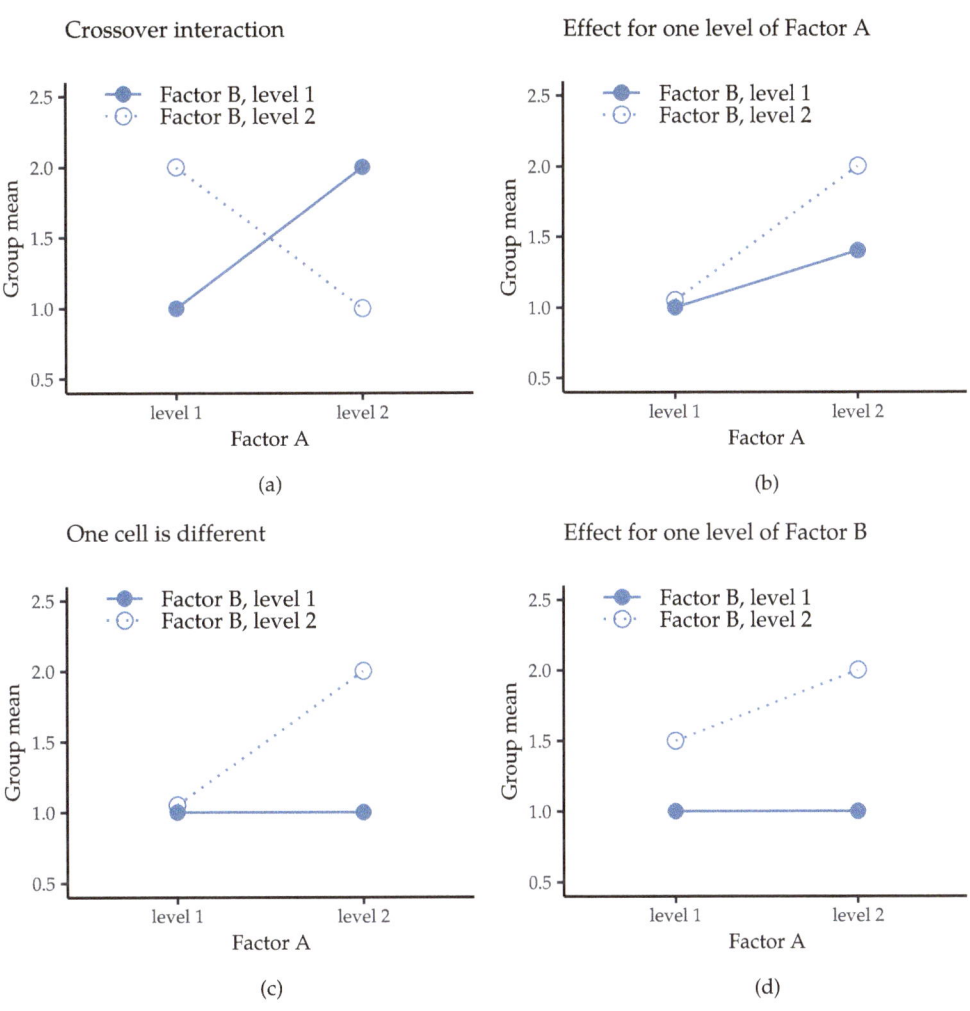

Figure 14.5: Qualitatively different interactions for a 2×2 ANOVA

Our main concern relates to the fact that the two lines aren't parallel. The effect of CBT (difference between solid line and dotted line) when the drug is Joyzepam (right side) appears to be near zero, even smaller than the effect of CBT when a placebo is used (left side). However, when Anxifree is administered, the effect of CBT is larger than the placebo (middle). Is this effect real, or is this just random variation due to chance? Our original ANOVA cannot answer this question, because we make no allowances for the idea that interactions even exist! In this section, we'll fix this problem.

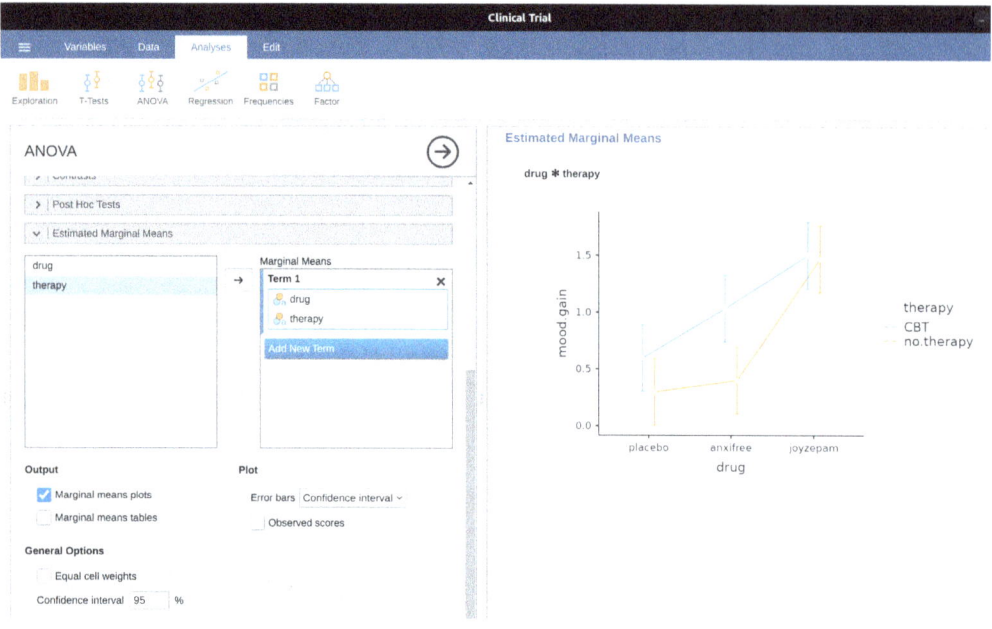

Figure 14.6: jamovi screen showing how to generate a descriptive interaction plot in ANOVA using the clinical trial data

14.2.1 What exactly is an interaction effect?

The key idea that we're going to introduce in this section is that of an interaction effect. In the ANOVA model we have looked at so far there are only two factors involved in our model (i.e., drug and therapy). But when we add an interaction we add a new component to the model: the combination of drug and therapy. Intuitively, the idea behind an interaction effect is fairly simple. It just means that the effect of Factor A is different, depending on which level of Factor B we're talking about. But what does that actually mean in terms of our data? The plot in Figure 14.5 depicts several different patterns that, although quite different to each other, would all count as an interaction effect. So it's not entirely straightforward to translate this qualitative idea into something mathematical that a statistician can work with.

[Additional technical detail[169]]

14.2.2 Degrees of freedom for the interaction

Calculating the degrees of freedom for the interaction is slightly trickier than the corresponding calculation for the main effects. Let's start by thinking about the ANOVA model as a whole. Once we include interaction effects in the model we're allowing every single group to have a unique mean, mu_{rc}. For an $R \times C$ factorial ANOVA, this means that there are $R \times C$ quantities of interest in the model and only the one constraint: all of the group means need to average out to the grand mean. So the model as a whole needs to have $(R \times C) - 1$ degrees of freedom. But the main effect of Factor

A has $R-1$ degrees of freedom, and the main effect of Factor B has $C-1$ degrees of freedom. Therefore the degrees of freedom for the interaction is:

$$\begin{aligned} df_{A:B} &= (R \times C - 1) - (R - 1) - (C - 1) \\ &= RC - R - C + 1 \\ &= (R-1)(C-1) \end{aligned}$$

which is just the product of the degrees of freedom associated with the row factor and the column factor.

What about the residual degrees of freedom? Because we've added interaction terms which absorb some degrees of freedom, there are fewer residual degrees of freedom left over. Specifically, note that if the model with interaction has a total of $(R \times C) - 1$, and there are N observations in your data set that are constrained to satisfy 1 grand mean, your residual degrees of freedom now become $N - (R \times C) - 1 + 1$, or just $N - (R \times C)$.

14.2.3 Running the ANOVA in jamovi

Adding interaction terms to the ANOVA model in jamovi is straightforward. In fact it is more than straightforward because it is the default option for ANOVA. This means that when you specify an ANOVA with two factors, e.g., drug and therapy then the interaction component – drug × therapy – is added automatically to the model.[170] When we run the ANOVA with the interaction term included, then we get the results shown in Figure 14.7.

ANOVA

ANOVA - mood.gain

	Sum of Squares	df	Mean Square	F	p	η²	η²p	ω²
drug	3.45	2	1.73	31.71	0.00002	0.71	0.84	0.68
therapy	0.47	1	0.47	8.58	0.01262	0.10	0.42	0.08
drug * therapy	0.27	2	0.14	2.49	0.12460	0.06	0.29	0.03
Residuals	0.65	12	0.05					

Figure 14.7: Results for the full factorial model, including the interaction component drug × therapy

As it turns out, while we do have a significant main effect of drug ($F_{2,12} = 31.7, p < .001$) and therapy type ($F_{1,12} = 8.6, p = .013$), there is no significant interaction between the two ($F_{2,12} = 2.5, p = 0.125$).

14.2.4 Interpreting the results

There's a couple of very important things to consider when interpreting the results of factorial ANOVA. First, there's the same issue that we had with one-way ANOVA, which is that if you obtain a significant main effect of (say) drug, it doesn't tell you anything about which drugs are different to one another. To find that out, you need to run additional analyses. We'll talk about some analyses that you can run in later Sections: Different ways to specify contrasts and Post hoc tests. The same is true for interaction effects. Knowing that there's a significant interaction doesn't tell you anything about what kind of interaction exists. Again, you'll need to run additional analyses.

Secondly, there's a very peculiar interpretation issue that arises when you obtain a significant interaction effect but no corresponding main effect. This happens sometimes. For instance, in the crossover interaction shown in Figure 14.5(a), this is exactly what you'd find. In this case, neither of the main effects would be significant, but the interaction effect would be. This is a difficult situation to interpret, and people often get a bit confused about it. The general advice that statisticians like to give in this situation is that you shouldn't pay much attention to the main effects when an interaction is present. The reason they say this is that, although the tests of the main effects are perfectly valid from a mathematical point of view, when there is a significant interaction effect the main effects rarely test interesting hypotheses. Recall from Section 14.1.1 that the null hypothesis for a main effect is that the marginal means are equal to each other, and that a marginal mean is formed by averaging across several different groups. But if you have a significant interaction effect then you know that the groups that comprise the marginal mean aren't homogeneous, so it's not really obvious why you would even care about those marginal means.

Here's what I mean. Again, let's stick with a clinical example. Suppose that we had a 2 × 2 design comparing two different treatments for phobias (e.g., systematic desensitisation vs flooding), and two different anxiety reducing drugs (e.g., Anxifree vs Joyzepam). Now, suppose what we found was that Anxifree had no effect when desensitisation was the treatment, and Joyzepam had no effect when flooding was the treatment. But both were pretty effective for the other treatment. This is a classic crossover interaction, and what we'd find when running the ANOVA is that there is no main effect of drug, but a significant interaction. Now, what does it actually mean to say that there's no main effect? Well, it means that if we average over the two different psychological treatments, then the average effect of Anxifree and Joyzepam is the same. But why would anyone care about that? When treating someone for phobias it is never the case that a person can be treated using an "average" of flooding and desensitisation. That doesn't make a lot of sense. You either get one or the other. For one treatment one drug is effective, and for the other treatment the other drug is effective. The interaction is the important thing and the main effect is kind of irrelevant.

This sort of thing happens a lot. The main effect are tests of marginal means, and when an interaction is present we often find ourselves not being terribly interested in marginal means because they imply averaging over things that the interaction tells us shouldn't be averaged! Of course, it's not always the case that a main effect is meaningless when an interaction is present. Often you can get a big main effect and a very small interaction, in which case you can still say things like "drug A is generally more effective than drug B" (because there was a big effect of drug), but you'd need to modify it a bit by adding that "the difference in effectiveness was different for different psychological treatments". In any case, the main point here is that whenever you get a significant interaction you should stop and think about what the main effect actually means in this context. Don't automatically assume that the main effect is interesting.

14.3 Effect size

The effect size calculation for a factorial ANOVA is pretty similar to those used in one-way ANOVA (see Effect size section). Specifically, we can use η^2 (eta-squared) as a simple way to measure how big the overall effect is for any particular term. As before, η^2 is defined by dividing the sum of squares associated with that term by the total sum of squares. For instance, to determine the size of the main effect of Factor A, we would use the following formula:

$$\eta_A^2 = \frac{SS_A}{SS_T}$$

As before, this can be interpreted in much the same way as R^2 in regression.[171] It tells you the proportion of variance in the outcome variable that can be accounted for by the main effect of Factor A. It is therefore a number that ranges from 0 (no effect at all) to 1 (accounts for all of the variability in the outcome). Moreover, the sum of all the η^2 values, taken across all the terms in the model, will sum to the the total R^2 for the ANOVA model. If, for instance, the ANOVA model fits perfectly (i.e., there is no within groups variability at all!), the η^2 values will sum to 1. Of course, that rarely if ever happens in real life.

However, when doing a factorial ANOVA, there is a second measure of effect size that people like to report, known as partial η^2. The idea behind partial η^2 (which is sometimes denoted $p\eta^2$ or η_p^2) is that, when measuring the effect size for a particular term (say, the main effect of Factor A), you want to deliberately ignore the other effects in the model (e.g., the main effect of Factor B). That is, you would pretend that the effect of all these other terms is zero, and then calculate what the η^2 value would have been. This is actually pretty easy to calculate. All you have to do is remove the sum of squares associated with the other terms from the denominator. In other words, if you want the partial η^2 for the main effect of Factor A, the denominator is just the sum of the SS values for Factor A and the residuals:

$$\text{partial } \eta_A^2 = \frac{SS_A}{SS_A + SS_R}$$

This will always give you a larger number than η^2, which the cynic in me suspects accounts for the popularity of partial η^2. And once again you get a number between 0 and 1, where 0 represents no effect. However, it's slightly trickier to interpret what a large partial η^2 value means. In particular, you can't actually compare the partial η^2 values across terms! Suppose, for instance, there is no within groups variability at all: if so, $SS_R = 0$. What that means is that every term has a partial η^2 value of 1. But that doesn't mean that all terms in your model are equally important, or indeed that they are equally large. All it mean is that all terms in your model have effect sizes that are large relative to the residual variation. It is not comparable across terms.

To see what I mean by this, it's useful to see a concrete example. First, let's have a look at the effect sizes for the original ANOVA (Table 14.6) without the interaction term from Figure 14.3.

Table 14.6: Effect sizes when the interaction term **is not** included in the ANOVA model

	eta.sq	partial.eta.sq
drug	0.71	0.79
therapy	0.10	0.34

Looking at the η^2 values first, we see that drug accounts for 71% of the variance (i.e. $\eta^2 = 0.71$) in mood.gain, whereas therapy only accounts for 10%. This leaves a total of 19% of the variation unaccounted for (i.e., the residuals constitute 19% of the variation in the outcome). Overall, this implies that we have a very large effect[172] of drug and a modest effect of therapy.

Now let's look at the partial η^2 values, shown in Figure 14.3. Because the effect of therapy isn't all that large, controlling for it doesn't make much of a difference, so the partial η^2 for drug doesn't increase very much, and we obtain a value of $p\eta^2 = 0.79$. In contrast, because the effect of drug was very large, controlling for it makes a big difference, and so when we calculate the partial η^2 for therapy you can see that it rises to $p\eta^2 = 0.34$. The question that we have to ask ourselves is, what do these partial η^2 values actually mean? The way I generally interpret the partial η^2 for the main effect of Factor A is to interpret it as a statement about a hypothetical experiment in which only Factor A was being varied. So, even though in this experiment we varied both A and B, we can easily imagine an experiment in which only Factor A was varied, and the partial η^2 statistic tells you how much of the variance in the outcome variable you would expect to see accounted for in that experiment. However, it should be noted that this interpretation, like many things associated with main effects, doesn't make a lot of sense when there is a large and significant interaction effect.

Speaking of interaction effects, Table 14.7 shows what we get when we calculate the effect sizes for the model that includes the interaction term, as in Figure 14.7. As you can see, the η^2 values for the main effects don't change, but the partial η^2 values do:

Table 14.7: Effect sizes when the interaction term **is** included in the ANOVA model

	eta.sq	partial.eta.sq
drug	0.71	0.84
therapy	0.10	0.42
drug*therapy	0.06	0.29

14.3.1 Estimated group means

In many situations you will want to report estimates of all the group means from the results of your ANOVA, as well as confidence intervals. You can use the 'Estimated Marginal Means' option in the jamovi ANOVA analysis to do this, as in Figure 14.8. If the ANOVA that you have run is a saturated model (i.e., contains all possible main effects and all possible interaction effects) then the estimates of the group means are actually identical to the sample means, though the confidence intervals will use a pooled estimate of the standard errors rather than use a separate one for each group.

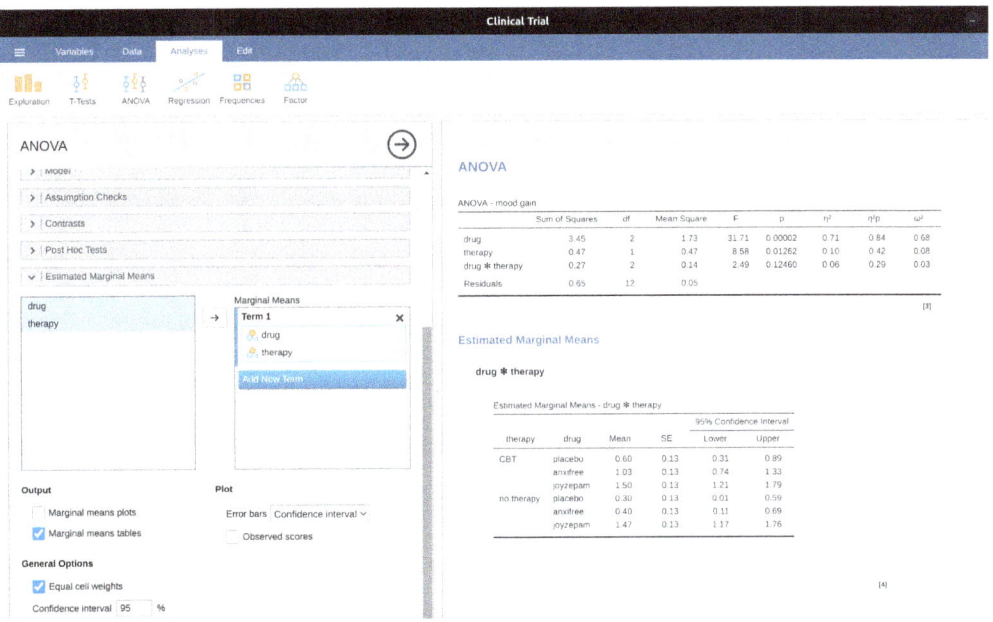

Figure 14.8: jamovi screenshot showing the marginal means for the saturated model, i.e. including the interaction component, with the clinical trial data set

In the output we see that the estimated mean mood gain for the placebo group with no therapy was 0.300, with a 95% confidence interval from 0.006 to 0.594. Note that these are not the same confidence intervals that you would get if you calculated them separately for each group, because of the fact that the ANOVA model assumes homogeneity of variance and therefore uses a pooled estimate of the standard deviation.

When the model doesn't contain the interaction term then the estimated group means will differ from the sample means. Instead of reporting the sample mean, jamovi calculates the value of the group means that would be expected on the basis of the marginal means (i.e., assuming no interaction).

Using the notation we developed earlier, the estimate for μ_{rc}, the mean for level r on the (row) Factor A and level c on the (column) Factor B would be $\mu_{..} + \alpha_r + \beta_c$. If there are genuinely no interactions between the two factors, this is actually a better estimate of the population mean than the raw sample mean. Removing the interaction term from the model, via the 'Model' options in the jamovi ANOVA analysis, provides the marginal means for the analysis shown in Figure 14.9.

ANOVA

ANOVA - mood.gain

	Sum of Squares	df	Mean Square	F	p	η²	η²p	ω²
drug	3.45	2	1.73	26.15	0.00002	0.71	0.79	0.68
therapy	0.47	1	0.47	7.08	0.01866	0.10	0.34	0.08
Residuals	0.92	14	0.07					

[3]

Estimated Marginal Means

drug ✱ therapy

Estimated Marginal Means - drug ✱ therapy

therapy	drug	Mean	SE	95% Confidence Interval	
				Lower	Upper
CBT	placebo	0.61	0.12	0.35	0.87
	anxifree	0.88	0.12	0.62	1.14
	joyzepam	1.64	0.12	1.38	1.90
no.therapy	placebo	0.29	0.12	0.03	0.55
	anxifree	0.56	0.12	0.30	0.82
	joyzepam	1.32	0.12	1.06	1.58

Figure 14.9: jamovi screenshot showing the marginal means for the unsaturated model, i.e. without the interaction component, with the clinical trial data set

14.4 Assumption checking

As with one-way ANOVA, the key assumptions of factorial ANOVA are homogeneity of variance (all groups have the same standard deviation), normality of the residuals, and independence of the observations. The first two are things we can check for. The third is something that you need to assess yourself by asking if there are any special relationships between different observations, for example repeated measures where the independent variable is time so there is a relationship between the observations at time one and time two: observations at different time points are from the same people. Additionally, if you aren't using a saturated model (e.g., if you've omitted the interaction terms) then you're also assuming that the omitted terms aren't important. Of course, you can check this last one by running an ANOVA with the omitted terms included and see if they're significant, so that's pretty easy. What about homogeneity of variance and normality of the residuals? As it turns out, these are pretty easy to check. It's no different to the checks we did for a one-way ANOVA.

14.4.1 Homogeneity of variance

As mentioned in Section 13.6.1 in the last chapter, it's a good idea to visually inspect a plot of the standard deviations compared across different groups / categories, and also see if the Levene test is consistent with the visual inspection. The theory behind the Levene test was discussed in Section 13.6.1, so I won't discuss it again. This test expects that you have a saturated model (i.e., including all of the relevant terms), because the test is primarily concerned with the within-group variance, and it doesn't really make a lot of sense to calculate this any way other than with respect to the full model. The Levene test can be specified under the ANOVA 'Assumption Checks' – 'Homogeneity Tests' option in jamovi, with the result shown as in Figure 14.10. The fact that the Levene test is non-significant means that, providing it is consistent with a visual inspection of the plot of standard deviations, we can safely assume that the homogeneity of variance assumption is not violated.

14.4.2 Normality of residuals

As with one-way ANOVA we can test for the normality of residuals is straightforward (see Section 13.6.4). Primarily it's a good idea to examine the residuals graphically using a QQ plot. See Figure 14.10.

14.5 Analysis of covariance (ANCOVA)

A variation in ANOVA is when you have an additional continuous variable that you think might be related to the dependent variable. This additional variable can be added to the analysis as a covariate, in the aptly named analysis of covariance (ANCOVA).

In ANCOVA the values of the dependent variable are "adjusted" for the influence of the covariate, and then the "adjusted" score means are tested between groups in the usual way. This technique can increase the precision of an experiment and provide a more "powerful" test of the equality of group means in the dependent variable. How does ANCOVA do this? Although the covariate itself is typically not of any experimental interest, adjustment for the covariate can decrease the estimate of experimental error. By reducing error variance precision is increased. This means that a false rejection of the null hypothesis (false negative or type II error) is less likely.

Despite this advantage, ANCOVA runs the risk of undoing real differences between groups and this should be avoided. Look at Figure 14.11 for example, which shows a plot of Statistics anxiety against age and shows two distinct groups – students who have either an Arts or Science background. ANCOVA with age as a covariate might lead to the conclusion that statistics anxiety does not differ in the two groups. Would this conclusion be reasonable – probably not because the ages of the two groups do not overlap and analysis of variance has essentially "extrapolated into a region with no data" (Everitt (1996), p. 68). Clearly, careful thought needs to be given to an analysis of covariance with distinct groups. This applies to both one-way and factorial designs, as ANCOVA can be used with both.

Assumption Checks

Homogeneity of Variances Test (Levene's)

F	df1	df2	p
0.22	5	12	0.94731

[3]

Normality Test (Shapiro-Wilk)

Statistic	p
0.96	0.53290

Q-Q Plot

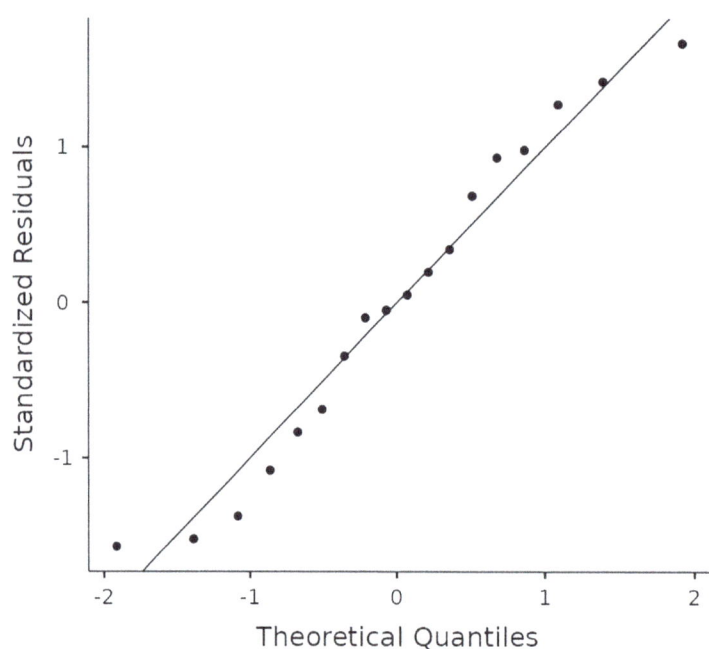

Figure 14.10: Checking assumptions in an ANOVA model

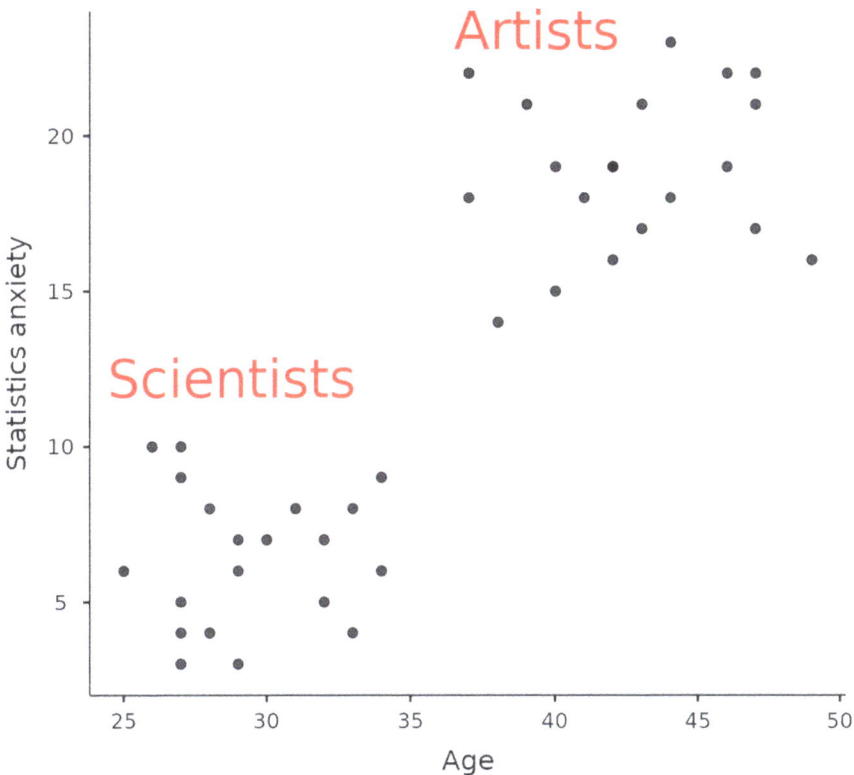

Figure 14.11: Plot of Statistics anxiety against age for two distinct groups

14.5.1 Running ANCOVA in jamovi

A health psychologist was interested in the effect of routine cycling and stress on happiness levels, with age as a covariate. You can find the data set in the file *ancova.csv*. Open this file in jamovi and then, to undertake an ANCOVA, select Analyses - ANOVA - ANCOVA to open the ANCOVA analysis window (Figure 14.12). Highlight the dependent variable 'happiness' and transfer it into the 'Dependent Variable' text box. Highlight the independent variables 'stress' and 'commute' and transfer them into the 'Fixed Factors' text box. Highlight the covariate 'age' and transfer it into the 'Covariates' text box. Then click on Estimated Marginal Means to bring up the plots and tables options.

An ANCOVA table showing 'Tests of Between Subjects Effects' is produced in the jamovi results window (Figure 14.13). The F-value for the covariate 'age' is significant at $p = .023$, suggesting that age is an important predictor of the dependent variable, happiness. When we look at the estimated marginal mean scores (Figure 14.14), adjustments have been made (compared to an analysis without the covariate) because of the inclusion of the covariate 'age' in this ANCOVA. A plot (Figure 14.15) is a good way of visualising and interpreting the significant effects.

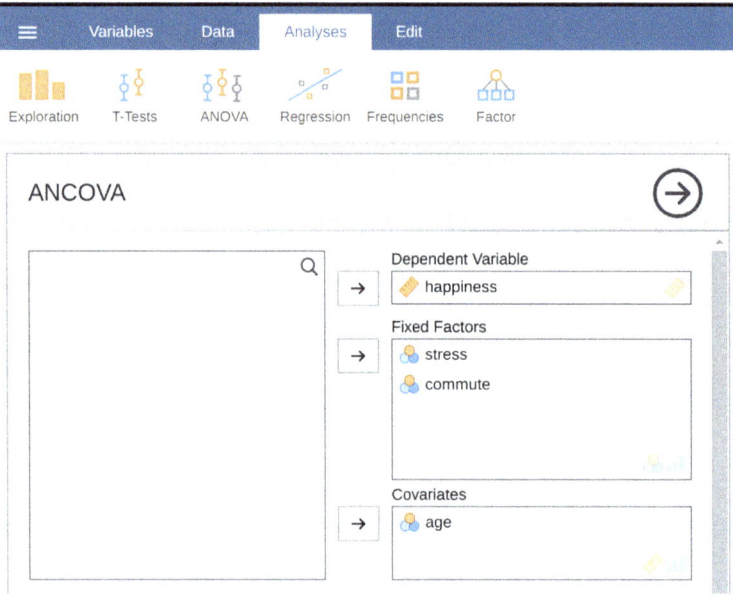

Figure 14.12: The jamovi ANCOVA analysis window

ANCOVA

ANCOVA - happiness

	Sum of Squares	df	Mean Square	F	p	η²	η²p	ω²
stress	2751.52	1	2751.52	52.61	< .00001	0.40	0.78	0.39
commute	2213.93	1	2213.93	42.33	< .00001	0.32	0.74	0.31
age	334.35	1	334.35	6.39	0.02316	0.05	0.30	0.04
stress ✶ commute	740.12	1	740.12	14.15	0.00188	0.11	0.49	0.10
Residuals	784.45	15	52.30					

Figure 14.13: jamovi ANCOVA output for happiness as a function of stress and commuting method, with age as a covariate

Estimated Marginal Means - stress ✶ commute

commute	stress	Mean	SE	95% Confidence Interval	
				Lower	Upper
drive	high	36.11	3.24	29.21	43.02
	low	51.09	3.26	44.13	58.04
cycle	high	43.58	3.84	35.40	51.76
	low	85.82	3.71	77.90	93.74

Figure 14.14: Table of mean happiness level as a function of stress and commuting method (adjusted for the covariate age) with 95% confidence intervals

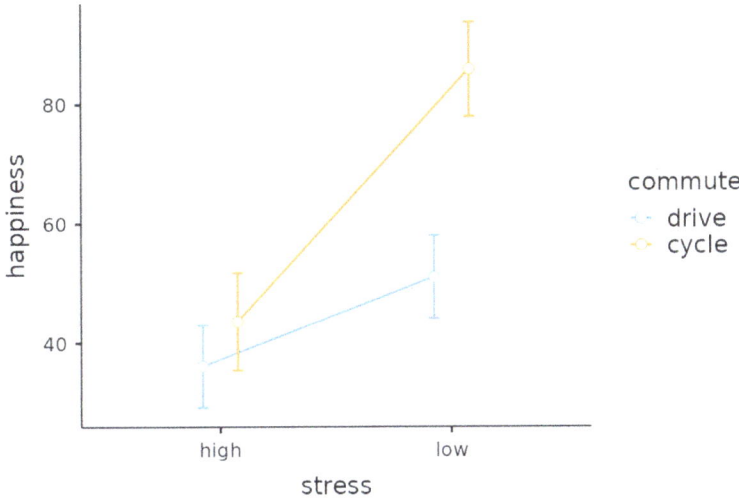

Figure 14.15: Plot of mean happiness level as a function of stress and commuting method

The F-value for the main effect 'stress' (52.61) has an associated probability of $p < .001$. The F-value for the main effect 'commute' (42.33) has an associated probability of $p < .001$. Since both of these are less than the probability that is typically used to decide if a statistical result is significant ($p < .05$) we can conclude that there was a significant main effect of stress ($F(1, 15) = 52.61, p < .001$) and a significant main effect of commuting method ($F(1, 15) = 42.33, p < .001$). A significant interaction between stress and commuting method was also found ($F(1, 15) = 14.15, p = .002$).

In Figure 14.15 we can see the adjusted, marginal, mean happiness scores when age is a covariate in an ANCOVA. In this analysis there is a significant interaction effect, whereby people with low stress who cycle to work are happier than people with low stress who drive and people with high stress regardless of whether they cycle or drive to work. There is also a significant main effect of stress – people with low stress are happier than those with high stress. And there is also a significant main effect of commuting behaviour – people who cycle are happier, on average, than those who drive to work.

One thing to be aware of is that, if you are thinking of including a covariate in your ANOVA, there is an additional assumption: the relationship between the covariate and the dependent variable should be similar for all levels of the independent variable. This can be checked by adding an interaction term between the covariate and each independent variable in the jamovi 'Model - Model' terms option. If the interaction effect is not significant it can be removed. If it is significant then a different and more advanced statistical technique might be appropriate (which is beyond the scope of this book so you might want to consult a friendly statistician).

14.6 ANOVA as a linear model

One of the most important things to understand about ANOVA and regression is that they're basically the same thing. On the surface of it, you maybe wouldn't think this is true. After all, the way that I've described them so far suggests that ANOVA is primarily concerned with testing for group differences, and regression is primarily concerned with understanding the correlations between variables. And, as far as it goes that's perfectly true. But when you look under the hood, so to speak, the underlying mechanics of ANOVA and regression are awfully similar. In fact, if you think about it, you've already seen evidence of this. ANOVA and regression both rely heavily on sums of squares (SS), both make use of F-tests, and so on. Looking back, it's hard to escape the feeling that Chapter 12 and Chapter 13 were a bit repetitive.

The reason for this is that ANOVA and regression are both kinds of **linear models**. In the case of regression, this is kind of obvious. The regression equation that we use to define the relationship between predictors and outcomes is the equation for a straight line, so it's quite obviously a linear model, with the equation:

$$Y_p = b_0 + b_1 X_{1p} + b_2 X_{2p} + \epsilon_p$$

where Y_p is the outcome value for the p-th observation (e.g., p-th person), X_{1p} is the value of the first predictor for the p-th observation, X_{2p} is the value of the second predictor for the p-th observation, the b_0, b_1, and b_2 terms are our regression coefficients, and ϵ_p is the p-th residual. If we ignore the residuals ϵ_p and just focus on the regression line itself, we get the following formula:

$$\hat{Y}_p = b_0 + b_1 X_{1p} + b_2 X_{2p}$$

where $\hat{Y}_p$ is the value of Y that the regression line predicts for person p, as opposed to the actually-observed value Y_p. The thing that isn't immediately obvious is that we can write ANOVA as a linear model as well. However, it's actually pretty straightforward to do this. Let's start with a really simple example, rewriting a 2×2 factorial ANOVA as a linear model.

14.6.1 Some data

To make things concrete, let's suppose that our outcome variable is the grade that a student receives in my class, a ratio-scale variable corresponding to a mark from 0 to 100. There are two predictor variables of interest: whether or not the student turned up to lectures (the attend variable) and whether or not the student actually read the textbook (the reading variable). We'll say that attend = 1 if the student attended class, and attend = 0 if they did not. Similarly, we'll say that reading = 1 if the student read the textbook, and reading = 0 if they did not.

Okay, so far that's simple enough. The next thing we need to do is to wrap some maths around this (sorry!). For the purposes of this example, let Y_p denote the grade of the p-th student in the class. This is not quite the same notation that we used earlier in this chapter. Previously, we've used the notation Y_{rci} to refer to the i-th person in the r-th group for predictor 1 (the row factor) and the c-th group for predictor 2 (the column factor). This extended notation was really handy for describing how the SS values are calculated, but it's a pain in the current context, so I'll switch notation here. Now, the Y_p notation is visually simpler than Y_{rci}, but it has the shortcoming that it doesn't actually keep track of the group memberships! That is, if I told you that $Y_{0,0,3} = 35$, you'd immediately know that we're talking about a student (the 3rd such student, in fact) who didn't attend the lectures (i.e., attend = 0) and didn't read the textbook (i.e. reading = 0), and who ended up failing the class (grade = 35). But if I tell you that $Y_p = 35$, all you know is that the p-th student didn't get a good grade. We've lost some key information here. Of course, it doesn't take a lot of thought to figure out how to fix this. What we'll do instead is introduce two new variables X_{1p} and X_{2p} that keep track of this information. In the case of our hypothetical student, we know that $X_{1p} = 0$ (i.e., attend = 0) and $X_{2p} = 0$ (i.e., reading = 0). So the data might look like Table 14.8.

Table 14.8: Data for grade, attendance and reading the textbook

person, p	grade, Y_p	attendance, X_{1p}	reading, X_{2p}
1	90	1	1
2	87	1	1
3	75	0	1
4	60	1	0
5	35	0	0
6	50	0	0
7	65	1	0
8	70	0	1

This isn't anything particularly special, of course. It's exactly the format in which we expect to see our data! See the data file *rtfm.csv*. We can use the jamovi 'Descriptives' analysis to confirm that this data set corresponds to a balanced design, with 2 observations for each combination of attend and reading. In the same way we can also calculate the mean grade for each combination. This is shown in Figure 14.16. Looking at the mean scores, one gets the strong impression that reading the text and attending the class both matter a lot.

14.6.2 ANOVA with binary factors as a regression model

Okay, let's get back to talking about the mathematics. We now have our data expressed in terms of three numeric variables: the continuous variable Y and the two binary variables X_1 and X_2. What I want you to recognise is that our 2×2 factorial ANOVA is exactly equivalent to the regression model:

$$Y_p = b_0 + b_1 X_{1p} + b_2 X_{2p} + \epsilon_p$$

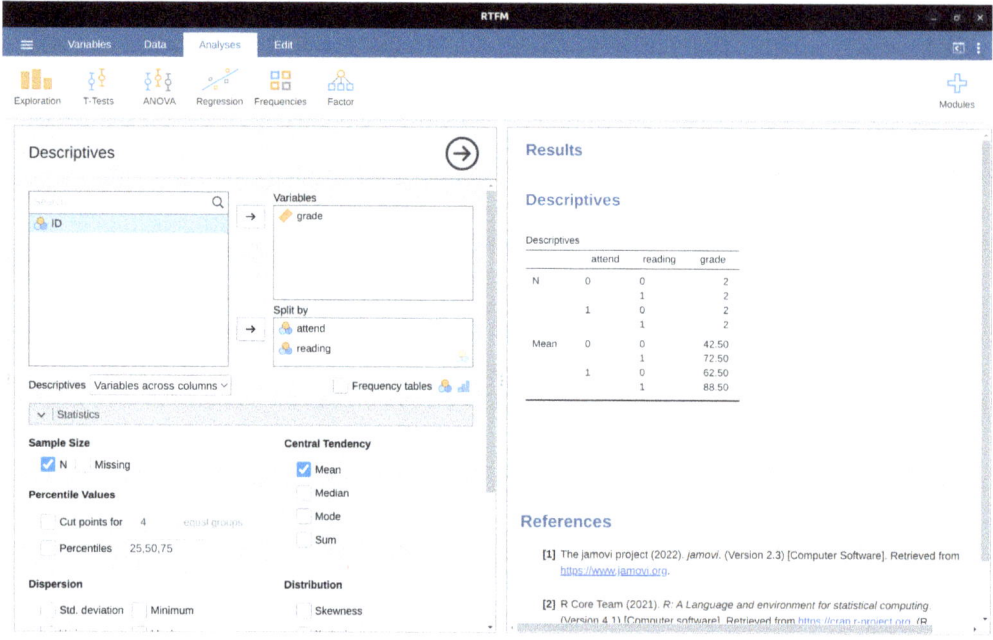

Figure 14.16: jamovi descriptives for the *rtfm.csv* data set

This is, of course, the exact same equation that I used earlier to describe a two-predictor regression model! The only difference is that X_1 and X_2 are now binary variables (i.e., values can only be 0 or 1), whereas in a regression analysis we expect that X_1 and X_2 will be continuous. There's a couple of ways I could try to convince you of this. One possibility would be to do a lengthy mathematical exercise proving that the two are identical. However, I'm going to go out on a limb and guess that most of the readership of this book will find that annoying rather than helpful. Instead, I'll explain the basic ideas and then rely on jamovi to show that ANOVA analyses and regression analyses aren't just similar, they're identical for all intents and purposes. Let's start by running this as an ANOVA. To do this, we'll use the rtfm data set, and Figure 14.17 shows what we get when we run the analysis in jamovi.

ANOVA

ANOVA - grade

	Sum of Squares	df	Mean Square	F	p	η²	η²p	ω²
attend	648.00	1	648.00	21.60	0.00559	0.27	0.81	0.26
reading	1568.00	1	1568.00	52.27	0.00079	0.66	0.91	0.64
Residuals	150.00	5	30.00					

Figure 14.17: ANOVA of the *rtfm.csv* data set in jamovi, without the interaction term

So, by reading the key numbers off the ANOVA table and the mean scores that we presented earlier, we can see that the students obtained a higher grade if they attended class ($F_{1,5} = 21.6, p = .0056$) and if they read the textbook ($F_{1,5} = 52.3, p = .0008$). Let's make a note of those p-values and those F-statistics.

Now let's think about the same analysis from a linear regression perspective. In the *rtfm.csv* data set, we have encoded attend and reading as if they were numeric predictors. In this case, this is perfectly acceptable. There really is a sense in which a student who turns up to class (i.e. attend = 1) has in fact done "more attendance" than a student who does not (i.e. attend = 0). So it's not at all unreasonable to include it as a predictor in a regression model. It's a little unusual, because the predictor only takes on two possible values, but it doesn't violate any of the assumptions of linear regression. And it's easy to interpret. If the regression coefficient for attend is greater than 0 it means that students that attend lectures get higher grades. If it's less than zero then students attending lectures get lower grades. The same is true for our reading variable.

Wait a second though. Why is this true? It's something that is intuitively obvious to everyone who has taken a few stats classes and is comfortable with the maths, but it isn't clear to everyone else at first pass. To see why this is true, it helps to look closely at a few specific students. Let's start by considering the 6th and 7th students in our data set (i.e. $p = 6$ and $p = 7$). Neither one has read the textbook, so in both cases we can set reading = 0. Or, to say the same thing in our mathematical notation, we observe $X_{2,6} = 0$ and $X_{2,7} = 0$. However, student number 7 did turn up to lectures (i.e., attend = 1, $X_{1,7} = 1$) whereas student number 6 did not (i.e., attend = 0, $X_{1,6} = 0$). Now let's look at what happens when we insert these numbers into the general formula for our regression line. For student number 6, the regression predicts that:

$$\begin{aligned}\hat{Y}_6 &= b_0 + b_1 X_{1,6} + b_2 X_{2,6} \\ &= b_0 + (b_1 \times 0) + (b_2 \times 0) \\ &= b_0\end{aligned}$$

So we're expecting that this student will obtain a grade corresponding to the value of the intercept term b_0. What about student 7? This time when we insert the numbers into the formula for the regression line, we obtain the following:

$$\begin{aligned}\hat{Y}_7 &= b_0 + b_1 X_{1,7} + b_2 X_{2,7} \\ &= b_0 + (b_1 \times 1) + (b_2 \times 0) \\ &= b_0 + b_1\end{aligned}$$

Because this student attended class, the predicted grade is equal to the intercept term b0 plus the coefficient associated with the attend variable, b_1. So, if b_1 is greater than zero, we're expecting that the students who turn up to lectures will get higher grades than those students who don't. If this coefficient is negative we're expecting the opposite: students who turn up at class end up performing much worse.

In fact, we can push this a little bit further. What about student number 1, who turned up to class ($X_{1,1} = 1$) and read the textbook ($X_{2,1} = 1$)? If we plug these numbers into the regression we get:

$$\begin{aligned}\hat{Y}_1 &= b_0 + b_1 X_{1,1} + b_2 X_{2,1} \\ &= b_0 + (b_1 \times 1) + (b_2 \times 1) \\ &= b_0 + b_1 + b_2\end{aligned}$$

So if we assume that attending class helps you get a good grade (i.e., $b_1 > 0$) and if we assume that reading the textbook also helps you get a good grade (i.e., $b_2 > 0$), then our expectation is that student 1 will get a grade that that is higher than student 6 and student 7.

And at this point you won't be at all suprised to learn that the regression model predicts that student 3, who read the book but didn't attend lectures, will obtain a grade of $b_2 + b_0$. I won't bore you with yet another regression formula. Instead, what I'll do is show you is Table 14.9 with the *expected grades*.

Table 14.9: Expected grades from the regression model

		read textbook	
		no	yes
attended?	no	β_0	$\beta_0 + \beta_2$
	yes	$\beta_0 + \beta_1$	$\beta_0 + \beta_1 + \beta_2$

As you can see, the intercept term b_0 acts like a kind of "baseline" grade that you would expect from those students who don't take the time to attend class or read the textbook. Similarly, b_1 represents the boost that you're expected to get if you come to class, and b_2 represents the boost that comes from reading the textbook. In fact, if this were an ANOVA you might very well want to characterise b_1 as the main effect of attendance, and b_2 as the main effect of reading! In fact, for a simple 2×2 ANOVA that's exactly how it plays out.

Okay, now that we're really starting to see why ANOVA and regression are basically the same thing, let's actually run our regression using the rtfm data and the jamovi regression analysis to convince ourselves that this is really true. Running the regression in the usual way gives the results shown in Figure 14.18.

There's a few interesting things to note here. First, notice that the intercept term is 43.5 which is close to the "group" mean of 42.5 observed for those two students who didn't read the text or attend class. Second, notice that we have the regression coefficient of $b_1 = 18.0$ for the attendance variable, suggesting that those students that attended class scored 18% higher than those who didn't. So our expectation would be that those students who turned up to class but didn't read the textbook would obtain a grade of $b_0 + b_1$, which is equal to $43.5 + 18.0 = 61.5$. You can verify for yourself that the same thing happens when we look at the students that read the textbook.

Model Coefficients - grade

Predictor	Estimate	SE	t	p
Intercept	43.50	3.35	12.97	0.00005
attend	18.00	3.87	4.65	0.00559
reading	28.00	3.87	7.23	0.00079

Figure 14.18: Regression analysis of the *rtfm.csv* data set in jamovi, without the interaction term

Actually, we can push a little further in establishing the equivalence of our ANOVA and our regression. Look at the p-values associated with the attend variable and the reading variable in the regression output. They're identical to the ones we encountered earlier when running the ANOVA. This might seem a little surprising, since the test used when running our regression model calculates a t-statistic and the ANOVA calculates an F-statistic. However, if you can remember all the way back to Chapter 7, I mentioned that there's a relationship between the t-distribution and the F-distribution. If you have some quantity that is distributed according to a t-distribution with k degrees of freedom and you square it, then this new squared quantity follows an F-distribution whose degrees of freedom are 1 and k. We can check this with respect to the t-statistics in our regression model. For the attend variable we get a t-value of 4.65. If we square this number we end up with 21.6, which matches the corresponding F-statistic in our ANOVA.

Finally, one last thing you should know. Because jamovi understands the fact that ANOVA and regression are both examples of linear models, it lets you extract the classic ANOVA table from your regression model using the 'Linear Regression' – 'Model Coefficients' – 'Omnibus Test' – 'ANOVA Test', and this will give you the table shown in Figure 14.19.

Omnibus ANOVA Test

	Sum of Squares	df	Mean Square	F	p
attend	648.00	1	648.00	21.60	0.00559
reading	1568.00	1	1568.00	52.27	0.00079
Residuals	150.00	5	30.00		

Note. Type 3 sum of squares

Figure 14.19: Omnibus ANOVA Test results from the jamovi regression analysis

14.6.3 How to encode non binary factors as contrasts

At this point, I've shown you how we can view a 2×2 ANOVA into a linear model. And it's pretty easy to see how this generalises to a $2 \times 2 \times 2$ ANOVA or a $2 \times 2 \times 2 \times 2$ ANOVA. It's the same thing, really. You just add a new binary variable for each of your factors. Where it begins to get trickier is when we consider factors that have more than two levels. Consider, for instance, the 3×2 ANOVA that we ran earlier in this chapter using the *clinicaltrial.csv* data. How can we convert the three-level drug factor into a numerical form that is appropriate for a regression?

The answer to this question is pretty simple, actually. All we have to do is realise that a three-level factor can be redescribed as two binary variables. Suppose, for instance, I were to create a new binary variable called druganxifree. Whenever the drug variable is equal to "anxifree" we set druganxifree = 1. Otherwise, we set druganxifree = 0. This variable sets up a **contrast**, in this case between anxifree and the other two drugs. By itself, of course, the druganxifree contrast isn't enough to fully capture all of the information in our drug variable. We need a second contrast, one that allows us to distinguish between joyzepam and the placebo. To do this, we can create a second binary contrast, called drugjoyzepam, which equals 1 if the drug is joyzepam and 0 if it is not. Taken together, these two contrasts allows us to perfectly discriminate between all three possible drugs. Table 14.10 illustrates this.

Table 14.10: Binary contrasts to discriminate between all three possible drugs

drug	druganxifree	drugjoyzepam
placebo	0	0
anxifree	1	0
joyzepam	0	1

If the drug administered to a patient is a placebo then both of the two contrast variables will equal 0. If the drug is Anxifree then the druganxifree variable will equal 1, and drugjoyzepam will be 0. The reverse is true for Joyzepam: drugjoyzepam is 1 and druganxifree is 0.

Creating contrast variables is not too difficult to do using the jamovi compute new variable command. For example, to create the druganxifree variable, write this logical expression in the compute new variable formula box:

IF(drug == 'anxifree', 1, 0)

Similarly, to create the new variable drugjoyzepam use this logical expression:

IF(drug == 'joyzepam', 1, 0)

Likewise for CBTtherapy:

IF(therapy == 'CBT', 1, 0)

You can see these new variables, and the corresponding logical expressions, in the jamovi data file *clinicaltrial2.omv* .

We have now recoded our three-level factor in terms of two binary variables, and we've already seen that ANOVA and regression behave the same way for binary variables. However, there are some additional complexities that arise in this case, which we'll discuss in the next section.

14.6.4 The equivalence between ANOVA and regression for non-binary factors

Now we have two different versions of the same data set. Our original data in which the drug variable from the *clinicaltrial.csv* file is expressed as a single three-level factor, and the expanded data *clinicaltrial2.omv* in which it is expanded into two binary contrasts. Once again, the thing that we want to demonstrate is that our original 3×2 factorial ANOVA is equivalent to a regression model applied to the contrast variables. Let's start by re-running the ANOVA, with results shown in Figure 14.20.

ANOVA

ANOVA - mood.gain

	Sum of Squares	df	Mean Square	F	p	η²	η²p	ω²
drug	3.45	2	1.73	26.15	0.00002	0.71	0.79	0.68
therapy	0.47	1	0.47	7.08	0.01866	0.10	0.34	0.08
Residuals	0.92	14	0.07					

Figure 14.20: jamovi ANOVA results, without interaction component

Obviously, there are no surprises here. That's the exact same ANOVA that we ran earlier. Next, let's run a regression using druganxifree, drugjoyzepam and CBTtherapy as the predictors. The results are shown in Figure 14.21.

Model Coefficients - mood.gain

Predictor	Estimate	SE	t	p
Intercept	0.29	0.12	2.38	0.03178
druganxifree	0.27	0.15	1.80	0.09386
drugjoyzepam	1.03	0.15	6.97	< .00001
CBTtherapy	0.32	0.12	2.66	0.01866

Figure 14.21: jamovi regression results, with contrast variables druganxifree and drugjoyzepam

Hmm. This isn't the same output that we got last time. Not surprisingly, the regression output prints out the results for each of the three predictors separately, just like it did every other time we conducted a regression analysis. On the one hand we can see that the p-value for the CBTtherapy variable is exactly the same as the one for the therapy factor in our original ANOVA, so we can be reassured that the regression model is doing the same thing as the ANOVA did. On the other hand, this regression model is testing the druganxifree contrast and the drugjoyzepam contrast *separately*, as if they were two completely unrelated variables. It's not surprising of course, because the poor regression analysis has no way of knowing that drugjoyzepam and druganxifree are actually the two different contrasts that we used to encode our three-level drug factor. As far as it knows, drugjoyzepam and druganxifree are no more related to one another than drugjoyzepam and CBTtherapy. However, you and I know better. At this stage we're not at all interested in determining whether these two contrasts are individually significant. We just want to know if there's an "overall" effect of drug. That is, what we want jamovi to do is to run some kind of "model comparison" test, one in which the two "drug related" contrasts are lumped together for the purpose of the test. Sound familiar? All we need to do is specify our null model, which in this case would include the CBTtherapy predictor, and omit both of the drug-related variables, as in Figure 14.22.

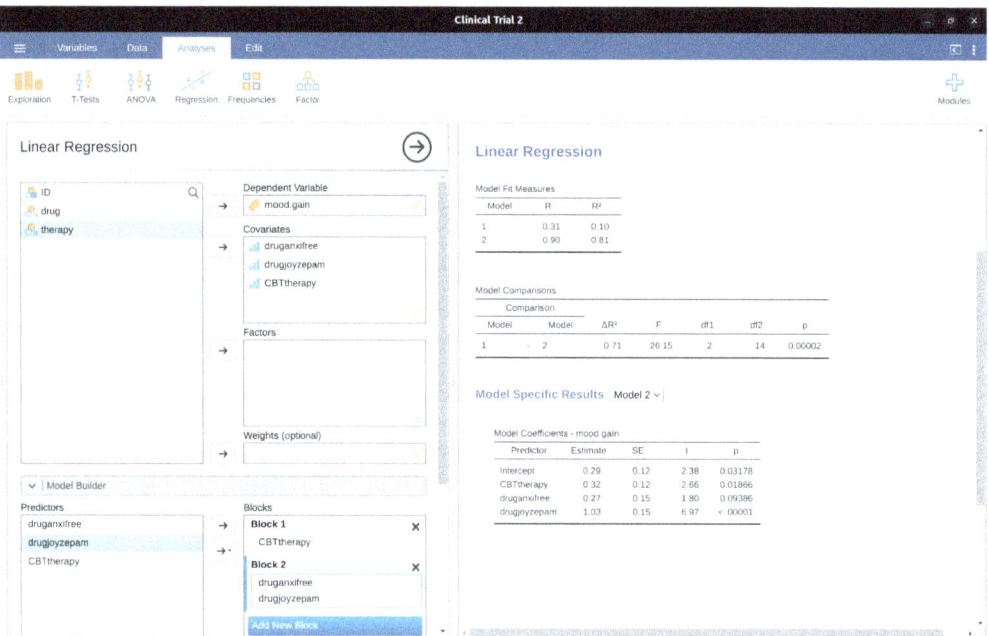

Figure 14.22: Model comparison in jamovi regression, null model 1 versus contrasts model 2

Ah, that's better. Our F-statistic is 26.15, the degrees of freedom are 2 and 14, and the p-value is 0.00002. The numbers are identical to the ones we obtained for the main effect of drug in our original ANOVA. Once again we see that ANOVA and regression are essentially the same. They are both linear models, and the underlying statistical machinery for ANOVA is identical to the machinery used in regression. The importance of this fact should not be understated. Throughout the rest of this chapter we're going to rely heavily on this idea.

Although we went through all the faff of computing new variables in jamovi for the contrasts druganxifree and drugjoyzepam, just to show that ANOVA and regression are essentially the same, in the jamovi linear regression analysis there is actually a nifty shortcut to get these contrasts, see Figure 14.23. What jamovi is doing here is allowing you to enter the predictor variables that are factors as, wait for it…factors! Smart, eh. You can also specify which group to use as the reference level, via the 'Reference Levels' option. We've changed this to 'placebo' and 'no.therapy', respectively, because this makes most sense.

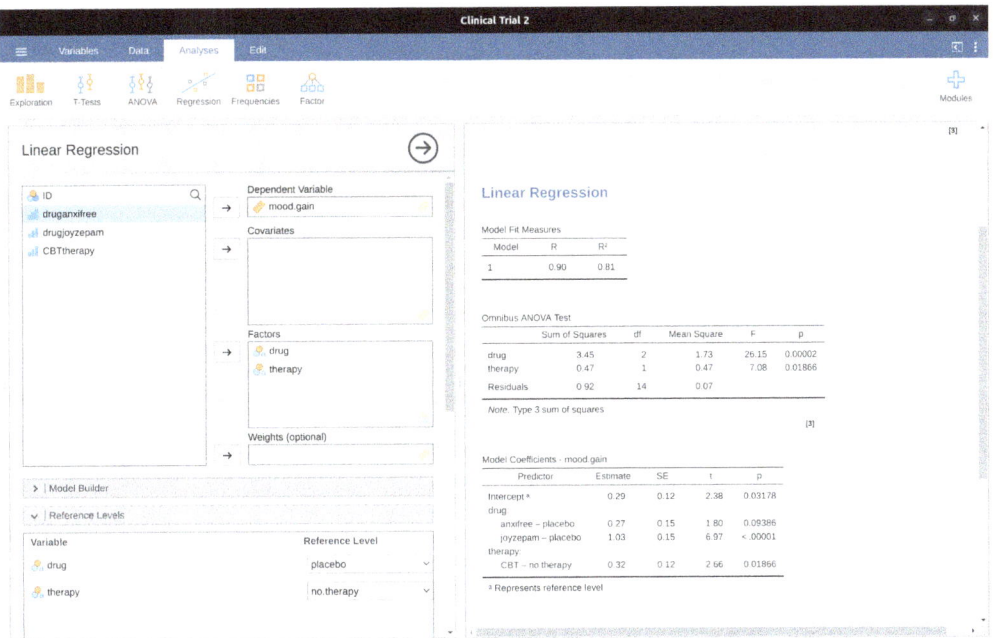

Figure 14.23: Regression analysis with factors and contrasts in jamovi, including omnibus ANOVA test results

If you also click on the 'ANOVA' test checkbox under the 'Model Coefficients' – 'Omnibus Test' option, we see that the F-statistic is 26.15, the degrees of freedom are 2 and 14, and the p-value is 0.00002 (Figure 14.23). The numbers are identical to those we obtained for the main effect of drug in our original ANOVA. Once again we see that ANOVA and regression are essentially the same. They are both linear models and the underlying statistical machinery for ANOVA and for regression is identical.

14.6.5 Degrees of freedom as parameter counting!

At long last, I can finally give a definition of degrees of freedom that I am happy with. Degrees of freedom are defined in terms of the number of parameters that have to be estimated in a model. For a regression model or an ANOVA, the number of parameters corresponds to the number of regression coefficients (i.e. b-values), including the intercept. Keeping in mind that any F-test is always a comparison between two models and the first df is the difference in the number of parameters. For example, in the model comparison above, the null model (mood.gain ~ CBTtherapy) has two parameters: there's one regression coefficient for the CBTtherapy variable, and a second one for the intercept. The alternative model (mood.gain ~ druganxifree + drugjoyzepam + CBTtherapy) has four parameters: one regression coefficient for each of the three contrasts, and one more for the intercept. So the degrees of freedom associated with the difference between these two models is $df_1 = 4 - 2 = 2$.

What about the case when there doesn't seem to be a null model? For instance, you might be thinking of the F-test that shows up when you select 'F Test' under the 'Linear

Regression' – 'Model Fit' options. I originally described that as a test of the regression model as a whole. However, that is still a comparison between two models. The null model is the trivial model that only includes 1 regression coefficient, for the intercept term. The alternative model contains $K + 1$ regression coefficients, one for each of the K predictor variables and one more for the intercept. So the df value that you see in this F-test is equal to $df_1 = K + 1 - 1 = K$.

What about the second df value that appears in the F-test? This always refers to the degrees of freedom associated with the residuals. It is possible to think of this in terms of parameters too, but in a slightly counter-intuitive way. Think of it like this. Suppose that the total number of observations across the study as a whole is N. If you wanted to perfectly describe each of these N values, you need to do so using, well... N numbers. When you build a regression model, what you're really doing is specifying that some of the numbers need to perfectly describe the data. If your model has K predictors and an intercept, then you've specified $K + 1$ numbers. So, without bothering to figure out exactly how this would be done, how many more numbers do you think are going to be needed to transform a $K + 1$ parameter regression model into a perfect re-description of the raw data? If you found yourself thinking that $(K+1)+(N-K-1) = N$, and so the answer would have to be $N-K-1$, well done! That's exactly right. In principle you can imagine an absurdly complicated regression model that includes a parameter for every single data point, and it would of course provide a perfect description of the data. This model would contain N parameters in total, but we're interested in the difference between the number of parameters required to describe this full model (i.e. N) and the number of parameters used by the simpler regression model that you're actually interested in (i.e., $K + 1$), and so the second degrees of freedom in the F test is $df_2 = N-K-1$, where K is the number of predictors (in a regression model) or the number of contrasts (in an ANOVA). In the example I gave above, there are ($N = 18$ observations in the data set and $K+1 = 4$ regression coefficients associated with the ANOVA model, so the degrees of freedom for the residuals is $df_2 = 18 - 4 = 14$.

14.7 Different ways to specify contrasts

In the previous section, I showed you a method for converting a factor into a collection of contrasts. In the method I showed you we specify a set of binary variables in which we defined a table like Table 14.11.

Table 14.11: Binary contrasts to discriminate between all three possible drugs

drug	druganxifree	drugjoyzepam
placebo	0	0
anxifree	1	0
joyzepam	0	1

Each row in the table corresponds to one of the factor levels, and each column corresponds to one of the contrasts. This table, which always has one more row than columns, has a special name. It is called a contrast matrix. However, there are lots of different ways to specify a contrast matrix. In this section I discuss a few of the standard contrast matrices that statisticians use and how you can use them in jamovi. If you're planning to read the section on Factorial ANOVA 3: unbalanced designs later on, it's worth reading this section carefully. If not, you can get away with skimming it, because the choice of contrasts doesn't matter much for balanced designs.

14.7.1 Treatment contrasts

In the particular kind of contrasts that I've described above, one level of the factor is special, and acts as a kind of "baseline" category (i.e., placebo in our example), against which the other two are defined. The name for these kinds of contrasts is treatment contrasts, also known as "dummy coding". In this contrast each level of the factor is compared to a base reference level, and the base reference level is the value of the intercept.

The name reflects the fact that these contrasts are natural and sensible when one of the categories in your factor really is special and actually does represent a baseline. That makes sense in our clinical trial example. The placebo condition corresponds to the situation where you don't give people any real drugs, and so it's special. The other two conditions are defined in relation to the placebo. In one case you replace the placebo with Anxifree, and in the other case your replace it with Joyzepam.

The table shown above is a matrix of treatment contrasts for a factor that has 3 levels. But suppose I want a matrix of treatment contrasts for a factor with 5 levels? You would set this out like Table 14.12.

Table 14.12: Matrix of treatment contrasts with 5 levels

Level	2	3	4	5
1	0	0	0	0
2	1	0	0	0
3	0	1	0	0
4	0	0	1	0
5	0	0	0	1

In this example, the first contrast is level 2 compared with level 1, the second contrast is level 3 compared with level 1, and so on. Notice that, by default, the *first* level of the factor is always treated as the baseline category (i.e., it's the one that has all zeros and doesn't have an explicit contrast associated with it). In jamovi you can change which category is the first level of the factor by manipulating the order of the levels of the variable shown in the 'Data Variable' window (double click on the name of the variable in the spreadsheet column to bring up the 'Data Variable' view.

14.7.2 Helmert contrasts

Treatment contrasts are useful for a lot of situations. However, they make most sense in the situation when there really is a baseline category, and you want to assess all the other groups in relation to that one. In other situations, however, no such baseline category exists, and it may make more sense to compare each group to the mean of the other groups. This is where we meet Helmert contrasts, generated by the 'helmert' option in the jamovi 'ANOVA' – 'Contrasts' selection box. The idea behind Helmert contrasts is to compare each group to the mean of the "previous" ones. That is, the first contrast represents the difference between group 2 and group 1, the second contrast represents the difference between group 3 and the mean of groups 1 and 2, and so on. This translates to a contrast matrix that looks like Table 14.13 for a factor with five levels.

Table 14.13: Matrix of helmert contrasts with 5 levels

1	-1	-1	-1	-1
2	1	-1	-1	-1
3	0	2	-1	-1
4	0	0	3	-1
5	0	0	0	4

With Helmert contrasts every contrast sums to zero (i.e., all the columns sum to zero). This means that, when we interpret the ANOVA as a regression, the intercept term corresponds to the grand mean $\mu_{..}$ if we are using Helmert contrasts. Compare this to treatment contrasts, in which the intercept term corresponds to the group mean for the baseline category. It doesn't matter very much if you have a balanced design, which we've assumed so far, but it will turn out to be important later when we consider Factorial ANOVA 3: unbalanced designs. In fact, the main reason why I've included this section is that contrasts become important if you want to understand unbalanced ANOVA.

14.7.3 Sum to zero contrasts

The third option that I should briefly mention are "sum to zero" contrasts, called "Simple" contrasts in jamovi, which are used to construct pairwise comparisons between groups. Specifically, each contrast encodes the difference between one of the groups and a baseline category, which in this case corresponds to the first group (Table 14.14).

Table 14.14: Matrix of "sum to" zero contrasts with 5 levels

1	-1	-1	-1	-1
2	1	0	0	0
3	0	1	0	0
4	0	0	1	0
5	0	0	0	1

Much like Helmert contrasts, we see that each column sums to zero, which means that the intercept term corresponds to the grand mean when ANOVA is treated as a regression model. When interpreting these contrasts, the thing to recognise is that each of these contrasts is a pairwise comparison between group 1 and one of the other four groups. Specifically, contrast 1 corresponds to a "group 2 minus group 1" comparison, contrast 2 corresponds to a "group 3 minus group 1" comparison, and so on.[173]

14.7.4 Optional contrasts in jamovi

There are options in jamovi that can generate different kinds of contrasts in ANOVA. See the 'Contrasts' option in the main ANOVA analysis window; Table 14.15 lists these contrast options.

Table 14.15: Contrasts available in the jamovi ANOVA analysis

Contrast type	
Deviation	Compares the mean of each level (except a reference category) to the mean of all of the levels (grand mean)
Simple	Like the treatment contrasts, the simple contrast compares the mean of each level to the mean of a specified level. This type of contrast is useful when there is a control group. By default the first category is the reference. However, with a simple contrast the intercept is the grand mean of all the levels of the factors.
Difference	Compares the mean of each level (except the first) to the mean of previous levels. (Sometimes called reverse Helmert contrasts)
Helmert	Compares the mean of each level of the factor (except the last) to the mean of subsequent levels
Repeated	Compares the mean of each level (except the last) to the mean of the subsequent level
Polynomial	Compares the linear effect and quadratic effect. The first degree of freedom contains the linear effect across all categories; the second degree of freedom, the quadratic effect. These contrasts are often used to estimate polynomial trends

14.8 Post hoc tests

Time to switch to a different topic. Rather than pre-planned comparisons that you have tested using contrasts, let's suppose you've done your ANOVA and it turns out that you obtained some significant effects. Because of the fact that the F-tests are "omnibus" tests that only really test the null hypothesis that there are no differences among groups, obtaining a significant effect doesn't tell you which groups are different to which other ones. We discussed this issue back in Chapter 13, and in that chapter our solution was to run t-tests for all possible pairs of groups, making corrections for multiple comparisons (e.g., Bonferroni, Holm) to control the type I error rate across all comparisons. The methods that we used back in Chapter 13 have the advantage of being relatively simple and being the kind of tools that you can use in a lot of different situations where you're testing multiple hypotheses, but they're not necessarily the best choices if you're interested in doing efficient post hoc testing in an ANOVA context. There are actually quite a lot of different methods for performing multiple comparisons in the statistics literature (Hsu, 1996), and it would be beyond the scope of an introductory text like this one to discuss all of them in any detail.

That being said, there's one tool that I do want to draw your attention to, namely Tukey's "Honestly Significant Difference", or **Tukey's HSD** for short. For once, I'll spare you the formulas and just stick to the qualitative ideas. The basic idea in Tukey's HSD is to examine all relevant pairwise comparisons between groups, and it's only really appropriate to use Tukey's HSD if it is pairwise differences that you're interested in.[174] For instance, earlier we conducted a factorial ANOVA using the *clinicaltrial.csv* data set, and where we specified a main effect for drug and a main effect of therapy we would be interested in the following four comparisons:

- The difference in mood gain for people given Anxifree versus people given the placebo.
- The difference in mood gain for people given Joyzepam versus people given the placebo.
- The difference in mood gain for people given Anxifree versus people given Joyzepam.
- The difference in mood gain for people treated with CBT and people given no therapy.

For any one of these comparisons, we're interested in the true difference between (population) group means. Tukey's HSD constructs simultaneous confidence intervals for all four of these comparisons. What we mean by 95% "simultaneous" confidence interval is that, if we were to repeat this study many times, then in 95% of the study results the confidence intervals would contain the relevant true value. Moreover, we can use these confidence intervals to calculate an adjusted p-value for any specific comparison.

The TukeyHSD function in jamovi is pretty easy to use. You simply specify the ANOVA model term that you want to run the post hoc tests for. For example, if we were looking to run post hoc tests for the main effects but not the interaction, we would open up the 'Post Hoc Tests' option in the ANOVA analysis screen, move the drug and therapy variables across to the box on the right, and then select the 'Tukey' checkbox in the list of possible post hoc corrections that could be applied. This, along with the corresponding results table, is shown in Figure 14.24.

The output shown in the 'Post Hoc Tests' results table is (I hope) pretty straightforward. The first comparison, for example, is the Anxifree versus placebo difference, and the first part of the output indicates that the observed difference in group means is .27. The next number is the standard error for the difference, from which we could calculate the 95% confidence interval if we wanted, though jamovi does not currently provide this option. Then there is a column with the degrees of freedom, a column with the t-value, and finally a column with the p-value. For the first comparison the adjusted p-value is .21. In contrast, if you look at the next line, we see that the observed difference between joyzepam and the placebo is 1.03, and this result is significant ($p < .001$).

So far, so good. What about the situation where your model includes interaction terms? For instance, the default option in jamovi is to allow for the possibility that there is an interaction between drug and therapy. If that's the case, the number of pairwise comparisons that we need to consider starts to increase. As before, we need to consider the three comparisons that are relevant to the main effect of drug and the one comparison that is relevant to the main effect of therapy.

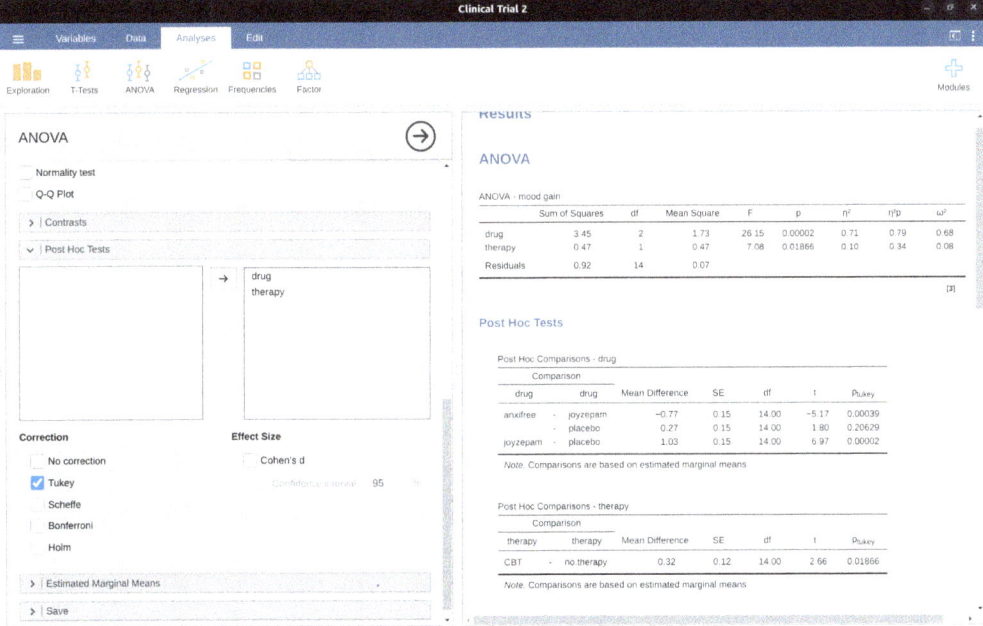

Figure 14.24: Tukey HSD post hoc test in jamovi factorial ANOVA, without an interaction term

But, if we want to consider the possibility of a significant interaction (and try to find the group differences that underpin that significant interaction), we need to include comparisons such as the following:

- The difference in mood gain for people given Anxifree and treated with CBT, versus people given the placebo and treated with CBT.
- The difference in mood gain for people given Anxifree and given no therapy, versus people given the placebo and given no therapy.
- etc.

There are quite a lot of these comparisons that you need to consider. So, when we run the Tukey post hoc analysis for this ANOVA model, we see that it has made a lot of pairwise comparisons (19 in total), as shown in Figure 14.25. You can see that it looks pretty similar to before, but with a lot more comparisons made.

Post Hoc Tests

Post Hoc Comparisons - drug

Comparison		Mean Difference	SE	df	t	p_tukey
drug	drug					
anxifree	joyzepam	−0.77	0.13	12.00	−5.69	0.00027
	placebo	0.27	0.13	12.00	1.98	0.15971
joyzepam	placebo	1.03	0.13	12.00	7.67	0.00002

Note. Comparisons are based on estimated marginal means

Post Hoc Comparisons - therapy

Comparison		Mean Difference	SE	df	t	p_tukey
therapy	therapy					
CBT	no.therapy	0.32	0.11	12.00	2.93	0.01262

Note. Comparisons are based on estimated marginal means

Post Hoc Comparisons - drug ✱ therapy

Comparison				Mean Difference	SE	df	t	p_tukey
drug	therapy	drug	therapy					
anxifree	CBT	anxifree	no.therapy	0.63	0.19	12.00	3.32	0.05298
		joyzepam	CBT	−0.47	0.19	12.00	−2.45	0.21392
		joyzepam	no.therapy	−0.43	0.19	12.00	−2.27	0.27506
		placebo	CBT	0.43	0.19	12.00	2.27	0.27506
		placebo	no.therapy	0.73	0.19	12.00	3.85	0.02187
	no.therapy	joyzepam	CBT	−1.10	0.19	12.00	−5.77	0.00096
		joyzepam	no.therapy	−1.07	0.19	12.00	−5.60	0.00126
		placebo	CBT	−0.20	0.19	12.00	−1.05	0.89172
		placebo	no.therapy	0.10	0.19	12.00	0.52	0.99401
joyzepam	CBT	joyzepam	no.therapy	0.03	0.19	12.00	0.17	0.99997
		placebo	CBT	0.90	0.19	12.00	4.72	0.00507
		placebo	no.therapy	1.20	0.19	12.00	6.30	0.00044
	no.therapy	placebo	CBT	0.87	0.19	12.00	4.55	0.00676
		placebo	no.therapy	1.17	0.19	12.00	6.12	0.00057
placebo	CBT	placebo	no.therapy	0.30	0.19	12.00	1.57	0.62800

Note. Comparisons are based on estimated marginal means

Figure 14.25: Tukey HSD post hoc test in jamovi factorial ANOVA with an interaction term

14.9 The method of planned comparisons

Following on from the previous sections on contrasts and post hoc tests in ANOVA, I think the method of planned comparisons is important enough to deserve a quick discussion. In our discussions of multiple comparisons, in the previous section and back in Chapter 13, I've been assuming that the tests you want to run are genuinely post hoc. For instance, in our drugs example above, maybe you thought that the drugs would all have different effects on mood (i.e., you hypothesised a main effect of drug), but you didn't have any specific hypothesis about how they would be different, nor did you have any real idea about which pairwise comparisons would be worth looking at. If that is the case, then you really have to resort to something like Tukey's HSD to do your pairwise comparisons.

The situation is rather different, however, if you genuinely did have real, specific hypotheses about which comparisons are of interest, and you never ever have any intention to look at any other comparisons besides the ones that you specified ahead of time. When this is true, and if you honestly and rigorously stick to your noble intentions to not run any other comparisons (even when the data look like they're showing you deliciously significant effects for stuff you didn't have a hypothesis test for), then it doesn't really make a lot of sense to run something like Tukey's HSD, because it makes corrections for a whole bunch of comparisons that you never cared about and never had any intention of looking at. Under those circumstances, you can safely run a (limited) number of hypothesis tests without making an adjustment for multiple testing. This situation is known as the method of planned comparisons, and it is sometimes used in clinical trials. However, further consideration is out of scope for this introductory book, but at least you know that this method exists!

14.10 Factorial ANOVA 3: unbalanced designs

Factorial ANOVA is a very handy thing to know about. It's been one of the standard tools used to analyse experimental data for many decades, and you'll find that you can't read more than two or three papers in psychology without running into an ANOVA in there somewhere. However, there's one huge difference between the ANOVAs that you'll see in a lot of real scientific articles and the ANOVAs that I've described so far. In in real life we're rarely lucky enough to have perfectly balanced designs. For one reason or another, it's typical to end up with more observations in some cells than in others. Or, to put it another way, we have an unbalanced design.

Unbalanced designs need to be treated with a lot more care than balanced designs, and the statistical theory that underpins them is a lot messier. It might be a consequence of this messiness, or it might be a shortage of time, but my experience has been that undergraduate research methods classes in psychology have a nasty tendency to ignore this issue completely. A lot of stats textbooks tend to gloss over it too. The net result of this, I think, is that a lot of active researchers in the field don't actually know that there's several different "types" of unbalanced ANOVAs, and they produce quite different answers. In fact, reading the psychological literature, I'm kind of amazed at the fact that most people who report the results of an unbalanced factorial ANOVA don't actually give you enough details to reproduce the analysis. I secretly suspect that most people don't even realise that their statistical software package is making a whole lot of substantive data analysis decisions on their behalf. It's actually a little terrifying when you think about it. So, if you want to avoid handing control of your data analysis to stupid software, read on.

14.10.1 The *coffee* data

As usual, it will help us to work with some data. The *coffee.csv* file contains a hypothetical data set that produces an unbalanced 3×2 ANOVA. Suppose we were interested in finding out whether or not the tendency of people to babble when they have too much coffee is purely an effect of the coffee itself, or whether there's some effect of the milk and sugar that people add to the coffee. Suppose we took 18 people and gave them some coffee to drink. The amount of coffee / caffeine was held constant, and we varied whether or not milk was added, so milk is a binary factor with two levels, "yes" and "no". We also varied the kind of sugar involved. The coffee might contain "real" sugar or it might contain "fake" sugar (i.e., artificial sweetener) or it might contain "none" at all, so the sugar variable is a three level factor. Our outcome variable is a continuous variable that presumably refers to some psychologically sensible measure of the extent to which someone is "babbling". The details don't really matter for our purpose. Take a look at the data in the jamovi spreadsheet view, as in Figure 14.26.

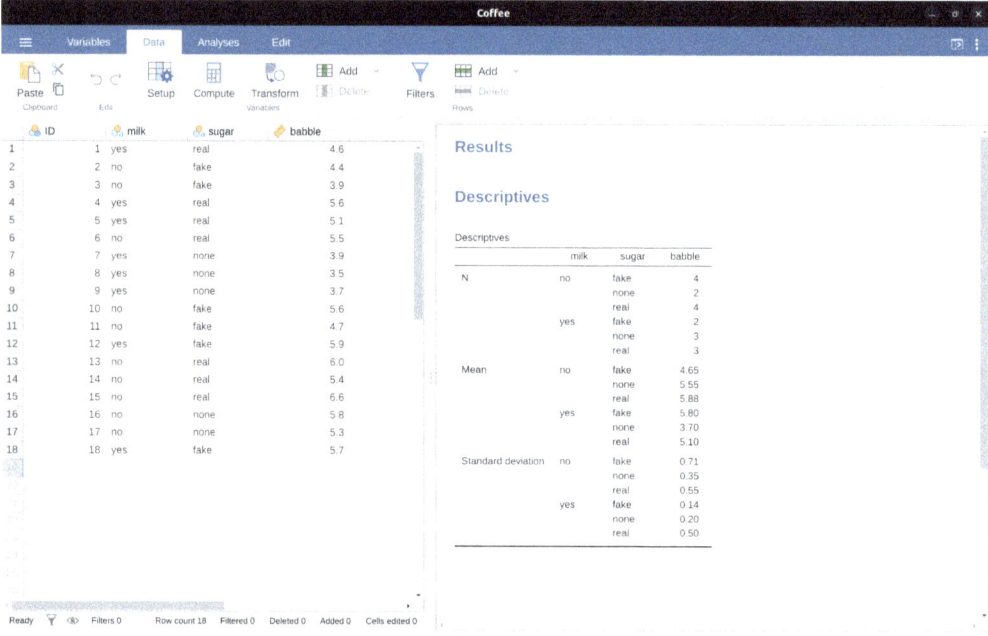

Figure 14.26: The *coffee.csv* data set in jamovi, with descriptive information aggregated by factor levels

Looking at the table of means in Figure 14.26 we get a strong impression that there are differences between the groups. This is especially true when we compare these means to the standard deviations for the babble variable. Across groups, this standard deviation varies from .14 to .71, which is fairly small relative to the differences in group means.[175] Whilst this at first may seem like a straightforward factorial ANOVA, a problem arises when we look at how many observations we have in each group. See the different Ns for different groups shown in Figure 14.26. This violates one of our original assumptions, namely that the number of people in each group is the same. We haven't really discussed how to handle this situation.

14.10.2 "Standard ANOVA" does not exist for unbalanced designs

Unbalanced designs lead us to the somewhat unsettling discovery that there isn't really any one thing that we might refer to as a standard ANOVA. In fact, it turns out that there are three fundamentally different ways[176] in which you might want to run an ANOVA in an unbalanced design. If you have a balanced design all three versions produce identical results, with the sums of squares, F-values, etc., all conforming to the formulas that I gave at the start of the chapter. However, when your design is unbalanced they don't give the same answers. Furthermore, they are not all equally appropriate to every situation. Some methods will be more appropriate to your situation than others. Given all this, it's important to understand what the different types of ANOVA are and how they differ from one another.

The first kind of ANOVA is conventionally referred to as **type I sum of squares**. I'm sure you can guess what the other two are called. The "sum of squares" part of the name was introduced by the SAS statistical software package and has become standard nomenclature, but it's a bit misleading in some ways. I think the logic for referring to them as different types of sum of squares is that, when you look at the ANOVA tables that they produce, the key difference in the numbers is the SS values. The degrees of freedom don't change, the MS values are still defined as SS divided by df, etc. However, what the terminology gets wrong is that it hides the reason *why* the SS values are different from one another. To that end, it's a lot more helpful to think of the three different kinds of ANOVA as three different *hypothesis testing strategies*. These different strategies lead to different SS values, to be sure, but it's the strategy that is the important thing here, not the SS values themselves. Recall from the section ANOVA as a linear model that any particular F-test is best thought of as a comparison between two linear models. So, when you're looking at an ANOVA table, it helps to remember that each of those F-tests corresponds to a pair of models that are being compared. Of course, this leads naturally to the question of which pair of models is being compared. This is the fundamental difference between ANOVA types I, II and III: each one corresponds to a different way of choosing the model pairs for the tests.

14.10.3 Type I sum of squares

The type I method is sometimes referred to as the "sequential" sum of squares, because it involves a process of adding terms to the model one at a time. Consider the *coffee* data, for instance. Suppose we want to run the full 3×2 factorial ANOVA, including interaction terms. The full model contains the outcome variable babble, the predictor variables sugar and milk, and the interaction term sugar × milk. This can be written as $babble \sim sugar + milk + sugar \times milk$. The type I strategy builds this model up sequentially, starting from the simplest possible model and gradually adding terms.

The simplest possible model for the data would be one in which neither milk nor sugar is assumed to have any effect on babbling. The only term that would be included in such a model is the intercept, written as babble ~ 1. This is our initial null hypothesis. The next simplest model for the data would be one in which only one of the two main effects is included. In the *coffee* data, there are two different possible choices here, because we could choose to add milk first or to add sugar first. The order actually turns out to matter, as we'll see later, but for now let's just make a choice arbitrarily and pick sugar. So, the second model in our sequence of models is babble ~ sugar, and it forms the alternative hypothesis for our first test. We now have our first hypothesis test (Table 14.16).

Table 14.16: Null and alternative hypotheses with the outcome variable "babble"

Null model:	$babble \sim 1$
Alternative model:	$babble \sim sugar$

This comparison forms our hypothesis test of the main effect of sugar. The next step in our model building exercise is to add the other main effect term, so the next model in our sequence is babble ~ sugar + milk. The second hypothesis test is then formed by comparing the following pair of models (Table 14.17).

Table 14.17: Further null and alternative hypotheses with the outcome variable "babble"

Null model:	$babble \sim sugar$
Alternative model:	$babble \sim sugar + milk$

This comparison forms our hypothesis test of the main effect of milk. In one sense, this approach is very elegant: the alternative hypothesis from the first test forms the null hypothesis for the second one. It is in this sense that the type I method is strictly sequential. Every test builds directly on the results of the last one. However, in another sense it's very inelegant, because there's a strong asymmetry between the two tests. The test of the main effect of sugar (the first test) completely ignores milk, whereas the test of the main effect of milk (the second test) does take sugar into account. In any case, the fourth model in our sequence is now the full model, babble ~ sugar + milk + sugar × milk, and the corresponding hypothesis test is shown in Table 14.18.

Table 14.18: And more possible null and alternative hypotheses with the outcome variable "babble"

Null model:	$babble \sim sugar + milk$
Alternative model:	$babble \sim sugar + milk + sugar * milk$

Type III sum of squares is the default hypothesis testing method used by jamovi ANOVA, so to run a type I sum of squares analysis we have to select 'Type 1' in the 'Sum of squares' selection box in the jamovi 'ANOVA' – 'Model' options. This gives us the ANOVA table shown in Figure 14.27.

The big problem with using type I sum of squares is the fact that it really does depend on the order in which you enter the variables. Yet, in many situations the researcher has no reason to prefer one ordering over another. This is presumably the case for our milk and sugar problem. Should we add milk first or sugar first? It feels exactly as arbitrary as a data analysis question as it does as a coffee-making question. There may in fact be some people with firm opinions about ordering, but it's hard to imagine a principled answer to the question. Yet, look what happens when we change the ordering, as in Figure 14.28.

ANOVA

ANOVA - babble

	Sum of Squares	df	Mean Square	F	p
sugar	3.56	2	1.78	6.75	0.01086
milk	0.96	1	0.96	3.63	0.08106
sugar ✱ milk	5.94	2	2.97	11.28	0.00175
Residuals	3.16	12	0.26		

Figure 14.27: ANOVA results table using type I sum of squares in jamovi

ANOVA

ANOVA - babble

	Sum of Squares	df	Mean Square	F	p
milk	1.44	1	1.44	5.48	0.03733
sugar	3.07	2	1.53	5.82	0.01708
milk ✱ sugar	5.94	2	2.97	11.28	0.00175
Residuals	3.16	12	0.26		

Figure 14.28: ANOVA results table using type I sum of squares in jamovi, but with factors entered in a different order (milk first)

The p-values for both main effect terms have changed, and fairly dramatically. Among other things, the effect of milk has become significant (though one should avoid drawing any strong conclusions about this, as I've mentioned previously). Which of these two ANOVAs should one report? It's not immediately obvious.

When you look at the hypothesis tests that are used to define the "first" main effect and the "second" one, it's clear that they're qualitatively different from one another. In our initial example, we saw that the test for the main effect of sugar completely ignores milk, whereas the test of the main effect of milk does take sugar into account. As such, the type I testing strategy really does treat the first main effect as if it had a kind of theoretical primacy over the second one. In my experience there is very rarely if ever any theoretically primacy of this kind that would justify treating any two main effects asymmetrically.

The consequence of all this is that type I tests are very rarely of much interest, and so we should move on to discuss type II tests and type III tests.

14.10.4 Type III sum of squares

Having just finished talking about type I tests, you might think that the natural thing to do next would be to talk about type II tests. However, I think it's actually a bit more natural to discuss type III tests (which are simple and the default in jamovi ANOVA) before talking about type II tests (which are trickier). The basic idea behind type III tests is extremely simple. Regardless of which term you're trying to evaluate, run the F-test in which the alternative hypothesis corresponds to the full ANOVA model as specified by the user, and the null model just deletes that one term that you're testing. For instance, in the coffee example, in which our full model was babble ~ sugar + milk + sugar × milk, the test for a main effect of sugar would correspond to a comparison between the following two models (Table 14.19).

Table 14.19: Null and alternative hypotheses with the outcome variable "babble", with type III sum of squares

Null model:	$babble \sim milk + sugar * milk$
Alternative model:	$babble \sim sugar+milk+sugar*milk$

Similarly the main effect of milk is evaluated by testing the full model against a null model that removes the milk term, like in Table 14.20.

Table 14.20: Further null and alternative hypotheses with the outcome variable 'babble', with type III sum of squares

Null model:	$babble \sim sugar + sugar * milk$
Alternative model:	$babble \sim sugar+milk+sugar*milk$

Finally, the interaction term sugar × milk is evaluated in exactly the same way. Once again, we test the full model against a null model that removes the sugar × milk interaction term, like in Table 14.21.

Table 14.21: Removing the interaction term from hypotheses with the outcome variable 'babble', with type III sum of squares

Null model:	$babble \sim sugar + milk$
Alternative model:	$babble \sim sugar+milk+sugar*milk$

The basic idea generalises to higher order ANOVAs. For instance, suppose that we were trying to run an ANOVA with three factors, A, B and C, and we wanted to consider all possible main effects and all possible interactions, including the three way interaction A × B × C. (Table 14.22) shows you what the Type III tests look like for this situation).

As ugly as that table looks, it's pretty simple. In all cases, the alternative hypothesis corresponds to the full model which contains three main effect terms (e.g., A), three two-way interactions (e.g., A*B) and one three-way interaction (i.e., A*B*C). The null model always contains 6 of these 7 terms, and the missing one is the one whose significance we're trying to test.

Table 14.22: Type III tests with three factors and all main effect and interaction term

Term being tested is	Null model is outcome ...	Alternative model is outcome ...
A	$B+C+A*B+A*C+B*C+A*B*C$	$A+B+C+A*B+A*C+B*C+A*B*C$
B	$A+C+A*B+A*C+B*C+A*B*C$	$A+B+C+A*B+A*C+B*C+A*B*C$
C	$A+B+A*B+A*C+B*C+A*B*C$	$A+B+C+A*B+A*C+B*C+A*B*C$
A*B	$A+B+C+A*C+B*C+A*B*C$	$A+B+C+A*B+A*C+B*C+A*B*C$
A*C	$A+B+C+A*B+B*C+A*B*C$	$A+B+C+A*B+A*C+B*C+A*B*C$
B*C	$A+B+C+A*B+A*C+A*B*C$	$A+B+C+A*B+A*C+B*C+A*B*C$
A*B*C	$A+B+C+A*B+A*C+B*C$	$A+B+C+A*B+A*C+B*C+A*B*C$

At first pass, type III tests seem like a nice idea. Firstly, we've removed the asymmetry that caused us to have problems when running type I tests. And because we're now treating all terms the same way, the results of the hypothesis tests do not depend on the order in which we specify them. This is definitely a good thing. However, there is a big problem when interpreting the results of the tests, especially for main effect terms. Consider the *coffee* data. Suppose it turns out that the main effect of milk is not significant according to the type III tests. What this is telling us is that babble ~ sugar + sugar*milk is a better model for the data than the full model. But what does that even mean? If the interaction term sugar*milk was also non significant, we'd be tempted to conclude that the data are telling us that the only thing that matters is sugar. But suppose we have a significant interaction term, but a non-significant main effect of milk. In this case, are we to assume that there really is an "effect of sugar", an "interaction between milk and sugar", but no "effect of milk"? That seems crazy. The right answer simply must be that it's meaningless[177] to talk about the main effect if the interaction is significant. In general, this seems to be what most statisticians advise us to do, and I think that's the right advice. But if it really is meaningless to talk about non-significant main effects in the presence of a significant interaction, then it's not at all obvious why type III tests should allow the null hypothesis to rely on a model that includes the interaction but omits one of the main effects that make it up. When characterised in this fashion, the null hypotheses really don't make much sense at all.

Later on, we'll see that type III tests can be redeemed in some contexts, but first let's take a look at the ANOVA results table using type III sum of squares, see Figure 14.29.

But be aware, one of the perverse features of the type III testing strategy is that typically the results turn out to depend on the contrasts that you use to encode your factors (see the Different ways to specify contrasts section if you've forgotten what the different types of contrasts are).[178]

ANOVA

ANOVA - babble

	Sum of Squares	df	Mean Square	F	p
milk	1.00	1	1.00	3.81	0.07467
sugar	2.13	2	1.07	4.04	0.04543
milk ✻ sugar	5.94	2	2.97	11.28	0.00175
Residuals	3.16	12	0.26		

Figure 14.29: ANOVA results table using type III sum of squares in jamovi

Okay, so if the p-values that typically come out of type III analyses (but not in jamovi) are so sensitive to the choice of contrasts, does that mean that type III tests are essentially arbitrary and not to be trusted? To some extent that's true, and when we turn to a discussion of type II tests we'll see that type II analyses avoid this arbitrariness entirely, but I think that's too strong a conclusion. Firstly, it's important to recognise that some choices of contrasts will always produce the same answers (ah, so this is what is happening in jamovi). Of particular importance is the fact that if the columns of our contrast matrix are all constrained to sum to zero, then the type III analysis will always give the same answers.

14.10.5 Type II sum of squares

Okay, so we've seen type I and III tests now, and both are pretty straightforward. Type I tests are performed by gradually adding terms one at a time, whereas type III tests are performed by taking the full model and looking to see what happens when you remove each term. However, both can have some limitations. Type I tests are dependent on the order in which you enter the terms, and type III tests are dependent on how you code up your contrasts. Type II tests are a little harder to describe, but they avoid both of these problems, and as a result they are a little easier to interpret.

Type II tests are broadly similar to type III tests. Start with a "full" model, and test a particular term by deleting it from that model. However, type II tests are based on the marginality principle which states that you should not omit a lower order term from your model if there are any higher order ones that depend on it. So, for instance, if your model contains the two-way interaction A × B (a 2nd order term), then it really ought to contain the main effects A and B (1st order terms). Similarly, if it contains a three-way interaction term A × B × C, then the model must also include the main effects A, B and C as well as the simpler interactions A × B, A × C and B × C. Type III tests routinely violate the marginality principle. For instance, consider the test of the main effect of A in the context of a three-way ANOVA that includes all possible interaction terms. According to type III tests, our null and alternative models are in Table 14.23.

Table 14.23: Type III tests for a main effect, A, in a three-way ANOVA with all possible interaction terms

Null model:	$outcome \sim$ $B+C+A*B+A*C+B*C+A*B*C$
Alternative model:	$outcome \sim A + B + C + A*B +$ $A*C + B*C + A*B*C$

Notice that the null hypothesis omits A, but includes A × B, A × C and A × B × C as part of the model. This, according to the type II tests, is not a good choice of null hypothesis. What we should do instead, if we want to test the null hypothesis that A is not relevant to our outcome, is to specify the null hypothesis that is the most complicated model that does not rely on A in any form, even as an interaction. The alternative hypothesis corresponds to this null model plus a main effect term of A. This is a lot closer to what most people would intuitively think of as a "main effect of A", and it yields the following as our type II test of the main effect of A (Table 14.24).[179]

Table 14.24: Type II tests for a main effect, A, in a three-way ANOVA with all possible interaction terms

Null model:	$outcome \sim B + C + B*C$
Alternative model:	$outcome \sim A + B + C + B*C$

Anyway, just to give you a sense of how the type II tests play out, see the full table (Table 14.25) of tests that would be applied in a three-way factorial ANOVA.

Table 14.25: Type II tests for a three-way factorial model

Term being tested is	Null model is outcome ...	Alternative model is outcome ...
A	$B + C + B*C$	$A + B + C + B*C$
B	$A + C + A*C$	$A + B + C + A*C$
C	$A + B + A*B$	$A + B + C + A*B$
A*B	$A + B + C + A*C + B*C$	$A + B + C + A*B + A*C + B*C$
A*C	$A + B + C + A*B + B*C$	$A + B + C + A*B + A*C + B*C$
B*C	$A + B + C + A*B + A*C$	$A + B + C + A*B + A*C + B*C$
A*B*C	$A + B + C + A*B + A*C + B*C$	$A+B+C+A*B+A*C+B*C+A*B*C$

In the context of the two way ANOVA that we've been using in the *coffee* data, the hypothesis tests are even simpler. The main effect of sugar corresponds to an F-test comparing these two models (Table 14.26). The test for the main effect of milk is in Table 14.27. Finally, the test for the interaction sugar × milk is in Table 14.28.

Table 14.26: Type II tests for the main effect of sugar in the *coffee* data

Null model:	$babble \sim milk$
Alternative model:	$babble \sim sugar + milk$

Table 14.27: Type II tests for the main effect of milk in the *coffee* data

Null model:	$babble \sim sugar$
Alternative model:	$babble \sim sugar + milk$

Table 14.28: Type II tests for the sugar × milk interaction term

Null model:	$babble \sim sugar + milk$
Alternative model:	$babble \sim sugar + milk + sugar*milk$

Running the tests are again straightforward. Just select 'Type 2' in the 'Sum of squares' selection box in the jamovi 'ANOVA' – 'Model' options, This gives us the ANOVA table shown in Figure 14.30.

ANOVA

ANOVA - babble

	Sum of Squares	df	Mean Square	F	p	η²	η²p	ω²
milk	0.96	1	0.96	3.63	0.08106	0.07	0.23	0.05
sugar	3.07	2	1.53	5.82	0.01708	0.23	0.49	0.19
milk ✻ sugar	5.94	2	2.97	11.28	0.00175	0.45	0.65	0.40
Residuals	3.16	12	0.26					

Figure 14.30: ANOVA results table using type II sum of squares in jamovi

Type II tests have some clear advantages over type I and type III tests. They don't depend on the order in which you specify factors (unlike type I), and they don't depend on the contrasts that you use to specify your factors (unlike type III). And although opinions may differ on this last point, and it will definitely depend on what you're trying to do with your data, I do think that the hypothesis tests that they specify are more likely to correspond to something that you actually care about. As a consequence, I find that it's usually easier to interpret the results of a type II test than the results of a type I or type III test. For this reason my tentative advice is that, if you can't think of any obvious model comparisons that directly map onto your research questions but you still want to run an ANOVA in an unbalanced design, type II tests are probably a better choice than type I or type III.[180]

14.10.6 Effect sizes (and non-additive sums of squares)

jamovi also provides the effect sizes η^2 and partial η^2 when you select these options, as in Figure 14.30. However, when you've got an unbalanced design there's a bit of extra complexity involved.

If you remember back to our very early discussions of ANOVA, one of the key ideas behind the sums of squares calculations is that if we add up all the SS terms associated with the effects in the model, and add that to the residual SS, they're supposed to add up to the total sum of squares. And, on top of that, the whole idea behind η^2 is that, because you're dividing one of the SS terms by the total SS value, an η^2 value can be interpreted as the proportion of variance accounted for by a particular term. But this is not so straightforward in unbalanced designs because some of the variance goes "missing".

This seems a bit odd at first, but here's why. When you have unbalanced designs your factors become correlated with one another, and it becomes difficult to tell the difference between the effect of Factor A and the effect of Factor B. In the extreme case, suppose that we'd run a 2×2 design in which the number of participants in each group had been as in Table 14.29.

Table 14.29: N participants in a 2 x 2 very (very!) unbalanced factorial design

	sugar	no sugar
milk	100	0
no milk	0	100

Here we have a spectacularly unbalanced design: 100 people have milk and sugar, 100 people have no milk and no sugar, and that's all. There are 0 people with milk and no sugar, and 0 people with sugar but no milk. Now suppose that, when we collected the data, it turned out there is a large (and statistically significant) difference between the "milk and sugar" group and the "no-milk and no-sugar" group. Is this a main effect of sugar? A main effect of milk? Or an interaction? It's impossible to tell, because the presence of sugar has a perfect association with the presence of milk. Now suppose the design had been a little more balanced (Table 14.30).

Table 14.30: N participants in a 2 x 2 still very unbalanced factorial design

	sugar	no sugar
milk	100	5
no milk	5	100

This time around, it's technically possible to distinguish between the effect of milk and the effect of sugar, because we have a few people that have one but not the other. However, it will still be pretty difficult to do so, because the association between sugar and milk is still extremely strong, and there are so few observations in two of the groups. Again, we're very likely to be in the situation where we *know* that the predictor variables (milk and sugar) are related to the outcome (babbling), but we don't know if the nature of that relationship is a main effect of one or the other predictor, or the interaction.

14.11 Summary

- Factorial ANOVA 1: balanced designs, focus on main effects and with interactions considered.
- Effect size, estimated means, and confidence intervals in a factorial ANOVA.
- Assumption checking in ANOVA.
- Analysis of Covariance (ANCOVA).
- Understanding ANOVA as a linear model, including Different ways to specify contrasts.
- Post hoc tests using Tukey's HSD and a brief commentary on The method of planned comparisons.
- Factorial ANOVA 3: unbalanced designs.

Chapter 15

Factor Analysis

Previous chapters have covered statistical tests for differences between two or more groups. However, sometimes when conducting research, we may wish to examine how multiple variables *co-vary*. That is, how they are related to each other and whether the patterns of relatedness suggest anything interesting and meaningful. For example, we are often interested in exploring whether there are any underlying unobserved latent factors that are represented by the observed, directly measured, variables in our data set. In statistics, latent factors are initially hidden variables that are not directly observed but are rather inferred (through statistical analysis) from other variables that are observed (directly measured). For a more technical and useful geometric explanation of the essentials of Factor Analysis take a look at Child (1990). Kline (1994) also provides a helpful introduction to the basic theory of Factor Analysis.

In this chapter we will cover how to undertake a number of different Factor Analysis and related techniques, starting with Exploratory Factor Analysis (EFA). EFA is a statistical technique for identifying underlying latent factors in a data set. Then we will cover Principal Component Analysis (PCA) which is a data reduction technique which, strictly speaking, does not identify underlying latent factors. Instead, PCA simply produces a linear combination of observed variables. Following this, the section on Confirmatory Factor Analysis (CFA) shows that, unlike EFA, with CFA you start with an idea – a model – of how the variables in your data are related to each other. You then test your model against the observed data and assess how good a fit the model is. A more sophisticated version of CFA is the so-called Multi-Trait Multi-Method CFA approach in which both latent factor and method variance are included in the model. This is useful when there are different methodological approaches used for measurement and therefore method variance is an important consideration. Finally, we will cover a related analysis: Internal consistency reliability analysis tests how consistently a scale measures a psychological construct.

15.1 Exploratory Factor Analysis

Exploratory Factor Analysis (EFA) is a statistical technique for revealing any hidden latent factors that can be inferred from our observed data. This technique calculates to what extent a set of measured variables, for example V_1, V_2, V_3, V_4, and V_5, can be represented as measures of an underlying latent factor. This latent factor cannot be measured through just one observed variable but instead is manifested in the relationships it causes in a set of observed variables.

In Figure 15.1 each observed variable V is 'caused' to some extent by the underlying latent factor (F), depicted by the coefficients b_1 to b_5 (also called factor loadings). Each observed variable also has an associated error term, e_1 to e_5. Each error term is the variance in the associated observed variable, V_i, that is unexplained by the underlying latent factor.

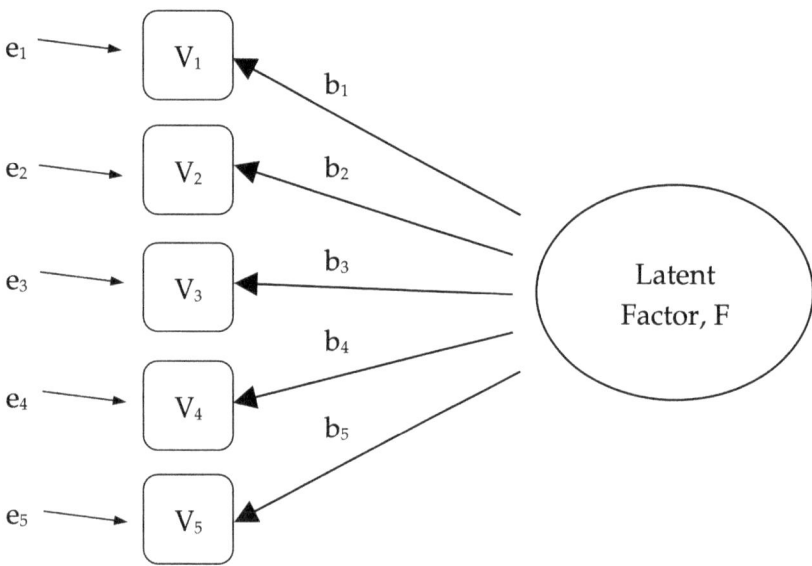

Figure 15.1: Latent factor underlying the relationship between several observed variables

In Psychology, latent factors represent psychological phenomena or constructs that are difficult to directly observe or measure. For example, personality, or intelligence, or thinking style. In the example in Figure 15.1 we may have asked people five specific questions about their behaviour or attitudes, and from that we are able to get a picture about a personality construct called, for example, extraversion. A different set of specific questions may give us a picture about an individual's introversion, or their conscientiousness.

Here's another example: we may not be able to directly measure statistics anxiety, but we can measure whether statistics anxiety is high or low with a set of questions in a questionnaire. For example, "$Q1$: Doing the assignment for a statistics course", "$Q2$: Trying to understand the statistics described in a journal article", and "$Q3$: Asking the lecturer for help in understanding something from the course", etc., each rated from low anxiety to high anxiety. People with high statistics anxiety will tend to give similarly high responses on these observed variables because of their high statistics anxiety. Likewise, people with low statistics anxiety will give similar low responses to these variables because of their low statistics anxiety.

In Exploratory Factor Analysis (EFA), we are essentially exploring the correlations between observed variables to uncover any interesting, important underlying (latent) factors that are identified when observed variables co-vary. We can use statistical software to estimate any latent factors and to identify which of our variables have a high loading[181] (e.g., loading > 0.5) on each factor, suggesting they are a useful measure, or indicator, of the latent factor. Part of this process includes a step called rotation, which to be honest is a pretty weird idea but luckily we don't have to worry about understanding it; we just need to know that it is helpful because it makes the pattern of loadings on different factors much clearer. As such, rotation helps with seeing more clearly which variables are linked substantively to each factor. We also need to decide how many factors are reasonable given our data, and helpful in this regard is something called Eigen values. We'll come back to this in a moment, after we have covered some of the main assumptions of EFA.

15.1.1 Checking assumptions

There are a couple of assumptions that need to be checked as part of the analysis. The first assumption is **sphericity**, which essentially checks that the variables in your data set are correlated with each other to the extent that they can potentially be summarised with a smaller set of factors. Bartlett's test for sphericity checks whether the observed correlation matrix diverges significantly from a zero (or null) correlation matrix. So, if Bartlett's test is significant ($p < .05$), this indicates that the observed correlation matrix is significantly divergent from the null, and is therefore suitable for EFA.

The second assumption is **sampling adequacy** and is checked using the Kaiser-Meyer-Olkin (KMO) Measure of Sampling Adequacy (MSA). The KMO index is a measure of the proportion of variance among observed variables that might be common variance. Using partial correlations, it checks for factors that load just two items. We seldom, if ever, want EFA producing a lot of factors loading just two items each. KMO is about sampling adequacy because partial correlations are typically seen with inadequate samples. If the KMO index is high (≈ 1), the EFA is efficient whereas if KMO is low (≈ 0), the EFA is not relevant. KMO values smaller than 0.5 indicates that EFA is not suitable and a KMO value of 0.6 should be present before EFA is considered suitable. Values between 0.5 and 0.7 are considered adequate, values between 0.7 and 0.9 are good and values between 0.9 and 1.0 are excellent.

15.1.2 What is EFA good for?

If the EFA has provided a good solution (i.e. factor model), then we need to decide what to do with our shiny new factors. Researchers often use EFA during psychometric scale development. They will develop a pool of questionnaire items that they think relate to one or more psychological constructs, use EFA to see which items "go together" as latent factors, and then they will assess whether some items should be removed because they don't usefully or distinctly measure one of the latent factors.[182]

In line with this approach, another consequence of EFA is to combine the variables that load onto distinct factors into a factor score, sometimes known as a scale score. There are two options for combining variables into a scale score:

- Create a new variable with a score weighted by the factor loadings for each item that contributes to the factor.
- Create a new variable based on each item that contributes to the factor, but weighting them equally.

In the first option each item's contribution to the combined score depends on how strongly it relates to the factor. In the second option we typically just average across all the items that contribute substantively to a factor to create the combined scale score variable. Which to choose is a matter of preference, though a disadvantage with the first option is that loadings can vary quite a bit from sample to sample, and in behavioural and health sciences we are often interested in developing and using composite questionnaire scale scores across different studies and different samples. In which case it is reasonable to use a composite measure that is based on the substantive items contributing equally rather than weighting by sample specific loadings from a different sample. In any case, understanding a combined variable measure as an average of items is simpler and more intuitive than using a sample specific optimally-weighted combination.

But let's not get ahead of ourselves; what we should really focus on now is how to do an EFA in jamovi.

15.1.3 EFA in jamovi

First, we need some data. Twenty-five personality self-report items (see Table 15.1) taken from the International Personality Item Pool (http://ipip.ori.org) were included as part of the Synthetic Aperture Personality Assessment (SAPA) web-based personality assessment (http://sapa-project.org) project. The 25 items are organized by five putative factors: Agreeableness, Conscientiousness, Extraversion, Neuroticism, and Openness.

The item data were collected using a 6-point response scale:

1. Very Inaccurate
2. Moderately Inaccurate
3. Slightly Inaccurate
4. Slightly Accurate
5. Moderately Accurate
6. Very Accurate

A sample of $N = 250$ responses is contained in the data set *bfi_sample.csv*. As researchers, we are interested in exploring the data to see whether there are some underlying latent factors that are measured reasonably well by the 25 observed variables in the *bfi_sample.csv* data file. Open up the data set and check that the 25 variables are coded as continuous variables (technically they are ordinal though for EFA in jamovi it mostly doesn't matter, except if you decide to calculate weighted factor scores in which case continuous variables are needed).

Table 15.1: Twenty-five observed variable items organised by five putative personality factors in the data set *bfi_sample.csv*

Variable name	Question / Item (short phrases that you should respond to by indicating how accurately the statement describes your typical behaviour or attitudes)	Coding (R: reverse)
A1	Am indifferent to the feelings of others	R
A2	Inquire about others' well-being	
A3	Know how to comfort others	
A4	Love children	
A5	Make people feel at ease	
C1	Am exacting in my work	
C2	Continue until everything is perfect	
C3	Do things according to a plan	
C4	Do things in a half-way manner	R
C5	Waste my time	R
E1	Don't talk a lot	R
E2	Find it difficult to approach others	R
E3	Know how to capitivate people	
E4	Make friends easily	
E5	Take charge	
N1	Get angry easily	
N2	Get irritated easily	
N3	Have frequent mood swings	
N4	Often feel blue	
N5	Panic easily	
O1	Am full of ideas	
O2	Avoid difficult reading material	R
O3	Carry the conversation to a higher level	
O4	Spend time reflecting on things	
O5	Will not probe deeply into a subject	R

To perform EFA in jamovi:

- Select 'Factor' – 'Exploratory Factor Analysis' from the main jamovi button bar to open the EFA analysis window (Figure 15.2).

- Select the 25 personality questions and transfer them into the 'Variables' box.

- Check appropriate options, including 'Assumption Checks', but also Rotation 'Method', 'Number of Factors' to extract, and 'Additional Output' options. See Figure 15.2 for suggested options for this illustrative EFA, and please note that the Rotation 'Method' and 'Number of Factors' extracted is typically adjusted by the researcher during the analysis to find the best result, as described below.

First, check the assumptions (Figure 15.3). You can see that (1) Bartlett's test of sphericity is significant, so this assumption is satisfied; and (2) the KMO measure of sampling adequacy (MSA) is 0.81 overall, suggesting good sampling adequacy. No problems here then!

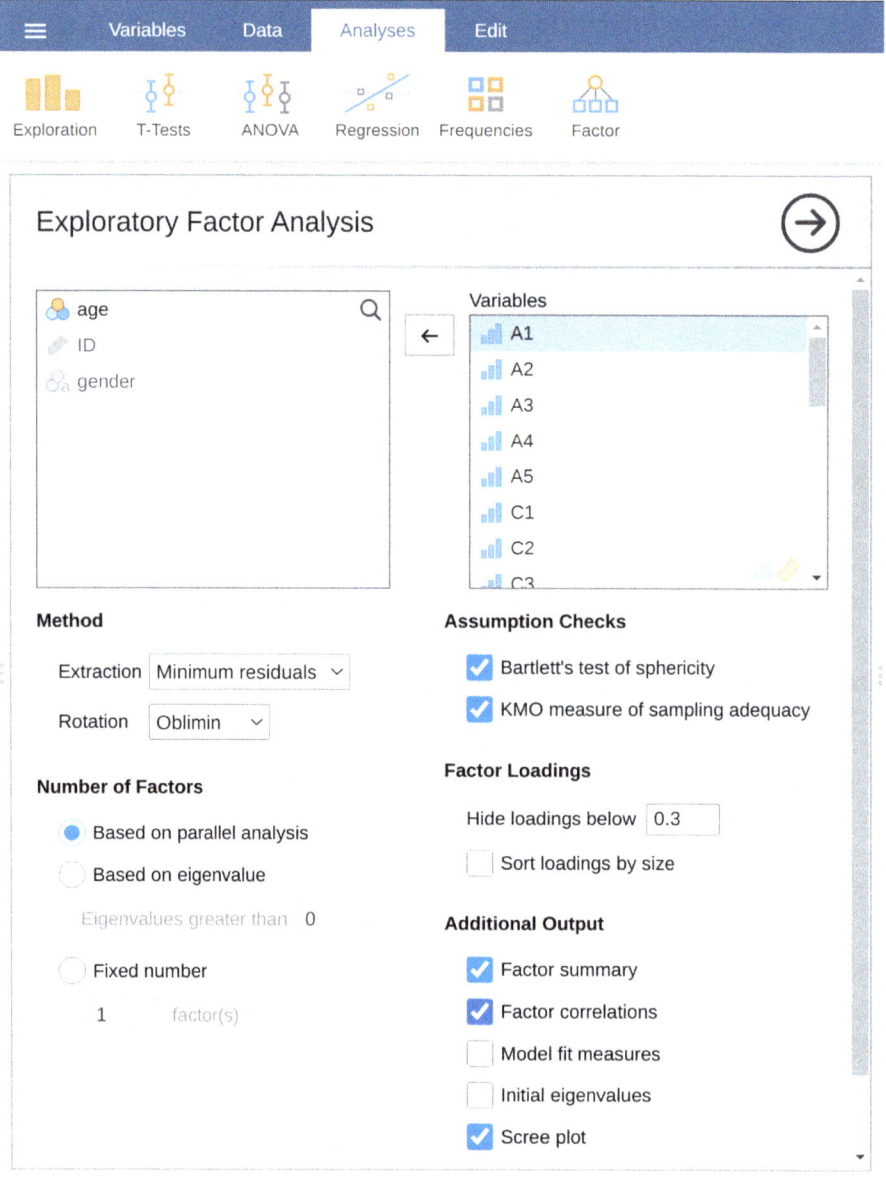

Figure 15.2: The jamovi EFA analysis window

Assumption Checks

Bartlett's Test of Sphericity

χ²	df	p
2204.28	300	< .00001

KMO Measure of Sampling Adequacy

	MSA
Overall	0.81
A1	0.59
A2	0.84
A3	0.82
A4	0.83
A5	0.86
C1	0.80
C2	0.81
C3	0.75
C4	0.79
C5	0.84
E1	0.83
E2	0.85
E3	0.83
E4	0.87
E5	0.91
N1	0.74
N2	0.71
N3	0.77
N4	0.83
N5	0.80
O1	0.84
O2	0.70
O3	0.82
O4	0.74
O5	0.76

Figure 15.3: jamovi EFA assumption checks for the personality questionnaire data

The next thing to check is how many factors to use (or "extract" from the data). Three different approaches are available:

- One convention is to choose all components with Eigen values greater than 1^{183}. This would give us four factors with our data (try it and see).

- Examination of the scree plot, as in Figure 15.4, lets you identify the "point of inflection". This is the point at which the slope of the scree curve clearly levels off, below the "elbow". This would give us five factors with our data. Interpreting scree plots is a bit of an art: in Figure 15.4 there is a noticeable step from 5 to 6 factors, but in other scree plots you look at it will not be so clear cut.

- Using a parallel analysis technique, the obtained Eigen values are compared to those that would be obtained from random data. The number of factors extracted is the number with Eigen values greater than what would be found with random data.

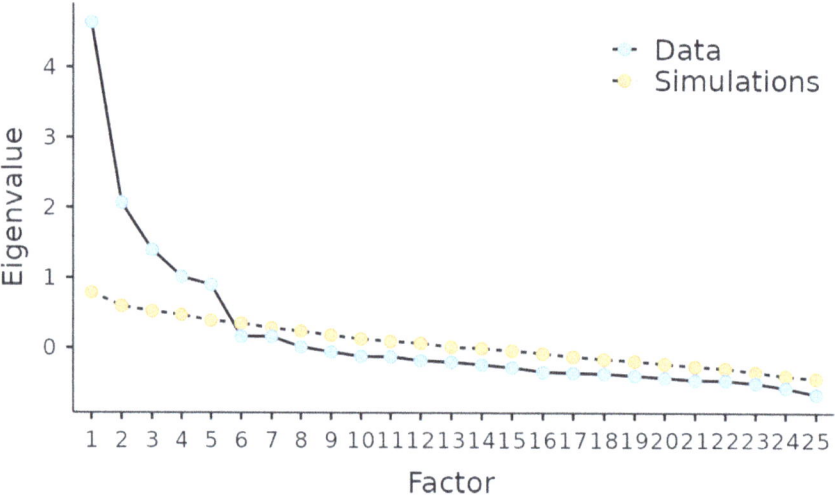

Figure 15.4: Scree plot of the personality data in jamovi EFA, showing a noticeable inflection and levelling off after point 5 (the "elbow")

The third approach is a good one according to Fabrigar et al. (1999), although in practice researchers tend to look at all three and then make a judgement about the number of factors that are most easily or helpfully interpreted. This can be understood as the "meaningfulness criterion", and researchers will typically examine, in addition to the solution from one of the approaches above, solutions with one or two more or fewer factors. They then adopt the solution which makes the most sense to them.

At the same time, we should also consider the best way to rotate the final solution. There are two main approaches to rotation: orthogonal (e.g., "varimax") rotation forces the selected factors to be uncorrelated, whereas oblique (e.g. "oblimin") rotation allows the selected factors to be correlated. Dimensions of interest to psychologists and behavioural scientists are not often dimensions we would expect to be orthogonal, so oblique solutions are arguably more sensible.[184]

Practically, if in an oblique rotation the factors are found to be substantially correlated (positive or negative, and > 0.3), as in Figure 15.5 where a correlation between two of the extracted factors is 0.31, then this would confirm our intuition to prefer oblique rotation. If the factors are, in fact, correlated, then an oblique rotation will produce a better estimate of the true factors and a better simple structure than will an orthogonal rotation. And, if the oblique rotation indicates that the factors have close to zero correlations between one another, then the researcher can go ahead and conduct an orthogonal rotation (which should then give about the same solution as the oblique rotation).

On checking the correlation between the extracted factors at least one correlation was greater than 0.3 (Figure 15.5), so an oblique ("oblimin") rotation of the five extracted factors is preferred. We can also see in Figure 15.5 that the proportion of overall variance in the data that is accounted for by the five factors is 46%. Factor 1 accounts for around 10% of the variance, factors 2 to 4 around 9% each, and factor 5 just over 7%. This isn't great; it would have been better if the overall solution accounted for a more substantive proportion of the variance in our data.

Factor Statistics

Summary

Factor	SS Loadings	% of Variance	Cumulative %
1	2.61	10.45	10.45
2	2.38	9.52	19.97
3	2.41	9.63	29.60
4	2.21	8.83	38.42
5	1.84	7.34	45.77

Inter-Factor Correlations

	1	2	3	4	5
1	—	−0.16	−0.16	0.02	−0.10
2		—	0.31	0.13	0.19
3			—	0.23	0.22
4				—	0.20
5					—

Figure 15.5: Factor summary statistics and correlations for a five factor solution in jamovi EFA

Exploratory Factor Analysis

Factor Loadings

	_____ Factor _____					
	1	2	3	4	5	Uniqueness
A1				−0.48		0.72
A2				0.71		0.48
A3				0.68		0.44
A4				0.43		0.67
A5				0.52		0.53
C1		0.72				0.48
C2		0.68				0.52
C3		0.51				0.75
C4		−0.64				0.47
C5		−0.58				0.52
E1			−0.53			0.70
E2			−0.70			0.35
E3			0.49		0.39	0.48
E4			0.59	0.36		0.43
E5			0.42			0.57
N1	0.79					0.35
N2	0.79					0.40
N3	0.69					0.47
N4	0.50		−0.45			0.45
N5	0.47					0.63
O1					0.57	0.63
O2					−0.49	0.68
O3					0.69	0.41
O4				0.33		0.75
O5					−0.52	0.68

Note. 'Minimum residual' extraction method was used in combination with a 'oblimin' rotation

Figure 15.6: Factor loadings for a five factor solution in jamovi EFA

Be aware that in every EFA you could potentially have the same number of factors as observed variables, but every additional factor you include will add a smaller amount of explained variance. If the first few factors explain a good amount of the variance in the original 25 variables, then those factors are clearly a useful, simpler substitute for the 25 variables. You can drop the rest without losing too much of the original variability. But if it takes 18 factors (for example) to explain most of the variance in those 25 variables, you might as well just use the original 25.

Figure 15.6 shows the factor loadings. That is, how the 25 different personality items load onto each of the five selected factors. We have hidden loadings less than 0.3 (set in the options shown in Figure 15.2).

For Factors 1, 2, 3 and 4 the pattern of factor loadings closely matches the putative factors specified in Table 15.1. Phew! And Factor 5 is pretty close, with four of the five observed variables that putatively measure "openness" loading pretty well onto the factor. Variable 04 doesn't quite seem to fit though, as the factor solution in Figure 15.6 suggests that it loads onto Factor 4 (albeit with a relatively low loading) but not substantively onto Factor 5.

The other thing to note is that those variables that were denoted as "R: reverse coding" in Table 15.1 are those that have negative factor loadings. Take a look at the items A1 ("Am indifferent to the feelings of others") and A2 ("Inquire about others' well-being"). We can see that a high score on 'A1' indicates low Agreeableness, whereas a high score on $A2$ (and all the other "A" variables for that matter) indicates high Agreeableness. Therefore A1 will be negatively correlated with the other "A" variables, and this is why it has a negative factor loading, as shown in Figure 15.6.

We can also see in Figure 15.6 the 'uniqueness' of each variable. Uniqueness is the proportion of variance that is 'unique' to the variable and not explained by the factors.[185] For example, 72% of the variance in 'A1' is not explained by the factors in the five factor solution. In contrast, 'N1' has relatively low variance not accounted for by the factor solution (35%). Note that the greater the 'uniqueness', the lower the relevance or contribution of the variable in the factor model.

To be honest, it's unusual to get such a neat solution in EFA. It's typically quite a bit more messy than this, and often interpreting the meaning of the factors is more challenging. It's not often that you have such a clearly delineated item pool. More often you will have a whole heap of observed variables that you think may be indicators of a few underlying latent factors, but you don't have such a strong sense of which variables are going to go where!

So, we seem to have a pretty good five factor solution, albeit accounting for a relatively low overall proportion of the observed variance. Let's assume we are happy with this solution and want to use our factors in further analysis. The straightforward option is to calculate an overall (average) score for each factor by adding together the score for each variable that loads substantively onto the factor and then dividing by the number of variables (in other words create a "mean score" for each person across the items for each scale). For each person in our data set that entails, for example for the Agreeableness factor, adding together $A1 + A2 + A3 + A4 + A5$, and then dividing by 5.[186] In essence, the factor score we have calculated is based on equally weighted scores from each of the included variables/items. We can do this in jamovi in two steps:

- Recode A1 into "A1R" by reverse scoring the values in the variable (i.e. $6 = 1$; $5 = 2$; $4 = 3$; $3 = 4$; $2 = 5$; $1 = 6$) using the jamovi transform variable command (see Figure 15.7).

- Compute a new variable, called "Agreeableness", by calculating the mean of A1R, A2, A3, A4 and A5. Do this using the jamovi compute new variable command (see Figure 15.8).

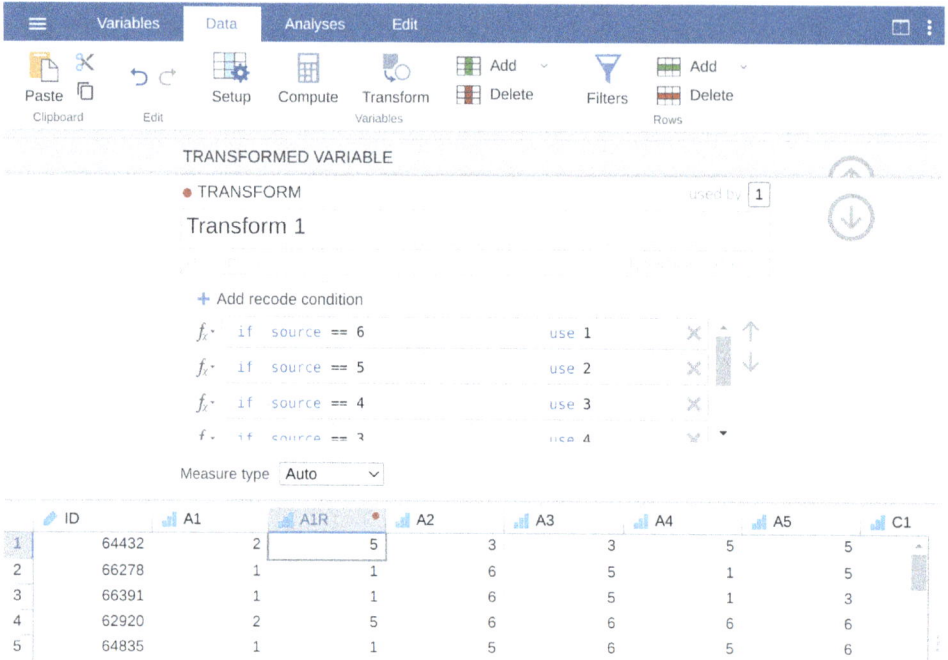

Figure 15.7: Recode variable using the jamovi Transform command

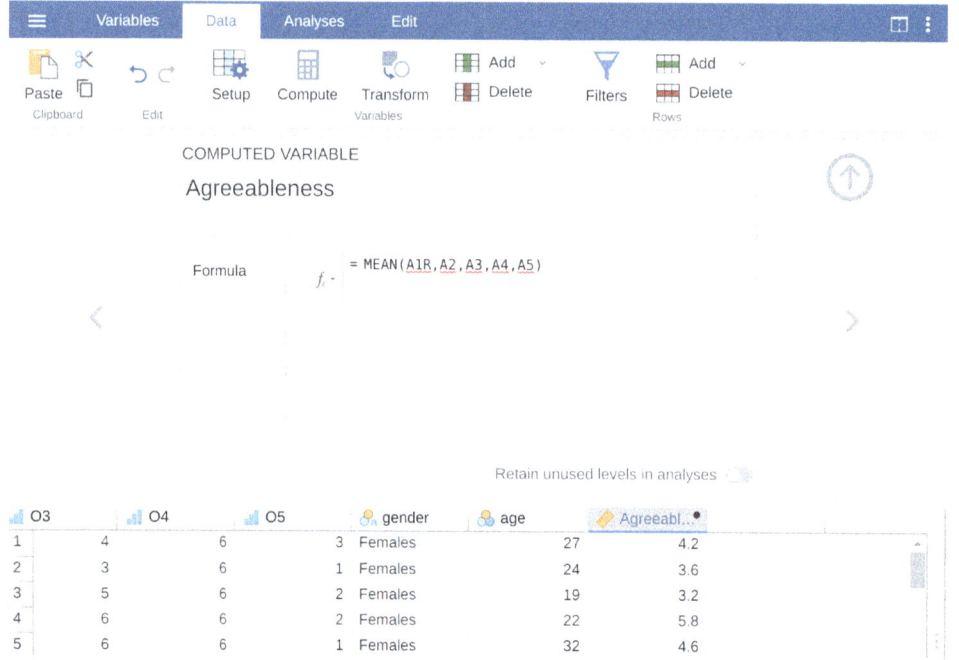

Figure 15.8: Compute new scale score variable using the jamovi Computed variable command

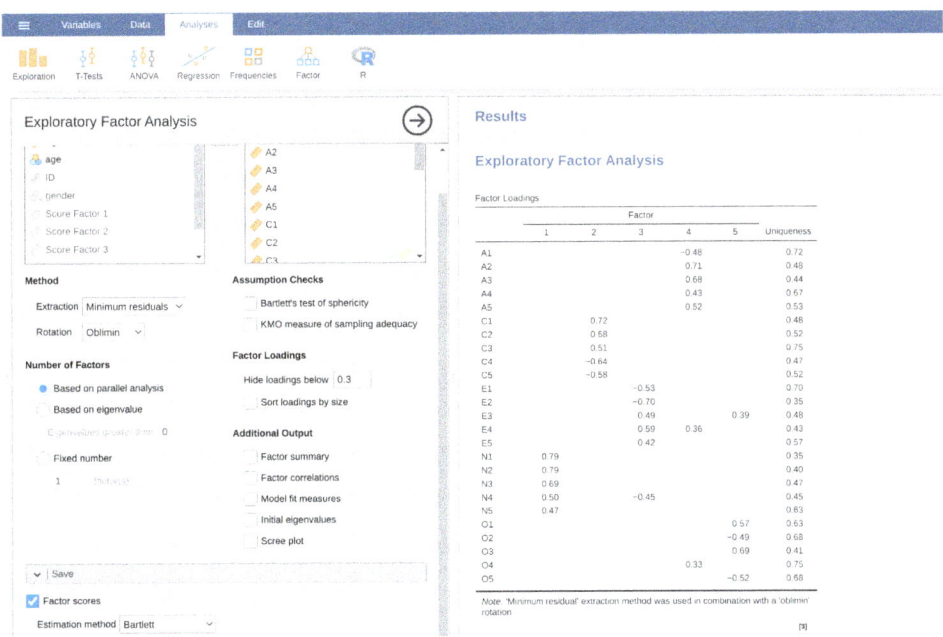

Figure 15.9: jamovi option for factor scores for the five factor solution, using the 'Bartlett' optimal weighting method

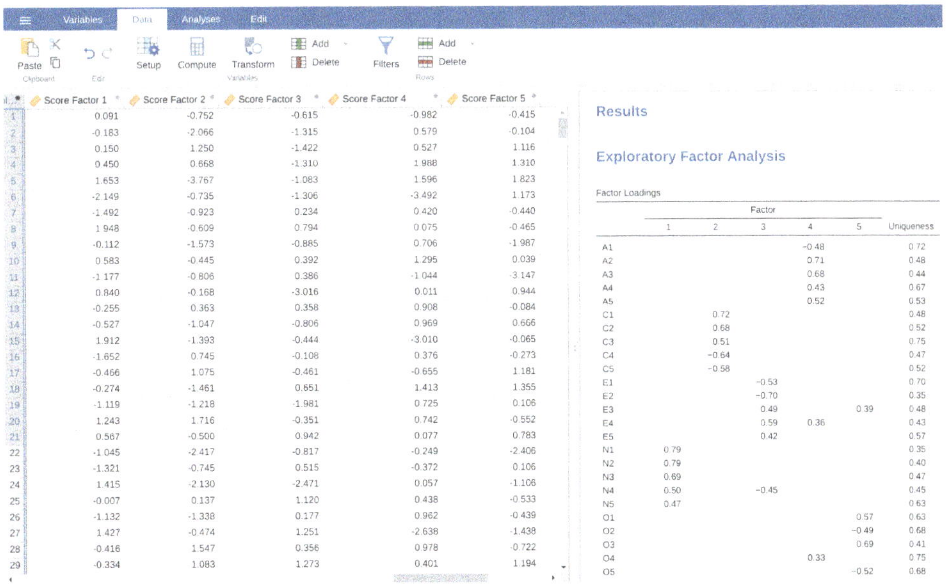

Figure 15.10: Data sheet view showing the five newly created factor score variables

Another option is to create an **optimally-weighted** factor score index. To do this, save the factor scores to the data set, using the 'Save' – 'Factor scores' checkbox. Once you have done this you will see that five new variables (columns) have been added to the data, one for each factor extracted. See Figure 15.9 and Figure 15.10.

389

Now you can go ahead and undertake further analyses, using either the mean score based factor scales (e.g., as in Figure 15.8) or using the optimally-weighted factor scores calculated by jamovi. Your choice! For example, one thing you might like to do is see whether there are any gender differences in each of our personality scales. We did this for the Agreeableness score that we calculated using the mean score approach, and although the t-test plot (Figure 15.11) showed that males were less agreeable than females, this was not a significant difference (Mann-Whitney $U = 5768, p = .075$).

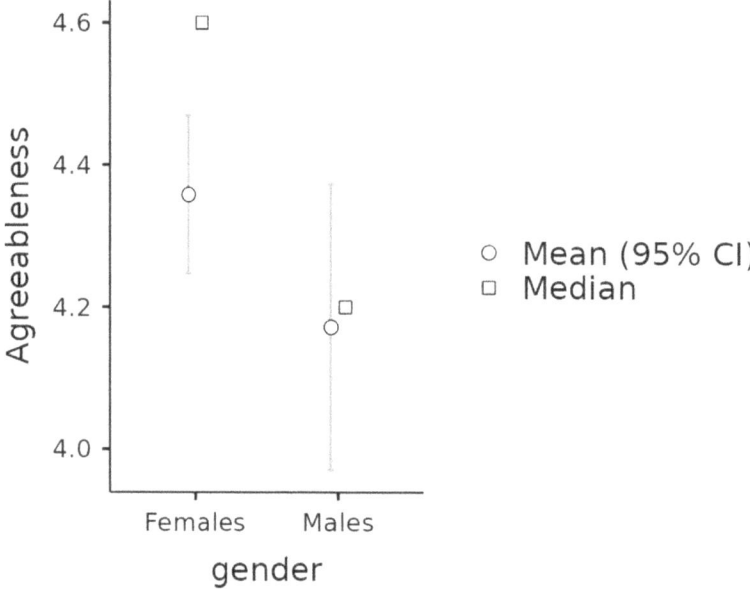

Figure 15.11: Comparing differences in Agreeableness factor-based scores between males and females

15.1.4 Writing up an EFA

Hopefully, so far we have given you some sense of EFA and how to undertake EFA in jamovi. So, once you have completed your EFA, how do you write it up? There is not a formal standard way to write up an EFA, and examples tend to vary by discipline and researcher. That said, there are some fairly standard pieces of information to include in your write-up:

1. What are the theoretical underpinnings for the area you are studying, and specifically for the constructs that you are interested in uncovering through EFA.

2. A description of the sample (e.g., demographic information, sample size, sampling method).

3. A description of the type of data used (e.g., nominal, continuous) and descriptive statistics.

4. Describe how you went about testing the assumptions for EFA. Details regarding sphericity checks and measures of sampling adequacy should be reported.

5. Explain what FA extraction method (e.g., 'Minimum residuals' or 'Maximum likelihood') was used.

6. Explain the criteria and process used for deciding how many factors were extracted in the final solution, and which items were selected. Clearly explain the rationale for key decisions during the EFA process.

7. Explain what rotation methods were attempted, the reasons why, and the results.

8. Final factor loadings should be reported in the results, in a table. This table should also report the uniqueness (or communality) for each variable (in the final column). Factor loadings should be reported with descriptive labels in addition to item numbers. Correlations between the factors should also be included, either at the bottom of this table, in a separate table.

9. Meaningful names for the extracted factors should be provided. You may like to use previously selected factor names, but on examining the actual items and factors you may think a different name is more appropriate.

15.2 Principal Component Analysis

In the previous section we saw that EFA works to identify underlying latent factors. And, as we saw, in one scenario the smaller number of latent factors can be used in further statistical analysis using some sort of combined factor scores.

In this way EFA is being used as a "data reduction" technique. Another type of data reduction technique, sometimes seen as part of the EFA family, is **Principal Component Analysis (PCA)**. However, PCA does not identify underlying latent factors. Instead it creates a linear composite score from a larger set of measured variables.

PCA simply produces a mathematical transformation to the original data with no assumptions about how the variables co-vary. The aim of PCA is to calculate a few linear combinations (components) of the original variables that can be used to summarize the observed data set without losing much information. However, if identification of underlying structure is a goal of the analysis, then EFA is to be preferred. And, as we saw, EFA produces factor scores that can be used for data reduction purposes just like principal component scores (Fabrigar et al., 1999).

PCA has been popular in Psychology for a number of reasons, and therefore it's worth mentioning, although nowadays EFA is just as easy to do given the power of desktop computers and can be less susceptible to bias than PCA, especially with a small number of factors and variables. Much of the procedure is similar to EFA, so although there are some conceptual differences, practically the steps are the same, and with large samples and a sufficient number of factors and variables, the results from PCA and EFA should be fairly similar.

To undertake PCA in jamovi, all you need to do is select 'Factor' – 'Principal Component Analysis' from the main jamovi button bar to open the PCA analysis window. Then you can follow the same steps from EFA in jamovi above.

15.3 Confirmatory Factor Analysis

So, our attempt to identify underlying latent factors using EFA with carefully selected questions from the personality item pool seemed to be pretty successful. The next step in our quest to develop a useful measure of personality is to check the latent factors we identified in the original EFA with a different sample. We want to see if the factors hold up, if we can confirm their existence with different data. This is a more rigorous check, as we will see. And it's called **Confirmatory Factor Analysis (CFA)** as we will, unsurprisingly, be seeking to confirm a pre-specified latent factor structure.[187]

In CFA, instead of doing an analysis where we see how the data goes together in an exploratory sense, we instead impose a structure, like in Figure 15.12, on the data and see how well the data fits our pre-specified structure. In this sense, we are undertaking a confirmatory analysis, to see how well a pre-specified **model** is confirmed by the observed data.

A straightforward Confirmatory Factor Analysis (CFA) of the personality items would therefore specify five latent factors as shown in Figure 15.12, each measured by five observed variables. Each variable is a measure of an underlying latent factor. For example, A1 is predicted by the underlying latent factor Agreeableness. And because A1 is not a perfect measure of the Agreeableness factor, there is an error term, e, associated with it. In other words, e represents the variance in A1 that is not accounted for by the Agreeableness factor. This is sometimes called **measurement error**.

The next step is to consider whether the latent factors should be allowed to correlate in our model. As mentioned earlier, in the psychological and behavioural sciences constructs are often related to each other, and we also think that some of our personality factors may be correlated with each other. So, in our model, we should allow these latent factors to co-vary, as shown by the double-headed arrows in Figure 15.12.

At the same time, we should consider whether there is any good, systematic reason for some of the error terms to be correlated with each other. One reason for this might be that there is a shared methodological feature for particular sub-sets of the observed variables such that the observed variables might be correlated for methodological rather than substantive latent factor reasons. We'll return to this possibility in a later section but, for now, there are no clear reasons that we can see that would justify correlating some of the error terms with each other.

Without any correlated error terms, the model we are testing to see how well it fits with our observed data is just as specified in Figure 15.12. Only parameters that are included in the model are expected to be found in the data, so in CFA all other possible parameters (coefficients) are set to zero. So, if these other parameters are not zero (for example there may be a substantial loading from A1 onto the latent factor Extraversion in the observed data, but not in our model) then we may find a poor fit between our model and the observed data.

Right, let's take a look at how we set this CFA analysis up in jamovi.

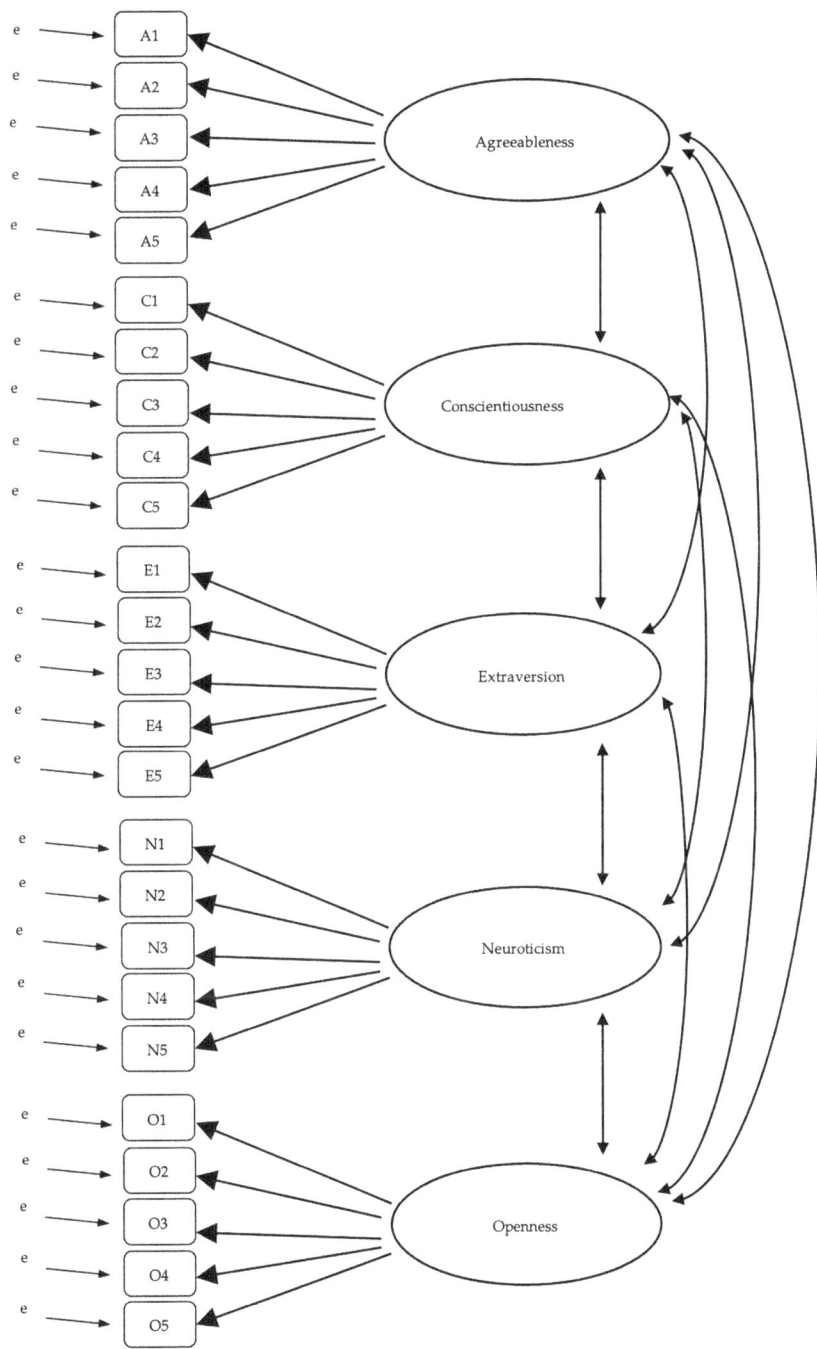

Figure 15.12: Initial pre-specification of latent factor structure for the five factor personality scales, for use in CFA

15.3.1 CFA in jamovi

Open up the *bfi_sample2.csv* file, check that the 25 variables are coded as ordinal (or continuous; it won't make any difference for this analysis). To perform CFA in jamovi:

- Select 'Factor - Confirmatory Factor Analysis' from the main jamovi button bar to open the CFA analysis window (Figure 15.13).

- Select the 5 A variables and transfer them into the 'Factors' box and give then the label "Agreeableness".

- Create a new Factor in the 'Factors' box and label it "Conscientiousness". Select the 5 C variables and transfer them into the 'Factors' box under the "Conscientiousness" label.

- Create another new Factor in the 'Factors' box and label it "Extraversion". Select the 5 E variables and transfer them into the 'Factors' box under the "Extraversion" label.

- Create another new Factor in the 'Factors' box and label it "Neuroticism". Select the 5 "N" variables and transfer them into the 'Factors' box under the "Neuroticism" label.

- Create another new Factor in the 'Factors' box and label it "Openness". Select the 5 O variables and transfer them into the 'Factors' box under the "Openness" label.

- Check other appropriate options, the defaults are ok for this initial work through, though you might want to check the "Path diagram" option under 'Plots' to see jamovi produce a (fairly) similar diagram to our Figure 15.12.

Once we have set up the analysis we can turn our attention to the jamovi results window and see what's what. The first thing to look at is **model fit** (Figure 15.14) as this tells us how good a fit our model is to the observed data. NB in our model only the pre-specified covariances are estimated, including the factor correlations by default. Everything else is set to zero.

There are several ways of assessing model fit. The first is a chi-square statistic that, if small, indicates that the model is a good fit to the data. However, the chi-squared statistic used for assessing model fit is pretty sensitive to sample size, meaning that with a large sample a good enough fit between the model and the data almost always produces a large and significant ($p < .05$) chi-square value.

So, we need some other ways of assessing model fit. In jamovi several are provided by default. These are the Comparative Fit Index (CFI), the Tucker Lewis Index (TLI) and the Root Mean Square Error of Approximation (RMSEA) together with the 90% confidence interval for the RMSEA. Some useful rules of thumb are that a satisfactory fit is indicated by CFI > 0.9, TLI > 0.9, and RMSEA of about 0.05 to 0.08. A good fit is CFI > 0.95, TLI > 0.95, and RMSEA and upper CI for RMSEA < 0.05.

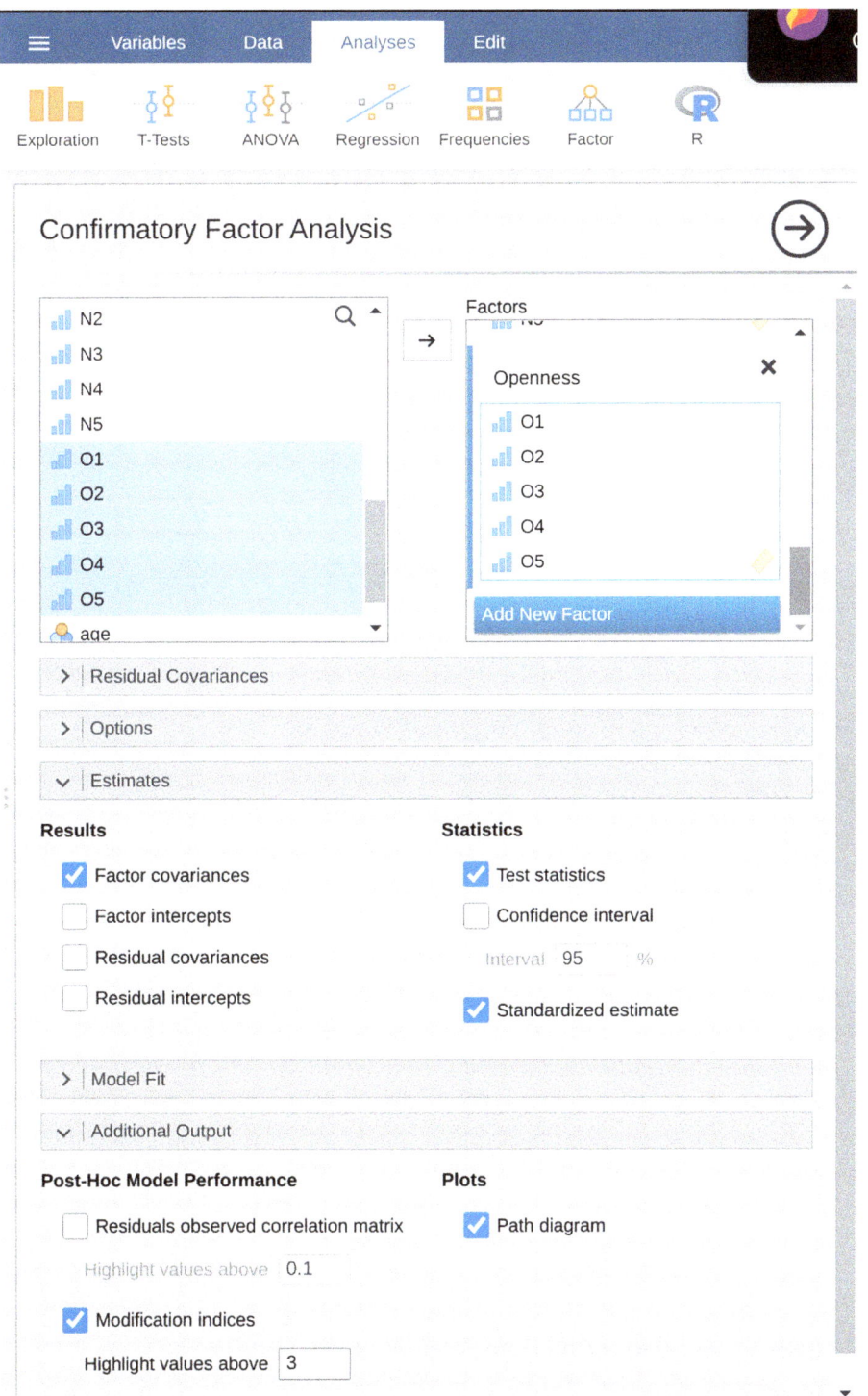

Figure 15.13: The jamovi CFA analysis window

Model Fit

Test for Exact Fit

χ²	df	p
739.73	265	<.00001

Fit Measures

CFI	TLI	RMSEA	RMSEA 90% CI Lower	RMSEA 90% CI Upper
0.76	0.73	0.08	0.08	0.09

Figure 15.14: The jamovi CFA Model Fit results for our CFA model

So, looking at Figure 15.14 we can see that the chi-square value is large and highly significant. Our sample size is not too large, so this possibly indicates a poor fit. The CFI is 0.762 and the TLI is 0.731, indicating poor fit between the model and the data. The RMSEA is 0.085 with a 90% confidence interval from 0.077 to 0.092, again this does not indicate a good fit.

Pretty disappointing, huh? But perhaps not too surprising given that in the earlier EFA, when we ran with a similar data set (see Exploratory Factor Analysis section), only around half of the variance in the data was accounted for by the five factor model.

Let's go on to look at the factor loadings and the factor covariance estimates, shown in Figure 15.15 and Figure 15.16. The Z-statistic and p-value for each of these parameters indicates they make a reasonable contribution to the model (i.e. they are not zero) so there doesn't appear to be any reason to remove any of the specified variable-factor paths, or factor-factor correlations from the model. Often the standardized estimates are easier to interpret, and these can be specified under the 'Estimates' option. These tables can usefully be incorporated into a written report or scientific article.

How could we improve the model? One option is to go back a few stages and think again about the items / measures we are using and how they might be improved or changed. Another option is to make some post hoc tweaks to the model to improve the fit. One way of doing this is to use "modification indices" (Figure 15.17), specified as an 'Additional output' option in jamovi.

What we are looking for is the highest modification index (MI) value. We would then judge whether it makes sense to add that additional term into the model, using a post hoc rationalisation. For example, we can see in Figure 15.17 that the largest MI for the factor loadings that are not already in the model is a value of 28.786 for the loading of N4 ("Often feel blue") onto the latent factor Extraversion. This indicates that if we add this path into the model then the chi-square value will reduce by around the same amount.

Confirmatory Factor Analysis

Factor Loadings

Factor	Indicator	Estimate	SE	Z	p	Stand. Estimate
Agreeableness	A1	0.47	0.10	4.73	<.00001	0.33
	A2	−0.77	0.08	−9.76	<.00001	−0.62
	A3	−1.06	0.08	−13.00	<.00001	−0.79
	A4	−0.75	0.10	−7.83	<.00001	−0.52
	A5	−0.94	0.08	−11.44	<.00001	−0.71
Conscientiousness	C1	0.80	0.08	10.53	<.00001	0.67
	C2	0.75	0.09	8.56	<.00001	0.57
	C3	0.54	0.09	6.19	<.00001	0.42
	C4	−1.07	0.09	−12.41	<.00001	−0.76
	C5	−1.12	0.10	−10.77	<.00001	−0.68
Extraversion	E1	0.90	0.11	8.51	<.00001	0.55
	E2	1.26	0.10	12.67	<.00001	0.76
	E3	−0.91	0.09	−9.89	<.00001	−0.63
	E4	−1.04	0.09	−11.28	<.00001	−0.69
	E5	−0.78	0.08	−9.37	<.00001	−0.60
Neuroticism	N1	1.34	0.09	14.17	<.00001	0.82
	N2	1.21	0.09	13.07	<.00001	0.76
	N3	1.22	0.10	12.49	<.00001	0.75
	N4	0.96	0.11	8.72	<.00001	0.57
	N5	0.84	0.11	7.63	<.00001	0.49
Openness	O1	0.67	0.08	8.12	<.00001	0.58
	O2	−0.65	0.12	−5.66	<.00001	−0.40
	O3	1.00	0.09	11.30	<.00001	0.83
	O4	0.23	0.09	2.73	0.00624	0.20
	O5	−0.51	0.09	−5.52	<.00001	−0.41

Figure 15.15: The jamovi CFA Factor Loadings table for our CFA model

But in our model adding this path arguably doesn't really make any theoretical or methodological sense, so it's not a good idea (unless you can come up with a persuasive argument that "Often feel blue" measures both Neuroticism and Extraversion). I can't think of a good reason. But, for the sake of argument, let's pretend it does make some sense and add this path into the model. Go back to the CFA analysis window (see Figure 15.13) and add N4 into the Extraversion factor. The results of the CFA will now change (not shown); the chi-square has come down to around 709 (a drop of around 30, roughly similar to the size of the MI) and the other fit indices have also improved, though only a bit. But it's not enough: it's still not a good fitting model.

If you do find yourself adding new parameters to a model using the MI values then always re-check the MI tables after each new addition, as the MIs are refreshed each time.

Factor Covariances

		Estimate	SE	Z	p	Stand. Estimate
Agreeableness	Agreeableness	1.00 [a]				
	Conscientiousness	−0.34	0.07	−4.51	< .00001	−0.34
	Extraversion	0.58	0.06	9.21	< .00001	0.58
	Neuroticism	0.16	0.08	2.14	0.03233	0.16
	Openness	−0.42	0.07	−5.95	< .00001	−0.42
Conscientiousness	Conscientiousness	1.00 [a]				
	Extraversion	−0.50	0.07	−7.76	< .00001	−0.50
	Neuroticism	−0.29	0.07	−3.92	0.00009	−0.29
	Openness	0.28	0.08	3.44	0.00058	0.28
Extraversion	Extraversion	1.00 [a]				
	Neuroticism	0.24	0.08	3.12	0.00179	0.24
	Openness	−0.53	0.07	−7.95	< .00001	−0.53
Neuroticism	Neuroticism	1.00 [a]				
	Openness	−0.18	0.08	−2.32	0.02033	−0.18
Openness	Openness	1.00 [a]				

[a] fixed parameter

Figure 15.16: The jamovi CFA Factor Covariances table for our CFA model

Factor Loadings – Modification Indices

	Agreeableness	Conscientiousness	Extraversion	Neuroticism	Openness
A1		11.14	14.21	1.53	0.39
A2		1.20	1.02	3.65	0.53
A3		1.76	4.70	2.95	0.00
A4		6.99	4.15	0.06	2.30
A5		3.84	11.41	8.18	3.50
C1	4.22		1.96	0.55	0.15
C2	1.40		0.01	12.08	0.01
C3	0.85		1.54	12.77	0.86
C4	0.96		1.05	2.71	0.21
C5	0.23		1.03	4.75	0.21
E1	13.55	0.50		1.36	3.14
E2	5.36	1.67		19.01	2.33
E3	4.40	5.85		6.20	28.02
E4	22.83	1.51		0.30	9.37
E5	2.40	8.78		2.51	2.89
N1	1.15	0.59	1.08		0.15
N2	1.03	9.83	7.58		3.18
N3	0.21	0.01	0.01		0.01
N4	1.65	14.03	28.79		0.65
N5	1.13	0.03	0.84		2.90
O1	0.07	0.19	1.89	0.28	
O2	6.43	2.43	8.06	6.85	
O3	2.71	2.09	7.70	1.04	
O4	1.85	13.39	10.54	8.87	
O5	1.58	4.49	2.97	2.19	

Figure 15.17: The jamovi CFA Factor Loadings Modification Indices

There is also a Table of Residual Covariance Modification Indices produced by jamovi (Figure 15.18). In other words, a table showing which correlated errors, if added to the model, would improve the model fit the most. It's a good idea to look across both MI tables at the same time, spot the largest MI, think about whether the addition of the suggested parameter can be reasonably justified and, if it can, add it to the model. And then you can start again looking for the biggest MI in the re-calculated results.

Residual Covariances – Modification Indices

	A1	A2	A3	A4	A5	C1	C2	C3	C4	C5	E1	E2	E3
A1		7.93	4.45	0.55	3.05	1.33	1.99	1.01	3.24	5.87	0.04	0.10	7.92
A2			0.08	0.96	1.84	1.15	3.10	3.42	3.74	2.67	0.01	3.32	0.84
A3				0.01	0.12	0.41	0.64	0.09	3.26	1.08	5.67	7.55	4.53
A4					2.19	0.93	6.77	0.40	2.34	12.73	0.35	0.32	0.10
A5						1.36	0.00	0.47	0.16	1.78	0.25	3.90	0.43
C1							13.88	0.19	0.34	10.69	0.09	2.14	3.04
C2								6.14	9.74	0.60	2.04	3.18	0.14
C3									2.06	0.85	0.01	0.01	0.32
C4										3.62	0.12	3.05	0.25
C5											0.00	1.41	0.45
E1												10.71	0.97
E2													2.10
E3													
E4													
E5													
N1													
N2													
N3													
N4													
N5													
O1													
O2													
O3													
O4													
O5													

	E4	E5	N1	N2	N3	N4	N5	O1	O2	O3	O4	O5
A1	0.06	6.91	1.28	11.77	1.24	1.63	2.13	6.01	0.85	0.80	8.11	1.05
A2	2.51	1.66	1.17	0.95	0.20	7.18	8.30	0.19	6.27	0.21	12.01	2.16
A3	3.07	0.82	0.27	0.58	1.07	0.65	0.98	0.03	0.50	0.42	1.25	0.14
A4	2.19	0.28	3.57	0.09	1.58	2.04	1.18	0.09	2.74	0.28	2.70	1.92
A5	1.66	2.28	1.93	0.15	0.39	0.19	0.03	0.04	2.98	0.25	0.06	0.17
C1	0.30	2.29	0.99	0.23	4.31	3.74	1.63	0.17	0.43	0.02	0.06	8.83
C2	0.56	5.96	0.15	1.89	0.04	2.60	6.32	0.25	9.81	0.69	4.28	0.09
C3	2.51	2.38	0.92	0.32	0.14	0.18	6.52	0.00	0.54	0.32	4.08	3.64
C4	1.07	0.07	7.97	1.73	6.53	1.52	4.20	0.01	11.85	5.32	5.08	6.37
C5	0.38	0.00	6.46	0.81	1.08	10.44	2.17	0.00	1.30	0.34	6.64	1.42
E1	0.04	0.36	2.78	1.63	1.40	4.29	0.06	2.56	0.50	0.03	0.93	0.04
E2	1.19	2.35	0.03	2.22	2.67	15.40	7.01	0.13	0.42	0.02	11.44	1.41
E3	0.93	1.23	1.50	0.05	1.21	1.55	2.72	3.34	2.67	16.90	2.86	0.60
E4		1.26	0.00	0.87	1.44	2.44	6.86	0.12	10.04	1.12	0.04	3.26
E5			1.36	6.54	1.54	1.10	0.00	5.97	0.90	0.42	0.33	0.23
N1				45.00	4.12	25.70	0.25	1.42	0.09	0.21	0.75	0.11
N2					4.37	3.48	6.25	3.23	2.53	0.82	0.04	4.68
N3						19.26	0.91	1.19	0.06	0.08	0.02	6.97
N4							8.57	0.01	0.25	0.74	10.89	0.65
N5								2.31	3.26	0.13	1.92	2.61
O1									0.04	4.69	0.33	0.95
O2										0.06	3.04	12.95
O3											2.93	3.24
O4												0.17
O5												

Figure 15.18: Residual Covariance Modification Indices produced by jamovi

You can keep going this way for as long as you like, adding parameters to the model based on the largest MI, and eventually you will achieve a satisfactory fit. But there will also be a strong possibility that in doing this you will have created a monster! A model that is ugly and deformed and doesn't have any theoretical sense or purity. In other words, be very careful!

So far, we have checked out the factor structure obtained in the EFA using a second sample and CFA. Unfortunately, we didn't find that the factor structure from the EFA was confirmed in the CFA, so it's back to the drawing board as far as the development of this personality scale goes.

Although we could have tweaked the CFA using modification indexes, there really were not any good reasons (that I could think of) for these suggested additional factor loadings or residual covariances to be included. However, sometimes there is a good reason for residuals to be allowed to co-vary (or correlate), and a good example of this is shown in the next section on Multi-Trait Multi-Method CFA. Before we do that, let's cover how to report the results of a CFA.

15.3.2 Reporting a CFA

There is not a formal standard way to write up a CFA, and examples tend to vary by discipline and researcher. That said, there are some fairly standard pieces of information to include in your write-up:

1. A theoretical and empirical justification for the hypothesized model.

2. A complete description of how the model was specified (e.g., the indicator variables for each latent factor, covariances between latent variables, and any correlations between error terms). A path diagram, like the one in Figure 15.12 would be good to include.

3. A description of the sample (e.g., demographic information, sample size, sampling method).

4. A description of the type of data used (e.g.,, nominal, continuous) and descriptive statistics.

5. Tests of assumptions and estimation method used.

6. A description of missing data and how the missing data were handled.

7. The software and version used to fit the model.

8. Measures, and the criteria used, to judge model fit.

9. Any alterations made to the original model based on model fit or modification indices.

10. All parameter estimates (i.e., loadings, error variances, latent (co)variances) and their standard errors, probably in a table.

15.4 Multi-Trait Multi-Method CFA

In this section we're going to consider how different measurement techniques or questions can be an important source of data variability, known as **method variance**. To do this, we'll use another psychological data set, one that contains data on "attributional style".

The Attributional Style Questionnaire (ASQ) was used (Hewitt et al., 2004) to collect psychological wellbeing data from young people in the United Kingdom and New Zealand. They measured attributional style for negative events, which is how people habitually explain the cause of bad things that happen to them (Peterson & Seligman, 1984). The ASQ measures three aspects of attributional style:

- Internality is the extent to which a person believes that the cause of a bad event is due to his/her own actions.

- Stability refers to the extent to which a person habitually believes the cause of a bad event is stable across time.

- Globality refers to the extent to which a person habitually believes that the cause of a bad event in one area will affect other areas of their lives.

There are six hypothetical scenarios and for each scenario respondents answer a question aimed at (a) internality, (b) stability and (c) globality. So there are $6 \times 3 = 18$ items overall. See Figure 15.19 for more details.

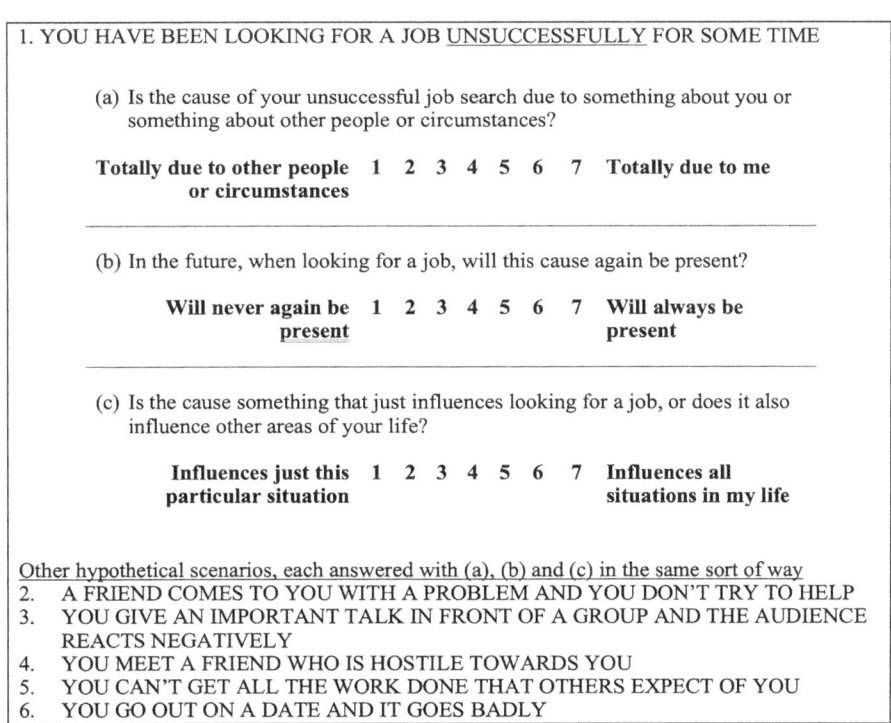

Figure 15.19: The Attributional Style Questionnaire (ASQ) for negative events

Researchers are interested in checking their data to see whether there are some underlying latent factors that are measured reasonably well by the 18 observed variables in the ASQ.

First, they try EFA with these 18 variables (not shown), but no matter how they extract or rotate, they can't find a good factor solution. Their attempt to identify underlying latent factors in the Attributional Style Questionnaire (ASQ) proved fruitless. If you get results like this then either your theory is wrong (there is no underlying latent factor structure for attributional style, which is possible), the sample is not relevant (which is unlikely given the size and characteristics of this sample of young adults from the United Kingdom and New Zealand), or the analysis was not the right tool for the job. We're going to look at this third possibility.

Remember that there were three dimensions measured in the ASQ: Internality, Stability and Globality, each measured by six questions as shown in Table 15.2.

What if, instead of doing an analysis where we see how the data goes together in an exploratory sense, we instead impose a structure, like in Table 15.2, on the data and see how well the data fits our pre-specified structure. In this sense, we are undertaking a confirmatory analysis, to see how well a pre-specified model is confirmed by the observed data.

A straightforward Confirmatory Factor Analysis (CFA) of the ASQ would therefore specify three latent factors as shown in the columns of Figure 15.24, each measured by six observed variables.

Table 15.2: Six questions on the ASQ for each of the Internality, Stability and Globality dimensions

Internality	Stability	Globality
Q1a	Q1b	Q1c
Q2a	Q2b	Q2c
Q3a	Q3b	Q3c
Q4a	Q4b	Q4c
Q5a	Q5b	Q5c
Q6a	Q6b	Q6c

We could depict this as in the diagram in Figure 15.20, which shows that each variable is a measure of an underlying latent factor. For example INT1 is predicted by the underlying latent factor Internality. And because INT1 is not a perfect measure of the Internality factor, there is an error term, e_1, associated with it. In other words, e_1 represents the variance in INT1 that is not accounted for by the Internality factor. This is sometimes called "measurement error".

The next step is to consider whether the latent factors should be allowed to correlate in our model. As mentioned earlier, in the psychological and behavioural sciences constructs are often related to each other, and we also think that Internality, Stability, and Globality might be correlated with each other, so in our model we should allow these latent factors to co-vary, as shown in Figure 15.21.

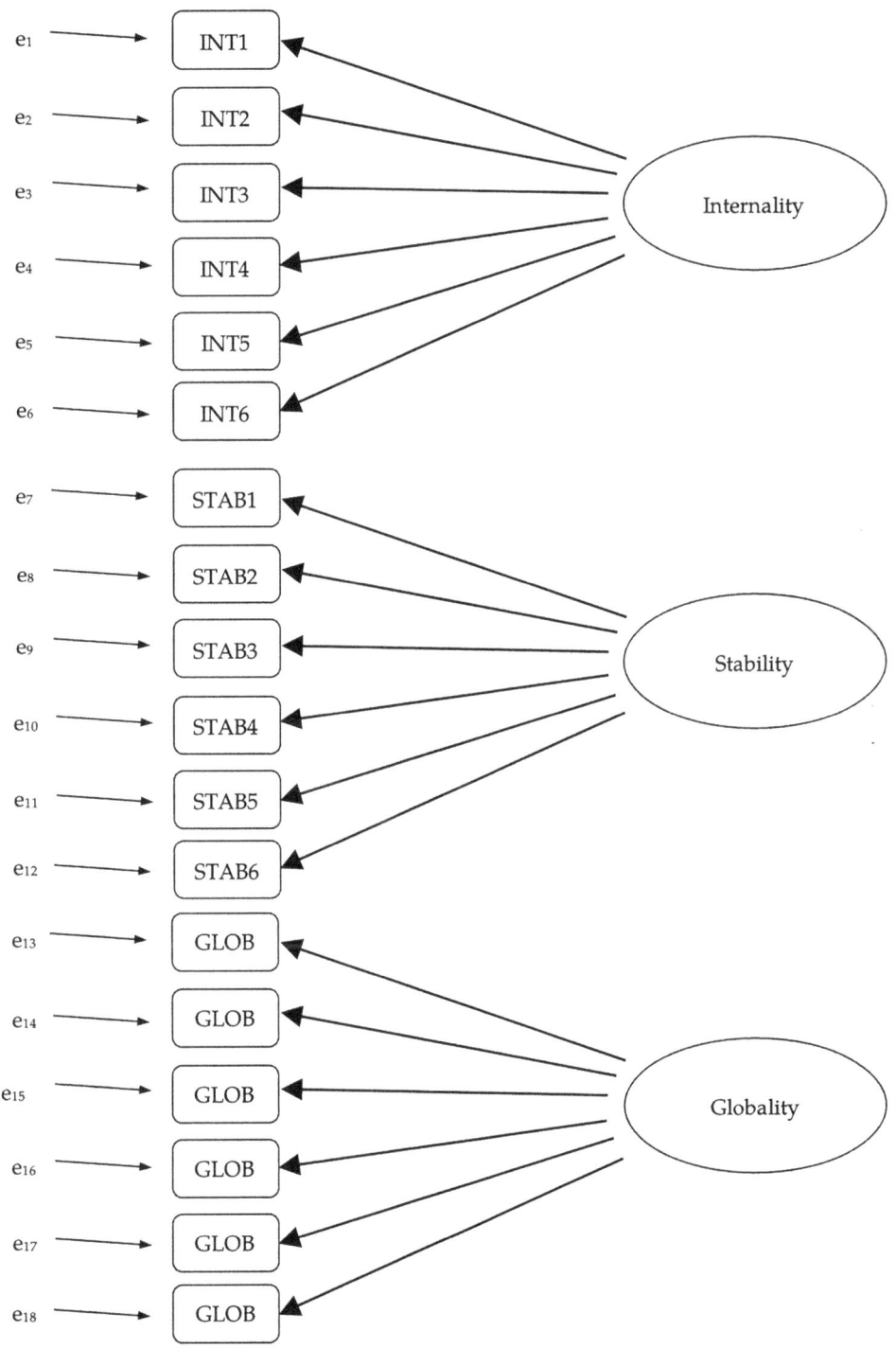

Figure 15.20: Initial pre-specification of latent factor structure for the ASQ

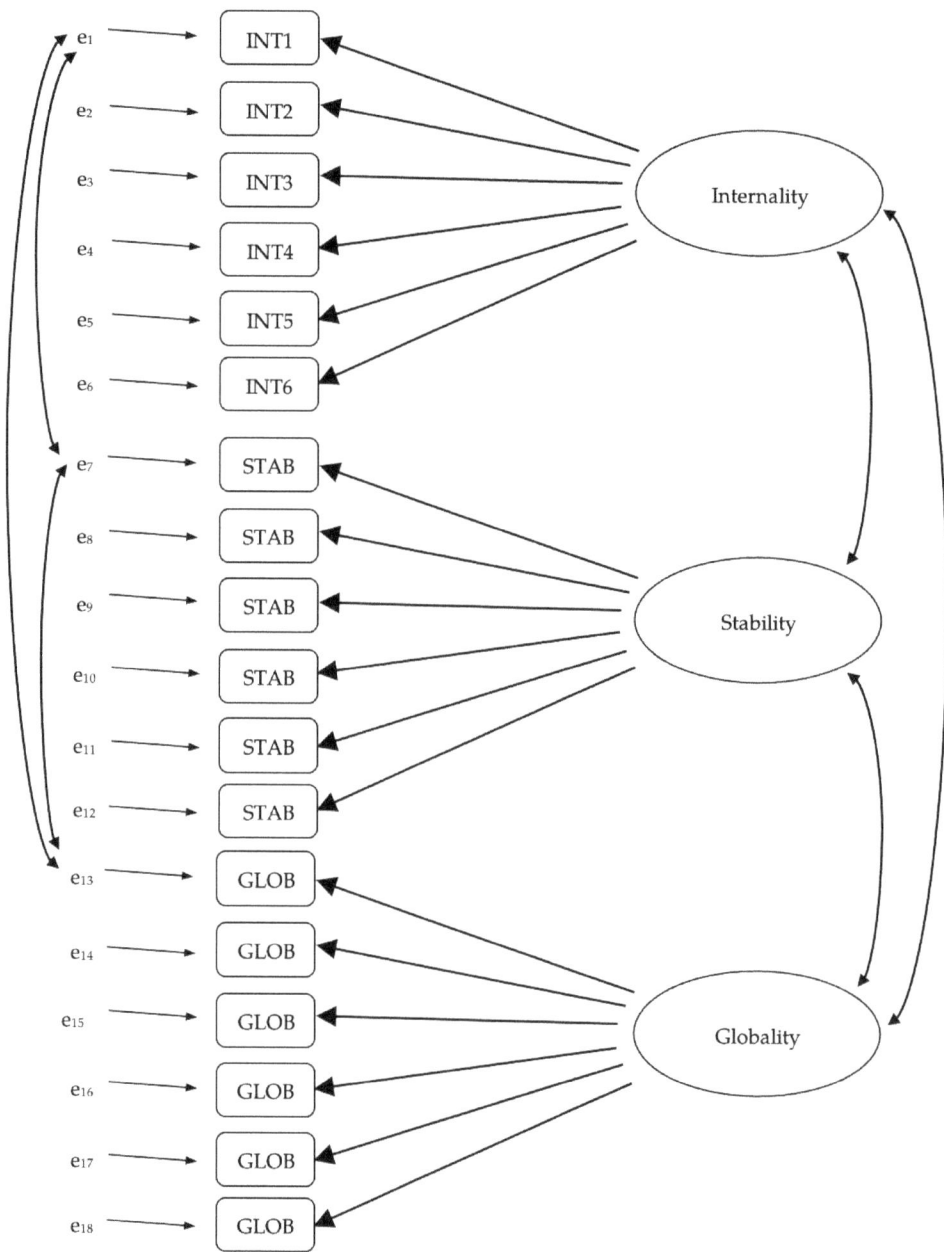

Figure 15.21: Final pre-specification of latent factor structure for the ASQ, including latent factor correlations, and shared method error term correlations for the observed variable INT1, STAB1 and GLOB1, in a CFA MTMM model. For clarity, other pre-specified error term correlations are not shown

At the same time, we should consider whether there is any good, systematic, reason for some of the error terms to be correlated with each other. Thinking back to the ASQ questions, there were three different sub-questions (a, b and c) for each main question (1-6). Q1 was about unsuccessful job hunting and it is plausible that this question has some distinctive artefactual or methodological aspects over and above the other questions (2-5), something to do with job hunting perhaps. Similarly, Q2 was about not helping a friend with a problem, and there may be some distinctive artefactual or methodological aspects to do with not helping a friend that is not present in the other questions (1, and 3-5).

So, as well as multiple factors, we also have multiple methodological features in the ASQ, where each of Questions 1-6 has a slightly different "method", but each "method" is shared across the sub-questions a, b and c. In order to incorporate these different methodological features into the model we can specify that certain error terms are correlated with each other. For example, the errors associated with INT1, STAB1 and GLOB1 should be correlated with each other to reflect the distinct and shared methodological variance of Q1a, Q1b and Q1c. Looking at Table 1.2, this means that as well as the latent factors represented by the columns, we will have correlated measurement errors for the variables in each row of the Table.

Whilst a basic CFA model like the one shown in Figure 15.20 could be tested against our observed data, we have in fact come up with a more sophisticated model, as shown in the diagram in Figure 15.21. This more sophisticated CFA model is known as a **Multi-Trait Multi-Method (MTMM)** model, and it is the one we will test in jamovi.

15.4.1 MTMM CFA in jamovi

Open up the *ASQ.csv* file and check that the 18 variables (six "Internality", six "Stability" and six "Globality" variables) are specified as continuous variables.

To perform MTMM CFA in jamovi:

- Select 'Factor' – 'Confirmatory Factor Analysis' from the main jamovi button bar to open the CFA analysis window (Figure 15.22).

- Select the 6 INT variables and transfer them into the 'Factors' box and give them the label "Internality".

- Create a new Factor in the 'Factors' box and label it "Stability". Select the 6 STAB variables and transfer them into the 'Factors' box under the "Stability" label.

- Create another new Factor in the 'Factors' box and label it "Globality". Select the 6 GLOB variables and transfer them into the 'Factors' box under the "Globality" label.

- Open up the Residual Covariances options, and for each of our pre-specified correlations move the associated variables across into the 'Residual Covariances' box on the right. For example, highlight both INT1 and STAB1 and then click the arrow to move these across. Now do the same for INT1 and GLOB1, for STAB1 and GLOB1, for INT2 and STAB2, for INT2 and GLOB2, for STAB2 and GLOB2, for INT3 and STAB3, and so on.

- Check other appropriate options. The defaults are ok for this initial work through, though you might want to check the "Path diagram" option under 'Plots' to see jamovi produce a (fairly) similar diagram to our Figure 15.21, and including all the error term correlations that we have added above.

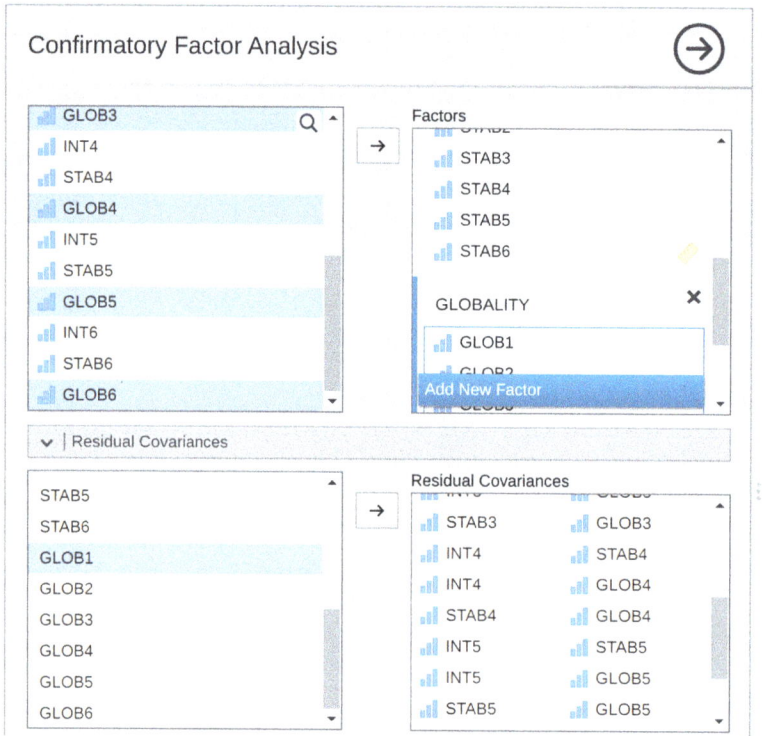

Figure 15.22: The jamovi CFA analysis window

Model Fit

Test for Exact Fit

χ^2	df	p
243.97	114	< .00001

Fit Measures

			RMSEA 90% CI	
CFI	TLI	RMSEA	Lower	Upper
0.98	0.98	0.02	0.02	0.02

Figure 15.23: The jamovi CFA Model Fit results for our CFA MTMM model

Once we have set up the analysis we can turn our attention to the jamovi results window and see what's what. The first thing to look at is "Model fit" as this tells us how good a fit our model is to the observed data (Figure 15.23). NB in our model only the pre-specified covariances are estimated, everything else is set to zero, so model fit is testing both whether the pre-specified "free" parameters are not zero, and conversely whether the other relationships in the data – the ones we have not specified in the model – can be held at zero.

Looking at Figure 15.23 we can see that the chi-square value is highly significant, which is not a surprise given the large sample size (N = 2748). The CFI is 0.98 and the TLI is also 0.98, indicating a very good fit. The RMSEA is 0.02 with a 90% confidence interval from 0.02 to 0.02 – pretty tight!

Overall, I think we can be satisfied that our pre-specified model is a very good fit to the observed data, lending support to our MTMM model for the ASQ.

We can now go on to look at the factor loadings and the factor covariance estimates, as in Figure 15.24. Often the standardized estimates are easier to interpret, and these can be specified under the 'Estimates' option. These tables can usefully be incorporated into a written report or scientific article.

You can see from Figure 15.24 that all of our pre-specified factor loadings and factor covariances are significantly different from zero. In other words, they all seem to be making a useful contribution to the model.

We've been pretty lucky with this analysis, getting a very good fit on our first attempt!

15.5 Internal consistency reliability analysis

After you have been through the process of initial scale development using EFA and CFA, you should have reached a stage where the scale holds up pretty well using CFA with different samples. One thing that you might also be interested in at this stage is to see how well the factors are measured using a scale that combines the observed variables.

In psychometrics we use reliability analysis to provide information about how consistently a scale measures a psychological construct (See earlier section on Section 2.3). **Internal consistency** is what we are concerned with here, and that refers to the consistency across all the individual items that make up a measurement scale. So, if we have $V1, V2, V3, V4$ and $V5$ as observed item variables, then we can calculate a statistic that tells us how internally consistent these items are in measuring the underlying construct.

A popular statistic used to check the internal consistency of a scale is **Cronbach's alpha** (Chronbach, 1951). Cronbach's alpha is a measure of equivalence (whether different sets of scale items would give the same measurement outcomes). Equivalence is tested by dividing the scale items into two groups (a "split-half") and seeing whether analysis of the two parts gives comparable results. Of course, there are many ways a set of items could be split, but if all possible splits are made then it is possible to produce a statistic that reflects the overall pattern of split-half coefficients. Cronbach's alpha (α) is such a statistic: a function of all the split-half coefficients for a scale. If a set of items that

Confirmatory Factor Analysis

Factor Loadings

Factor	Indicator	Estimate	SE	Z	p	Stand. Estimate
INTERNALITY	INT1	0.55	0.05	12.28	< .00001	0.34
	INT2	0.50	0.05	10.52	< .00001	0.28
	INT3	0.61	0.04	13.95	< .00001	0.38
	INT4	0.64	0.05	13.45	< .00001	0.36
	INT5	0.54	0.04	12.52	< .00001	0.33
	INT6	0.66	0.04	16.50	< .00001	0.45
STABILITY	STAB1	0.53	0.04	14.97	< .00001	0.35
	STAB2	0.48	0.03	14.54	< .00001	0.34
	STAB3	0.69	0.03	20.20	< .00001	0.46
	STAB4	0.65	0.03	20.72	< .00001	0.47
	STAB5	0.67	0.03	21.47	< .00001	0.49
	STAB6	0.66	0.03	22.17	< .00001	0.51
GLOBALITY	GLOB1	0.71	0.04	18.16	< .00001	0.40
	GLOB2	0.73	0.04	20.32	< .00001	0.44
	GLOB3	0.93	0.04	25.10	< .00001	0.54
	GLOB4	0.83	0.03	24.99	< .00001	0.53
	GLOB5	0.76	0.03	22.71	< .00001	0.48
	GLOB6	0.96	0.03	27.66	< .00001	0.59

[3]

Factor Estimates

Factor Covariances

		Estimate	SE	Z	p	Stand. Estimate
INTERNALITY	INTERNALITY	1.00 [a]				
	STABILITY	0.52	0.03	17.10	< .00001	0.52
	GLOBALITY	0.45	0.03	14.96	< .00001	0.45
STABILITY	STABILITY	1.00 [a]				
	GLOBALITY	0.70	0.02	35.47	< .00001	0.70
GLOBALITY	GLOBALITY	1.00 [a]				

[a] fixed parameter

Figure 15.24: The jamovi CFA Factor Loadings and Covariances tables for our CFA MTMM model

measure a construct (e.g., an Extraversion scale) has an α of 0.80, then the proportion of error variance in the scale is 0.20. In other words, a scale with an α of 0.80 includes approximately 20% error.

BUT, and that's a BIG "BUT", Cronbach's alpha is not a measure of unidimensionality (i.e. an indicator that a scale is measuring a single factor or construct rather than multiple related constructs). Scales that are multidimensional will cause alpha to be under-estimated if not assessed separately for each dimension, but high values for alpha are not necessarily indicators of unidimensionality. So, an α of 0.80 does not mean that 80% of a single underlying construct is accounted for. It could be that the 80% comes from more than one underlying construct. That's why EFA and CFA are useful to do first.

Further, another feature of α is that it tends to be sample specific: it is not a characteristic of the scale, but rather a characteristic of the sample in which the scale has been used. A biased, unrepresentative, or small sample could produce a very different α coefficient than a large, representative sample. α can even vary from large sample to large sample. Nevertheless, despite these limitations, Cronbach's α has been popular in Psychology for estimating internal consistency reliability. It's pretty easy to calculate, understand and interpret, and therefore it can be a useful initial check on scale performance when you administer a scale with a different sample, from a different setting or population, for example.

An alternative is **McDonald's omega** (ω), and jamovi also provides this statistic. Whereas α makes the following assumptions: (a) no residual correlations, (b) items have identical loadings, and (c) the scale is unidimensional, ω does not and is therefore a more robust reliability statistic. If these assumptions are not violated then α and ω will be similar, but if they are then ω is to be preferred.

Sometimes a threshold for α or ω is provided, suggesting a "good enough" value. This might be something like αs of 0.70 or 0.80 representing "acceptable" and "good" reliability, respectively. However, this does depend on what exactly the scale is supposed to be measuring, so thresholds like this should be used cautiously. It could be better to simply state that an α or ω of 0.70 is associated with 30% error variance in a scale, and an α or ω of 0.80 is associated with 20%.

Can α be too high? Probably: if you are getting an α coefficient above 0.95 then this indicates high inter-correlations between the items and that there might be too much overly redundant specificity in the measurement, with a risk that the construct being measured is perhaps overly narrow.

15.5.1 Reliability analysis in jamovi

We have a third sample of personality data to use to undertake reliability analysis: in the *bfi_sample3.csv* file. Once again, check that the 25 personality item variables are coded as continuous. To perform reliability analysis in jamovi:

- Select 'Factor' – 'Reliability Analysis' from the main jamovi button bar to open the reliability analysis window (Figure 15.25).

- Select the 5 A variables and transfer them into the 'Items' box.

- Under the "Reverse Scaled Items" option, select variable A1 in the "Normal Scaled Items" box and move it across to the "Reverse Scaled Items" box.

- Check other appropriate options, as in Figure 15.25.

Once done, look across at the jamovi results window. You should see something like Figure 15.26. This tells us that the Cronbach's α coefficient for the Agreeableness scale is 0.72. This means that just under 30% of the Agreeableness scale score is error variance. McDonald's ω is also given, and this is 0.74, not much different from α.

We can also check how α or ω can be improved if a specific item is dropped from the scale. For example, α would increase to 0.74 and ω to 0.75 if we dropped item A1. This isn't a big increase, so probably not worth doing.

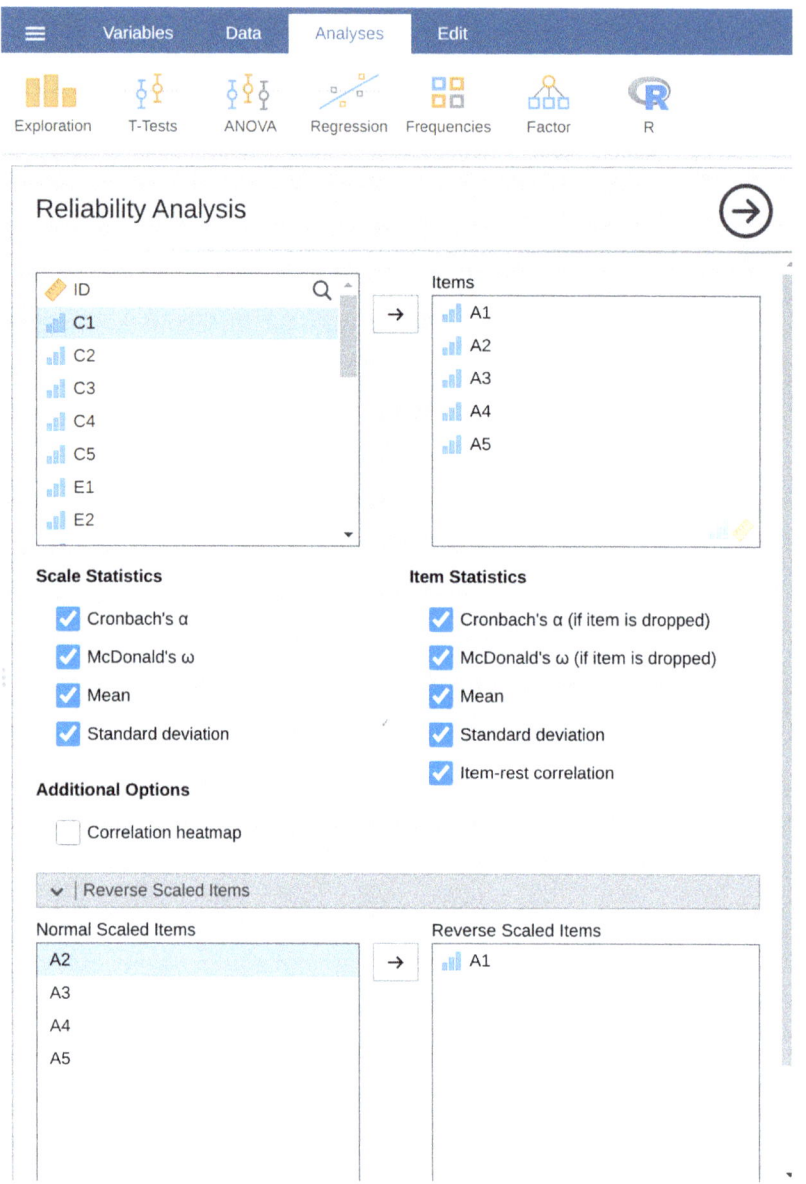

Figure 15.25: The jamovi 'Reliability Analysis' window

The process of calculating and checking scale statistics (α and ω) is the same for all the other scales, and they all had similar reliability estimates apart from Openness. For Openness, the amount of error variance in the Scale score is around 40%, which is high and indicates that Openness is substantially less consistent as a reliable measure of a personality attribute than the other personality scales.

Reliability Analysis

Scale Reliability Statistics

	Mean	SD	Cronbach's α	McDonald's ω
scale	4.65	0.90	0.70	0.72

[3]

Item Reliability Statistics

				If item dropped	
	Mean	SD	Item-rest correlation	Cronbach's α	McDonald's ω
A1 [a]	4.63	1.37	0.29	0.72	0.73
A2	4.68	1.24	0.54	0.62	0.65
A3	4.64	1.30	0.61	0.59	0.61
A4	4.83	1.42	0.40	0.68	0.70
A5	4.48	1.33	0.47	0.64	0.66

[a] reverse scaled item

Figure 15.26: The jamovi Reliability Analysis results for the Agreeableness factor

15.6 Summary

In this chapter on Factor Analysis and related techniques we have introduced and demonstrated statistical analyses that assess the pattern of relationships in a data set. Specifically, we have covered:

- Exploratory Factor Analysis (EFA). EFA is a statistical technique for identifying underlying latent factors in a data set. Each observed variable is conceptualised as representing the latent factor to some extent, indicated by a factor loading. Researchers also use EFA as a way of data reduction, i.e. identifying observed variables than can be combined into new factor variables for subsequent analysis.

- Principal Component Analysis (PCA) is a data reduction technique which, strictly speaking, does not identify underlying latent factors. Instead, PCA simply produces a linear combination of observed variables.

- Confirmatory Factor Analysis (CFA). Unlike EFA, with CFA you start with an idea – a model – of how the variables in your data are related to each other. You then test your model against the observed data and assess how good a fit the model is to the data.

- In Multi-Trait Multi-Method CFA (MTMM CFA), both latent factor and method variance are included in the model in an approach that is useful when there are different methodological approaches used and therefore method variance is an important consideration.

- Internal consistency reliability analysis. This form of reliability analysis tests how consistently a scale measures a measurement (psychological) construct.

Part VI

Endings, alternatives and prospects

Chapter 16

Bayesian statistics

In our reasonings concerning matter of fact, there are all imaginable degrees of assurance, from the highest certainty to the lowest species of moral evidence. A wise man, therefore, proportions his belief to the evidence.
– David Hume[188]

The ideas I've presented to you in this book describe inferential statistics from the frequentist perspective. I'm not alone in doing this. In fact, almost every textbook given to undergraduate psychology students presents the opinions of the frequentist statistician as *the* theory of inferential statistics, the one true way to do things. I have taught this way for practical reasons. The frequentist view of statistics dominated the academic field of statistics for most of the 20th century, and this dominance is even more extreme among applied scientists. It was and is current practice among psychologists to use frequentist methods. Because frequentist methods are ubiquitous in scientific papers, every student of statistics needs to understand those methods, otherwise they will be unable to make sense of what those papers are saying! Unfortunately, in my opinion at least, the current practice in psychology is often misguided, and the reliance on frequentist methods is partly to blame. In this chapter I explain why I think this and provide an introduction to Bayesian statistics, an approach that I think is generally superior to the orthodox approach.

This chapter comes in two parts. In the first three sections I talk about what Bayesian statistics are all about, covering the basic mathematical rules for how it works as well as an explanation for why I think the Bayesian approach is so useful. Afterwards, I provide a brief overview of how you can do Bayesian *t*-tests.

16.1 Probabilistic reasoning by rational agents

From a Bayesian perspective statistical inference is all about *belief revision*. I start out with a set of candidate hypotheses, h, about the world. I don't know which of these hypotheses is true, but do I have some beliefs about which hypotheses are plausible and which are not. When I observe the data, d, I have to revise those beliefs. If the data are consistent with a hypothesis, my belief in that hypothesis is strengthened. If the data are inconsistent with the hypothesis, my belief in that hypothesis is weakened. That's it! At the end of this section I'll give a precise description of how Bayesian reasoning works, but first I want to work through a simple example in order to introduce the key ideas. Consider the following reasoning problem:

I'm carrying an umbrella. Do you think it will rain?

In this problem I have presented you with a single piece of data (d = I'm carrying the umbrella), and I'm asking you to tell me your belief or hypothesis about whether it's raining. You have two alternatives, h: either it will rain today or it will not. How should you solve this problem?

16.1.1 Priors: what you believed before

The first thing you need to do is ignore what I told you about the umbrella, and write down your pre-existing beliefs about rain. This is important. If you want to be honest about how your beliefs have been revised in the light of new evidence (data) then you must say something about what you believed before those data appeared! So, what might you believe about whether it will rain today? You probably know that I live in Australia and that much of Australia is hot and dry. The city of Adelaide where I live has a Mediterranean climate, very similar to southern California, southern Europe or northern Africa. I'm writing this in January and so you can assume it's the middle of summer. In fact, you might have decided to take a quick look on Wikipedia[189] and discovered that Adelaide gets an average of 4.4 days of rain across the 31 days of January. Without knowing anything else, you might conclude that the probability of January rain in Adelaide is about 15%, and the probability of a dry day is 85% (see Table 16.1). If this is really what you believe about Adelaide rainfall (and now that I've told it to you I'm betting that this really is what you believe) then what I have written here is your **prior distribution**, written $P(h)$.

Table 16.1: How likely is it to rain in Adelaide – pre-existing beliefs based on knowledge of average January rainfall

Hypothesis	Degree of Belief
Rainy day	0.15
Dry day	0.85

16.1.2 Likelihoods: theories about the data

To solve the reasoning problem you need a theory about my behaviour. When does Danielle carry an umbrella? You might guess that I'm not a complete idiot,[190] and I try to carry umbrellas only on rainy days. On the other hand, you also know that I have young kids, and you wouldn't be all that surprised to know that I'm pretty forgetful about this sort of thing. Let's suppose that on rainy days I remember my umbrella about 30% of the time (I really am awful at this). But let's say that on dry days I'm only about 5% likely to be carrying an umbrella. So you might write this out as in Table 16.2.

Table 16.2: How likely am I to be carrying an umbrella on rainy and dry days

Hypothesis	Data Umbrella	Data No umbrella
Rainy day	0.30	0.70
Dry day	0.05	0.95

It's important to remember that each cell in this table describes your beliefs about what data d will be observed, *given* the truth of a particular hypothesis h. This "conditional probability" is written $P(d|h)$, which you can read as "the probability of d given h". In Bayesian statistics, this is referred to as the **likelihood** of the data d given the hypothesis h.[191]

16.1.3 The joint probability of data and hypothesis

At this point all the elements are in place. Having written down the priors and the likelihood, you have all the information you need to do Bayesian reasoning. The question now becomes how do we use this information? As it turns out, there's a very simple equation that we can use here, but it's important that you understand why we use it, so I'm going to try to build it up from more basic ideas.

Let's start out with one of the rules of probability theory. I listed it way back in Table 7.1, but I didn't make a big deal out of it at the time, and you probably ignored it. The rule in question is the one that talks about the probability that two things are true. In our example, you might want to calculate the probability that today is rainy (i.e., hypothesis h is true) and I'm carrying an umbrella (i.e., data d is observed). The **joint probability** of the hypothesis and the data is written $P(d, h)$, and you can calculate it by multiplying the prior $P(h)$ by the likelihood $P(d|h)$. Mathematically, we say that:

$$P(d, h) = P(d|h)P(h)$$

So, what is the probability that today is a rainy day *and* I remember to carry an umbrella? As we discussed earlier, the prior tells us that the probability of a rainy day is 15%, and the likelihood tells us that the probability of me remembering my umbrella on a rainy day is 30%. So the probability that both of these things are true is calculated by multiplying the two:

$$P(rainy, umbrella) = P(umbrella|rainy) \times P(rainy)$$
$$= 0.30 \times 0.15$$
$$= 0.045$$

In other words, before being told anything about what actually happened, you think that there is a 4.5% probability that today will be a rainy day and that I will remember an umbrella. However, there are of course four possible things that could happen, right? So let's repeat the exercise for all four. If we do that, we end up with Table 16.3.

This table captures all the information about which of the four possibilities are likely. To really get the full picture, though, it helps to add the row totals and column totals. That gives us Table 16.4.

Table 16.3: Four possibilities combining rain (or not) and umbrella carrying (or not)

	Umbrella	No-umbrella
Rainy	0.045	0.105
Dry	0.0425	0.807

Table 16.4: Four possibilities combining rain (or not) and umbrella carrying (or not), with row and column totals

	Umbrella	No-umbrella	Total
Rainy	0.045	0.105	0.15
Dry	0.0425	0.807	0.85
Total	0.0875	0.912	1

This is a very useful table, so it's worth taking a moment to think about what all these numbers are telling us. First, notice that the row sums aren't telling us anything new at all. For example, the first row tells us that if we ignore all this umbrella business, the chance that today will be a rainy day is 15%. That's not surprising, of course, as that's our prior.[192] The important thing isn't the number itself. Rather, the important thing is that it gives us some confidence that our calculations are sensible! Now take a look at the column sums and notice that they tell us something that we haven't explicitly stated yet. In the same way that the row sums tell us the probability of rain, the column sums tell us the probability of me carrying an umbrella. Specifically, the first column tells us that on average (i.e., ignoring whether it's a rainy day or not) the probability of me carrying an umbrella is 8.75%. Finally, notice that when we sum across all four logically-possible events, everything adds up to 1. In other words, what we have written down is a proper probability distribution defined over all possible combinations of data and hypothesis.

Now, because this table is so useful, I want to make sure you understand what all the elements correspond to and how they written (Table 16.5).

Table 16.5: Four possibilities combining rain (or not) and umbrella carrying (or not), expressed as conditional probabilities

	Umbrella	No-umbrella	
Rainy	P(Umbrella, Rainy)	P(No-umbrella, Rainy)	P(Rainy)
Dry	P(Umbrella, Dry)	P(No-umbrella, Dry)	P(Dry)
	P(Umbrella)	P(No-umbrella)	

Finally, let's use "proper" statistical notation. In the rainy day problem, the data corresponds to the observation that I do or do not have an umbrella. So we'll let d_1 refer to the possibility that you observe me carrying an umbrella, and d_2 refers to you observing me not carrying one. Similarly, h_1 is your hypothesis that today is rainy, and h_2 is the hypothesis that it is not. Using this notation, the table looks like Table 16.6.

Table 16.6: Four possibilities combining rain (or not) and umbrella carrying (or not), expressed in hypothetical terms as conditional probabilities

	d_1	d_2	
h_1	$P(h_1, d_1)$	$P(h_1, d_2)$	$P(h_1)$
h_2	$P(h_2, d_1)$	$P(h_2, d_2)$	$P(h_2)$
	$P(d_1)$	$P(d_2)$	

16.1.4 Updating beliefs using Bayes' rule

The table we laid out in the last section is a very powerful tool for solving the rainy day problem, because it considers all four logical possibilities and states exactly how confident you are in each of them before being given any data. It's now time to consider what happens to our beliefs when we are actually given the data. In the rainy day problem, you are told that I really am carrying an umbrella. This is something of a surprising event. According to our table, the probability of me carrying an umbrella is only 8.75%. But that makes sense, right? A woman carrying an umbrella on a summer day in a hot dry city is pretty unusual, and so you really weren't expecting that. Nevertheless, the data tells you that it is true. No matter how unlikely you thought it was, you must now adjust your beliefs to accommodate the fact that you now *know* that I have an umbrella.[193] To reflect this new knowledge, our *revised* table must have the following numbers. (see Table 16.7).

Table 16.7: Revising beliefs given new data about umbrella carrying

	Umbrella	No-umbrella
Rainy		0
Dry		0
Total	1	0

In other words, the facts have eliminated any possibility of "no umbrella", so we have to put zeros into any cell in the table that implies that I'm not carrying an umbrella. Also, you know for a fact that I am carrying an umbrella, so the column sum on the left must be 1 to correctly describe the fact that $P(umbrella) = 1$.

What two numbers should we put in the empty cells? Again, let's not worry about the maths, and instead think about our intuitions. When we wrote out our table the first time, it turned out that those two cells had almost identical numbers, right? We worked out that the joint probability of "rain and umbrella" was 4.5%, and the joint probability of "dry and umbrella" was 4.25%. In other words, before I told you that I am in fact carrying an umbrella, you'd have said that these two events were almost identical in probability, yes? But notice that both of these possibilities are consistent with the fact that I actually am carrying an umbrella. From the perspective of these two possibilities, very little has changed. I hope you'd agree that it's still true that these two possibilities are equally plausible. So what we expect to see in our final table is some numbers that preserve the fact that "rain and umbrella" is *slightly* more plausible than "dry and umbrella", while still ensuring that numbers in the table add up. Something like Table 16.8, perhaps?

Table 16.8: Revising probabilities given new data about umbrella carrying

	Umbrella	No-umbrella
Rainy	0.514	0
Dry	0.486	0
Total	1	0

What this table is telling you is that, after being told that I'm carrying an umbrella, you believe that there's a 51.4% chance that today will be a rainy day, and a 48.6% chance that it won't. That's the answer to our problem! The **posterior probability** of rain $P(h|d)$ given that I am carrying an umbrella is 51.4%.

How did I calculate these numbers? You can probably guess. To work out that there was a 0.514 probability of "rain", all I did was take the 0.045 probability of "rain and umbrella" and divide it by the 0.0875 chance of "umbrella". This produces a table that satisfies our need to have everything sum to 1, and our need not to interfere with the relative plausibility of the two events that are actually consistent with the data. To say the same thing using fancy statistical jargon, what I've done here is divide the joint probability of the hypothesis and the data $P(d, h)$ by the **marginal probability** of the data $P(d)$, and this is what gives us the posterior probability of the hypothesis given the data that have been observed. To write this as an equation:[194]

However, remember what I said at the start of the last section, namely that the joint probability, $P(d, h)$, is calculated by multiplying the prior, $P(h)$, by the likelihood, $P(d|h)$. In real life, the things we actually know how to write down are the priors and the likelihood, so let's substitute those back into the equation. This gives us the following formula for the posterior probability:

$$P(h|d) = \frac{P(d|h)P(h)}{P(d)}$$

And this formula, folks, is known as **Bayes' rule**. It describes how a learner starts out with prior beliefs about the plausibility of different hypotheses, and tells you how those beliefs should be revised in the face of data. In the Bayesian paradigm, all statistical inference flows from this one simple rule.

16.2 Bayesian hypothesis tests

In Chapter 9 I described the orthodox approach to hypothesis testing. It took an entire chapter to describe, because null hypothesis testing is a very elaborate contraption that people find very hard to make sense of. In contrast, the Bayesian approach to hypothesis testing is incredibly simple. Let's pick a setting that is closely analogous to the orthodox scenario. There are two hypotheses that we want to compare, a null hypothesis, h_0, and an alternative hypothesis, h_1. Prior to running the experiment we have some beliefs, $P(h)$, about which hypotheses are true. We run an experiment and obtain data, d. Unlike frequentist statistics, Bayesian statistics does allow us to talk about the probability that the null hypothesis is true. Better yet, it allows us to calculate the **posterior probability of the null hypothesis**, using Bayes' rule:

$$P(h_0|d) = \frac{P(d|h_0)P(h_0)}{P(d)}$$

This formula tells us exactly how much belief we should have in the null hypothesis after having observed the data, d. Similarly, we can work out how much belief to place in the alternative hypothesis using essentially the same equation. All we do is change the subscript:

$$P(h_1|d) = \frac{P(d|h_1)P(h_1)}{P(d)}$$

It's all so simple that I feel like an idiot even bothering to write these equations down, since all I'm doing is copying Bayes' rule from the previous section.[195]

16.2.1 The Bayes factor

In practice, most Bayesian data analysts tend not to talk in terms of the raw posterior probabilities $P(h_0|d)$ and $P(h_1|d)$. Instead, we tend to talk in terms of the **posterior odds** ratio. Think of it like betting. Suppose, for instance, the posterior probability of the null hypothesis is 25%, and the posterior probability of the alternative is 75%. The alternative hypothesis is three times as probable as the null, so we say that the odds are 3:1 in favour of the alternative. Mathematically, all we have to do to calculate the posterior odds is divide one posterior probability by the other:

$$\frac{P(h_1|d)}{P(h_0|d)} = \frac{0.75}{0.25} = 3$$

Or, to write the same thing in terms of the equations above:

$$\frac{P(h_1|d)}{P(h_0|d)} = \frac{P(d|h_1)}{P(d|h_0)} \times \frac{P(h_1)}{P(h_0)}$$

Actually, this equation is worth expanding on. There are three different terms here that you should know. On the left-hand side, we have the posterior odds, which tells you what you believe about the relative plausibilty of the null hypothesis and the alternative hypothesis after seeing the data. On the right-hand side, we have the **prior odds**, which indicates what you thought before seeing the data. In the middle, we have the **Bayes factor**, which describes the amount of evidence provided by the data (Table 16.9).

The Bayes factor (sometimes abbreviated as BF) has a special place in Bayesian hypothesis testing, because it serves a similar role to the p-value in orthodox hypothesis testing. The Bayes factor quantifies the strength of evidence provided by the data, and as such it is the Bayes factor that people tend to report when running a Bayesian hypothesis test.

Table 16.9: Posterior odds given the Bsyes factor and prior odds

$$\frac{P(h_1|d)}{h_0|d} = \frac{P(d|h_1)}{d|h_0} \times \frac{P(h_1)}{h_0}$$

⇑ ⇑ ⇑

Posterior odds | Bayes factor | Prior odds

The reason for reporting Bayes factors rather than posterior odds is that different researchers will have different priors. Some people might have a strong bias to believe the null hypothesis is true, others might have a strong bias to believe it is false. Because of this, the polite thing for an applied researcher to do is report the Bayes factor. That way, anyone reading the paper can multiply the Bayes factor by their own personal prior odds, and they can work out for themselves what the posterior odds would be. In any case, by convention we like to pretend that we give equal consideration to both the null hypothesis and the alternative, in which case the prior odds equals 1, and the posterior odds becomes the same as the Bayes factor.

16.2.2 Interpreting Bayes factors

One of the really nice things about the Bayes factor is the numbers are inherently meaningful. If you run an experiment and you compute a Bayes factor of 4, it means that the evidence provided by your data corresponds to betting odds of 4:1 in favour of the alternative. However, there have been some attempts to quantify the standards of evidence that would be considered meaningful in a scientific context. The two most widely used are from Jeffreys (1961) and Kass & Raftery (1995). Of the two, I tend to prefer the Kass & Raftery (1995) table because it's a bit more conservative. So here it is (Table 16.10).

Table 16.10: Bayes factors and strength of evidence

Bayes factor	Interpretation
1 - 3	Negligible evidence
3-20	Positive evidence
20-150	Strong evidence
> 150	Very strong evidence

And to be perfectly honest, I think that even the Kass & Raftery (1995) standards are being a bit charitable. If it were up to me, I'd have called the "positive evidence" category "weak evidence". To me, anything in the range 3:1 to 20:1 is "weak" or "modest" evidence at best. But there are no hard and fast rules here. What counts as strong or weak evidence depends entirely on how conservative you are and upon the standards that your community insists upon before it is willing to label a finding as "true".

In any case, note that all the numbers listed above make sense if the Bayes factor is greater than 1 (i.e., the evidence favours the alternative hypothesis). However, one big practical advantage of the Bayesian approach relative to the orthodox approach is that it also allows you to quantify evidence for the null. When that happens, the Bayes factor will be less than 1. You can choose to report a Bayes factor less than 1, but to be honest I find it confusing. For example, suppose that the likelihood of the data under the null hypothesis $P(d|h_0)$ is equal to 0.2, and the corresponding likelihood $P(d|h_1)$ under the alternative hypothesis is 0.1. Using the equations given above, Bayes factor here would be:

$$BF = \frac{P(d|h_1)}{P(d|h_0)} = \frac{0.1}{0.2} = 0.5$$

Read literally, this result tells is that the evidence in favour of the alternative is 0.5 to 1. I find this hard to understand. To me, it makes a lot more sense to turn the equation "upside down", and report the amount op evidence in favour of the null. In other words, what we calculate is this:

$$BF' = \frac{P(d|h_0)}{P(d|h_1)} = \frac{0.2}{0.1} = 2$$

And what we would report is a Bayes factor of 2:1 in favour of the null. Much easier to understand, and you can interpret this using the table above.

16.3 Why be a Bayesian?

Up to this point I've focused exclusively on the logic underpinning Bayesian statistics. We've talked about the idea of "probability as a degree of belief", and what it implies about how a rational agent should reason about the world. The question that you have to answer for yourself is this: how do you want to do your statistics? Do you want to be an orthodox statistician, relying on sampling distributions and p-values to guide your decisions? Or do you want to be a Bayesian, relying on things like prior beliefs, Bayes factors and the rules for rational belief revision? And to be perfectly honest, I can't answer this question for you. Ultimately it depends on what you think is right. It's your call and your call alone. That being said, I can talk a little about why I prefer the Bayesian approach.

16.3.1 Statistics that mean what you think they mean

> *You keep using that word. I do not think it means what you think it means*
> – Inigo Montoya, *The Princess Bride*[196]

To me, one of the biggest advantages to the Bayesian approach is that it answers the right questions. Within the Bayesian framework, it is perfectly sensible and allowable to refer to "the probability that a hypothesis is true". You can even try to calculate this probability. Ultimately, isn't that what you want your statistical tests to tell you? To an actual human being, this would seem to be the whole point of doing statistics, i.e., to determine what is true and what isn't. Any time that you aren't exactly sure about what the truth is, you should use the language of probability theory to say things like "there is an 80% chance that Theory A is true, but a 20% chance that Theory B is true instead".

This seems so obvious to a human, yet it is explicitly forbidden within the orthodox framework. To a frequentist, such statements are a nonsense because "the theory is true" is not a repeatable event. A theory is true or it is not, and no probabilistic statements are allowed, no matter how much you might want to make them. There's a reason why, back in Section 9.5, I repeatedly warned you not to interpret the p-value as the probability that the null hypothesis is true. There's a reason why almost every textbook on statistics is forced to repeat that warning. It's because people desperately want that to be the correct interpretation. Frequentist dogma notwithstanding, a lifetime of experience of teaching undergraduates and of doing data analysis on a daily basis suggests to me that most actual humans think that "the probability that the hypothesis is true" is not only meaningful, it's the thing we care most about. It's such an appealing idea that even trained statisticians fall prey to the mistake of trying to interpret a p-value this way. For example, here is a quote from an official Newspoll report in 2013, explaining how to interpret their (frequentist) data analysis:[197]

> *Throughout the report, where relevant, statistically significant changes have been noted. All significance tests have been based on the 95 percent level of confidence.* **This means that if a change is noted as being statistically significant, there is a 95 percent probability that a real change has occurred**, *and is not simply due to chance variation.* (emphasis added)

Nope! That's not what p < .05 means. That's not what 95% confidence means to a frequentist statistician. The bolded section is just plain wrong. Orthodox methods cannot tell you that "there is a 95% chance that a real change has occurred", because this is not the kind of event to which frequentist probabilities may be assigned. To an ideological frequentist, this sentence should be meaningless. Even if you're a more pragmatic frequentist, it's still the wrong definition of a p-value. It is simply not an allowed or correct thing to say if you want to rely on orthodox statistical tools.

On the other hand, let's suppose you are a Bayesian. Although the bolded passage is the wrong definition of a p-value, it's pretty much exactly what a Bayesian means when they say that the posterior probability of the alternative hypothesis is greater than 95%. And here's the thing. If the Bayesian posterior is actually the thing you want to report, why are you even trying to use orthodox methods? If you want to make Bayesian claims, all you have to do is be a Bayesian and use Bayesian tools.

Speaking for myself, I found this to be the most liberating thing about switching to the Bayesian view. Once you've made the jump, you no longer have to wrap your head around counter-intuitive definitions of p-values. You don't have to bother remembering why you can't say that you're 95% confident that the true mean lies within some interval. All you have to do is be honest about what you believed before you ran the study and then report what you learned from doing it. Sounds nice, doesn't it? To me, this is the big promise of the Bayesian approach. You do the analysis you really want to do, and express what you really believe the data are telling you.

16.3.2 Evidentiary standards you can believe

> *If p is below .02 it is strongly indicated that the null hypothesis fails to account for the whole of the facts. We shall not often be astray if we draw a conventional line at .05 and consider that smaller values of p indicate a real discrepancy.*
> – Sir Ronald Fisher (Fisher, 1925, p. 79)

Consider the quote above by Sir Ronald Fisher, one of the founders of what has become the orthodox approach to statistics. If anyone has ever been entitled to express an opinion about the intended function of *p*-values, it's Fisher. In this passage, taken from his classic guide, *Statistical Methods for Research Workers*, he's pretty clear about what it means to reject a null hypothesis at $p < .05$. In his opinion, if we take $p < .05$ to mean there is "a real effect", then "we shall not often be astray". This view is hardly unusual. In my experience, most practitioners express views very similar to Fisher's. In essence, the $p < .05$ convention is assumed to represent a fairly stringent evidential standard.

Well, how true is that? One way to approach this question is to try to convert *p*-values to Bayes factors, and see how the two compare. It's not an easy thing to do because a *p*-value is a fundamentally different kind of calculation to a Bayes factor, and they don't measure the same thing. However, there have been some attempts to work out the relationship between the two, and it's somewhat surprising. For example, Johnson (2013) presents a pretty compelling case that (for *t*-tests at least) the $p < .05$ threshold corresponds roughly to a Bayes factor of somewhere between 3:1 and 5:1 in favour of the alternative. If that's right, then Fisher's claim is a bit of a stretch. Let's suppose that the null hypothesis is true about half the time (i.e., the prior probability of H_0 is 0.5), and we use those numbers to work out the posterior probability of the null hypothesis given that it has been rejected at $p < .05$. Using the data from Johnson (2013), we see that if you reject the null at $p < .05$, you'll be correct about 80% of the time. I don't know about you but, in my opinion, an evidential standard that ensures you'll be wrong on 20% of your decisions isn't good enough. The fact remains that, quite contrary to Fisher's claim, if you reject at $p < .05$ you shall quite often go astray. It's not a very stringent evidential threshold at all.

16.3.3 The *p*-value is a lie.

The cake is a lie.
The cake is a lie.
The cake is a lie.
The cake is a lie.
– Portal[198]

Okay, at this point you might be thinking that the real problem is not with orthodox statistics, just the $p < .05$ standard. In one sense, that's true. The recommendation that Johnson (2013) gives is not that "everyone must be a Bayesian now". Instead, the suggestion is that it would be wiser to shift the conventional standard to something like a $p < .01$ level. That's not an unreasonable view to take, but in my view the problem is a little more severe than that. In my opinion, there's a fairly big problem built into the way most (but not all) orthodox hypothesis tests are constructed. They are grossly naive about how humans actually do research, and because of this most *p*-values are wrong.

Sounds like an absurd claim, right? Well, consider the following scenario. You've come up with a really exciting research hypothesis and you design a study to test it. You're very diligent, so you run a power analysis to work out what your sample size should be, and you run the study. You run your hypothesis test and out pops a *p*-value of 0.072. Really bloody annoying, right?

What should you do? Here are some possibilities:

1. You conclude that there is no effect and try to publish it as a null result
2. You guess that there might be an effect and try to publish it as a "borderline significant" result.
3. You give up and try a new study.
4. You collect some more data to see if the p-value goes up or (preferably!) drops below the "magic" criterion of $p < .05$.

Which would you choose? Before reading any further, I urge you to take some time to think about it. Be honest with yourself. But don't stress about it too much, because you're screwed no matter what you choose. Based on my own experiences as an author, reviewer and editor, as well as stories I've heard from others, here's what will happen in each case:

- Let's start with option 1. If you try to publish it as a null result, the paper will struggle to be published. Some reviewers will think that $p = .072$ is not really a null result. They'll argue it's borderline significant. Other reviewers will agree it's a null result, but will claim that even though some null results are publishable, yours isn't. One or two reviewers might even be on your side, but you'll be fighting an uphill battle to get it through.

- Okay, let's think about option number 2. Suppose you try to publish it as a borderline significant result. Some reviewers will claim that it's a null result and should not be published. Others will claim that the evidence is ambiguous, and that you should collect more data until you get a clear significant result. Again, the publication process does not favour you.

- Given the difficulties in publishing an "ambiguous" result like $p = .072$, option number 3 might seem tempting: give up and do something else. But that's a recipe for career suicide. If you give up and try a new project every time you find yourself faced with ambiguity, your work will never be published. And if you're in academia without a publication record, you can lose your job. So that option is out.

- It looks like you're stuck with option 4. You don't have conclusive results, so you decide to collect some more data and re-run the analysis. Seems sensible, but unfortunately for you, if you do this all of your p-values are now incorrect. All of them. Not just the p-values that you calculated for this study. All of them. All the p-values you calculated in the past and all the p-values you will calculate in the future. Fortunately, no-one will notice. You'll get published, and you'll have lied.

Wait, what? How can that last part be true? I mean, it sounds like a perfectly reasonable strategy, doesn't it? You collected some data, the results weren't conclusive, so now what you want to do is collect more data until the the results are conclusive. What's wrong with that?

Honestly, there's nothing wrong with it. It's a reasonable, sensible and rational thing to do. In real life, this is exactly what every researcher does. Unfortunately, the theory of null hypothesis testing as I described it in Chapter 9 forbids you from doing this.[199]

The reason is that the theory assumes that the experiment is finished and all the data are in. And because it assumes the experiment is over, it only considers two possible decisions. If you're using the conventional $p < .05$ threshold, those decisions are shown in Table 16.11.

Table 16.11: Conventional null hypothesis signicance testing (NHST) with $p < .05$)

Outcome	Action
p less than .05	Reject the null
p greater than .05	Retain the null

What you're doing is adding a third possible action to the decision making problem. Specifically, what you're doing is using the p-value itself as a reason to justify continuing the experiment. And as a consequence you've transformed the decision-making procedure into one that looks more like Table 16.12.

Table 16.12: Carrying on data collecting based on p-values obtained in preliminary testing

Outcome	Action
p less than .05	Stop the experiment and reject the null
p between .05 and .1	Continue the experiment
p greater than .1	Stop the experiment and retain the null

The "basic" theory of null hypothesis testing isn't built to handle this sort of thing, not in the form I described in Chapter 9. If you're the kind of person who would choose to "collect more data" in real life, it implies that you are not making decisions in accordance with the rules of null hypothesis testing. Even if you happen to arrive at the same decision as the hypothesis test, you aren't following the decision process it implies, and it's this failure to follow the process that is causing the problem.[200] Your p-values are a lie.

Worse yet, they're a lie in a dangerous way, because they're all *too small*. To give you a sense of just how bad it can be, consider the following (worst case) scenario. Imagine you're a really super-enthusiastic researcher on a tight budget who didn't pay any attention to my warnings above. You design a study comparing two groups. You desperately want to see a significant result at the $p < .05$ level, but you really don't want to collect any more data than you have to (because it's expensive). In order to cut costs you start collecting data but every time a set of observations arrive you run a t-test on your data. If the t-test says $p < .05$, then you stop the experiment and report a significant result. If not, you keep collecting data. You keep doing this until you reach your pre-defined spending limit for this experiment. Let's say that limit kicks in at $N = 1000$ observations. As it turns out, the truth of the matter is that there is no real effect to be found: the null hypothesis is true. So, what's the chance that you'll make it to the end of the experiment and (correctly) conclude that there is no effect? In an ideal world, the answer here should be 95%. After all, the whole point of the $p < .05$ criterion is to control the type I error rate at 5%, so what we'd hope is that there's only a 5% chance of falsely rejecting the null hypothesis in this situation. However, there's no guarantee that will be true. You're breaking the rules. Because you're running tests repeatedly, "peeking" at your data to see if you've gotten a significant result, all bets are off.

So how bad is it? The answer from a simulation study is shown as the solid line in Figure 16.1, and it's astoundingly bad.

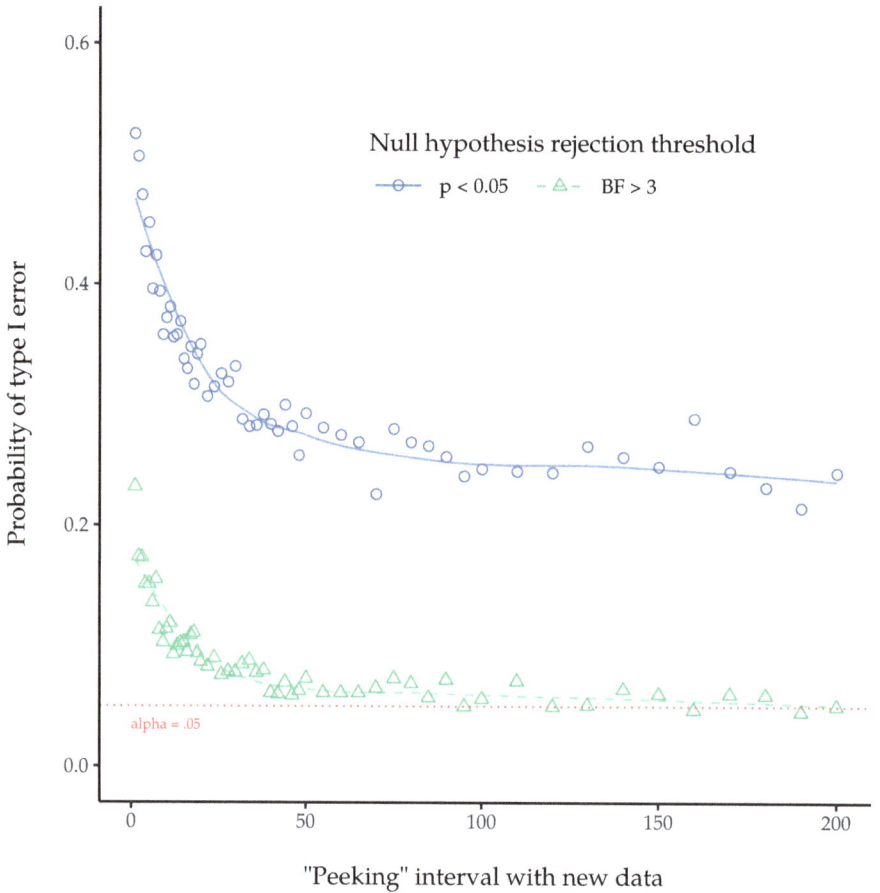

Figure 16.1: Probability of type I error in an experiment with target N of 1000 per group and **peeking** at different intervals:- things can go badly wrong if you **peek** at your data and re-run your tests as new data arrives. If you are a frequentist, this is *very wrong* (blue circles and solid line). If you are a bayesian, it is not so bad (green triangles and dashed line). The alpha level was set at 0.05 (red dotted line) in this simulation.

If you peek at your data after every single observation, there is a 52% chance that you will make a type I error. That's, um, quite a bit bigger than the 5% that it's supposed to be. And it doesn't improve much with less frequent peeking: if you only peek every 10, or every 50 observations, then the type I error rates are still way too high: 37% and 29%, respectively. By way of comparison, imagine that you had used the following strategy. Start collecting data. Every single time an observation arrives, run Bayesian t-tests and look at the Bayes factor. I'll assume that Johnson (2013) is right, and I'll treat a Bayes factor of 3:1 as roughly equivalent to a p-value of .05.[201] This time around, our trigger-happy researcher uses the following procedure. If the Bayes factor is 3:1 or more in favour of the null, stop the experiment and retain the null. If it is 3:1 or more in favour of the alternative, stop the experiment and reject the null. Otherwise continue testing. Now, just like last time, let's assume that the null hypothesis is true. What happens? As it happens, I ran the simulations for this scenario too, and the results are shown as the dashed line in Figure 16.1. It turns out that the type I error rate for peeking every time a new observation arrives is 23%, much much lower than the 52% rate that we were getting by using the orthodox t-test. And for peeking every 10 or 50 observations, the rates are 11% and 7%, respectively.

16.3.4 Is it really this bad?

The example I gave in the previous section is a pretty extreme situation. In real life, people don't run hypothesis tests every time a new observation arrives. So it's not fair to say that the $p < .05$ threshold "really" corresponds to a 52% type I error rate (i.e., $p = 0.52$). But the fact remains that if you want your p-values to be honest, then you either have to switch to a completely different way of doing hypothesis tests or enforce a strict rule of no peeking. You are not allowed to use the data to decide when to terminate the experiment. You are not allowed to look at a "borderline" p-value and decide to collect more data. You aren't even allowed to change your data analyis strategy after looking at data. You are strictly required to follow these rules. Otherwise the p-values you calculate will be nonsense.

And yes, these rules are surprisingly strict. As a class exercise a couple of years back, I asked students to think about this scenario. Suppose you started running your study with the intention of collecting $N = 80$ people. When the study starts out you follow the rules, refusing to look at the data or run any tests. But when you reach $N = 50$ your willpower gives in… and you take a peek. Guess what? You've got a significant result! Now, sure, you know you said that you'd keep running the study out to a sample size of $N = 80$, but it seems sort of pointless now, right? The result is significant with a sample size of $N = 50$, so wouldn't it be wasteful and inefficient to keep collecting data? Aren't you tempted to stop? Just a little? Well, keep in mind that if you do, your type I error rate at $p < .05$ just ballooned out to 8%. When you report $p < .05$ in your paper, what you're really saying is $p < .08$. That's how bad the consequences of "just one peek" can be.

Now consider this. The scientific literature is filled with t-tests, ANOVAs, regressions and chi-square tests. When I wrote this book I didn't pick these tests arbitrarily. The reason why these four tools appear in most introductory statistics texts is that these are the bread-and-butter tools of science. None of these tools include a correction to deal with "data peeking": they all assume that you're not doing it. But how realistic is that assumption? In real life, how many people do you think have "peeked" at their data before the experiment was finished and adapted their subsequent behaviour after seeing what the data looked like? Except when the sampling procedure is fixed by an external constraint, I'm guessing the answer is "most people have done it". If that has happened, you can infer that the reported p-values are wrong. Worse yet, because we don't know what decision process they actually followed, we have no way to know what the p-values should have been. You can't compute a p-value when you don't know the decision-making procedure that the researcher used. And so the reported p-value remains a lie.

Given all of the above, what is the take home message? It's not that Bayesian methods are foolproof. If a researcher is determined to cheat, they can always do so. Bayes' rule cannot stop people from lying, nor can it stop them from rigging an experiment. That's not my point here. My point is the same one I made at the very beginning of the book in Section 1.1: the reason why we run statistical tests is to protect us from ourselves. And the reason why "data peeking" is such a concern is that it's so tempting, even for honest researchers. A theory for statistical inference has to acknowledge this. Yes, you might try to defend p-values by saying that it's the fault of the researcher for not using them properly, but to my mind that misses the point. A theory of statistical inference that is so completely naive about humans that it doesn't even consider the possibility that the researcher might look at their own data isn't a theory worth having. In essence, my point is this:

Good laws have their origins in bad morals.
– Ambrosius Macrobius[202]

Good rules for statistical testing have to acknowledge human frailty. None of us are without sin. None of us are beyond temptation. A good system for statistical inference should still work even when it is used by actual humans. Orthodox null hypothesis testing does not.[203]

16.4 Bayesian t-tests

An important type of statistical inference problem discussed in this book is comparing two means, discussed in some detail in Chapter 11 on t-tests. If you can remember back that far, you'll recall that there are several versions of the t-test. I'll talk a little about Bayesian versions of the independent samples t-tests and the paired samples t-test in this section.

16.4.1 Independent samples t-test

The most common type of t-test is the independent samples t-test, and it arises when you have data as in the *harpo.csv* data set that we used in Chapter 11 on t-tests. In this data set, we have two groups of students, those who received lessons from Anastasia and those who took their classes with Bernadette. The question we want to answer is whether there's any difference in the grades received by these two groups of students. Back in Chapter 11 I suggested you could analyse this kind of data using the Independent Samples t-test in jamovi, which gave us the results in Figure 16.2. As we obtain a p-value less than 0.05, we reject the null hypothesis.

What does the Bayesian version of the t-test look like? We can get the Bayes factor analysis by selecting the 'Bayes factor' checkbox under the 'Tests' option, and accepting the suggested default value for the 'Prior'. This gives the results shown in the table in Figure 16.3. What we get in this table is a Bayes factor statistic of 1.75, meaning that the evidence provided by these data are about 1.8:1 in favour of the alternative hypothesis.

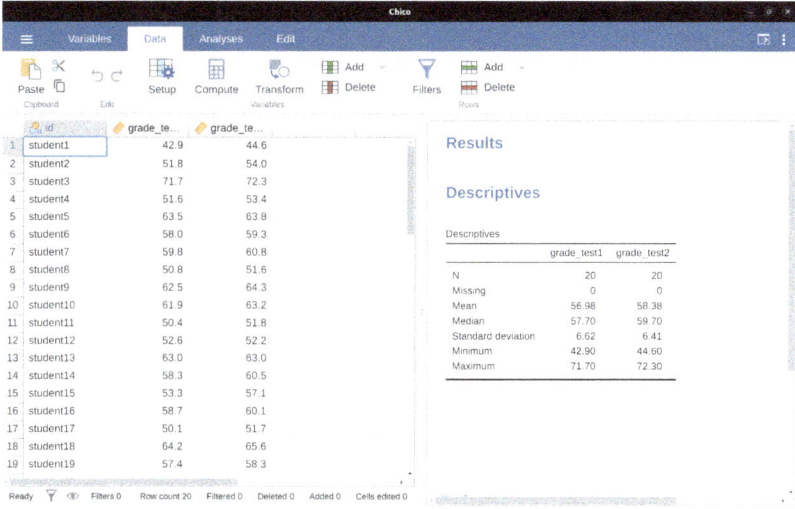

Figure 16.2: Independent samples *t*-test result in jamovi

Independent Samples T-Test

		Statistic	±%	df	p	Mean difference	SE difference		Effect Size
grade	Student's t	2.12		31.00	0.04253	5.48	2.59	Cohen's d	0.74
	Bayes factor₁₀	1.75	0.00						
	Welch's t	2.03		23.02	0.05361	5.48	2.69	Cohen's d	0.72

Figure 16.3: Bayes factor analysis alongside independent samples *t*-test

Before moving on, it's worth highlighting the difference between the orthodox test results and the Bayesian one. According to the orthodox test, we obtained a significant result, though only barely. Nevertheless, many people would happily accept $p = .043$ as reasonably strong evidence for an effect. In contrast, notice that the Bayesian test doesn't even reach 2:1 odds in favour of an effect, and would be considered very weak evidence at best. In my experience that's a pretty typical outcome. Bayesian methods usually require more evidence before rejecting the null.

16.4.2 Paired samples *t*-test

Back in Section 11.5 I discussed the *chico.csv* data set in which student grades were measured on two tests, and we were interested in finding out whether grades went up from test 1 to test 2. Because every student did both tests, the tool we used to analyse the data was a paired samples *t*-test. Figure 16.4 shows the jamovi results table for the conventional paired *t*-test alongside the Bayes factor analysis. At this point, I hope you can read this output without any difficulty. The data provide evidence of about 6000:1 in favour of the alternative. We could probably reject the null with some confidence!

Paired Samples T-Test									95% Confidence Interval			
			Statistic	±%	df	p	Mean difference	SE difference	Lower	Upper		Effect Size
grade_test2	grade_test1	Student's t	6.48		19.00	<.00001	1.40	0.22	0.95	1.86	Cohen's d	1.45
		Bayes factor₁₀	5991.58	0.00								

Figure 16.4: Paired samples *t*-test and Bayes factor result in jamovi

16.5 Summary

The first half of this chapter was focused primarily on the theoretical underpinnings of Bayesian statistics. I introduced the mathematics for how Bayesian inference works in the section on Probabilistic reasoning by rational agents, and gave a very basic overview of Bayesian hypothesis tests. Finally, I devoted some space to talking about why I think Bayesian methods are worth using.

Then I gave a practical example, with Bayesian *t*-tests. If you're interested in learning more about the Bayesian approach, there are many good books you could look into. John Kruschke's book, *Doing Bayesian Data Analysis*, is a pretty good place to start (Kruschke, 2011) and is a nice mix of theory and practice. His approach is a little different to the "Bayes factor" approach that I've discussed here, so you won't be covering the same ground. If you're a cognitive psychologist, you might want to check out Lee & Wagenmakers (2014). I picked these two because I think they're especially useful for people in my discipline, but there's a lot of good books out there, so look around!

Epilogue

"Begin at the beginning", the King said, very gravely, "and go on till you come to the end: then stop"
– Lewis Carroll, *Alice in Wonderland*

The undiscovered statistics

First, I'm going to talk a bit about some of the content that I wish I'd had the chance to cram into this book, just so that you can get a sense of what other ideas are out there in the world of statistics. One thing that students often fail to realise is that their introductory statistics classes are just that, an introduction. If you want to go out into the wider world and do real data analysis, you have to learn a whole lot of new tools that extend the content of your undergraduate lectures in all sorts of different ways. Don't assume that something can't be done just because it wasn't covered in undergrad. Don't assume that something is the right thing to do just because it was covered in an undergrad class. To stop you from falling victim to that trap, I think it's useful to give a bit of an overview of some of the other ideas out there.

Omissions within the topics covered

Even within the topics that I have covered in the book, there are a lot of omissions that I'd like to redress in the future version. Just sticking to things that are purely about statistics (rather than things associated with jamovi), the following is a representative but not exhaustive list of topics that I'd like to expand on at some time:

- **Other types of correlations.** In Chapter 12 I talked about two types of correlation: Pearson and Spearman. Both of these methods of assessing correlation are applicable to the case where you have two continuous variables and want to assess the relationship between them. What about the case where your variables are both nominal scale? Or when one is nominal scale and the other is continuous? There are actually methods for computing correlations in such cases (e.g., polychoric correlation), and it would be good to see these included.

- **More detail on effect sizes.** In general, I think the treatment of effect sizes throughout the book is a little more cursory than it should be. In almost every instance, I've tended just to pick one measure of effect size (usually the most popular one) and describe that. However, for almost all tests and models there are multiple ways of thinking about effect size, and I'd like to go into more detail in the future.

- **Dealing with violated assumptions.** In a number of places in the book I've talked about some things you can do when you find that the assumptions of your test (or model) are violated, but I think that I ought to say more about this. In particular, I think it would have been nice to talk in a lot more detail about how you can tranform variables to fix problems. I talked a bit about this in Chapter 6, but the discussion isn't detailed enough I think.

- **Interaction terms for regression.** In Chapter 14 I talked about the fact that you can have interaction terms in an ANOVA, and I also pointed out that ANOVA can be interpreted as a kind of linear regression model. Yet, when talking about regression in Chapter 12 I made no mention of interactions at all. However, there's nothing stopping you from including interaction terms in a regression model. It's just a little more complicated to figure out what an "interaction" actually means when you're talking about the interaction between two continuous predictors, and it can be done in more than one way. Even so, I would have liked to talk a little about this.

- **Method of planned comparison.** As I mentioned this in Chapter 14, it's not always appropriate to be using a post hoc correction like Tukey's HSD when doing an ANOVA, especially when you had a very clear (and limited) set of comparisons that you cared about ahead of time. I would like to talk more about this in the future.

- **Multiple comparison methods.** Even within the context of talking about post hoc tests and multiple comparisons, I would have liked to talk about the methods in more detail, and talk about what other methods exist besides the few options I mentioned.

Statistical models missing from the book

Statistics is a huge field. The core tools that I've described in this book (chi-square tests, t-tests, regression and ANOVA) are basic tools that are widely used in everyday data analysis, and they form the core of most introductory stats books. However, there are a lot of other tools out there. There are so very many data analysis situations that these tools don't cover, and it would be great to give you a sense of just how much more there is, for example:

- **Nonlinear regression.** When discussing regression in Chapter 12, we saw that regression assumes that the relationship between predictors and outcomes is linear. On the other hand, when we talked about the simpler problem of correlation in Chapter 4, we saw that there exist tools (e.g., Spearman correlations) that are able to assess non-linear relationships between variables. There are a number of tools in statistics that can be used to do non-linear regression. For instance, some non-linear regression models assume that the relationship between predictors and outcomes is monotonic (e.g., isotonic regression), while others assume that it is smooth but not necessarily monotonic (e.g., Lowess regression), while others assume that the relationship is of a known form that happens to be nonlinear (e.g., polynomial regression).

- **Logistic regression.** Yet another variation on regression occurs when the outcome variable is binary, but the predictors are continuous. For instance, suppose you're investigating social media, and you want to know if it's possible to predict whether or not someone is on Twitter as a function of their income, their age, and a range of other variables. This is basically a regression model, but you can't use regular linear regression because the outcome variable is binary (you're either on Twitter or you're not). Because the outcome variable is binary, there's no way that the residuals could possibly be normally distributed. There are a number of tools that statisticians can apply to this situation, the most prominent of which is logistic regression.

- **The General Linear Model (GLM).** The GLM is actually a family of models that includes logistic regression, linear regression, (some) nonlinear regression, ANOVA and many others. The basic idea in the GLM is essentially the same idea that underpins linear models, but it allows for the idea that your data might not be normally distributed, and allows for nonlinear relationships between predictors and outcomes. There are a lot of very handy analyses that you can run that fall within the GLM, so it's a very useful thing to know about.

- **Survival analysis.** In Chapter 2 I talked about "differential attrition", the tendency for people to leave the study in a non-random fashion. Back then, I was talking about it as a potential methodological concern, but there are a lot of situations in which differential attrition is actually the thing you're interested in. Suppose, for instance, you're interested in finding out how long people play different kinds of computer games in a single session. Do people tend to play RTS (real time strategy) games for longer stretches than FPS (first person shooter) games? You might design your study like this. People come into the lab, and they can play for as long or as little as they like. Once they're finished, you record the time they spent playing. However, due to ethical restrictions, let's suppose that you cannot allow them to keep playing longer than two hours. A lot of people will stop playing before the two-hour limit, so you know exactly how long they played. But some people will run into the two-hour limit, and so you don't know how long they would have kept playing if you'd been able to continue the study. As a consequence, your data are systematically censored: you're missing all of the very long times. How do you analyse this data sensibly? This is the problem that survival analysis solves. It is specifically designed to handle this situation, where you're systematically missing one "side" of the data because the study ended. It's very widely used in health research, and in that context it is often literally used to analyse survival. For instance, you may be tracking people with a particular type of cancer, some who have received treatment A and others who have received treatment B, but you only have funding to track them for five years. At the end of the study period some people are alive, others are not. In this context, survival analysis is useful for determining which treatment is more effective, and telling you about the risk of death that people face over time.

- **Mixed models.** Repeated measures ANOVA is often used in situations where you have observations clustered within experimental units. A good example of this is when you track individual people across multiple time points. Let's say you're tracking happiness over time, for two people. Aaron's happiness starts at 10, then drops to 8, and then to 6. Belinda's happiness starts at 6, then rises to 8 and then to 10. Both of these two people have the same "overall" level of happiness (the average across the three time points is 8), so a repeated measures ANOVA analysis would treat Aaron and Belinda the same way. But that's clearly wrong. Aaron's happiness is decreasing, whereas Belinda's is increasing. If you want to optimally analyse data from an experiment where people can change over time, then you need a more powerful tool than repeated measures ANOVA. The tools that people use to solve this problem are called "mixed" models, because they are designed to learn about individual experimental units (e.g., happiness of individual people over time) as well as overall effects (e.g., the effect of money on happiness over time). Repeated measures ANOVA is perhaps the simplest example of a mixed model, but there's a lot you can do with mixed models that you can't do with repeated measures ANOVA.

- **Multidimensional scaling.** Factor Analysis is an example of an "unsupervised learning" model. What this means is that, unlike most of the "supervised learning" tools I've mentioned, you can't divide up your variables into predictors and outcomes. Regression is supervised learning whereas Factor Analysis is unsupervised learning. It's not the only type of unsupervised learning model however. For example, in Factor Analysis one is concerned with the analysis of correlations between variables. However, there are many situations where you're actually interested in analysing similarities or dissimilarities between objects, items or people. There are a number of tools that you can use in this situation, the best known of which is multidimensional scaling (MDS). In MDS, the idea is to find a "geometric" representation of your items. Each item is "plotted" as a point in some space, and the distance between two points is a measure of how dissimilar those items are.

- **Clustering.** Another example of an unsupervised learning model is clustering (also referred to as classification), in which you want to organise all of your items into meaningful groups, such that similar items are assigned to the same groups. A lot of clustering is unsupervised, meaning that you don't know anything about what the groups are, you just have to guess. There are other "supervised clustering" situations where you need to predict group memberships on the basis of other variables, and those group memberships are actually observables. Logistic regression is a good example of a tool that works this way. However, when you don't actually know the group memberships, you have to use different tools (e.g., k-means clustering). There are even situations where you want to do something called "semi-supervised clustering", in which you know the group memberships for some items but not others. As you can probably guess, clustering is a pretty big topic, and a pretty useful thing to know about.

- **Causal models.** One thing that I haven't talked about much in this book is how you can use statistical modelling to learn about the causal relationships between variables. For instance, consider the following three variables which might be of interest when thinking about how someone died in a firing squad. We might want to measure whether or not an execution order was given (variable A), whether or not a marksman fired their gun (variable B), and whether or not the person got hit with a bullet (variable C). These three variables are all correlated with one another (e.g., there is a correlation between guns being fired and people getting hit with bullets), but we actually want to make stronger statements about them than merely talking about correlations. We want to talk about causation. We want to be able to say that the execution order (A) causes the marksman to fire (B) which causes someone to get shot (C). We can express this by a directed arrow notation: we write it as $A \rightarrow B \rightarrow C$. This "causal chain" is a fundamentally different explanation for events than one in which the marksman fires first, which causes the shooting $B \rightarrow C$, and then causes the executioner to "retroactively" issue the execution order, $B \rightarrow A$. This "common effect" model says that A and C are both caused by B. You can see why these are different. In the first causal model, if we had managed to stop the executioner from issuing the order (intervening to change A), then no shooting would have happened. In the second model, the shooting would have happened any way because the marksman was not following the execution order. There is a big literature in statistics on trying to understand the causal relationships between variables, and a number of different tools exist to help you test different causal stories about your data. The most widely used of these tools (in psychology at least) is structural equations modelling (SEM), and at some point I'd like to extend the book to talk about it.

Of course, even this listing is incomplete. I haven't mentioned time series analysis, item response theory, market basket analysis, classification and regression trees, or any of a huge range of other topics. However, the list that I've given above is essentially my wish list for this book. Sure, it would double the length of the book, but it would mean that the scope has become broad enough to cover most things that applied researchers in psychology would need to use.

Other ways of doing inference

A different sense in which this book is incomplete is that it focuses pretty heavily on a very narrow and old-fashioned view of how inferential statistics should be done. In Chapter 8 I talked a little bit about the idea of unbiased estimators, sampling distributions and so on. In Chapter 9 I talked about the theory of null hypothesis significance testing and p-values. These ideas have been around since the early 20th century, and the tools that I've talked about in the book rely very heavily on the theoretical ideas from that time. I've felt obligated to stick to those topics because the vast majority of data analysis in science is also reliant on those ideas. However, the theory of statistics is not restricted to those topics and, whilst everyone should know about them because of their practical importance, in many respects those ideas do not represent best practice for contemporary data analysis. One of the things that I'm especially happy with is that I've been able to go a little beyond this. Chapter 16 now presents the Bayesian perspective in a reasonable amount of detail, but the book overall is still pretty heavily weighted towards the frequentist orthodoxy. Additionally, there are a number of other approaches to inference that are worth mentioning:

- Bootstrapping. Throughout the book, whenever I've introduced a hypothesis test, I've had a strong tendency just to make assertions like "the sampling distribution for BLAH is a t-distribution" or something like that. In some cases, I've actually attempted to justify this assertion. For example, when talking about χ^2 tests in Chapter 10 I made reference to the known relationship between normal distributions and χ^2 distributions (see Chapter 7) to explain how we end up assuming that the sampling distribution of the goodness-of-fit statistic is χ^2. However, it's also the case that a lot of these sampling distributions are, well, wrong. The χ^2 test is a good example. It is based on an assumption about the distribution of your data, an assumption which is known to be wrong for small sample sizes! Back in the early 20th century, there wasn't much you could do about this situation. Statisticians had developed mathematical results that said that "under assumptions BLAH about the data, the sampling distribution is approximately BLAH", and that was about the best you could do. A lot of times they didn't even have that. There are lots of data analysis situations for which no-one has found a mathematical solution for the sampling distributions that you need. And so up until the late 20th century, the corresponding tests didn't exist or didn't work. However, computers have changed all that now. There are lots of fancy tricks, and some not-so-fancy, that you can use to get around it. The simplest of these is bootstrapping, and in it's simplest form it's incredibly simple. What you do is simulate the results of your experiment lots and lots of times, under the twin assumptions that (a) the null hypothesis is true and (b) the unknown population distribution actually looks pretty similar to your raw data. In other words, instead of assuming that the data are (for instance) normally distributed, just assume that the population looks the same as your sample, and then use computers to simulate the sampling distribution for your test statistic if that assumption holds. Despite relying on a somewhat dubious assumption (i.e., the population distribution is the same as the sample!) bootstrapping is quick and easy method that works remarkably well in practice for lots of data analysis problems.

- Cross validation. One question that pops up in my stats classes every now and then, usually by a student trying to be provocative, is "Why do we care about inferential statistics at all? Why not just describe your sample?" The answer to the question is usually something like this, "Because our true interest as scientists is not the specific sample that we have observed in the *past*, we want to make predictions about data we might observe in the future". A lot of the issues in statistical inference arise because of the fact that we always expect the future to be similar to but a bit different from the past. Or, more generally, new data won't be quite the same as old data. What we do, in a lot of situations, is try to derive mathematical rules that help us to draw the inferences that are most likely to be correct for new data, rather than to pick the statements that best describe old data. For instance, given two models A and B, and a data set X you collected today, try to pick the model that will best describe a new data set Y that you're going to collect tomorrow. Sometimes it's convenient to simulate the process, and that's what cross-validation does. What you do is divide your data set into two subsets, X_1 and X_2. Use the subset X_1 to train the model (e.g., estimate regression coefficients, let's say), but then assess the model performance on the other one X_2. This gives you a measure of how well the model generalises from an old data set to a new one, and is often a better measure of how good your model is than if you just fit it to the full data set X.

- Robust statistics. Life is messy, and nothing really works the way it's supposed to. This is just as true for statistics as it is for anything else, and when trying to analyse data we're often stuck with all sorts of problems in which the data are just messier than they're supposed to be. Variables that are supposed to be normally distributed are not actually normally distributed, relationships that are supposed to be linear are not actually linear, and some of the observations in your data set are almost certainly junk (i.e., not measuring what they're supposed to). All of this messiness is ignored in most of the statistical theory I developed in this book. However, ignoring a problem doesn't always solve it. Sometimes, it's actually okay to ignore the mess, because some types of statistical tools are "robust", i.e., if the data don't satisfy your theoretical assumptions they nevertheless still work pretty well. Other types of statistical tools are not robust, and even minor deviations from the theoretical assumptions cause them to break. Robust statistics is a branch of stats concerned with this question, and they talk about things like the "breakdown point" of a statistic. That is, how messy does your data have to be before the statistic cannot be trusted? I touched on this in places. The mean is not a robust estimator of the central tendency of a variable, but the median is. For instance, suppose I told you that the ages of my five best friends are 34, 39, 31, 43 and 4003 years. How old do you think they are on average? That is, what is the true population mean here? If you use the sample mean as your estimator of the population mean, you get an answer of 830 years. If you use the sample median as the estimator of the population mean, you get an answer of 39 years. Notice that, even though you're "technically" doing the wrong thing in the second case (using the median to estimate the mean!) you're actually getting a better answer. The problem here is that one of the observations is clearly, obviously, a lie. I don't have a friend aged 4003 years. It's probably a typo, I probably meant to type 43. But what if I had typed 53 instead of 43, or 34 instead of 43? Could you be sure if this was a typo or not? Sometimes the errors in the data are subtle, so you can't detect them just by eyeballing the sample, but they're still errors that contaminate your data, and they still affect your conclusions. Robust statistics is concerned with how you can make safe inferences even when faced with contamination that you don't know about. It's pretty cool stuff.

Miscellaneous topics

- Suppose you're doing a survey, and you're interested in exercise and weight. You send data to four people. Adam says he exercises a lot and is not overweight. Briony says she exercises a lot and is not overweight. Carol says she does not exercise and is overweight. Tim says he does not exercise and refuses to answer the question about his weight. Elaine does not return the survey. You now have a missing data problem. There is one entire survey missing, and one question missing from another one, What do you do about it? Ignoring missing data is not, in general, a safe thing to do. Let's think about Tim's survey here. Firstly, notice that, on the basis of his other responses, he appear to be more similar to Carol (neither of us exercise) than to Adam or Briony. So if you were forced to guess his weight, you'd guess that he is closer to her than to them. Maybe you'd make some correction for the fact that Adam and Tim are males and Briony and Carol are females. The statistical name for this kind of guessing is "imputation". Doing imputation safely is hard, but it's important, especially when the missing data are missing in a systematic way. Because of the fact that people who are overweight are often pressured to feel poorly about their weight (often thanks to public health campaigns), we actually have reason to suspect that the people who are not responding are more likely to be overweight than the people who do respond. Imputing a weight to Tim means that the number of overweight people in the sample will probably rise from 1 out of 3 (if we ignore Tim), to 2 out of 4 (if we impute Tim's weight). Clearly this matters. But doing it sensibly is more complicated than it sounds. Earlier, I suggested you should treat Tim like Carol, since they gave the same answer to the exercise question. But that's not quite right. There is a systematic difference between them. She answered the question, and Tim didn't. Given the social pressures faced by overweight people, isn't it likely that Tim is *more* overweight than Carol? And of course this is still ignoring the fact that it's not sensible to impute a *single* weight to Tim, as if you actually knew his weight. Instead, what you need to do it is impute a range of plausible guesses (referred to as multiple imputation), in order to capture the fact that you're more uncertain about Tim's weight than you are about Carol's. And let's not get started on the problem posed by the fact that Elaine didn't send in the survey. As you can probably guess, dealing with missing data is an increasingly important topic. In fact, I've been told that a lot of journals in some fields will not accept studies that have missing data unless some kind of sensible multiple imputation scheme is followed.

- Power analysis. In Chapter 9 I discussed the concept of power (i.e., how likely are you to be able to detect an effect if it actually exists) and referred to power analysis, a collection of tools that are useful for assessing how much power your study has. Power analysis can be useful for planning a study (e.g., figuring out how large a sample you're likely to need), but it also serves a useful role in analysing data that you already collected. For instance, suppose you get a significant result, and you have an estimate of your effect size. You can use this information to estimate how much power your study actually had. This is kind of useful, especially if your effect size is not large. For instance, suppose you reject the null hypothesis at $p < .05$, but you use power analysis to figure out that your estimated power was only .08. The significant result means that, if the null hypothesis was in fact true, there was a 5% chance of getting data like this. But the low power means that, even if the null hypothesis is false and the effect size was really as small as it looks, there was only an 8% chance of getting data like you did. This suggests that you need to be pretty cautious, because luck seems to have played a big part in your results, one way or the other!

- Data analysis using theory-inspired models. In a few places in this book I've mentioned response time (RT) data, where you record how long it takes someone to do something (e.g., make a simple decision). I've mentioned that RT data are almost invariably non-normal, and positively skewed. Additionally, there's a thing known as the speed / accuracy trade-off: if you try to make decisions too quickly (low RT) then you're likely to make poorer decisions (lower accuracy). So if you measure both the accuracy of a participant's decisions and their RT, you'll probably find that speed and accuracy are related. There's more to the story than this, of course, because some people make better decisions than others regardless of how fast they're going. Moreover, speed depends on both cognitive processes (i.e., time spent thinking) but also physiological ones (e.g., how fast can you move your muscles). It's starting to sound like analysing this data will be a complicated process. And indeed it is, but one of the things that you find when you dig into the psychological literature is that there already exist mathematical models (called "sequential sampling models") that describe how people make simple decisions, and these models take into account a lot of the factors I mentioned above. You won't find any of these theoretically-inspired models in a standard statistics textbook. Standard stats textbooks describe standard tools, tools that could meaningfully be applied in lots of different disciplines, not just psychology. ANOVA is an example of a standard tool that is just as applicable to psychology as to pharmacology. Sequential sampling models are not, they are psychology-specific, more or less. This doesn't make them less powerful tools. In fact, if you're analysing data where people have to make choices quickly you should really be using sequential sampling models to analyse the data. Using ANOVA or regression or whatever won't work as well, because the theoretical assumptions that underpin them are not well-matched to your data. In contrast, sequential sampling models were explicitly designed to analyse this specific type of data, and their theoretical assumptions are extremely well-matched to the data.

Learning the basics, and learning them in jamovi

Okay, that was a long list. And even that listing is massively incomplete. There really are a lot of big ideas in statistics that I haven't covered in this book. It can seem pretty depressing to finish an almost 500-page textbook only to be told that this is only the beginning, especially when you start to suspect that half of the stuff you've been taught is wrong. For instance, there are a lot of people in the field who would strongly argue against the use of the classical ANOVA model, yet I've devoted two whole chapters to it! Standard ANOVA can be attacked from a Bayesian perspective, or from a robust statistics perspective, or even from a "it's just plain wrong" perspective (people very frequently use ANOVA when they should actually be using mixed models). So why learn it at all?

As I see it, there are two key arguments. Firstly, there's the pure pragmatism argument. Rightly or wrongly, ANOVA is widely used. If you want to understand the scientific literature, you need to understand ANOVA. And secondly, there's the "incremental knowledge" argument. In the same way that it was handy to have seen one-way ANOVA before trying to learn factorial ANOVA, understanding ANOVA is helpful for understanding more advanced tools, because a lot of those tools extend on or modify the basic ANOVA setup in some way. For instance, although mixed models are way more useful than ANOVA and regression, I've never heard of anyone learning how mixed models work without first having worked through ANOVA and regression. You have to learn to crawl before you can climb a mountain.

Actually, I want to push this point a bit further. One thing that I've done a lot of in this book is talk about fundamentals. I spent a lot of time on probability theory. I talked about the theory of estimation and hypothesis tests in more detail than I needed to. Why did I do all this? Looking back, you might ask whether I really needed to spend all that time talking about what a probability distribution is, or why there was even a section on probability density. If the goal of the book was to teach you how to run a t-test or an ANOVA, was all that really necessary? Was this all just a huge waste of everyone's time???

The answer, I hope you'll agree, is no. The goal of an introductory stats is not to teach ANOVA. It's not to teach t-tests, or regressions, or histograms, or p-values. The goal is to start you on the path towards becoming a skilled data analyst. And in order for you to become a skilled data analyst, you need to be able to do more than ANOVA, more than t-tests, regressions and histograms. You need to be able to think properly about data. You need to be able to learn the more advanced statistical models that I talked about in the last section, and to understand the theory upon which they are based. And you need to have access to software that will let you use those advanced tools. And this is where, in my opinion at least, all that extra time I've spent on the fundamentals pays off. If you understand probability theory, you'll find it much easier to switch from frequentist analyses to Bayesian ones.

In short, I think that the big payoff for learning statistics this way is extensibility. For a book that only covers the very basics of data analysis, this book has a massive overhead in terms of learning probability theory and so on. There's a whole lot of other things that it pushes you to learn besides the specific analyses that the book covers. So if your goal had been to learn how to run an ANOVA in the minimum possible time, well, this book wasn't a good choice. But as I say, I don't think that is your goal. I think you want to learn how to do data analysis. And if that really is your goal, you want to make sure that the skills you learn in your introductory stats class are naturally and cleanly extensible to the more complicated models that you need in real world data analysis. You want to make sure that you learn to use the same tools that real data analysts use, so that you can learn to do what they do. And so yeah, okay, you're a beginner right now (or you were when you started this book), but that doesn't mean you should be given a dumbed-down story, a story in which I don't tell you about probability density, or a story where I don't tell you about the nightmare that is factorial ANOVA with unbalanced designs. And it doesn't mean that you should be given baby toys instead of proper data analysis tools. Beginners aren't dumb, they just lack knowledge. What you need is not to have the complexities of real-world data analysis hidden from from you. What you need are the skills and tools that will let you handle those complexities when they inevitably ambush you in the real world.

And what I hope is that this book is able to help you with that.

Author's note – If you see anything clever sounding in this book that doesn't seem to have a reference, I can absolutely promise you that the idea was someone else's. This is an introductory textbook: none of the ideas are original. I'll take responsibility for all the errors, but I can't take credit for any of the good stuff. Everything smart in this book came from someone else.

References

Adair, G. (1984). The hawthorne effect: A reconsideration of the methodological artifact. *Journal of Applied Psychology*, *69*, 334–345. https://doi.org/10.1037/0021-9010.69.2.334

Agresti, A. (1996). *An introduction to categorical data analysis*. Wiley. https://doi.org/10.1002/0470114754

Agresti, A. (2002). *Categorical data analysis* (2nd ed.). Wiley. https://doi.org/10.1002/0471249688

Akaike, H. (1974). A new look at the statistical model identification. *IEEE Transactions on Automatic Control*, *19*, 716–723. https://doi.org/10.1109/TAC.1974.1100705

Anscombe, F. J. (1973). Graphs in statistical analysis. *American Statistician*, *27*, 17–21. https://doi.org/10.1080/00031305.1973.10478966

Bickel, P. J., Hammel, E. A., & O'Connell, J. W. (1975). Sex bias in graduate admissions: Data from Berkeley. *Science*, *187*, 398–404. https://doi.org/10.1126/science.187.4175.398

Box, G. E. P. (1953). Non-normality and tests on variances. *Biometrika*, *40*, 318–335. https://doi.org/10.2307/2333350

Box, G. E. P. (1976). Science and statistics. *Journal of the American Statistical Association*, *71*, 791–799. https://doi.org/10.1080/01621459.1976.10480949

Box, J. F. (1987). Guinness, gosset, fisher, and small samples. *Statistical Science*, *2*, 45–52. https://doi.org/10.1214/ss/1177013437

Brown, M. B., & Forsythe, A. B. (1974). Robust tests for equality of variances. *Journal of the American Statistical Association*, *69*, 364–367. https://doi.org/10.2307/2285659

Campbell, D. T., & Stanley, J. C. (1963). *Experimental and quasi-experimental designs for research*. Houghton Mifflin.

Child, D. (1990). *The essentials of factor analysis* (2nd ed.). Cassell Educational.

Chronbach, L. J. (1951). Coefficient alpha and the internal structure of tests. *Psychometrika*, *16(3)*, 297–334. https://doi.org/10.1007/BF02310555

Cochran, W. G. (1954). The χ^2 test of goodness of fit. *The Annals of Mathematical Statistics*, *23*, 315–345. https://doi.org/10.1214/aoms/1177729380

Cohen, J. (1988). *Statistical power analysis for the behavioral sciences* (2nd ed.). Lawrence Erlbaum. https://doi.org/10.4324/9780203771587

Cramer, H. (1946). *Mathematical methods of statistics*. Princeton University Press. https://doi.org/10.1515/9781400883868

Dunn, O. J. (1961). Multiple comparisons among means. *Journal of the American Statistical Association*, *56*, 52–64. https://doi.org/10.1080/01621459.1961.10482090

Ellis, P. D. (2010). *The essential guide to effect sizes: Statistical power, meta-analysis, and the interpretation of research results*. Cambridge University Press. https://doi.org/10.1017/CBO9780511761676

Evans, J. St. B. T., Barston, J. L., & Pollard, P. (1983). On the conflict between logic and belief in syllogistic reasoning. *Memory and Cognition*, *11*, 295–306. https://doi.org/10.3758/BF03196976

Evans, M., Hastings, N., & Peacock, B. (2011). *Statistical distributions (3rd ed)*. Wiley. https://doi.org/10.1002/9780470627242

Everitt, B. S. (1996). *Making sense of statistics in psychology. A second-level course*. Oxford University Press.

Fabrigar, L. R., Wegener, D. T., MacCallum, R. C., & Strahan, E. J. (1999). Evaluating the use of exploratory factor analysis in psychological research. *Psychological Methods*, *4*, 272–299. https://doi.org/10.1037/1082-989X.4.3.272

Fisher, R. A. (1922a). On the interpretation of χ^2 from contingency tables, and the calculation of *p*. *Journal of the Royal Statistical Society*, *84*, 87–94. https://doi.org/10.1111/j.2397-2335.1922.tb00768.x

Fisher, R. A. (1922b). On the mathematical foundation of theoretical statistics. *Philosophical Transactions of the Royal Society A*, *222*, 309–368. https://doi.org/10.1098/rsta.1922.0009

Fisher, R. A. (1925). *Statistical methods for research workers*. Oliver & Boyd.

Fox, J., & Weisberg, S. (2011). *An R companion to applied regression* (2nd ed.). Sage.

Gelman, A., & Stern, H. (2006). The difference between "significant" and "not significant" is not itself statistically significant. *The American Statistician*, *60*, 328–331. https://doi.org/10.1198/000313006X152649

Geschwind, N. (1972). Language and the brain. *Scientific American*, *226(4)*, 76–83. https://doi.org/10.1038/scientificamerican0472-76

Hays, W. L. (1994). *Statistics* (5th ed.). Harcourt Brace.

Hedges, L. V. (1981). Distribution theory for glass's estimator of effect size and related estimators. *Journal of Educational Statistics*, *6*, 107–128. https://doi.org/10.2307/1164588

Hedges, L. V., & Olkin, I. (1985). *Statistical methods for meta-analysis*. Academic Press. https://doi.org/10.1016/C2009-0-03396-0

Hewitt, A. K., Foxcroft, D. R., & MacDonald, J. (2004). Multitrait-multimethod confirmatory factor analysis of the attributional style questionnaire. *Personality and Individual Differences*, *37(7)*, 1483–1491. https://doi.org/10.1016/j.paid.2004.02.005

Hogg, R. V., McKean, J. V., & Craig, A. T. (2005). *Introduction to mathematical statistics* (6th ed.). Pearson.

Holm, S. (1979). A simple sequentially rejective multiple test procedure. *Scandinavian Journal of Statistics*, *6*, 65–70. https://doi.org/10.2307/4615733

Hróbjartsson, A., & Gøtzsche, P. (2010). Placebo interventions for all clinical conditions. *Cochrane Database of Systematic Reviews*, *1*. https://doi.org/10.1002/14651858.cd003974.pub3

Hsu, J. C. (1996). *Multiple comparisons: Theory and methods*. Chapman & Hall. https://doi.org/10.1201/b15074

Ioannidis, J. P. A. (2005). Why most published research findings are false. *PLoS Med*, *2(8)*, 697–701. https://doi.org/10.1371/journal.pmed.1004085

Jeffreys, H. (1961). *The theory of probability* (3rd ed.). Oxford. https://doi.org/10.1093/oso/9780198503682.001.0001

Johnson, V. E. (2013). Revised standards for statistical evidence. *Proceedings of the National Academy of Sciences*, *48*, 19313–19317. https://doi.org/10.1073/pnas.1313476110

Kahneman, D., & Tversky, A. (1973). On the psychology of prediction. *Psychological*

Review, *80*, 237–251. https://doi.org/10.1037/h0034747

Kass, R. E., & Raftery, A. E. (1995). Bayes factors. *Journal of the American Statistical Association*, *90*, 773–795. https://doi.org/10.1080/01621459.1995.10476572

Keynes, J. M. (1923). *A tract on monetary reform*. Macmillan & Company.

Kline, P. (1994). *An easy guide to factor analysis*. Routledge. https://doi.org/10.4324/9781315788135

Kruschke, J. K. (2011). *Doing Bayesian data analysis: A tutorial with R and BUGS*. Academic Press.

Kruskal, W. H., & Wallis, W. A. (1952). Use of ranks in one-criterion variance analysis. *Journal of the American Statistical Association*, *47*, 583–621. https://doi.org/10.1080/01621459.1952.10483441

Kühberger, A., Fritz, A., & Scherndl, T. (2014). Publication bias in psychology: A diagnosis based on the correlation between effect size and sample size. *Public Library of Science One*, *9*, 1–8. https://doi.org/10.1371/journal.pone.0105825

Larntz, K. (1978). Small-sample comparisons of exact levels for chi-squared goodness-of-fit statistics. *Journal of the American Statistical Association*, *73*, 253–263. https://doi.org/10.1080/01621459.1978.10481567

Lee, M. D., & Wagenmakers, E.-J. (2014). *Bayesian cognitive modeling: A practical course*. Cambridge University Press. https://doi.org/10.1017/CBO9781139087759

Lehmann, E. L. (2011). *Fisher, Neyman, and the creation of classical statistics*. Springer.

Levene, H. (1960). Robust tests for equality of variances. In Olkin, I. and others (Ed.), *Contributions to probability and statistics: Essays in honor of harold hotelling* (pp. 278–292). Stanford University Press.

McGrath, R. E., & Meyer, G. J. (2006). When effect sizes disagree: The case of r and d. *Psychological Methods*, *11*, 386–401. https://doi.org/10.1037/1082-989x.11.4.386

Meehl, P. H. (1967). Theory testing in psychology and physics: A methodological paradox. *Philosophy of Science*, *34*, 103–115. https://doi.org/10.1086/288135

Pearson, K. (1900). On the criterion that a given system of deviations from the probable in the case of a correlated system of variables is such that it can be reasonably supposed to have arisen from random sampling. *Philosophical Magazine*, *50*, 157–175. https://doi.org/10.1080/14786440009463897

Peterson, C., & Seligman, M. (1984). Causal explanations as a risk factor for depression: Theory and evidence. *Psychological Review*, *91*, 347–374. https://doi.org/10.1037/0033-295X.91.3.347

Pfungst, O. (1911). *Clever hans (the horse of mr. Von osten): A contribution to experimental animal and human psychology* (C. L. Rahn, Trans.). Henry Holt.

Rosenthal, R. (1966). *Experimenter effects in behavioral research*. Appleton.

Sahai, H., & Ageel, M. I. (2000). *The analysis of variance: Fixed, random and mixed models*. Birkhauser.

Shaffer, J. P. (1995). Multiple hypothesis testing. *Annual Review of Psychology*, *46*, 561–584. https://doi.org/10.1146/annurev.ps.46.020195.003021

Shapiro, S. S., & Wilk, M. B. (1965). An analysis of variance test for normality (complete samples). *Biometrika*, *52*, 591–611. https://doi.org/10.1093/biomet/52.3-4.591

Sokal, R. R., & Rohlf, F. J. (1994). *Biometry: The principles and practice of statistics in biological research* (3rd ed.). Freeman.

Stevens, S. S. (1946). On the theory of scales of measurement. *Science*, *103*, 677–680. https://doi.org/10.1126/science.103.2684.677

Stigler, S. M. (1986). *The history of statistics*. Harvard University Press.

Student, A. (1908). The probable error of a mean. *Biometrika*, *6*, 1–2. https://doi.org/

10.1093/biomet/6.1.1

Tversky, A., & Kahneman, D. (1974). Judgment under uncertainty: Heuristics and biases. *Science*, *185*(4157), 1124–1131. https://doi.org/10.1126/science.185.4157.1124

Welch, B. L. (1947). The generalization of "Student's" problem when several different population variances are involved. *Biometrika*, *34*, 28–35. https://doi.org/10.1093/biomet/34.1-2.28

Welch, B. L. (1951). On the comparison of several mean values: An alternative approach. *Biometrika*, *38*, 330–336. https://doi.org/10.1093/biomet/38.3-4.330

Wilkinson, L., Wills, D., Rope, D., Norton, A., & Dubbs, R. (2006). *The grammar of graphics*. Springer.

Yates, F. (1934). Contingency tables involving small numbers and the χ^2 test. *Supplement to the Journal of the Royal Statistical Society*, *1*, 217–235. https://doi.org/10.2307/2983604

Chapter notes

Notes for chapter 1

1. The quote comes from Auden's 1946 poem *Under Which Lyre: A Reactionary Tract for the Times*, delivered as part of a commencement address at Harvard University. The history of the poem is kind of interesting, see Adam Kirsch's analysis in the *Harvard Magazine*, https://www.harvardmagazine.com/2007/11/a-poets-warning.html

2. Including the suggestion that common sense is in short supply among scientists.

3. In my more cynical moments I feel like this fact alone explains 95% of what I read on the internet.

4. Earlier versions of these notes incorrectly suggested that they actually were sued. But that's not true. There's a nice commentary on this by Alex Reinhart here: https://www.refsmmat.com/posts/2016-05-08-simpsons-paradox-berkeley.html A big thank you to Wilfried Van Hirtum for pointing this out to me.

5. Which might explain why physics is just a teensy bit further advanced as a science than we are.

Notes for chapter 2

6. Presidential Address to the First Indian Statistical Congress, 1938. Source: https://en.wikiquote.org/wiki/Ronald_Fisher

7. Actually, I've been informed by readers with greater physics knowledge than I that temperature isn't strictly an interval scale, in the sense that the amount of energy required to heat something up by 3° depends on its current temperature. So in the sense that physicists care about, temperature isn't actually an interval scale. But it still makes a cute example so I'm going to ignore this little inconvenient truth.

8. Ah, psychology... never an easy answer to anything!

9. Annoyingly though, there's a lot of different names used out there. I won't list all of them – there would be no point in doing that – other than to note that "response variable" is sometimes used where I've used "outcome". Sigh. This sort of terminological confusion is very common, I'm afraid.

10. The reason why I say that it's unmeasured is that if you have measured it, then you can use some fancy statistical tricks to deal with the confounder. Because of the existence of these statistical solutions to the problem of confounders, we often refer to a confounder that we have measured and dealt with as a covariate. Dealing with covariates is a more advanced topic, but I thought I'd mention it in passing since it's kind of comforting to at least know that this stuff exists.

11. Some people might argue that if you're not honest then you're not a real scientist. Which does have some truth to it I guess, but that's disingenuous (look up the "No true Scotsman" fallacy). The fact is that there are lots of people who are employed ostensibly as scientists, and whose work has all of the trappings of science, but who are outright fraudulent. Pretending that they don't exist by saying that they're not scientists is just muddled thinking.

12. Clearly, the real effect is that only insane people would even try to read *Finnegans Wake*.

Notes for chapter 3

13. Source: *Dismal Light* (1968).

14. At the time of first writing this in August 2018. Later versions of this book will use later versions of jamovi.

15. Although jamovi is updated frequently it doesn't usually make much of a difference for the sort of work we'll do in this book. In fact, during the writing of the book I upgraded several times and it didn't make much difference at all to what is in this book.

16. From now on, we'll use single quote marks to signify a label, command, option, or outputs in the jamovi interface.

17. In later versions of jamovi there is a pre-defined function 'Z' to compute z-scores, which is much easier!

18. R is a powerful statistical programming language. In fact, jamovi is just a user-friendly interface that sits on top of the R engine.

19. You can change the default value for missing values in jamovi from the top right menu (three vertical dots), but this only works at the time of importing data files into jamovi. The default missing value in the data set should not be a valid number associated with any of the variables, e.g., you could use -9999 as this is unlikely to be a valid value.

20. I know this is a bit of a fudge, but it does work and hopefully this will be fixed in a later version of jamovi.

Notes for chapter 4

21. Note for non-Australians: the AFL is an Australian rules football competition. You don't need to know anything about Australian rules in order to follow this section.

22. Okay, now let's try to write a formula for the mean. By tradition, we use $\bar{X}$ as the notation for the mean. So the calculation for the mean could be expressed using the following formula:

$$\bar{X} = \frac{X_1 + X_2 ... + X_{N-1} + X_N}{N}$$

This formula is entirely correct but it's terribly long, so we make use of the **summation symbol** $\sum$ to shorten it. If I want to add up the first five observations I could write out the sum the long way, $X_1 + X_2 + X_3 + X_4 + X_5$ or I could use the summation symbol to shorten it to this:

$$\sum_{i=1}^{5} X_i$$

Taken literally, this could be read as "the sum, taken over all i values from 1 to 5, of the value X_i". But basically what it means is "add up the first five observations". In any case, we can use this notation to write out the formula for the mean, which looks like this:

$$\bar{X} = \frac{1}{N} \sum_{i=1}^{N} X_i$$

In all honesty, I can't imagine that all this mathematical notation helps clarify the concept of the mean at all. In fact, it's really just a fancy way of writing out the same thing I said in words: add all the values up and then divide by the total number of items. However, that's not really the reason I went into all that detail. My goal was to try to make sure that everyone reading this book is clear on the notation that we'll be using throughout the book: $\bar{X}$ for the mean, $\sum$ for the idea of summation, X_i for the ith observation, and N for the total number of observations. We're going to be re-using these symbols a fair bit so it's important that you understand them well enough to be able to "read" the equations, and to be able to see that it's just saying "add up lots of things and then divide by another thing". *The choice to use $\sum$ to denote summation isn't arbitrary. It's the Greek upper case letter sigma, which is the analogue of the letter S in that alphabet. Similarly, there's an equivalent symbol used to denote the multiplication of lots of numbers, because multiplications are also called "products" we use the $\prod$ symbol for this (the Greek upper case pi, which is the analogue of the letter P.*

23. www.abc.net.au/news/2010-09-24/housing-bubble-debate-boils-over/2273406

24. However, whilst our calculations for this little example are at an end, we do have a couple of things left to talk about. First, we should really try to write down a proper mathematical formula. But in order to do this I need some mathematical notation to refer to the mean absolute deviation. Irritatingly, "mean absolute deviation" and "median absolute deviation" have the same acronym (MAD), which leads to a certain amount of ambiguity so I'd better come up with something different for the mean absolute deviation. Sigh. What I'll do is use AAD instead, short for average absolute deviation. Now that we have some unambiguous notation, here's the formula that describes what we just calculated:

$$AAD(X) = \frac{1}{N} \sum_{i=1}^{N} |X_i - \bar{X}| = 15.52$$

25. Well, I will very briefly mention the one that I think is coolest, for a very particular definition of "cool", that is. Variances are additive. Here's what that means. Suppose I have two variables X and Y, whose variances are $Var(X)$ and $Var(Y)$ respectively. Now imagine I want to define a new variable Z that is the sum of the two, $Z = X + Y$. As it turns out, the variance of Z is equal to $Var(X) + Var(Y)$. This is a very useful property, but it's not true of the other measures that I talk about in this section.

26. The formula that we use to calculate the variance of a set of observations is as follows:

$$VAR(X) = \frac{1}{N} \sum_{i=1}^{N} (X_i - \bar{X})^2$$

As you can see, it's basically the same formula that we used to calculate the average absolute deviation, except that instead of using "absolute deviations" we use "squared deviations". It is for this reason that the variance is sometimes referred to as the "mean square deviation".

27. With the possible exception of the third question.

28. In other words, the formula that jamovi is using is this one:

$$\frac{1}{N-1} \sum_{i=1}^{N} (X_i - \bar{X})^2$$

29. Because the standard deviation is equal to the square root of the variance, you probably won't be surprised to see that the formula is:

$$s = \sqrt{\frac{1}{N} \sum_{i=1}^{N} (X_i - \bar{X})^2}$$

and in jamovi there is a check box for 'Std. deviation' right above the check box for 'Variance'. Selecting this gives a value of 26.07 for the standard deviation.

30. For reasons that will make sense when we return to this topic in Chapter 8 I'll refer to this new quantity as $\hat{\sigma}$ (read as: "sigma hat"), and the formula for this is:

$$\hat{\sigma} = \sqrt{\frac{1}{N-1}\sum_{i=1}^{N}(X_i - \bar{X})^2}$$

31. One formula for the skewness of a data set is:

$$skewness(X) = \frac{1}{N\hat{\sigma}^3}\sum_{i=1}^{N}(X_i - \bar{X})^3$$

where N is the number of observations, $\bar{X}$ is the sample mean, and $\hat{\sigma}$ is the standard deviation (the "divide by $N-1$" version, that is).

32. The equation for kurtosis is pretty similar in spirit to the formulas we've seen already for the variance and the skewness. Except that where the variance involved squared deviations and the skewness involved cubed deviations, the kurtosis involves raising the deviations to the fourth power:[a]

$$kurtosis(X) = \frac{1}{N\hat{\sigma}^4}\sum_{i=1}^{N}(X_i - \bar{X})^4 - 3$$

I know, it's not terribly interesting to me either. — [a] The "-3" part is something that statisticians tack on to ensure that the normal curve has kurtosis zero. It looks a bit stupid, just sticking a "-3" at the end of the formula, but there are good mathematical reasons for doing this.

33. Sometimes jamovi will also present numbers in an unusual way. If a number is very small, or very large, then jamovi switches to an exponential form for numbers. For example 6.51e-4 is the same as saying that the decimal point is moved 4 places to the left, so the actual number is 0.000651. If there is a plus sign (i.e. 6.51e+4 then the decimal point is moved to the right, i.e. 65,100.00. Usually only very small or very large numbers are expressed in this way, for example 6.51e-16, which would be quite unwieldy to write out in the normal way.

34. In actual maths, the equation for the z-score is:

$$z_i = \frac{X_i - \bar{X}}{\hat{\sigma}}$$

35. Though some caution is usually warranted. It's not always the case that one standard deviation on variable A corresponds to the same "kind" of thing as one standard deviation on variable B. Use common sense when trying to determine whether or not the z-scores of two variables can be meaningfully compared.

Notes for chapter 5

36. The origin of this quote is Tufte's lovely book *The Visual Display of Quantitative Information*.

37. This altered version of the AFL Margins By Year data set isn't available to open / load into jamovi. I simply changed a couple of the values of margin in the data set so that they were over 300. You can do this yourself if you want.

38. jamovi uses the symbol "==" here to mean "matches".

Notes for chapter 6

39. The quote comes from *Home is the Hangman*, published in 1975.

40. I offer up my teenage attempts to be "cool" as evidence that some things just can't be done.

41. You can do this in the 'Compute new variable' screen, though just calculating 2 + 2 for every cell of a new variable is not very useful!

42. Note that this is a very different operator to the equals operator =. A common typo that people make when trying to write logical commands in jamovi (or other languages, since the "= versus ==" distinction is important in many computer and statistical programs) is to accidentally type = when you really mean ==. Be especially cautious with this, I've been programming in various languages since I was a teenager and I still screw this up a lot.

43. Now, here's a quirk in jamovi. When you have simple logical expressions like the ones we have already met, e.g., $2 + 2 == 5$ then jamovi neatly states 'false' (or 'true') in the corresponding spreadsheet column. Underneath the hood, jamovi stores 'false' as 0 and 'true' as 1. When we have more complex logical expressions, such as $(2 + 2 == 4)$ or $(2 + 2 == 5)$, then jamovi just displays either 0 or 1, depending whether the logical expression is evaluated as false, or true.

44. The absolute value of a number is its distance from zero, regardless of whether its sign is negative or positive.

45. The reason we have to use the 'IF' command and keep zero as zero is that you cannot just use 'likert.centred / opinion.strength' to calculate the sign of 'likert.centred', because mathematically dividing zero by zero does not work. Try it and see.

46. If you've read further into the book, and are re-reading this section, then a good example of this would be someone choosing to do an ANOVA using 'AgeCats' as the grouping variable, instead of running a regression using Age as a predictor. There are sometimes good reasons for doing this. For instance, if the relationship between Age and your outcome variable is highly non-linear and you aren't comfortable with trying to run non-linear regression! However, unless you really do have a good rationale for doing this, it's best not to. It tends to introduce all sorts of other problems (e.g., the data will probably violate the normality assumption) and you can lose a lot of statistical power.

47. We'll leave the box-cox function until later on.

Notes for chapter 7

48. This doesn't mean that frequentists can't make hypothetical statements, of course. It's just that if you want to make a statement about probability then it must be possible to redescribe that statement in terms of a sequence of potentially observable events, together with the relative frequencies of different outcomes that appear within that sequence.

49. Note that the term "success" is pretty arbitrary and doesn't actually imply that the outcome is something to be desired. If θ referred to the probability that any one passenger gets injured in a bus crash I'd still call it the success probability, but that doesn't mean I want people to get hurt in bus crashes!

50. For those readers who know a little calculus, I'll give a slightly more precise explanation. In the same way that probabilities are non-negative numbers that must sum to 1, probability densities are non-negative numbers that must integrate to 1 (where the integral is taken across all possible values of X). To calculate the probability that X falls between a and b we calculate the definite integral of the density function over the corresponding range, $\int_a^b p(x)dx$. If you don't remember or never learned calculus, don't worry about this. It's not needed for this book.

51. In the equation for the binomial, $X!$ is the factorial function (i.e., multiply all whole numbers from 1 to X):

$$P(X|\theta, N) = \frac{N!}{X!(N-X)!} \theta^X (1-\theta)^{N-X}$$

If this equation doesn't make a lot of sense to you, don't worry too much about it.

52. As was the case with the binomial distribution, I have included the formula for the normal distribution in this book, because I think it's important enough that everyone who learns statistics should at least look at it, but since this is an introductory text I don't want to focus on it, so I've tucked it away in this footnote:

$$p(X|\mu, \sigma) = \frac{1}{\sigma\sqrt{2\pi}} e^{-\frac{(X-\mu)^2}{2\sigma^2}}$$

53. There is a subtle and somewhat frustrating characteristic of continuous distributions that makes the y-axis behave a bit oddly - the height of the curve here is not actually the probability of observing a particular x value. On the other hand, it is true that the heights of the curve tells you which x values are more likely (the higher ones!). (see Probability density section for all the annoying details)

54. In practice, the normal distribution is so handy that people tend to use it even when the variable isn't actually continuous. As long as there are enough categories (e.g., Likert-scale responses to a questionnaire), it's pretty standard practice to use the normal distribution as an approximation. This works out much better in practice than you'd think.

55. For those readers who know a little calculus, I'll give a slightly more precise explanation. In the same way that probabilities are non-negative numbers that must sum to 1, probability densities are non-negative numbers that must integrate to 1 (where the integral is taken across all possible values of X). To calculate the probability that X falls between a and b we calculate the definite integral of the density function over the corresponding range, $\int_a^b p(x)dx$. If you don't remember or never learned calculus, don't worry about this. It's not needed for this book.

Notes for chapter 8

56. The proper mathematical definition of randomness is extraordinarily technical, and way beyond the scope of this book. We'll be non-technical here and say that a process has an element of randomness to it whenever it is possible to repeat the process and get different answers each time.

57. Nothing in life is that simple. There's not an obvious division of people into binary categories like "schizophrenic" and "not schizophrenic". But this isn't a clinical psychology text so please forgive me a few simplifications here and there.

58. Technically, the law of large numbers pertains to any sample statistic that can be described as an average of independent quantities. That's certainly true for the sample mean. However, it's also possible to write many other sample statistics as averages of one form or another. The variance of a sample, for instance, can be rewritten as a kind of average and so is subject to the law of large numbers. The minimum value of a sample, however, cannot be written as an average of anything and is therefore not governed by the law of large numbers.

59. As usual, I'm being a bit sloppy here. The central limit theorem is a bit more general than this section implies. Like most introductory stats texts I've discussed one situation where the central limit theorem holds: when you're taking an average across lots of independent events drawn from the same distribution. However, the central limit theorem is much broader than this. There's a whole class of things called "U-statistics" for instance, all of which satisfy the central limit theorem and therefore become normally distributed for large sample sizes. The mean is one such statistic, but it's not the only one.

60. Please note that if you were actually interested in this question you would need to be a lot more careful than I'm being here. You can't just compare IQ scores in Whyalla to Port Pirie and assume that any differences are

due to lead poisoning. Even if it were true that the only differences between the two towns corresponded to the different refineries (and it isn't, not by a long shot), you need to account for the fact that people already believe that lead pollution causes cognitive deficits. If you recall back to Chapter 2, this means that there are different demand effects for the Port Pirie sample than for the Whyalla sample. In other words, you might end up with an illusory group difference in your data, caused by the fact that people think that there is a real difference. I find it pretty implausible to think that the locals wouldn't be well aware of what you were trying to do if a bunch of researchers turned up in Port Pirie with lab coats and IQ tests, and even less plausible to think that a lot of people would be pretty resentful of you for doing it. Those people won't be as co-operative in the tests. Other people in Port Pirie might be more motivated to do well because they don't want their home town to look bad. The motivational effects that would apply in Whyalla are likely to be weaker, because people don't have any concept of "iron ore poisoning" in the same way that they have a concept for "lead poisoning". Psychology is hard.

61. I should note that I'm hiding something here. Unbiasedness is a desirable characteristic for an estimator, but there are other things that matter besides bias. However, it's beyond the scope of this book to discuss this in any detail. I just want to draw your attention to the fact that there's some hidden complexity here.

62. Dividing by $N-1$ gives us an unbiased estimate of the population variance:

$$\hat{\sigma}^2 = \frac{1}{N-1}\sum_{i=1}^{N}(X_i - \bar{X})^2$$

and similarly for standard deviation:

$$\hat{\sigma} = \sqrt{\frac{1}{N-1}\sum_{i=1}^{N}(X_i - \bar{X})^2}$$

Okay, I'm hiding something else here. In a bizarre and counter-intuitive twist, since $\hat{\sigma}^2$ is an unbiased estimator of σ^2, you'd assume that taking the square root would be fine and $\hat{\sigma}$ would be an unbiased estimator of σ. Right? Weirdly, it's not. There's actually a subtle, tiny bias in $\hat{\sigma}$. This is just bizarre: $\hat{\sigma}^2$ is an unbiased estimate of the population variance σ^2, but when you take the square root, it turns out that $\hat{\sigma}$ is a biased estimator of the population standard deviation σ. Weird, weird, weird, right? So, why is $\hat{\sigma}$ biased? The technical answer is "because non-linear transformations (e.g., the square root) don't commute with expectation", but that just sounds like gibberish to everyone who hasn't taken a course in mathematical statistics. Fortunately, it doesn't matter for practical purposes. The bias is small, and in real life everyone uses $\hat{\sigma}$ and it works just fine. Sometimes mathematics is just annoying.

63. This quote appears on a great many t-shirts and websites, and even gets a mention in a few academic papers (e.g., https://jse.amstat.org/v10n3/friedman.html, but I've never found the original source.

64. Mathematically, we write this as:

$$\mu - (1.96 \times SEM) \leq \bar{X} \leq \mu + (1.96 \times SEM)$$

where the SEM is equal to $\frac{\sigma}{\sqrt{N}}$ and we can be 95% confident that this is true. However, that's not answering the question that we're actually interested in. The equation above tells us what we should expect about the sample mean given that we know what the population parameters are. What we want is to have this work the other way around. We want to know what we should believe about the population parameters, given that we have observed a particular sample. However, it's not too difficult to do this. Using a little high school algebra, a sneaky way to rewrite our equation is like this:

$$\bar{X} - (1.96 \times SEM) \leq \mu \leq \bar{X} + (1.96 \times SEM)$$

What this is telling is is that the range of values has a 95% probability of containing the population mean μ. We refer to this range as a **95% confidence interval**, denoted CI_{95}. In short, as long as N is sufficiently large (large enough for us to believe that the sampling distribution of the mean is normal), then we can write this as our formula for the 95% confidence interval:

$$CI_{95} = \bar{X} \pm (1.96 \times \frac{\sigma}{\sqrt{N}})$$

Notes for chapter 9

65. The quote comes from Wittgenstein's (1922) text, *Tractatus Logico-Philosphicus*.

66. A technical note. The description below differs subtly from the standard description given in a lot of introductory texts. The orthodox theory of null hypothesis testing emerged from the work of Sir Ronald Fisher and Jerzy Neyman in the early 20th century; but Fisher and Neyman actually had very different views about how it should work. The standard treatment of hypothesis testing that most texts use is a hybrid of the two approaches. The treatment here is a little more Neyman-style than the orthodox view, especially as regards the meaning of the *p*-value.

67. My apologies to anyone who actually believes in this stuff, but on my reading of the literature on ESP it's just not reasonable to think this is real. To be fair, though, some of the studies are rigorously designed, so it's actually an interesting area for thinking about psychological research design. And of course it's a free country so you can spend your own time and effort proving me wrong if you like, but I wouldn't think that's a terribly practical use of your intellect.

68. This analogy only works if you're from an adversarial legal system like UK/US/Australia. As I understand these things, the French inquisitorial system is quite different.

69. An aside regarding the language you use to talk about hypothesis testing. First, one thing you really want to avoid is the word "prove". A statistical test really doesn't prove that a hypothesis is true or false. Proof implies certainty and, as the saying goes, statistics means never having to say you're certain. On that point almost everyone would agree. However, beyond that there's a fair amount of confusion. Some people argue that you're only allowed to make statements like "rejected the null", "failed to reject the null", or possibly "retained the null". According to this line of thinking you can't say things like "accept the alternative" or "accept the null". Personally I think this is too strong. In my opinion, this conflates null hypothesis testing with Karl Popper's falsificationist view of the scientific process. Whilst there are similarities between falsificationism and null hypothesis testing, they aren't equivalent. However, whilst I personally think it's fine to talk about accepting a hypothesis (on the proviso that "acceptance" doesn't actually mean that it's necessarily true, especially in the case of the null hypothesis), many people will disagree. And more to the point, you should be aware that this particular weirdness exists so that you're not caught unawares by it when writing up your own results.

70. Strictly speaking, the test I just constructed has $\alpha = .057$, which is a bit too generous. However, if I'd chosen 39 and 61 to be the boundaries for the critical region then the critical region only covers 3.5% of the distribution. I figured that it makes more sense to use 40 and 60 as my critical values, and be willing to tolerate a 5.7% type I error rate, since that's as close as I can get to a value of $\alpha = .05$.

71. The internet seems fairly convinced that Ashley said this, though I can't for the life of me find anyone willing to give a source for the claim.

72. That's $p = .000000000000000000000000136$ for folks that don't like scientific notation!

73. Note that the *p* here has nothing to do with a *p*-value. The *p* argument in the jamovi binomial test corresponds to the probability of making a correct response, according to the null hypothesis. In other words, it's the θ value.

74. Although in practice a very small effect size is worrying, because even very minor methodological flaws might be responsible for the effect; and in practice no experiment is perfect, so there are always methodological issues to worry about.

75. Notice that the true population parameter θ doesn't necessarily correspond to an immutable fact of nature. In this context θ is just the true probability that people would correctly guess the colour of the card in the other room. As such the population parameter can be influenced by all sorts of things. Of course, this is all on the assumption that ESP actually exists!

76. One possible exception to this is when researchers study the effectiveness of a new medical treatment and they specify in advance what an important effect size would be to detect, for example over and above any existing treatment. In this way some information about the potential value of a new treatment can be obtained.

77. Although this book describes both Neyman's and Fisher's definition of the *p*-value, most don't. Most introductory textbooks will only give you the Fisher version.

78. In this case, the Pearson chi-square test of independence (see Chapter 10).

Notes for chapter 10

79. Also sometimes referred to as "chi-squared".

80. A vector is a sequence of data elements of the same basic type:

81. If we let k refer to the total number of categories (i.e., $k = 4$ for our cards data), then the χ^2 statistic is given by:
$$\chi^2 = \sum_{i=1}^{k} \frac{(O_i - E_i)^2}{E_i}$$
Intuitively, it's clear that if chi^2 is small, then the observed data O_i are very close to what the null hypothesis predicted E_i, so we're going to need a large χ^2 statistic in order to reject the null.

82. If you rewrite the equation for the goodness-of-fit statistic as a sum over $k-1$ independent things you get the "proper" sampling distribution, which is chi-square with $k-1$ degrees of freedom. It's beyond the scope of an introductory book to show the maths in that much detail. All I wanted to do is give you a sense of why the goodness-of-fit statistic is associated with the chi-square distribution.

83. I feel obliged to point out that this is an over-simplification. It works nicely for quite a few situations, but every now and then we'll come across degrees of freedom values that aren't whole numbers. Don't let this worry you too much; when you come across this just remind yourself that "degrees of freedom" is actually a bit of a messy concept, and that the nice simple story that I'm telling you here isn't the whole story. For an introductory class it's usually best to stick to the simple story, but I figure it's best to warn you to expect this simple story to fall apart. If I didn't give you this warning you might start getting confused when you see $df = 3.4$ or something, (incorrectly) thinking that you had misunderstood something that I've taught you rather than (correctly) realising that there's something that I haven't told you.

84. In practice, the sample size isn't always fixed. For example, we might run the experiment over a fixed period of time and the number of people participating depends on how many people show up. That doesn't matter for the current purposes.

85. Well, sort of. The conventions for how statistics should be reported tend to differ somewhat from discipline to discipline. I've tended to stick with how things are done in psychology, since that's what I do. But the general principle of providing enough information to the reader to allow them to check your results is pretty universal, I think.

86. To some people, this advice might sound odd, or at least in conflict with the "usual" advice on how to write a technical report. Very typically, students are told that the "results" section of a report is for describing the data and reporting statistical analysis, and the "discussion" section is for providing interpretation. That's true as far as it goes, but I think people often interpret it way too literally. The way I usually approach it is to provide a quick and simple interpretation of the data in the results section, so that my reader understands what the data are telling us. Then, in the discussion, I try to tell a bigger story about how my results fit with the rest of the scientific literature. In short, don't let the "interpretation goes in the discussion" advice turn your results section into incomprehensible garbage. Being understood by your reader is much more important.

87. If you've been reading very closely, and are as much of a mathematical pedant as I am, there is one thing about the way I wrote up the chi-square test in the last section that might be bugging you a little bit. There's something that feels a bit wrong with writing "$\chi^2(3) = 8.44$", you might be thinking. After all, it's the goodness-of-fit statistic that is equal to 8.44, so shouldn't I have written $X^2 = 8.44$ or maybe $GOF = 8.44$? This seems to be conflating the sampling distribution (i.e., χ^2 with df = 3) with the test statistic (i.e., X^2). Odds are you figured it was a typo, since χ and X look pretty similar. Oddly, it's not. Writing $\chi^2(3)= 8.44$ is essentially a highly condensed way of writing "the sampling distribution of the test statistic is $\chi^2(3)$, and the value of the test statistic is 8.44" In one sense, this is kind of stupid. There are lots of different test statistics out there that turn out to have a chi-square sampling distribution. The X^2 statistic that we've used for our goodness-of-fit test is only one of many (albeit one of the most commonly encountered ones). In a sensible, perfectly organised world we'd always have a separate name for the test statistic and the sampling distribution. That way, the stat block itself would tell you exactly what it was that the researcher had calculated. Sometimes this happens. For instance, the test statistic used in the Pearson goodness-of-fit test is written X^2, but there's a closely related test known as the G-test[a] (Sokal & Rohlf, 1994), in which the test statistic is written as G. As it happens, the Pearson goodness-of-fit test and the G-test both test the same null hypothesis, and the sampling distribution is exactly the same (i.e., chi-square with $k-1$ degrees of freedom). If I'd done a G-test for the cards data rather than a goodness-of-fit test, then I'd have ended up with a test statistic of $G = 8.65$, which is slightly different from the $X^2 = 8.44$ value that I got earlier and which produces a slightly smaller p-value of $p = .034$. Suppose that the convention was to report the test statistic, then the sampling distribution, and then the p-value. If that were true, then these two situations would produce different stat blocks: my original result would be written $X^2 = 8.44$, $\chi^2(3)$, $p = .038$, whereas the new version using the G-test would be written as $G = 8.65$, $\chi^2(3)$, $p = .034$. However, using the condensed reporting standard, the original result is written $\chi^2(3) = 8.44, p = .038$, and the new one is written $\chi^2(3) = 8.65, p = .034$, and so it's actually unclear which test I actually ran. So why don't we live in a world in which the contents of the stat block uniquely specifies what tests were ran? The deep reason is that life is messy. We (as users of statistical tools) want it to be nice and neat and organised. We want it to be designed, as if it were a product, but that's not how life works. Statistics is an intellectual discipline just as much as any other one, and as such it's a massively distributed, partly-collaborative and partly-competitive project that no-one really understands completely. The things that you and I use as data analysis tools weren't created by an Act of the Gods of Statistics. They were invented by lots of different people, published as papers in academic journals, implemented, corrected and modified by lots of other people, and then explained to students in textbooks by someone else. As a consequence, there's a lot of test statistics that don't even have names, and as a consequence they're just given the same name as the corresponding sampling distribution. As we'll see later, any test statistic that follows a χ^2 distribution is commonly called a "chi-square statistic", anything that follows a t distribution is called a "t-statistic", and so on. But, as the χ^2 versus G example illustrates, two different things with the same sampling distribution are still, well, different. As a consequence, it's sometimes a good idea to be clear about what the actual test was that you ran, especially if you're doing something unusual. If you just say "chi-square test" it's not actually clear what test you're talking about. Although, since the two most common chi-square tests are the goodness-of-fit test and the independence test, most readers with stats training can probably guess. Nevertheless, it's something to be aware of. – [a] Complicating matters, the G-test is a special case of a whole class of tests that are known as likelihood ratio tests (LRTs). I don't cover LRTs in this book, but they are quite handy things to know about.

88. A technical note. The way I've described the test pretends that the column totals are fixed (i.e., the researcher intended to survey 87 robots and 93 humans) and the row totals are random (i.e., it just turned out that 28 people chose the puppy). To use the terminology from my mathematical statistics textbook (Hogg et al., 2005) I should technically refer to this situation as a chi-square test of homogeneity and reserve the term chi-square test of independence for the situation where both the row and column totals are random outcomes of the experiment. In the initial drafts of this book that's exactly what I did. However, it turns out that these two tests are identical, and so I've collapsed them together.

89. Technically, E_{ij} here is an estimate, so I should probably write it $\hat{E}_{ij}$, but since no-one else does, I won't either. Now that we've figured out how to calculate the expected frequencies, it's straightforward to define a test statistic, following the exact same strategy that we used in the goodness-of-fit test. In fact, it's pretty much the same statistic. For a contingency table with r rows and c columns, the equation that defines our X^2 statistic is:

$$X^2 = \sum_{i=1}^{r} \sum_{j=1}^{c} \frac{(E_{ij} - O_{ij})^2}{E_{ij}}$$

The only difference is that I have to include two summation signs (i.e., $\sum$) to indicate that we're summing over both rows and columns.

90. A problem many of us worry about in real life.

91. Yates (1934) suggested a simple fix, in which you redefine the goodness-of-fit statistic as:
$$\chi^2 = \sum_i \frac{(|E_i - O_i| - 0.5)^2}{E_i}$$
Basically, he just subtracts off 0.5 everywhere.

92. Mathematically, they're very simple. To calculate the ϕ statistic, you just divide your X^2 value by the sample size, and take the square root:
$$\phi = \sqrt{\frac{X^2}{N}}$$
The idea here is that the ϕ statistic is supposed to range between 0 (no association at all) and 1 (perfect association), but it doesn't always do this when your contingency table is bigger than 2×2, which is a total pain. For bigger tables it's actually possible to obtain $\phi > 1$, which is pretty unsatisfactory. So, to correct for this, people usually prefer to report the V statistic proposed by Cramer (1946). It's a pretty simple adjustment to ϕ. If you've got a contingency table with r rows and c columns, then define $k = min(r, c)$ to be the smaller of the two values. If so, then Cramér's V statistic is:
$$V = \sqrt{\frac{X^2}{N(k-1)}}$$

93. This example is based on a joke article published in the *Journal of Irreproducible Results*

94. Not surprisingly, the Fisher exact test is motivated by Fisher's interpretation of a *p*-value, not Neyman's! See Section 9.5.

Notes for chapter 11

95. Informal experimentation in my garden suggests that yes, it does. Australian natives are adapted to low phosphorus levels relative to everywhere else on Earth, so if you've bought a house with a bunch of exotics and you want to plant natives, keep them separate; nutrients to European plants are poison to Australian ones.

96. In order to do this I had to change the measurement level for X to 'Continuous', as during the opening / import of the csv file jamovi made this a nominal level variable, which isn't right for my analysis.

97. Adopting the notation from Section 7.5, a statistician might write this as:
$$X \sim Normal(\mu_0, \sigma^2)$$

98. In other words, if the null hypothesis is true then the sampling distribution of the mean can be written as follows:
$$\bar{X} \sim Normal(\mu_0, SE(\bar{X}))$$

99. Again, see Section 4.5 if you've forgotten why this is true.

100. Actually this is too strong. Strictly speaking the z-test only requires that the sampling distribution of the mean be normally distributed. If the population is normal then it necessarily follows that the sampling distribution of the mean is also normal. However, as we saw when talking about the central limit theorem, it's quite possible (even commonplace) for the sampling distribution to be normal even if the population distribution itself is nonnormal. However, in light of the sheer ridiculousness of the assumption that the true standard deviation is known, there really isn't much point in going into details on this front!

101. Well, sort of. As I understand the history, Gosset only provided a partial solution; the general solution to the problem was provided by Sir Ronald Fisher.

102. More seriously, I tend to think the reverse is true. I get very suspicious of technical reports that fill their results sections with nothing except the numbers. It might just be that I'm an arrogant jerk, but I often feel like an author that makes no attempt to explain and interpret their analysis to the reader either doesn't understand it themselves, or is being a bit lazy. Your readers are smart, but not infinitely patient. Don't annoy them if you can help it.

103. A technical comment. In the same way that we can weaken the assumptions of the z-test so that we're only talking about the sampling distribution, we can weaken the t-test assumptions so that we don't have to assume normality of the population. However, for the t-test it's trickier to do this. As before, we can replace the assumption of population normality with an assumption that the sampling distribution of $\bar{X}$ is normal. However, remember that we're also relying on a sample estimate of the standard deviation, and so we also require the sampling distribution of $\hat{\sigma}$ to be chi-square. That makes things nastier, and this version is rarely used in practice. Fortunately, if the population distribution is normal, then both of these two assumptions are met.

104. Although it is the simplest, which is why I started with it.

105. A funny question almost always pops up at this point: what the heck is the population being referred to in this case? Is it the set of students actually taking Dr Harpo's class (all 33 of them)? The set of people who might take the class (an unknown number of them)? Or something else? Does it matter which of these we pick? It's traditional in an introductory behavioural stats class to mumble a lot at this point, but since I get asked this question every year by my students, I'll give a brief answer. Technically yes, it does matter. If you change your definition of what the "real-world" population actually is, then the sampling distribution of your observed mean $\bar{X}$ changes too. The t-test relies on an assumption that the observations are sampled at random from an infinitely large population and, to the extent that real life isn't like that, then the t-test can be wrong. In practice, however, this isn't usually a big deal. Even though the assumption is almost always wrong, it doesn't lead to a lot of pathological behaviour from the test, so we tend to just ignore it.

106. Mathematically, we can write this as:
$$w_1 = N_1 - 1$$
$$w_2 = N_2 - 1$$
Now that we've assigned weights to each sample we calculate the pooled estimate of the variance by taking the weighted average of the two variance estimates, $\hat{sigma}_1^2$ and $\hat{sigma}_2^2$:
$$\hat{\sigma}_p^2 = \frac{w_1 \hat{\sigma}_1^2 + w_2 \hat{\sigma}_2^2}{w_1 + w_2}$$
Finally, we convert the pooled variance estimate to a pooled standard deviation estimate, by taking the square root:
$$\hat{\sigma}_p = \sqrt{\frac{w_1 \hat{\sigma}_1^2 + w_2 \hat{\sigma}_2^2}{w_1 + w_2}}$$
And if you mentally substitute ($w_1 = N_1 - 1$) and $w_2 = N_2 - 1$ into this equation you get a very ugly looking formula. A very ugly formula that actually seems to be the "standard" way of describing the pooled standard deviation estimate. It's not my favourite way of thinking about pooled standard deviations, however. I prefer to think about it like this. Our data set actually corresponds to a set of N observations which are sorted into two groups. So let's use the notation X_{ik} to referto the grade received by the i-th student in the k-th tutorial group. That is, X_{11} is the grade received by the first student in Anastasia's class, X_{21} is her second student, and so on. And we have two separate group means $\bar{X}_1$ and $\bar{X}_2$, which we could "generically" refer to using the notation $\bar{X}_k$, i.e., the mean grade for the k-th tutorial group. So far, so good. Now, since every single student falls into one of the two tutorials, we can describe their deviation from the group mean as the difference:
$$X_{ik} - \bar{X}_k$$
So why not just use these deviations (i.e., the extent to which each student's grade differs from the mean grade in their tutorial)? Remember, a variance is just the average of a bunch of squared deviations, so let's do that. Mathematically, we could write it like this:
$$\frac{\sum_{ik}(X_{ik} - \bar{X}_k)^2}{N}$$

where the notation "$\sum_{ik}$" is a lazy way of saying "calculate a sum by looking at all students in all tutorials", since each "ik" corresponds to one student.[a] But, as we saw in Chapter 8, calculating the variance by dividing by N produces a biased estimate of the population variance. And previously we needed to divide by $(N-1)$ to fix this. However, as I mentioned at the time, the reason why this bias exists is because the variance estimate relies on the sample mean, and to the extent that the sample mean isn't equal to the population mean it can systematically bias our estimate of the variance. But this time we're relying on two sample means! Does this mean that we've got more bias? Yes, yes it does. And does this mean we now need to divide by $(N-2)$ instead of $(N-1)$, in order to calculate our pooled variance estimate? Why, yes:

$$\hat{\sigma}_p^2 = \frac{\sum_{ik}(X_{ik} - \bar{X}_k)^2}{N-2}$$

Oh, and if you take the square root of this then you get $\hat{\sigma}_p$, the pooled standard deviation estimate. In other words, the pooled standard deviation calculation is nothing special. It's not terribly different to the regular standard deviation calculation. — [a] A more correct notation will be introduced in Chapter 13.

107. As long as the two variables really do have the same standard deviation, then our estimate for the standard error is:

$$SE(\bar{X}_1 - \bar{X}_2) = \hat{\sigma}\sqrt{\frac{1}{N_1} + \frac{1}{N_2}}$$

and our t-statistic is therefore:

$$t = \frac{\bar{X}_1 - \bar{X}_2}{SE(\bar{X}_1 - \bar{X}_2)}$$

108. Strictly speaking, it is the **difference** in the means that should be normally distributed, but if both groups have normally distributed data then the difference in means will also be normally distributed. In practice, the central limit theorem assures us that, generally, the distributions of the two sample means being tested will themselves approach normal distributions as the sample sizes get large, regardless of the distributions of the underlying data.

109. Well, I guess you can average apples and oranges, and what you end up with is a delicious fruit smoothie. But no one really thinks that a fruit smoothie is a very good way to describe the original fruits, do they?

110. But you can still estimate the standard error of the difference between sample means, it just ends up looking different:

$$SE(\bar{X}_1 - \bar{X}_2) = \sqrt{\frac{\hat{\sigma}_1^2}{N_1} + \frac{\hat{\sigma}_2^2}{N_2}}$$

The reason why it's calculated this way is beyond the scope of this book. What matters for our purposes is that the t-statistic that comes out of the Welch t-test is actually somewhat different to the one that comes from the Student t-test.

111. This design is very similar to the one that motivated the McNemar test (Section 10.7). This should be no surprise. Both are standard repeated measures designs involving two measurements. The only difference is that this time our outcome variable is interval scale (working memory capacity) rather than a binary, nominal scale variable (a yes-or-no question).

112. At this point we have Drs Harpo, Chico and Zeppo. No prizes for guessing who Dr Groucho is.

113. They introduce a small correction by multiplying the usual value of d by $\frac{(N-3)}{(N-2.25)}$.

114. If you are interested, you can look at how this was done in the *chico2.omv* file

115. This is a massive oversimplification.

116. Either that, or the Kolmogorov-Smirnov test, which is probably more traditional than the Shapiro-Wilk. Although most things I've read seem to suggest Shapiro-Wilk is the better test of normality, the Kolomogorov

Smirnov is a general purpose test of distributional equivalence that can be adapted to handle other kinds of distribution tests. In jamovi the Shapiro-Wilk test is preferred.

117. The test statistic that it calculates is conventionally denoted as W, and it's calculated as follows. First, we sort the observations in order of increasing size, and let $\bar{X}_1$ be the smallest value in the sample, X_2 be the second smallest and so on. Then the value of W is given by:

$$W = \frac{(\sum_{i=1}^{N} a_i X_i)^2}{\sum_{i=1}^{N} (X_i - \bar{X})^2}$$

where $\bar{X}$ is the mean of the observations, and the a_i values are ...mumble, mumble... something complicated that is a bit beyond the scope of an introductory text.

118. Actually, there are two different versions of the test statistic that differ from each other by a constant value. The version that I've described is the one that jamovi calculates.

Notes for chapter 12

119. I've noticed that in jamovi you can also specify an 'ID' variable type, but for our purposes it does not matter how we specify the ID variable as we won't be including it in any analyses.

120. Actually, even that table is more than I'd bother with. In practice most people pick one measure of central tendency, and one measure of variability only.

121. The formula for the Pearson's correlation coefficient can be written in several different ways. I think the simplest way to write down the formula is to break it into two steps. Firstly, let's introduce the idea of a **covariance**. The covariance between two variables X and Y is a generalisation of the notion of the variance and is a mathematically simple way of describing the relationship between two variables that isn't terribly informative to humans:

$$Cov(X,Y) = \frac{1}{N-1} \sum_{i=1}^{N} (X_i - \bar{X})(Y_i - \bar{Y})$$

Because we're multiplying (i.e., taking the "product" of) a quantity that depends on X by a quantity that depends on Y and then averaging,[a] you can think of the formula for the covariance as an "average cross product" between X and Y. The covariance has the nice property that, if X and Y are entirely unrelated, then the covariance is exactly zero. If the relationship between them is positive (in the sense shown in Figure 12.4 then the covariance is also positive, and if the relationship is negative then the covariance is also negative. In other words, the covariance captures the basic qualitative idea of correlation. Unfortunately, the raw magnitude of the covariance isn't easy to interpret as it depends on the units in which X and Y are expressed and, worse yet, the actual units that the covariance itself is expressed in are really weird. For instance, if X refers to the dani.sleep variable (units: hours) and Y refers to the dani.grump variable (units: grumps), then the units for their covariance are $hours \times grumps$. And I have no freaking idea what that would even mean. The Pearson correlation coefficient r fixes this interpretation problem by standardising the covariance, in pretty much the exact same way that the z-score standardises a raw score, by dividing by the standard deviation. However, because we have two variables that contribute to the covariance, the standardisation only works if we divide by both standard deviations.[b] In other words, the correlation between X and Y can be written as follows:

$$r_{XY} = \frac{Cov(X,Y)}{\hat{\sigma}_X \hat{\sigma}_Y}$$

—[a] Just like we saw with the variance and the standard deviation, in practice we divide by $N-1$ rather than N. [b] This is an oversimplification, but it'll do for our purposes.

122. Also sometimes written as $y = mx + c$ where m is the slope coefficient and c is the intercept (constant) coefficient:

$$\hat{Y}_i = b_0 + b_1 X_i$$

123. The ϵ symbol is the Greek letter epsilon. It's traditional to use ϵ_i or e_i to denote a residual:
$$\epsilon_i = Y_i - \hat{Y}_i$$
which in turn means that we can write down the complete linear regression model as:
$$Y_i = b_0 + b_1 X_i + \epsilon_i$$

124. Or at least, I'm assuming that it doesn't help most people. But on the off chance that someone reading this is a proper kung fu master of linear algebra (and to be fair, I always have a few of these people in my intro stats class), it will help you to know that the solution to the estimation problem turns out to be $\hat{b} = (X'X)^{-1}X'y$, where $\hat{b}$ is a vector containing the estimated regression coefficients, X is the "design matrix" that contains the predictor variables (plus an additional column containing all ones; strictly X is a matrix of the regressors, but I haven't discussed the distinction yet), and y is a vector containing the outcome variable. For everyone else, this isn't exactly helpful and can be downright scary. However, since quite a few things in linear regression can be written in linear algebra terms, you'll see a bunch of footnotes like this one in this chapter. If you can follow the maths in them, great. If not, ignore it.

125. The formula for the general case: The equation that I gave in the main text shows you what a multiple regression model looks like when you include two predictors. Not surprisingly then, if you want more than two predictors all you have to do is add more X terms and more b coefficients. In other words, if you have K predictor variables in the model then the regression equation look like this:
$$Y_i = b_0 + \left(\sum_{k=1}^{K} b_k X_{ik}\right) + \epsilon_i$$

126. And by "sometimes" I mean "almost never". In practice everyone just calls it "R-squared".

127. The adjusted R^2 value introduces a slight change to the calculation, as follows. For a regression model with K predictors, fit to a data set containing N observations, the adjusted R^2 is:
$$\text{adj.} R^2 = 1 - \left(\frac{SS_{res}}{SS_{tot}} \times \frac{N-1}{N-K-1}\right)$$

128. Formally, our "null model" corresponds to the fairly trivial "regression" model in which we include 0 predictors and only include the intercept term b_0: $H_0 : Y_0 = b_0 + \epsilon_i$ If our regression model has K predictors, the "alternative model" is described using the usual formula for a multiple regression model:
$$H_1 : Y_i = b_0 + \left(\sum_{k=1}^{K} b_k X_{ik}\right) + \epsilon_i$$

How can we test these two hypotheses against each other? The trick is to understand that it's possible to divide up the total variance SS_{tot} into the sum of the residual variance SS_{res} and the regression model variance SS_{mod}. I'll skip over the technicalities, since we'll get to that later when we look at ANOVA in Chapter 13. But just note that $SS_{mod} = SS_{tot} - SS_{res}$ And we can convert the sums of squares into mean squares by dividing by the degrees of freedom:
$$MS_{mod} = \frac{SS_{mod}}{df_{mod}}$$
$$MS_{res} = \frac{SS_{res}}{df_{res}}$$

So, how many degrees of freedom do we have? As you might expect the df associated with the model is closely tied to the number of predictors that we've included. In fact, it turns out that $df_{mod} = K$. For the residuals the total degrees of freedom is $df_{res} = N - K - 1$. Now that we've got our mean square values we can calculate an F-statistic like this:
$$F = \frac{MS_{mod}}{MS_{res}}$$
and the degrees of freedom associated with this are K and $N - K - 1$.

129. For advanced readers only. The vector of residuals is $\epsilon = y - X\hat{b}$. For K predictors plus the intercept, the estimated residual variance is $\hat{\sigma}^2 = \frac{\epsilon'\epsilon}{(N-K-1)}$. The estimated covariance matrix of the coefficients is $\hat{\sigma}^2 (X'X)^{-1}$, the main diagonal of which is $se(\hat{b})$, our estimated standard errors.

130. Note that, although jamovi has done multiple tests here, it hasn't done a Bonferroni correction or anything (see Chapter 13). These are standard one-sample t-tests with a two-sided alternative. If you want to make corrections for multiple tests, you need to do that yourself.

131. Fortunately, confidence intervals for the regression weights can be constructed in the usual fashion $CI(b) = \hat{b} \pm (t_{crit} \times SE(\hat{b}))$ where $se(\hat{b})$ is the standard error of the regression coefficient, and t_crit is the relevant critical value of the appropriate t-distribution. For instance, if it's a 95% confidence interval that we want, then the critical value is the 97.5th quantile of a t-distribution with $N - K - 1$ degrees of freedom. In other words, this is basically the same approach to calculating confidence intervals that we've used throughout.

132. Strictly, you standardise all the *regressors*. That is, every "thing" that has a regression coefficient associated with it in the model. For the regression models that I've talked about so far, each predictor variable maps onto exactly one regressor, and vice versa. However, that's not actually true in general and we'll see some examples of this later in Chapter 14. But, for now we don't need to care too much about this distinction.

133. Leaving aside the interpretation issues, let's look at how it's calculated. What you could do is standardise all the variables yourself and then run a regression, but there's a much simpler way to do it. As it turns out, the β coefficient for a predictor X and outcome Y has a very simple formula, namely $\beta_X = b_X \times \frac{\sigma_X}{\sigma_Y}$ where σ_X is the standard deviation of the predictor, and σ_Y is the standard deviation of the outcome variable Y. This makes matters a lot simpler.

134. The way we calculate these is to divide the ordinary residual by an estimate of the (population) standard deviation of these residuals. For technical reasons, the formula for this is:

$$\epsilon'_i = \frac{\epsilon_i}{\hat{\sigma}\sqrt{1-h_i}}$$

where $\hat{\sigma}$ in this context is the estimated population standard deviation of the ordinary residuals, and h_i is the "hat value" of the ith observation. I haven't explained hat values to you yet, so this won't make a lot of sense. For now, it's enough to interpret the standardised residuals as if we'd converted the ordinary residuals to z-scores.

135. The formula for doing the calculations this time is subtly different $\epsilon^*_i = \frac{\epsilon_i}{\hat{\sigma}_{(-i)}\sqrt{1-h_i}}$. Notice that our estimate of the standard deviation here is written $\hat{\sigma}_{(-i)}$. What this corresponds to is the estimate of the residual standard deviation that you would have obtained if you just deleted the ith observation from the data set. This sounds like the sort of thing that would be a nightmare to calculate, since it seems to be saying that you have to run N new regression models (even a modern computer might grumble a bit at that, especially if you've got a large data set). Fortunately, this standard deviation estimate is actually given by the following equation: $\hat{\sigma}_{(-i)} = \hat{\sigma}\sqrt{\frac{N-K-1-\epsilon'^2_i}{N-K-2}}$.

136. $f(x, \lambda) = \frac{x^\lambda - 1}{\lambda}$ for all values of λ except $\lambda = 0$. When $\lambda = 0$ we just take the natural logarithm (i.e., $\ln(x)$).

137. In jamovi, you can compute this new variable using the formula 'SQRT(ABS(Residuals))'.

138. It's a bit beyond the scope of this chapter to talk about how to deal with violations of homogeneity of variance, but I'll give you a quick sense of what you need to consider. The **main** thing to worry about, if homogeneity of variance is violated, is that the standard error estimates associated with the regression coefficients are no longer entirely reliable, and so your t-tests for the coefficients aren't quite right either. A simple fix to the problem is to make use of a "heteroscedasticity corrected covariance matrix" when estimating the standard errors. These are often called **sandwich estimators**, and these can be estimated in R (but not directly in jamovi).

139. The formula for the k-th VIF is: $VIF_k = \frac{1}{1-R^2_{(-k)}}$ where $R^2_{(-k)}$ refers to R-squared value you would get if you ran a regression using X_k as the outcome variable, and all the other X variables as the predictors. The idea here is that $R^2_{(-k)}$ is a very good measure of the extent to which X_k is correlated with all the other variables in the model. Better yet, the square root of the VIF is pretty interpretable: it tells you how much wider the confidence interval for the corresponding coefficient b_k is, relative to what you would have expected if the predictors are all nice and uncorrelated with one another.

140. Again, for the linear algebra fanatics: the "hat matrix" is defined to be that matrix H that converts the vector of observed values y into a vector of predicted values $\hat{y}$, such that $\hat{y} = Hy$. The name comes from the fact that this is the matrix that "puts a hat on y". The hat value of the i-th observation is the i-th diagonal element of this matrix (so technically I should be writing it as h_{ii} rather than h_i). And here's how it's calculated: $H = X(X'X)^1 X'$.

141. $D_i = \frac{\epsilon_i^{*2}}{K+1} \times \frac{h_i}{1-h_i}$ Notice that this is a multiplication of something that measures the outlier-ness of the observation (the bit on the left), and something that measures the leverage of the observation (the bit on the right).

142. In jamovi you can save the Cook's distance values to the data set, then draw a boxplot of the Cook's distance values to identify the specific outliers. Or you could use a more powerful regression program such as the "car" package in R which has more options for advanced regression diagnostic analysis.

143. In the context of a linear regression model (and ignoring terms that don't depend on the model in any way!), the AIC for a model that has K predictor variables plus an intercept is $AIC = \frac{SS_{res}}{\hat{\sigma}^2} + 2K$.

144. While I'm on this topic I should point out that the empirical evidence suggests that BIC is a better criterion than AIC. In most simulation studies that I've seen, BIC does a much better job of selecting the correct model.

145. We can fit both models to the data and obtain a residual sum of squares for both models. I'll denote these as: $SS_{res}^{(1)}$ and $SS_{res}^{(2)}$ respectively. The superscripting here just indicates which model we're talking about. Then our F statistic is:

$$F = \frac{\frac{SS_{res}^{(1)} - SS_{res}^{(2)}}{k}}{\frac{SS_{res}^2}{N-p-1}}$$

where N is the number of observations, p is the number of predictors in the full model (not including the intercept), and k is the difference in the number of parameters between the two models.[d] The degrees of freedom here are k and $N-p-1$. Note that it's often more convenient to think about the difference between those two SS values as a sum of squares in its own right. That is:

$$SS_\Delta = SS_{res}^{(1)} - SS_{res}^{(2)}$$

The reason why this is helpful is that we can express SS_Δ as a measure of the extent to which the two models make different predictions about the the outcome variable. Specifically:

$$SS_\Delta = \sum_i (\hat{y}_i^{(2)} - \hat{y}_i^{(1)})^2$$

where $\hat{y}_{i(1)}$ is the predicted value for y_i according to model M_1 and $\hat{y}_{i(2)}$ is the predicted value for y_i according to model M_2. — [d] It's worth noting in passing that this same F statistic can be used to test a much broader range of hypotheses than those that I'm mentioning here. Very briefly, notice that the nested model M_1 corresponds to the full model M_2 when we constrain some of the regression coefficients to zero. It is sometimes useful to construct sub-models by placing other kinds of constraints on the regression coefficients. For instance, maybe two different coefficients might have to sum to zero. You can construct hypothesis tests for those kind of constraints too, but it is somewhat more complicated and the sampling distribution for F can end up being something known as the non-central F-distribution, which is way beyond the scope of this book! All I want to do is alert you to this possibility.

Notes for chapter 13

146. When all groups have the same number of observations, the experimental design is said to be "balanced". Balance isn't such a big deal for one-way ANOVA, which is the topic of this chapter. It becomes more important when you start doing more complicated ANOVAs.

147. So the formula for the total sum of squares is almost identical to the formula for the variance:

$$SS_{tot} = \sum_{k=1}^{G} \sum_{i=1}^{N_k} (Y_{ik} - \bar{Y})^2$$

148. One very nice thing about the total sum of squares is that we can break it up into two different kinds of variation First, we can talk about the within-group sum of squares, in which we look to see how different each individual person is from their own group mean:

$$SS_w = \sum_{k=1}^{G} \sum_{i=1}^{N_k} (Y_{ik} - \bar{Y}_k)^2$$

where $\bar{Y}_k$ is a group mean. In our example, $\bar{Y}_k$ would be the average mood change experienced by those people given the k-th drug. So, instead of comparing individuals to the average of all people in the experiment, we're only comparing them to those people in the the same group. As a consequence, you'd expect the value of SS_w to be smaller than the total sum of squares, because it's completely ignoring any group differences, i.e., whether the drugs will have different effects on people's moods.

149. In order to quantify the extent of this variation, what we do is calculate the between-group sum of squares:

$$SS_b = \sum_{k=1}^{G} \sum_{i=1}^{N_k} (\bar{Y}_k - \bar{Y})^2$$
$$= \sum_{k=1}^{G} N_k (\bar{Y}_k - \bar{Y})^2$$

150. SS_w is also referred to in an independent ANOVA as the error variance, or SS_{error}.

151. At a fundamental level ANOVA is a competition between two different statistical models, H_0 and H_1. When I described the null and alternative hypotheses at the start of the section, I was a little imprecise about what these models actually are. I'll remedy that now, though you probably won't like me for doing so. If you recall, our null hypothesis was that all of the group means are identical to one another. If so, then a natural way to think about the outcome variable Y_{ik} is to describe individual scores in terms of a single population mean μ, plus the deviation from that population mean. This deviation is usually denoted ϵ_{ik} and is traditionally called the error or residual associated with that observation. Be careful though. Just like we saw with the word "significant", the word "error" has a technical meaning in statistics that isn't quite the same as its everyday English definition. In everyday language, "error" implies a mistake of some kind, but in statistics it doesn't (or at least, not necessarily). With that in mind, the word "residual" is a better term than the word "error". In statistics both words mean "leftover variability", that is "stuff" that the model can't explain. In any case, here's what the null hypothesis looks like when we write it as a statistical model:

$$Y_{ik} = \mu + \epsilon_{ik}$$

where we make the assumption (discussed later) that the residual values ϵ_{ik} are normally distributed, with mean 0 and a standard deviation σ that is the same for all groups. To use the notation that we introduced in the Introduction to probability we would write this assumption like this:

$$\epsilon_{ik} \sim Normal(0, \sigma^2)$$

What about the alternative hypothesis, H_1? The only difference between the null hypothesis and the alternative hypothesis is that we allow each group to have a different population mean. So, if we let μ_k denote the population mean for the k-th group in our experiment, then the statistical model corresponding to H_1 is:

$$Y_{ik} = \mu_k + \epsilon_{ik}$$

where, once again, we assume that the error terms are normally distributed with mean 0 and standard deviation σ. That is, the alternative hypothesis also assumes that $\epsilon \sim Normal(0, \sigma^2)$ Okay, now that we've described the statistical models underpinning H_0 and H_1 in more detail, it's now pretty straightforward to say what the mean square values are measuring, and what this means for the interpretation of F. I won't bore you with the proof of this but it turns out that the within-groups mean square, MS_w, can be viewed as an estimator of the error variance σ^2. The between-groups mean square MS_b is also an estimator, but what it estimates is the error variance plus a quantity that depends on the true differences among the group means. If we call this quantity Q, then we can see that the F statistic is basically:[a]

$$F = \frac{\hat{Q} + \hat{\sigma}^2}{\hat{\sigma}^2}$$

where the true value $Q = 0$ if the null hypothesis is true, and $Q < 0$ if the alternative hypothesis is true (e.g., Hays (1994), ch. 10). Therefore, at a bare minimum the *F value must be larger than* 1 to have any chance of rejecting the null hypothesis. Note that this doesn't mean that it's impossible to get an F-value less than 1. What it means is that if the null hypothesis is true the sampling distribution of the F-ratio has a mean of 1,[b] and so we need to see F-values larger than 1 in order to safely reject the null. To be a bit more precise about the sampling distribution, notice that if the null hypothesis is true, both MS_b and MS_w are estimators of the variance of the residuals ϵ_{ik}. If those residuals are normally distributed, then you might suspect that the estimate of the variance of ϵ_{ik} is chi-square distributed, because (as discussed in the Section 7.6) that's what a chi-square distribution is: it's what you get when you square a bunch of normally-distributed things and add them up. And since the F distribution is (again, by definition) what you get when you take the ratio between two things that are χ^2 distributed, we have our sampling distribution. Obviously, I'm glossing over a whole lot of stuff when I say this, but in broad terms, this really is where our sampling distribution comes from. — [a]If you read ahead to Chapter 14 and look at how the "treatment effect" at level k of a factor is defined in terms of the α_k values (see [Factorial ANOVA 2: balanced designs, interactions allowed]), it turns out that Q refers to a weighted mean of the squared treatment effects, $Q = \frac{(\sum_{k=1}^{G} N_k \alpha_k^2)}{(G-1)}$. — [b]Or, if we want to be sticklers for accuracy, $1 + \frac{2}{df_2 - 2}$.

152. Or, to be precise, party like "it's 1899 and we've got no friends and nothing better to do with our time than do some calculations that wouldn't have made any sense in 1899 because ANOVA didn't exist until about the 1920s".

153. In the Excel *clinicaltrial-anova.xls* the value for SS_b worked out to be very slightly different, 3.45, than that shown in the text above (rounding errors!).

154. The jamovi results are more accurate than the ones in the text above, due to rounding errors.

155. https://daniellakens.blogspot.com/2015/06/why-you-should-use-omega-squared.html

156. If you *do* have some theoretical basis for wanting to investigate some comparisons but not others, it's a different story. In those circumstances you're not really running "post hoc" analyses at all, you're making "planned comparisons". I do talk about this situation later in the book - Section 14.9, but for now I want to keep things simple.

157. It's worth noting in passing that not all adjustment methods try to do this. What I've described here is an approach for controlling "family wise type I error rate". However, there are other post hoc tests that seek to control the "false discovery rate", which is a somewhat different thing.

158. If you remember back to A worked example, which I hope you at least skimmed even if you didn't read the whole thing, I described the statistical models underpinning ANOVA in this way:

$$H_0 : Y_{ik} = \mu + \epsilon_{ik}$$
$$H_1 : Y_{ik} = \mu_k + \epsilon_{ik}$$

In these equations μ refers to a single grand population mean which is the same for all groups, and μk is the population mean for the k-th group. Up to this point we've been mostly interested in whether our data are best described in terms of a single grand mean (the null hypothesis) or in terms of different group-specific means (the alternative hypothesis). This makes sense, of course, as that's actually the important research

question! However, all of our testing procedures have, implicitly, relied on a specific assumption about the residuals, ϵ_ik, namely that:
$$\epsilon_{ik} \sim Normal(0, \sigma^2)$$
None of the maths works properly without this bit. Or, to be precise, you can still do all the calculations and you'll end up with an F-statistic, but you have no guarantee that this F-statistic actually measures what you think it's measuring, and so any conclusions that you might draw on the basis of the F-test might be wrong.

159. The Levene test is shockingly simple. Suppose we have our outcome variable Y_{ik}. All we do is define a new variable, which I'll call Z_{ik}, corresponding to the absolute deviation from the group mean:
$$Z_{ik} = Y_{ik} - \bar{Y}_k$$
Okay, what good does this do us? Well, let's take a moment to think about what Z_{ik} actually is and what we're trying to test. The value of Z_{ik} is a measure of how the i-th observation in the k-th group deviates from its group mean. And our null hypothesis is that all groups have the same variance, i.e., the same overall deviations from the group means! So the null hypothesis in a Levene test is that the population means of Z are identical for all groups. Hmm. So what we need now is a statistical test of the null hypothesis that all group means are identical. Where have we seen that before? Oh right, that's what ANOVA is, and so all that the Levene test does is run an ANOVA on the new variable Z_{ik}. What about the Brown-Forsythe test? Does that do anything particularly different? Nope. The only change from the Levene test is that it constructs the transformed variable Z in a slightly different way, using deviations from the group medians rather than deviations from the group means. That is, for the Brown-Forsythe test:
$$Z_{ik} = Y_{ik} - median_k(Y)$$
where $median_k(Y)$ is the median for group k.

160. So let's let R_ik refer to the ranking given to the ith member of the kth group. Now, let's calculate $\bar{R}_k$, the average rank given to observations in the kth group:
$$\bar{R}_k = \frac{1}{N_k} \sum_i R_{ik}$$
and let's also calculate $\bar{R}$, the grand mean rank:
$$\bar{R} = \frac{1}{N} \sum_i \sum_k R_{ik}$$
Now that we've done this, we can calculate the squared deviations from the grand mean rank $\bar{R}$. When we do this for the individual scores, i.e., if we calculate $(R_{ik} - \bar{R})^2$, what we have is a "nonparametric" measure of how far the ik-th observation deviates from the grand mean rank. When we calculate the squared deviation of the group means from the grand means, i.e., if we calculate $(R_{ik} - \bar{R})^2$, then what we have is a nonparametric measure of how much the group deviates from the grand mean rank. With this in mind, we'll follow the same logic that we did with ANOVA and define our ranked sums of squares measures, much like we did earlier. First, we have our "total ranked sums of squares":
$$RSS_{tot} = \sum_k \sum_i (R_{ik} - \bar{R})^2$$
and we can define the "between groups ranked sums of squares" like this:
$$RSS_b = \sum_k k \sum_i (\bar{R}_k - \bar{R})^2$$
$$= \sum_k N_k (\bar{R}_k - \bar{R})^2$$
So, if the null hypothesis is true and there are no true group differences at all, you'd expect the between group rank sums RSS_b to be very small, much smaller than the total rank sums RSS_{tot}. Qualitatively this is very much the same as what we found when we went about constructing the ANOVA F-statistic, but for technical reasons the Kruskal-Wallis test statistic, usually denoted K, is constructed in a slightly different way:
$$K = (N - 1) \times \frac{RSS_b}{RSS_{tot}}$$
and if the null hypothesis is true, then the sampling distribution of K is approximately chi-square with $G - 1$ degrees of freedom (where G is the number of groups). The larger the value of K, the less consistent the data are with the null hypothesis, so this is a one-sided test. We reject H_0 when K is sufficiently large.

161. However, from a purely mathematical perspective it's needlessly complicated. I won't show you the derivation, but you can use a bit of algebraic jiggery-pokery[a] to show that the equation for K can be:

$$K = \frac{12}{N(N-1)} \sum_k N_k \bar{R}_k^2 - 3(N+1)$$

It's this last equation that you sometimes see given for K. This is way easier to calculate than the version I described in the previous section, but it's just that it's totally meaningless to actual humans. It's probably best to think of K the way I described it earlier, as an analogue of ANOVA based on ranks. But keep in mind that the test statistic that gets calculated ends up with a rather different look to it than the one we used for our original ANOVA. — [a] A technical term.

162. More to the point, in the mathematical notation I introduced above, this is telling us that $f_3 = 2$. Yay. So, now that we know this, the tie correction factor (TCF) is:

$$TCF = 1 - \frac{\sum_j f_j^3 - f_j}{N^3 - N}$$

The tie-corrected value of the Kruskal-Wallis statistic is obtained by dividing the value of K by this quantity. It is this tie-corrected version that jamovi calculates.

163. $(n - k)$: (number of subjects - number of groups)

164. As with all of the chapters in this book, there are quite a few different sources that I've relied upon, but the one stand-out text that I've been most heavily influenced by is Sahai & Ageel (2000). It's not a good book for beginners, but it's an excellent book for more advanced readers who are interested in understanding the mathematics behind ANOVA.

Notes for chapter 14

165. The nice thing about the subscript notation is that it generalises nicely. If our experiment had involved a third factor, then we could just add a third subscript. In principle, the notation extends to as many factors as you might care to include, but in this book we'll rarely consider analyses involving more than two factors, and never more than three.

166. Technically, marginalising isn't quite identical to a regular mean. It's a weighted average where you take into account the frequency of the different events that you're averaging over. However, in a balanced design, all of our cell frequencies are equal by definition so the two are equivalent. We'll discuss unbalanced designs later, and when we do so you'll see that all of our calculations become a real headache. But let's ignore this for now.

167. NB There are some rounding errors here, the value of the mean square, to 5 decimal places, is 1.72667. And the value of the residual mean square to 5 decimal places, is 0.05444. jamovi actually uses many more decimal places in its calculations, but the figures shown in the results tables are rounded for clarity. Though you can change the number of decimal places displayed by jamovi if you want.

168. Now that we've got our notation straight, we can compute the sum of squares values for each of the two factors in a relatively familiar way. For Factor A, our between group sum of squares is calculated by assessing the extent to which the (row) marginal means $\bar{Y}_{1.}$, $\bar{Y}_{2.}$ etc, are different from the grand mean $\bar{Y}_{..}$ We do this in the same way that we did for one-way ANOVA: calculate the sum of squared difference between the $\bar{Y}_{i.}$ values and the $\bar{Y}_{..}$ values. Specifically, if there are N people in each group, then we calculate this:

$$SS_A = (N \times C) \sum_{r=1}^{R} (\bar{Y}_{r.} - \bar{Y}_{..})^2$$

As with one-way ANOVA, the most interesting[a] part of this formula is the bit, which corresponds to the squared deviation associated with level r. All that this formula does is calculate this squared deviation for all R levels of the factor, add them up, and then multiply the result by $N \times C$. The reason for this last part is that there are multiple cells in our design that have level r on Factor A. In fact, there are C of them, one corresponding to each possible level of Factor B. For instance, in our example there are two different cells in the design corresponding to the anxifree drug: one for people with no.therapy and one for the CBT group. Not only that, within each of these cells there are N observations. So, if we want to convert our SS value into a quantity that calculates the between groups sum of squares on a "per observation" basis, we have to multiply by $N \times C$. The formula for Factor B is of course the same thing, just with some subscripts shuffled around:

$$SS_B = (N \times R) \sum_{c=1}^{C} (\bar{Y}_{.c} - \bar{Y}_{..})^2$$

Now that we have these formulas we can check them against the jamovi output from the earlier section. Once again, a dedicated spreadsheet programme is helpful for these sorts of calculations, so please have a go yourself. You can also take a look at the version I did in Excel in the file *clinicaltrial_factorialanova.xls*. First, let's calculate the sum of squares associated with the main effect of drug. There are a total of $N = 3$ people in each group and $C = 2$ different types of therapy. Or, to put it another way, there are $3 \times 2 = 6$ people who received any particular drug. When we do these calculations in a spreadsheet programme, we get a value of 3.45 for the sum of squares associated with the main effect of drug. Not surprisingly, this is the same number that you get when you look up the SS value for the drugs factor in the ANOVA table that I presented earlier, in Figure 14.3. We can repeat the same kind of calculation for the effect of therapy. Again there are $N = 3$ people in each group, but since there are $R = 3$ different drugs, this time around we note that there are $3 \times 3 = 9$ people who received CBT and an additional 9 people who received the placebo. So our calculation in this case gives us a value of 0.47 for the sum of squares associated with the main effect of therapy. Once again, we are not surprised to see that our calculations are identical to the ANOVA output in Figure 14.3. So that's how you calculate the SS values for the two main effects. These SS values are analogous to the between group sum of squares values that we calculated when doing one-way ANOVA in Chapter 13. However, it's not a good idea to think of them as between groups SS values anymore, just because we have two different grouping variables and it's easy to get confused. In order to construct an F test, however, we also need to calculate the within groups sum of squares. In keeping with the terminology that we used in Chapter 12 and the terminology that jamovi uses when printing out the ANOVA table, I'll start referring to the within groups SS value as the residual sum of squares SS_R. The easiest way to think about the residual SS values in this context, I think, is to think of it as the leftover variation in the outcome variable after you take into account the differences in the marginal means (i.e., after you remove SS_A and SS_B). What I mean by that is we can start by calculating the total sum of squares, which I'll label SS_T. The formula for this is pretty much the same as it was for one-way ANOVA. We take the difference between each observation Y_{rci} and the grand mean $\hat{Y}_{..}$, square the differences, and add them all:

$$SS_T = \sum_{r=1}^{R} \sum_{c=1}^{C} \sum_{i=1}^{N} (Y_{rci} - \bar{Y}_{..})^2$$

The "triple summation" here looks more complicated than it is. In the first two summations, we're summing across all levels of Factor A (i.e., over all possible rows r in our table) and across all levels of Factor B (i.e., all possible columns c). Each rc combination corresponds to a single group and each group contains N people, so we have to sum across all those people (i.e., all i values) too. In other words, all we're doing here is summing across all observations in the data set (i.e., all possible rci combinations). At this point, we know the total variability of the outcome variable SST, and we know how much of that variability can be attributed to Factor A (SS_A) and how much of it can be attributed to Factor B (SS_B). The residual sum of squares is thus defined to be the variability in Y that can't be attributed to either of our two factors (or their interaction if you also calculate the interaction effect, which is the default in jamovi). In other words:

$$SS_R = SS_T - (SS_A + SS_B)$$

Of course, there is a formula that you can use to calculate the residual SS directly, but I think that it makes more conceptual sense to think of it like this. The whole point of calling it a residual is that it's the leftover variation, and the formula above makes that clear. I should also note that, in keeping with the terminology used in the regression chapter, it is commonplace to refer to $SS_A + SS_B$ as the variance attributable to the "ANOVA model", denoted SSM, and so we often say that the total sum of squares is equal to the model sum of squares plus the residual sum of squares. Later on in this chapter we'll see that this isn't just a surface similarity: ANOVA and regression are actually the same thing under the hood. In any case, it's probably worth taking a moment to check that we can calculate SS_R using this formula and verify that we do obtain the same answer that jamovi produces in its ANOVA table. The calculations are pretty straightforward when done in a spreadsheet (see the clinicaltrial_factorialanova.xls file). — [a]English translation: "least tedious".

169. As a consequence, the way that the idea of an interaction effect is formalised in terms of null and alternative hypotheses is slightly difficult, and I'm guessing that a lot of readers of this book probably won't be all that

interested. Even so, I'll try to give the basic idea here. To start with, we need to be a little more explicit about our main effects. Consider the main effect of Factor A (drug in our running example). We originally formulated this in terms of the null hypothesis that the two marginal means μ_r. are all equal to each other. Obviously, if all of these are equal to each other, then they must also be equal to the grand mean $\mu_{..}$ as well, right? So what we can do is define the effect of Factor A at level r to be equal to the difference between the marginal mean μ_r. and the grand mean $\mu_{..}$. Let's denote this effect by α_r, and note that:

$$\alpha_r = \mu_{r.} - \mu_{..}$$

Now, by definition all of the α_r values must sum to zero, for the same reason that the average of the marginal means μ_c must be the grand mean $\mu_{..}$. We can similarly define the effect of Factor B at level i to be the difference between the column marginal mean $\mu_{.c}$ and the grand mean $\mu_{..}$:

$$\beta_c = \mu_{.c} - \mu_{..}$$

and once again, these β_c values must sum to zero. The reason that statisticians sometimes like to talk about the main effects in terms of these α_r and β_c values is that it allows them to be precise about what it means to say that there is no interaction effect. If there is no interaction at all, then these α_r and β_c values will perfectly describe the group means μ_{rc}. Specifically, it means that:

$$\mu_{rc} = \mu_{..} + \alpha_r + \beta_c$$

That is, there's nothing special about the group means that you couldn't predict perfectly by knowing all the marginal means. And that's our null hypothesis, right there. The alternative hypothesis is that:

$$\mu_{rc} \neq \mu_{..} + \alpha_r + \beta_c$$

for at least one group rc in our table. However, statisticians often like to write this slightly differently. They'll usually define the specific interaction associated with group rc to be some number, awkwardly referred to as $(\alpha\beta)_{rc}$, and then they will say that the alternative hypothesis is that:

$$\mu_{rc} = \mu_{..} + \alpha_r + \beta_c + (\alpha\beta)_{rc}$$

where $(\alpha\beta)_{rc}$ is non-zero for at least one group. This notation is kind of ugly to look at, but it is handy as we'll see when discussing how to calculate the sum of squares. How should we calculate the sum of squares for the interaction terms, $SS_{A:B}$? Well, first off, it helps to notice how we have just defined the interaction effect in terms of the extent to which the actual group means differ from what you'd expect by just looking at the marginal means. Of course, all of those formulas refer to population parameters rather than sample statistics, so we don't actually know what they are. However, we can estimate them by using sample means in place of population means. So for Factor A, a good way to estimate the main effect at level r is as the difference between the sample marginal mean $\bar{Y}_{rc}$ and the sample grand mean $\bar{Y}_{..}$. That is, we would use this as our estimate of the effect:

$$\hat{\alpha}_r = bar Y_{r.} - \bar{Y}_{..}$$

Similarly, our estimate of the main effect of Factor B at level c can be defined as:

$$\hat{\beta}_c = \hat{Y}_{.c} - \bar{Y}_{..}$$

Now, if you go back to the formulas that I used to describe the SS values for the two main effects, you'll notice that these effect terms are exactly the quantities that we were squaring and summing! So, what's the analog of this for interaction terms? The answer to this can be found by first rearranging the formula for the group means μ_{rc} under the alternative hypothesis, so that we get this:

$$\begin{aligned}(\alpha\beta)_{rc} &= \mu_{rc} - \mu_{..} - \alpha_r - \beta_c \\ &= \mu_{rc} - \mu_{..} - (\mu_{r.} - \mu_{..}) - (\mu_{.c} - \mu_{..}) \\ &= \mu_{rc} - \mu_{r.} - \mu_{.c} + \mu_{..}\end{aligned}$$

So, once again if we substitute our sample statistics in place of the population means, we get the following as our estimate of the interaction effect for group rc, which is:

$$\hat{(\alpha\beta)}_{rc} = \bar{Y}_{rc} - \hat{Y}_{r.} - \bar{Y}_{.c} + \bar{Y}_{..}$$

Now all we have to do is sum all of these estimates across all R levels of Factor A and all C levels of Factor B, and we obtain the following formula for the sum of squares associated with the interaction as a whole:

$$SS_{A:B} = N \sum_{r=1}^{R} \sum_{c=1}^{C} (\bar{Y}_{rc} - \bar{Y}_{r.} - \bar{Y}_{.c} + \bar{Y}_{..})^2$$

where we multiply by N because there are N observations in each of the groups, and we want our SS values to reflect the variation among observations accounted for by the interaction, not the variation among groups. Now that we have a formula for calculating $SS_{A:B}$, it's important to recognise that the interaction term is part of the model (of course), so the total sum of squares associated with the model, SSM, is now equal to the sum of the three relevant SS values, $SS_A + SS_B + SS_{A:B}$. The residual sum of squares SSR is still defined as the leftover variation, namely $SS_T - SS_M$, but now that we have the interaction term this becomes:

$$SS_R = SS_T - (SS_A + SS_B + SS_{A:B})$$

As a consequence, the residual sum of squares SS_R will be smaller than in our original ANOVA that didn't include interactions.

170. You may have spotted this already when looking at the main effects analysis in jamovi that we described earlier. For the purpose of the explanation in this book I removed the interaction component from the earlier model to keep things clean and simple.

171. This chapter seems to be setting a new record for the number of different things that the letter R can stand for. So far we have R referring to the software package, the number of rows in our table of means, the residuals in the model, and now the correlation coefficient in a regression. Sorry. We clearly don't have enough letters in the alphabet. However, I've tried pretty hard to be clear on which thing R is referring to in each case.

172. Implausibly large, I would think. The artificiality of this data set is really starting to show!

173. What's the difference between treatment and simple contrasts, I hear you ask? Well, as a basic example consider a gender main effect, with $m = 0$ and $f = 1$. The coefficient corresponding to the treatment contrast will measure the difference in mean between females and males, and the intercept would be the mean of the males. However, with a simple contrast, i.e., $m = -1$ and $f = 1$, the intercept is the average of the means and the main effect is the difference of each group mean from the intercept.

174. If, for instance, you actually find yourself interested to know if Group A is significantly different from the mean of Group B and Group C, then you need to use a different tool (e.g., Scheffe's method, which is more conservative, and beyond the scope of this book). However, in most cases you probably are interested in pairwise group differences, so Tukey's HSD is a pretty useful thing to know about.

175. This discrepancy in standard deviations might (and should) make you wonder if we have a violation of the homogeneity of variance assumption. I'll leave it as an exercise for the reader to double check this using the Levene test option.

176. Actually, this is a bit of a lie. ANOVAs can vary in other ways besides the ones I've discussed in this book. For instance, I've completely ignored the difference between fixed-effect models in which the levels of a factor are "fixed" by the experimenter or the world, and random-effect models in which the levels are random samples from a larger population of possible levels (this book only covers fixed-effect models). Don't make the mistake of thinking that this book, or any other one, will tell you "everything you need to know" about statistics, any more than a single book could possibly tell you everything you need to know about psychology, physics or philosophy. Life is too complicated for that to ever be true. This isn't a cause for despair, though. Most researchers get by with a basic working knowledge of ANOVA that doesn't go any further than this book does. I just want you to keep in mind that this book is only the beginning of a very long story, not the whole story.

177. Or, at the very least, rarely of interest.

178. However, in jamovi the results for type III sum of squares ANOVA are the same regardless of the contrast selected, so jamovi is obviously doing something different!

179. Note, of course, that this does depend on the model that the user specified. If the original ANOVA model doesn't contain an interaction term for $B \times C$, then obviously it won't appear in either the null or the alternative. But that's true for types I, II and III. They never include any terms that you didn't include, but they make different choices about how to construct tests for the ones that you did include.

180. I find it amusing to note that the default in R is type I and the default in SPSS and jamovi is type III. Neither of these appeals to me all that much. Relatedly, I find it depressing that almost nobody in the psychological literature ever bothers to report which type is used either. The only way I can ever make any sense of what people typically report is to try to guess from auxiliary cues which software they were using, and to assume that they never changed the default settings. Please don't do this! Now that you know about these issues make sure you indicate what software you used, and if you're reporting ANOVA results for unbalanced data, then specify what Type of tests you ran, specify order information if you've done type I tests and specify contrasts if you've done type III tests. Or, even better, do hypotheses tests that correspond to things you really care about and then report those!

Notes for chapter 15

181. Quite helpfully, factor loadings can be interpreted like standardized regression coefficients.

182. A more advanced statistical technique, one which is beyond the scope of this book, undertakes regression modelling where latent factors are used in prediction models of other latent factors. This is called "structural equation modelling" and there are specific software programs and R packages dedicated to this approach.

183. An Eigen value indicates how much of the variance in the observed variables a factor accounts for. A factor with an Eigen value > 1 accounts for more variance than a single observed variable.

184. Oblique rotations provide two factor matrices, one called a structure matrix and one called a pattern matrix. In jamovi just the pattern matrix is shown in the results as this is typically the most useful for interpretation, though some experts suggest that both can be helpful. In a structure matrix coefficients show the relationship between the variable and the factors whilst ignoring the relationship of that factor with all the other factors (i.e. a zero-order correlation). Pattern matrix coefficients show the unique contribution of a factor to a variable whilst controlling for the effects of other factors on that variable (akin to standardized partial regression coefficient). Under orthogonal rotation, structure and pattern coefficients are the same.

185. Sometimes reported in Factor Analysis is "communality" which is the amount of variance in a variable that is accounted for by the factor solution. Uniqueness is equal to $(1 \sim \text{communality})$.

186. Remembering to first reverse score some variables if necessary.

187. As an aside, given that we had a pretty firm idea from our initial "putative" factors, we could just have gone straight to CFA and skipped the EFA step. Whether you use EFA and then go on to CFA, or go straight to CFA, is a matter of judgement and how confident you are initially that you have the model about right (in terms of number of factors and variables). Earlier on in the development of scales, or the identification of underlying latent constructs, researchers tend to use EFA. Later on, as they get closer to a final scale, or if they want to check an established scale in a new sample, then CFA is a good option.

Notes for chapter 16

188. https://en.wikiquote.org/wiki/David%20Hume

189. https://en.wikipedia.org/wiki/Climate_of_Adelaide

190. It's a leap of faith, I know, but let's run with it okay?

191. Um. I hate to bring this up, but some statisticians would object to me using the word "likelihood" here. The problem is that the word "likelihood" has a very specific meaning in frequentist statistics, and it's not quite the same as what it means in Bayesian statistics. As far as I can tell Bayesians didn't originally have any

192. Just to be clear, "prior" information is pre-existing knowledge or beliefs, before we collect or use any data to improve that information.

193. If we were being a bit more sophisticated, we could extend the example to accommodate the possibility that I'm lying about the umbrella. But let's keep things simple, shall we?

194. You might notice that this equation is actually a restatement of the same basic rule I listed at the start of the last section. If you multiply both sides of the equation by $P(d)$, then you get $P(d)P(h|d) = P(d,h)$, which is the rule for how joint probabilities are calculated. So I'm not actually introducing any "new" rules here, I'm just using the same rule in a different way:
$$P(h|d) = \frac{P(d,h)}{P(d)}$$

195. Obviously, this is a highly simplified story. All the complexity of real-life Bayesian hypothesis testing comes down to how you calculate the likelihood, $P(d|h)$, when the hypothesis h is a complex and vague thing. I'm not going to talk about those complexities in this book, but I do want to highlight that although this simple story is true as far as it goes, real life is messier than I'm able to cover in an introductory stats textbook.

196. https://www.imdb.com/title/tt0093779/quotes I should note in passing that I'm not the first person to use this quote to complain about frequentist methods. Rich Morey and colleagues had the idea first. I'm shamelessly stealing it because it's such an awesome pull quote to use in this context and I refuse to miss any opportunity to quote *The Princess Bride*.

197. https://about.abc.net.au/reports-publications/appreciation-survey-summary-report-2013/

198. https://knowyourmeme.com/memes/the-cake-is-a-lie

199. In the interests of being completely honest, I should acknowledge that not all orthodox statistical tests rely on this silly assumption. There are a number of sequential analysis tools that are sometimes used in clinical trials and the like. These methods are built on the assumption that data are analysed as they arrive, and these tests aren't horribly broken in the way I'm complaining about here. However, sequential analysis methods are constructed in a very different fashion to the "standard" version of null hypothesis testing. They don't make it into any introductory textbooks, and they're not very widely used in the psychological literature. The concern I'm raising here is valid for every single orthodox test I've presented so far and for almost every test I've seen reported in the papers I read.

200. A related problem: https://xkcd.com/1478/

201. Some readers might wonder why I picked 3:1 rather than 5:1, given that Johnson (2013) suggests that $p = .05$ lies somewhere in that range. I did so in order to be charitable to the p-value. If I'd chosen a 5:1 Bayes factor instead, the results would look even better for the Bayesian approach. In some ways, this is remarkable. The entire *point* of orthodox null hypothesis testing is to control the type I error rate. Bayesian methods aren't actually designed to do this at all. Yet, as it turns out, when faced with a "trigger happy" researcher who keeps running hypothesis tests as the data come in, the Bayesian approach is much more effective. Even the 3:1 standard, which most Bayesians would consider unacceptably lax, is much safer than the $p < .05$ rule.

202. https://www.quotationspage.com/quotes/Ambrosius_Macrobius/

203. Okay, I just know that some knowledgeable frequentists will read this and start complaining about this section. Look, I'm not dumb. I absolutely know that if you adopt a sequential analysis perspective, you can avoid

these errors within the orthodox framework. I also know that you can explictly design studies with interim analyses in mind. So yes, in one sense I'm attacking a "straw man" version of orthodox methods. However, the straw man that I'm attacking is the one that *has been used by most practitioners*. If it ever reaches the point where sequential methods become the norm among experimental psychologists and I'm no longer forced to read 20 extremely dubious ANOVAs a day, I promise I'll rewrite this section and dial down the vitriol. But until that day arrives, I stand by my claim that default Bayes factor methods are much more robust in the face of data analysis practices as they exist in the real world. *Default* orthodox methods suck, and we all know it.

About the team

Alessandra Tosi was the managing editor for this book.

Tricia de Souza and Adèle Kreager proof-read this manuscript.

The cover was designed by Jeevanjot Kaur Nagpal, and produced in InDesign using the Fontin and Calibri fonts.

David Foxcroft and Cameron Craig produced the printed PDF editions.

David Foxcroft produced the HTML edition.

Raegan Allen was in charge of marketing.

This book was peer-reviewed by two referees. Experts in their field, these readers give their time freely to help ensure the academic rigour of our books. We are grateful for their generous and invaluable contributions.